Proceedings of the 4th International Conference "Advanced Composite Materials Engineering" COMAT 2012

Transilvania University of Brasov

18- 20 October 2012, Brasov, Romania

Petre P. Teodorescu, Chairman
Sorin Vlase, President
Michael M. Dediu, Editor

DERC Publishing House
Tewksbury (Boston), Massachusetts, U. S. A.

Published and printed in the
United States of America

Library of Congress Cataloging in Publication Data

Advanced Composite Materials Engineering, the 4th International Conference, 18- 20 October 2012, Brasov, Romania / Petre P. Teodorescu, Chairman, Sorin Vlase, President
Michael M. Dediu, Editor
        p. cm. – (Proceedings of the 4th International Conference "Advanced Composite Materials Engineering" COMAT 2012)
        Includes bibliographical references

ISBN-13: 978-0981730059
ISBN-10: 0981730051

Preface

Advanced composite materials engineering is one of the most important area of research and development at the present time. Composite materials are a mixture of two or more different components, whose properties complement each other, and the result is a material with superior properties to those specific to each component. The utilization of composite materials is growing fast in strategic fields such as the aircraft, automotive, biomedical, naval and space industries, as well as in textile, paints, magnetic fluids, high-quality paper coatings, and microelectronics.

These Proceedings of the 4th International Conference "Advanced Composite Materials Engineering" COMAT 2012 include 67 papers, which analyse many important practical applications. The topics range from design optimization of composite stiffened structures for aerospace applications, wood engineering, sonic composite visualization, asphalt mixtures with polypropylene fibers and dental filling materials, to new lignocellulosic composite materials, incorporation of different inorganic nanoparticles in polymer matrix, and carbon fiber reinforced polymer-matrix composites.

We thank Mrs. Sophia Dediu for her assistance in preparing this volume.

There is, certainly, much more that can be said about advanced composite materials engineering, than we presented here. We hope that the papers included here will provide ideas for our audience, and will stimulate more research, development and applications.

We look forward to receiving comments and suggestions from our readers.

Michael M. Dediu

The 4th International Conference
"Advanced Composite Materials Engineering "
COMAT 2012
18- 20 October 2012, Brasov, Romania

YOUNG MODULI EVALUATION OF THE SOFT-WOOD COMPONENTS BASED ON THE GLOBAL COMPRESSION TESTS'

Ioan Száva[1], Botond-Pál Gálfi[2], Sorin Vlase[3]
[1]Prof. dr. eng., Transilvania University of Brasov, Romania
[2]PhD. stud. eng., Transilvania University of Brasov, Romania
[1]Prof. dr. eng., Transilvania University of Brasov, Romania

1. ANALYTICAL APPROACH OF THE PROBLEM

Usually, all experimental investigations are focused on the global mechanical behavior; only a few numbers of them aiming on the individual components' elastic properties.

The following innovative proposal consists of in keeping the soft-wood specimen intact and to evaluate the individual elastic properties of early and late rings [2] under these conditions. Soft-wood specimens (pine-wood) with dimensions of 20 *mm* x 20 *mm* x 50 *mm* (*R*∗*T*∗*L*) were tested under compression. These specimens contain several annual rings. The elastic properties of the specimen are determined by the elastic properties of the different rings. In a soft-wood specimen having a tangential-radial (*T-R*) cross-section, the early-, and late annual rings (the corresponding early-, and late-wood segments/regions, respectively) are parallel-connected (Figure 1). When the cross-section of the specimen is longitudinal-tangential (*L-T*), the adequate regions are serial-connected. Starting from this fact, it became useful to establish their individual elastic properties for a more accurate and easier finite element modeling.

In the authors' opinion, each specimen set has to be consisted of 3 specimens worked-out identically and made of the same wood-item. The number of sets had to be large enough for a further statistical evaluation. In each set, two specimens (denoted by *k* and *m*) were used to determine the adequate Young modules, and the third one (noted with *g*) served for validation of the obtained results. The specimens were compressed in a longitudinal direction. Based on the results measured (force-displacement curves) we developed and then used a simple analytical procedure (described bellow) to obtain the individual properties of the early-, and late segments/regions. In the authors opinion, the soft-wood specimen *k* (see Figure 1,a), can be considered as a unit of two parallel-connected elements: *1* - early-wood, *2* - late-wood, each having different cross-sectional area. They will work together and for a given cross-section, they will offer a nominal, global mechanical behavior. If one change only their cross section's ratio, consequently one will obtain, of course, other global mechanical behaviors. For a given mechanical system (Figure 2), regarding on the tested specimen *k*, one can use a static equilibrium equation and the Bernoulli hypothesis to obtain the corresponding load-ratios $N_{k,1}^j$, $N_{k,2}^j$ (from the applied load F_k^j) for these characteristic elements (*1* and *2*):

$$\begin{cases} F_k^j = N_{k,1}^j + N_{k,2}^j; \\ \Delta \ell_{k,1}^j = \Delta \ell_{k,2}^j = \Delta \ell_k^j; \end{cases} \text{ or } \begin{cases} F_k^j = N_{k,1}^j + N_{k,2}^j; \\ \varepsilon_{k,1}^j = \varepsilon_{k,2}^j = \varepsilon_k^j . \end{cases} \quad (1)$$

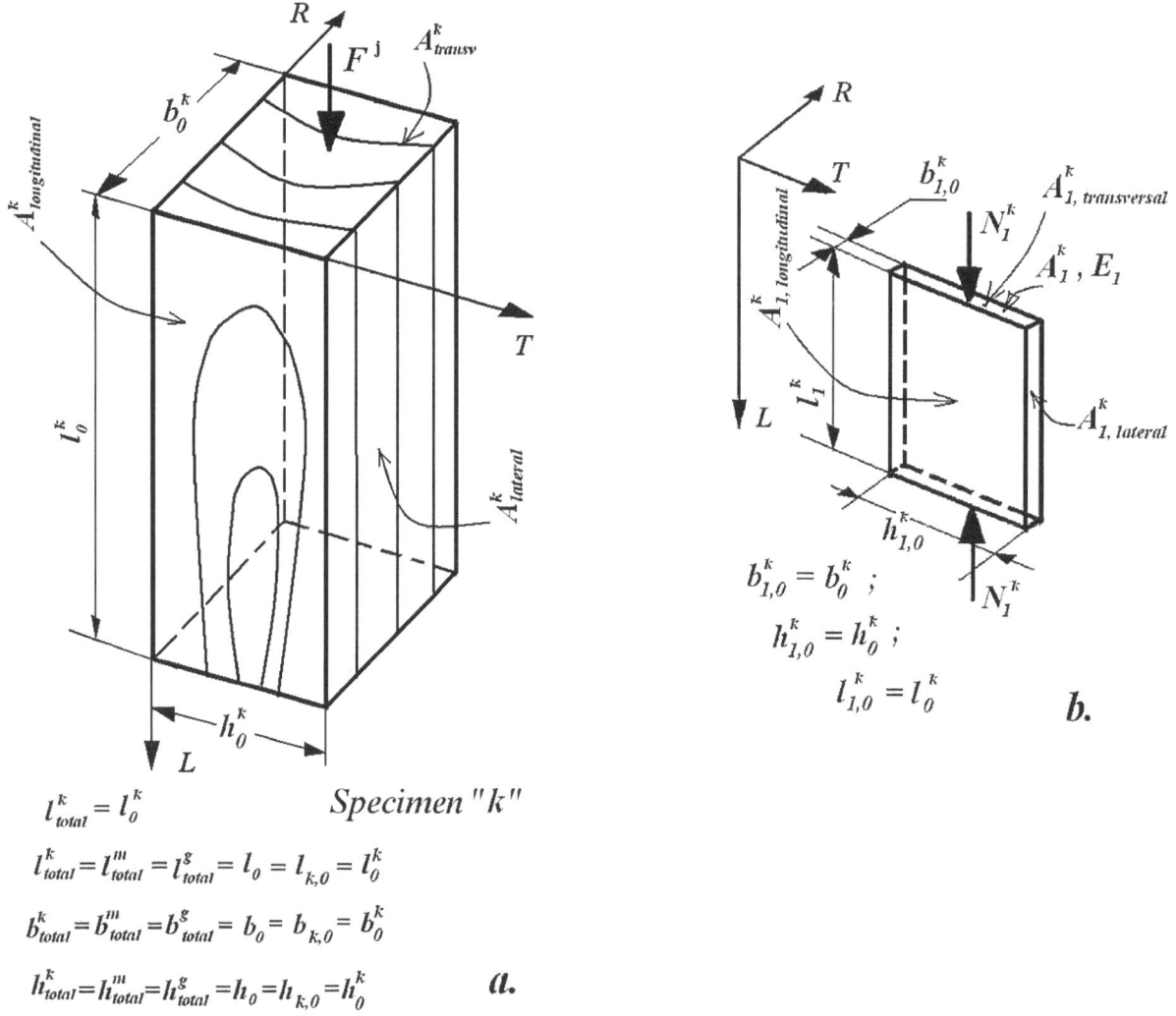

Figure 1
Geometrical parameters for the soft-wood specimen tested and one of the representative elements (here: *1* - early-wood)

The shortening of these two characteristic elements are:

$$\begin{cases} \dfrac{N_{k,1}^{j}\, \ell_{k,0}}{E_1\, A_{k,1}} = \Delta \ell_{k,1}^{j}\,; \\[3mm] \dfrac{N_{k,2}^{j}\, \ell_{k,0}}{E_2\, A_{k,2}} = \Delta \ell_{k,2}^{j}\,, \end{cases}$$ (2)

from where we get the loads that belong to these elements:

$$\begin{cases} N_{k,1}^{j} = \dfrac{\Delta \ell_{k,1}^{j}}{\ell_{k,0}}\, E_1\, A_{k,1} = \varepsilon_{k,1}^{j}\, E_1\, A_{k,1} = \varepsilon_{k}^{j}\, E_1\, A_{k,1}\,; \\[3mm] N_{k,2}^{j} = \dfrac{\Delta \ell_{k,2}^{j}}{\ell_{k,0}}\, E_2\, A_{k,2} = \varepsilon_{k,2}^{j}\, E_2\, A_{k,2} = \varepsilon_{k}^{j}\, E_2\, A_{k,2}\,. \end{cases}$$ (3)

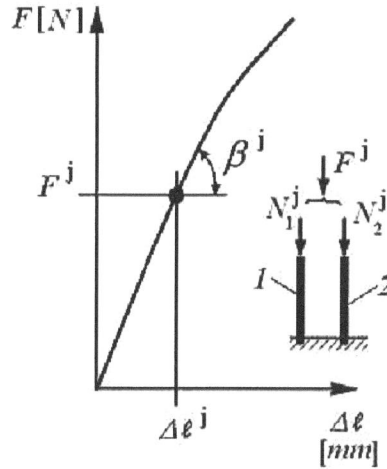

Figure 2
Calculus schema for parallel-connected elements

By substituting them into the static equilibrium equation, finally one can obtain a relationship between the measured values (where we have among others: the real cross-sectional areas and the global strains of the specimen tested) and the Young-modules E_1, E_2 which belong to these two characteristic elements.

Specimen **k**:

$$A_{k,1}\, E_1 + A_{k,2}\, E_2 = F_k^j / \varepsilon_k^j, \tag{4}$$

Specimen **m**:

$$A_{m,1}\, E_1 + A_{m,2}\, E_2 = F_m^j / \varepsilon_m^j. \tag{5}$$

By solving this system, will result the unknown Young moduli E_1, E_2 for the early-, and late wood parts, respectively.

The third specimen (denoted by **g**) has also the same dimensions, but the ratio between the early and late rings' area ($A_{g,1}, A_{g,2}$) is different.

One can determine for this specimen a *global Young modulus*

$$E_g = \frac{\sigma_g}{\varepsilon_g} = \frac{F_g}{A_g\, \varepsilon_g}, \tag{6}$$

and, as for the first two specimens, one can write:

$$A_{g1}\, E_1 + A_{g2}\, E_2 = F_g / \varepsilon_g. \tag{7}$$

and

$$A_{g1}\, E_1 + A_{g2}\, E_2 = A_g\, E_g, \tag{9}$$

which yields

$$E_g = \frac{A_{g1}\, E_1 + A_{g2}\, E_2}{A_g}. \tag{10}$$

By using the global Young modulus E_g, it becomes possible to draw (to predict) the linear zone of the stress-strain curve $\sigma - \varepsilon$, and to compare it to the real, experimentally-obtained, stress-strain curve.

2. EXPERIMENTAL RESULTS

This used investigation method (**V**ideo **I**mage **C**orrelation, VIC-3D) is a full-field one and its 3D version practically eliminates all disadvantages or limitations of the other experimental methods.

In principle, the system consists of two high-resolution video cameras, mounted on a tripod by means of a high-precision connecting rod (see Figure 3).

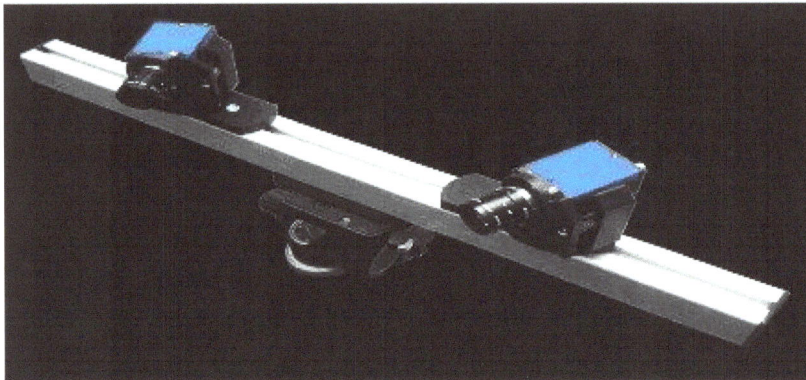

Figure 3. The VIC-3D setup

The system allows measurements in normal working conditions, due to the fact, that *on can eliminate the rigid body movements from the displacements field*. This fact represents one of its main advantages.

The tested object will be sprayed in advance with a water-soluble paint, in order to obtaining a non-uniform dotted surface. The sizes of dotes depend on the surface sizes. In this way one can assure different grey-intensity of each pixel from the analysed surface.

The cameras, after an adequate calibration, made the image acquisition in an [$n.m$] matrix of pixels, firstly for the unloaded tested specimen.

Each captured image (from these two cameras) will be analysed step-by-step, based on the principle schema from Figure 4.

The program allows the pre-selecting of a primarily cell sizes (in this case 5.5=25 *pixels*, see Figure 4.).

For this primarily cell (or sub-set), the program will establish/determine an unique grey-code, correlated to the median pixel, respectively its 3D high-accuracy positioning, too.

By analysing of the whole image (by crossing over it with a pre-selected step: a number of pixels), each primarily cells will obtain a nominated spatial positioning and also an unique grey-code, too.

After loading of the tested specimen, for the new captured images, the program will identify the new positions of these primarily cells.

From these information, based on the high-accuracy calibration, one can obtain both the displacements' field (in space, 3-D), respectively the strains' field, too.

The accuracy of the tested VIC-3D system, from the ISI-SYS GmbH is approximately 1 *micrometer* (1 *micron*), but it depends also on the specimen sizes, optical montage, and displacements magnitudes etc.

The system is very flexible, allowing to evaluate a relatively large field of displacements (starting form some *microns* up to several *mm*), even point-by-point, even along a line, a circle etc.

These results are offered even in colour graph (similarly with the FEM analysis results), even can be exported in Excel-files, and destined to draw-up several useful graphs.

In order to offer validation opportunities for numerical analysis of different structural elements (I mean: to validate their numerical models), the VIC-3D results (the displacement field and the strain field from the structural element's surface) can serve as input data for Boundary Element Method (BEM).

The presented method of Video Image Correlation seems to satisfy in very good conditions also all requirements of the nondestructive testing of the composite materials, where, the wood-based ones occupy an important place.

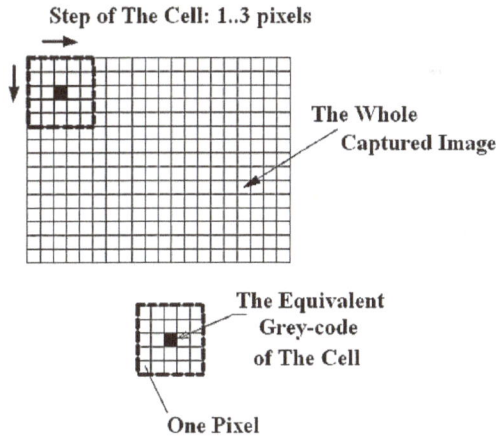

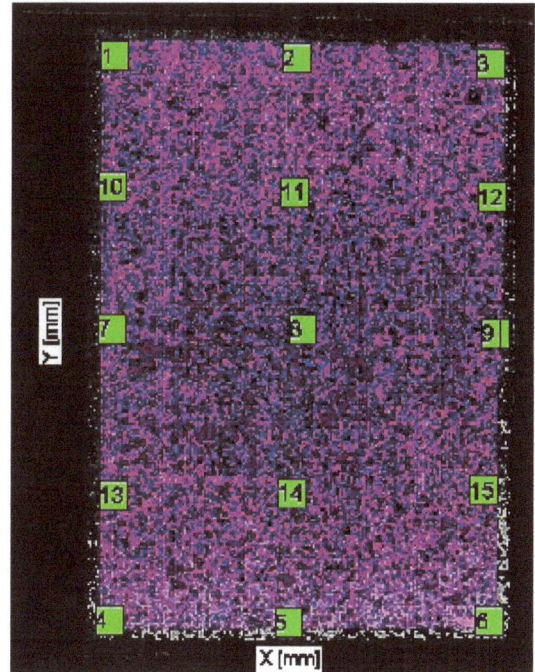

Figure 4.
The measuring principle based on the scanning procedure

Figure 5. The specimen's prepared surface and its selected points for calculus

Based on the mentioned procedure, each soft-wood specimen were prepared (sprayed in advance with a water-soluble paint, in order to obtaining a non-uniform dotted surface).

During the compression tests, the captured images-pairs (by the 2 cameras) are stored and analysed by the Vic-3D software. In this sense, on the analysed surface were pre-selected an adequate number of points, marked in Figure 5 by numbers (1...15).

The obtained information in displacements for each of these selected points allow to determine with a good accuracy their relative displacements, respectively the corresponding shortening/elongations and from them: the adequate strains.

During the compression tests, the force applied and the images captured by VIC-3D were correlated. Finally, one has got the graphs $F - \Delta \ell$ for specimens **k** and **m** during the loading phase (Figure 6), respectively, by making use of the adequate $\sigma - \varepsilon$ curves. Based on an original image analyzer program, it became possible to determine the early-, and late-wood cross-sectional areas.

We have got $A_1 = 364.86 \, mm^2$; $A_2 = 35.14 \, mm^2$ for specimen $L8 \equiv k$. As regards specimen $L6 \equiv m$ we have $A_1 = 352.25 \, mm^2$; $A_2 = 47.75 \, mm^2$.

Finally $A_1 = 364.86 \, mm^2$, $A_2 = 35.14 \, mm^2$

are the results for specimen $L7 \equiv g$. The corresponding Young moduli are as follows:

$$E_{1,average} = 2858.59 \, N / mm^2,$$

$$E_{2,average} = 14486.36 \, N / mm^2.$$

Calculations for the third specimen have been carried out in the same way. Utilizing both the area ratio for the third specimen ($A_{g,1}, A_{g,2}$), and the average values of the Young moduli $E_{1,average}$, $E_{2,average}$ was drawn the analytical $\sigma - \varepsilon$ stress-strain curve as well as that obtained experimentally (Figure 7).

One can observe that these analytical values approximate very well the data obtained experimentally. This fact validates the method we have proposed.

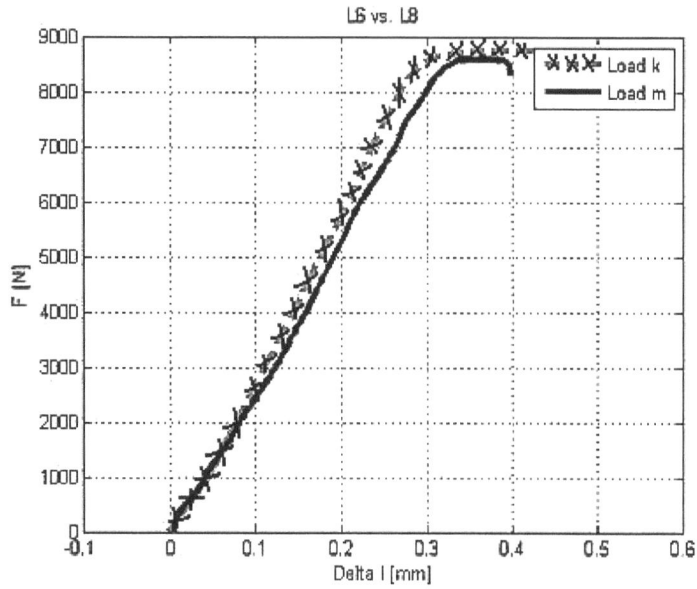

Fig. 6 Force- shortening $F - \Delta\ell$ curves for the specimens k and m

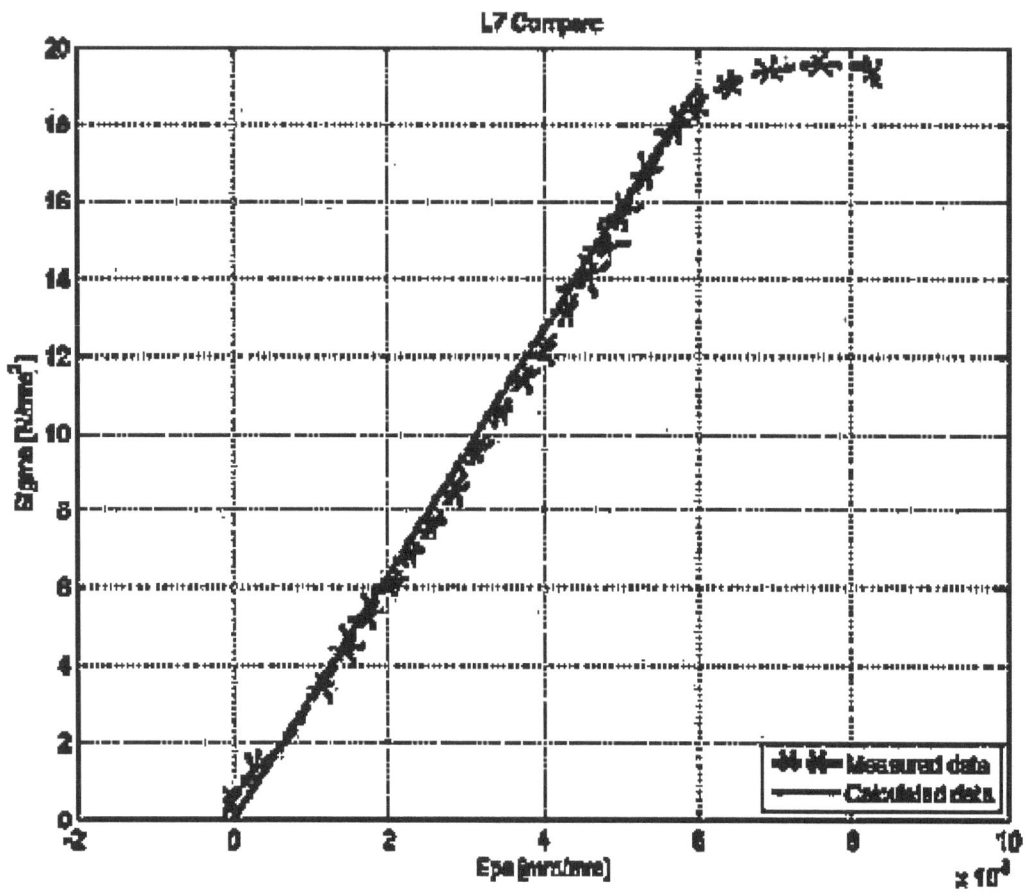

Figure 7
A comparison of the predicted and measured values concerning specimen g

3. CONCLUSIONS AND FURTHER GOALS

The authors have developed/elaborated a new and simple method to establish the individual contribution of the annual rings and their elastic properties.

The main advantage of this new method consists in providing a better approach to the elastic properties of the individual rings, which can also be included in a more accurate finite element modeling.

On the basis of the measurement strategies presented one can determine the elastic constants of the soft-wood (for early and late rings and various types of wood material) with an acceptable accuracy provided that we carry out a representative series of statistical measurements.

Accordingly, the designer has to take only the area ratio of the early and late annual rings of the wood specimen used and he can also predict by analytical or numerical calculus the real loading capacity of that structure.

Determination of the shear moduli which belong to the same main directions needs more complex and very expensive equipment.

ACKNOWLEDGEMENT

The authors, from the "Transilvania" University of Brasov have had a chance to use the VIC-3D (2009), Video Image Correlation system in their experimental investigations for a limited period. We would like to thank the generosity of ISI-Sys GmbH, Germany (system producer) and the Correlated Solution Company, USA (software producer).

REFERENCES

[1]   BODIG, J., JAYNE, B, A.: *Mechanics of Wood and Wood Composites.* Van Nostrand Reinhold   Company, 1982.

[2]   GÁLFI, B., SZÁVA, I.: *Experimental Method to Establish the Individual Fibres' Mechanical   Properties of the Hard-wood Specimens.* The 26th Symposium on Advances in Experimental   Mechanics,   23-26   September,   2009, Montanuniversität Leoben, Austria, Volume of the Symposium,   ISBN: 978-3-902544- 02-5, pp. 63-64.

[3]   SZALAI, J.: *Strength of Materials and Elasticity of Wood and Wood-based Materials-*, vol. 1, West University of Hungary, Sopron, 1994. (in Hungarian)

ON THE SONIC COMPOSITE VISUALIZATION WITH HAPTIC INTERFACES

V. Chiroiu[1], Cornel Brişan[2], Ş. Donescu[3], L. Munteanu[4]

[1]Institute of Solid Mechanics, veturiachiroiu@yahoo.com
[2]Technical University of Cluj-Napoca, Cornel.Brisan@mmfm.utcluj.ro
[3]Technical University of Civil Engineering, stefania.donescu@yahoo.com
[4]Institute of Solid Mechanics, ligia_munteanu@hotmail.com

Abstract: *A model for the virtual reality system configuration using a haptic interface for physically manipulation in order to display tactual information to the user is introduced in this paper. The idea is to have a correct visual feedback of the interface and to use it in controlling the transition between the atomistic and continuum regions for the sonic composites.*

Keywords: *Haptics, Sonic composite, Full band-gap.*

1. INTRODUCTION

A sonic composite is a 3D finite size array composed of scatterers embedded into the matrix. The scatterers are local resonators which scater, diffuse or disperse energy, such as spheres, bars, chains of various geometries made of functionally graded, piezoelectric materials, steel or other materials [1]. Figure 1 shows a scatterer made of carbon nanotube rope segments. Figure 1 shows such a carbon nanotube rope made from 6 subropes, each subrope being composed from 7 groups of single wall carbon nanotubes. Each group contains 25 carbon nanotubes with two different radii (zigzag and armchair 6.26A, $h = 0.617A$ and 16.33A, $h = 0.998A$), and the core group consists of 49 chiral carbon nanotube with the same radius (3.22A and $h = 0.6A$), into a polymeric matrix [1, 2]

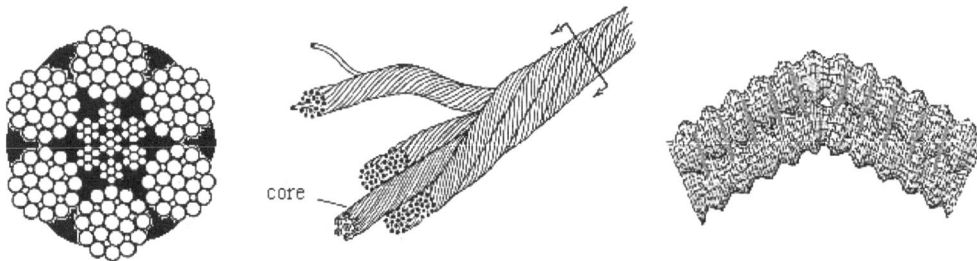

Figure 1 Scatterer made of carbon nanotube nanoropes [1, 2].

The sonic compositer is a sonic version of a photonic crystal and is architectured such that the sound is not allowed to propagate in certain full band-gaps due to complete reflections. We must add that commercially available sound absorbing materials cannot selectively attenuate sound in a desired frequency band. The design and understanding of new architectures of such materials as scatterers have many applications such as acoustic filters, sonic panels, sound shields, transducers, acoustic wave guides for stopping of noise and vibrations in closed spaces (especially for achieving the acoustic comfort control in engineering and applied aeronautics).

The haptic technologies provide new tools to grow the human capacity to manipulate the matter in order to create new materials with desired properties. Haptic interfaces are devices that can be applied for manual interaction of matter with the virtual environments or tele-operated remote systems [3, 4]. The transition or boundary region between the atomistic and continuum regions can be manipulated via the haptic visualization.

2. METHOD

The transition or boundary region between the atomistic and continuum regions (Figure2) is the key of the modeling [5]. At the interface between the atoms and the nodes of the continuum region, we construct a one-to-one correspondence with the aid of the Chebyshev polynomials of the second kind. The polynomial Chebyshev of second kind on the interval $[-1,1]$ are given by [6-8]

$$U_n(x) = [1/(n+1)]T'_{n+1}(x), \; -1 \leq x \leq 1, \; n \in N_0. \tag{1}$$

In (1), $T_n(x) = \cos(n \arccos x)$ are the Chebyshev polynomials of the first kind, and N_0 is a set of natural numbers. The shifted Chebyshev polynomials are defined as

$$U_n^*(t) = U_n(2t-1) = \{1/[2\sqrt{t(1-t)}]\}\sin[(n+1)\arccos(2t-1)], \; 0 \leq t \leq 1.$$

The functions

$$F^*(r,t) = 1/[1+r)^2 - 4rt], \; h^*(x) = 2\sqrt{2(1-t)}, \; |r| \leq 1, \; 0 \leq t \leq 1, \tag{2}$$

are the generating and weighting functions for these polynomials .

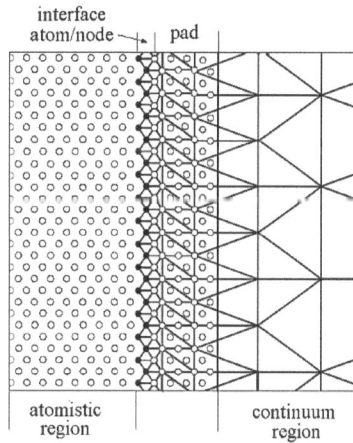

interface
atom/node ⌐⌐ pad

atomistic region   continuum region

Figure 2 The atomistic/continuum transition region [6].

To obtain the basic recurrence relations, the first can be derived for $U_n(x)$ and then for $U_n^*(t)$ by substituting $2t-1$, $0 \leq t \leq 1$ for x. As a result, it is obtained

$$4tU_n^*(t) = U_{n-1}^*(t) + 2U_n^*(t) + U_{n+1}^*(t),$$

$$2tU_n^{*\prime}(t) = 2nU_n^*(t) + 2U_{n-1}^{*\prime}(t) + U_n^{*\prime}(t), \; n \geq 1, \tag{3}$$

$$U_n^{*\prime}(t) = 4nU_{n-1}^*(t) + U_{n-2}^{*\prime}(t), \; n \geq 2 .$$

At the macrosopic, the carbon nanorope is modeled as a rod of length l, with circular cross section of radius less than its length $a << l$. A new parametrization is applied to the region occupied by the body $x \in [-a,a]$, $s \in [0,l]$, for a given time interval $t \in [0,T]$. Instead of $x \in [-a,a]$ we can use $x \in [0,1]$. For any integrable function $A(x,s)$, $x \in [0,1]$, $s \in [0,l]$, $t \in [0,T]$, we consider an expansion of the form

$$. A(x,s,t) = \sum_{k=0}^{\infty} A^{(k)}(s,t)\widehat{U}_k^*(x), \; x \in [0,1], \; s \in [0,l], \; t \in [0,T], \tag{4}$$

where $A^{(k)}(s,t)$ is the kth coefficient in the expansion of $A(x,s,t)$ in the orthonormal Chebyshev polynomials $\{\widehat{U}_k^*(x)\}_{k=0}^{\infty}$ of the second kind. Let us to consider the integral

$$I^{(k)}(A) = \int_0^1 A(x,s,t)\widehat{U}_k^*(x)h(x)\mathrm{d}x, \quad k \in \mathrm{N}_0 \ s \in [0,l], \tag{5}$$

Where $h(s)$ is a properly chosen weighting function. This integral verifies the property of linearity

$$I^{(k)}[\alpha(s)A + \beta(s)B] = \alpha(s)I^k(A) + \beta(s)I^k(B), \tag{6}$$

for any functions A and B of the form (4). Also, it is easily to show that $I^{(k)}(A)$ is equal to the kth coefficient in the expansion of A in these polynomials with respect to x

$$I^{(k)}(A) = \int_0^1 A(x,s,t)\widehat{U}_k^*(x)h(x)\mathrm{d}x = A^{(k)}(s,t), \text{ with } k \in \mathrm{N}_0. \tag{7}$$

The external moments fix the ends of the tube. We suppose the rope deforms by bending and torsion. At the macroscopic scale, the motion of the rod between $t = 0$ and r_{ij} is known from the given mapping

$$S(0,t), \quad \forall t \in [0,t_1], \tag{8}$$

which takes a material point in the initial domain at $t = 0$ to a spatial position at $t = t_1$.

We take s to be the coordinate along the central line of the natural state. The orthonormal basis of the Lagrange coordinate system is denoted by (e_1, e_2, e_3), and the orthonormal basis of the Euler coordinate system by (d_1, d_2, d_3). The basis $\{d_k\}$, $k = 1,2,3$ is related to $\{e_k\}$, $k = 1,2,3$ by the Euler angles θ, ψ and φ. These angles determine the orientation of the Euler axes relative to the Lagrange axes. The curvature C, in the longitudinal direction, the nondimensional curvature c, the deformation parameter ζ, are defined by

$$C = \frac{2ch}{\sqrt{3}R^2(1-v^2)}, \quad c = \frac{\sqrt{3\zeta}}{2}, \quad \zeta = \frac{R - R_c}{R}, \tag{9}$$

where R and R_c, are the radius before and after deformation.

The energy field $E^{tr}(x,s)$ of the transition region verifies the conditions

$$E^{tr}(x,s) \to E^c(x,s) \text{ for } x_{tr} \to x_c, \ s_{tr} \to s_c \tag{10}$$

and

$$E^{tr}(x,s) \to E^a(x,s) \text{ for } x_{tr} \to x_a, \ s_{tr} \to s_a, \tag{11}$$

where

$$E^{tr}(x,s) = \sum_{k=0}^{\infty} E^{tr(k)}(s)\widehat{U}_k^*(x), \tag{12}$$

with $x_{tr} \in [0,1]$, $s_{tr} \in [0,l]$. In (12) $E^{tr(k)}(s)$ is the kth coefficient in the expansion of $E^{tr}(x,s)$ in the orthonormal Chebyshev polynomials $\{\widehat{U}_k^*(x)\}_{k=0}^{\infty}$ of the second kind. These coefficients can be calculated as

$$E^{(k)tr}(s) \to \int_0^1 E^a(x,s)\widehat{U}_k^*(x)h_a(x)\mathrm{d}x, \quad k \in \mathrm{N}_0, \text{ for } x_{tr} \to x_a, \ s_{tr} \to s_a,$$

and

$$E^{(k)tr}(s) \to \int_0^1 E^c(x,s)\widehat{U}_k^*(x)h_c(x)\mathrm{d}x, \quad k \in \mathrm{N}_0, \text{ for } x_{tr} \to x_c, \ s_{tr} \to s_c,$$

with unknown functions $h_a(x)$ and $h_c(x)$ that are determined from an inverse problem such that (10) and (11) hold.

The total potential energy of the coupled atomistic-continuum model is obtained by summing the energies associated with the atomistic, continuum and transition regions as

$$E(x,s) = E^c(x_c,s_c) + E^a(x_a,s_a) + E^{tr}(x_{tr},s_{tr}), \ (x_c,s_c) \in I_c, \ (x_a,s_a) \in I_a,$$
$$(x_{tr},s_{tr}) \in I_{tr} \tag{13}$$

$$I_c \cup I_a \cup I_{tr} = [0,1] \times [0,l], \ I_i \cap I_j = O, \ i,j = a,c,tr.$$

3. RESULTS and conclusions

The matrix resolution is important for the quality of the results [9-12]. There are some techniques to create the perception of depth such as active or passive stereo [9]. The use of active and passive actuators together could make possible to simulate various virtual environments with the stable, realistic illusion [10, 11].

The tool that changes the structure is a sphere. Before the compression the interface is represented in Figure 3. Figure 4 shows different cross sections of the nanorope segments for different values of ζ after compression.

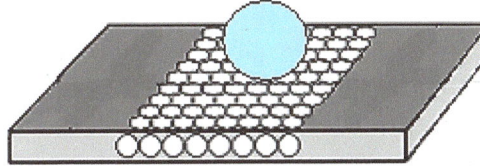

Figure 3. The interface before the compression.

The haptic feedback is calculated from the depth of the indentation of the sphere into the interface and from the neighboring field. The perturbed surface normal n_p and the gradient of the cells height ∇h are given by

$$n_p - n = (\nabla h \cdot n)n - \nabla h, \tag{14}$$

where n is the original normal.

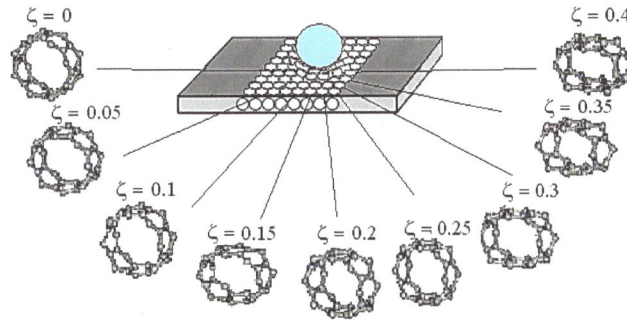

Figure 4. The interface after the compression

The calculations were carried for a sonic plate of length $l = 18$cm and width $d = 2$cm, whith the scatterers made of carbon nanotube rope segments.

The first observed thing is that the macroscopic modeling is valid up to the point of local buckling at $\vartheta = 25.58°$. For larger angles, the equation must be coupled with an atomistic theory. When the external bending moment increases, the axial compression in the tube increases too, and when the compressive stress reaches a critical value, the tube will locally buckle. The value of ζ given by (9) at the point of local buckling is around 0.14.

With the increase in the bending angle ϑ, the top and bottom parts of the kink get closer to each other, and at a certain stage, the distance between them reaches the critical equilibrium distance. Upon additional bending, this distance remains unchanged because there are no external normal loads applied on the walls to prevail over the repulsive van der Waals forces.

For $\vartheta > 25.58°$, the atomistic theory put into evidence a region in which a specific mechanism of deformation appears.

The guided waves are accompanied by evanescent waves which extend to the periodic array of the scatterers surrounding the wave-guide. It is strongly expected that mode coupling waves arise between adjacent wave-guides. The output of the coupled modes is compared with the input waves, as shown in Figure5 for $\zeta = 0.35$. The remarkable result is that the ratio of the coupled and input waves is −3 to −4 dB around the frequency of 7kHz to 7.8kHz in the band-gap of the sonic material.

The visualization shows that a portion of the wall flattens and forms a domain that rotates about a central hinge line. The nanotube becomes a mechanism and the macroscopic modeling is no longer valid in this domain. The major advances in haptics and dynamic contacts can be found in [13, 14].

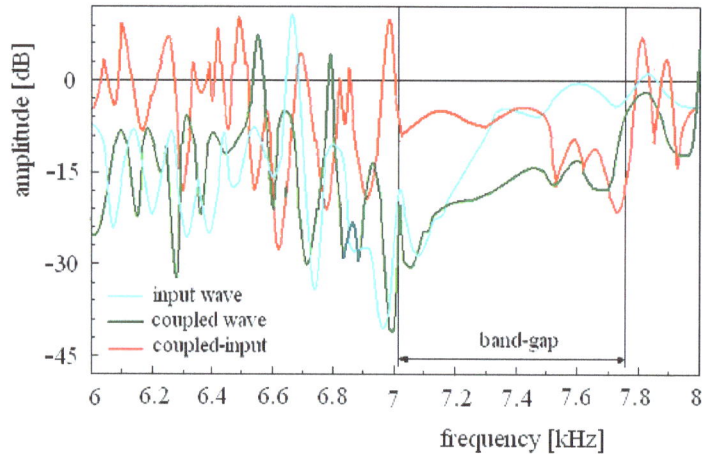

Figure 5 The input and coupled waves for sonic composite.

ACKNOLEDGEMENT

This research was elaborated through the PN-II-PT-PCCA-2011-3.1-0190 Project of the National Authority for Scientific Research (ANCS, UEFISCSU), Romania.

REFERENCES

[1] Munteanu, L., *Nanocomposites*, Editura Academiei, 2012.

[2] Chiroiu, V., Munteanu, L., V.P.Paun, P.P.Teodorescu, *On the bending and torsion of carbon nanotubes ropes* New Applications of Micro-and Nanotechnologies, Series of Micro and Nanoengineering, vol.14, 26-44, Publishing House of the Romanian Academy (eds. M.Zaharescu, M.Ciurea, I.Kleps, D.Dascalu) 2009.

[3] Chiroiu, V., Brisan, C., Donescu, St., Munteanu, L., *On the material visualization system with haptic feedback,* International Semiconductor Conference- CAS 2012, Sinaia, 2012.

[4] Dorjgotov, E., Beneš, B., Madhavan, K., *An immersive granular material visualization system with haptic feedback*, in Theory and Practice of Computer Graphics (Ik Soo Lim and David Duce (eds.) 2007

[5] Curtin, W.A., Miller, F., *Atomistic/continuum coupling in computational material science*, Modelling and Simulation in Materials Science and Engineering, 11, R33–R68, 2003.

[6] Nikabadze, M.U., *A system of equations of the thin-body theory*, Vestn. Mosk. Univ., Ser.1: Mat.Mekh., 1,   30–35, 2006.

[7] Nikabadze, M.U., *Application of Chebyshev polynomials to the theory of thin bodies*, Moscow University Mechanics Bulletin, 62(5), 56–63, 2007.

[8] Nikabadze, M.U., *The unit tensors of second and fourth ranks under a new parametrization of a shell space*, Vestn. Mosk. Univ., Ser.1: Mat.Mekh., 6, 25–28, 2000.

[9] Munteanu, L., Brisan, C., Donescu, St., Chiroiu, V., *On the compression viewed as a geometric transformation*, CMC: Computers, Materials & Continua, 2012 (in press).

[10]   Evans, K.E., Alderson, A., *Auxetic materials: Functional materials and structures from lateral thinking*, Advanced materials, 12(9), pp.617–628, 2000.

[11]   Brisan, C., Csiszar, A., *Computation and Analysis of the Workspace of a Reconfigurable Parallel Robotic System,* Mechanism and Machine Theory journal, vol.46, pp.1647–1668, 2011.

[12]   Pacurari, R., Csizar, A., Brisan, **C.**, *Basic aspects concerning modular design of reconfigurable parallel*

16

manipulators for assembly tasks at nanoscale, Mechanika, 2(76), pp.69–76, 2009.

[13] Srinivasan, M.A., Basdogan, C., *Haptic in virtual environments taxonomy, research status, and challenges,* Comput. & Graphics, vol.21, no.4, pp.393–404, 1997.

[14] Gilardi, G., Sharf, I., *Literature survey of contact dynamics modeling*, Mechanism and Machine Theory, vol.37, pp.1213–1239, 2002.

DENTAL FILLING MATERIALS' YOUNG MODULI DETERMINATION USING ESPI/SHEAROGRAPHY METHOD - PRELIMINARY RESULTS -

Dániel Száva[1], Ioan Száva[2], Sorin Vlase[3], Botond-Pál Gálfi[4],

[1] Assist., PhD stud., University of Medicine and Pharmaceutics Târgu-Mureş, Romania
[2] Prof. dr .eng.,"Transilvania"University of Brasov, Romania
[3] Prof. dr .eng.,"Transilvania"University of Brasov, Romania
[4] Dr .eng.,"Transilvania"University of Brasov, Romania

1. PRELIMINARIES

In order to select the better of the filling materials one should take into consideration not only their biocompatibility but also their mechanical characteristics.

The following two main aspects should be taken into account:

- the complex phenomenon of the curing process, when it is produced certain polymerisation shrinkage of the filling materials and
- the phenomenon of the mastication itself.

Most of the dental practitioners believe that polymerization shrinkage is the causing mechanism for the poor marginal adaptation between the two surfaces (tooth-filling material), ultimately leading to micro leakage and appearance of secondary caries lesions. Although this misconception is in clear contradiction with the facts experienced in dental practice it is widely spread.

The goals of the described investigations consist in proving that the poor marginal adaptation is caused mainly by the different mechanical characteristics (e.g.: Young modulus, Poisson ratio) of the tooth and filling materials, respectively offering an objective selecting procedure/method for the most widely spread dental filling materials in our region from this point of view.

It is a well-known fact that during the mastication, the filling material will be subjected to compression and consequently lateral shrinkage will appear. This shrinkage produces practically a horizontal load on the lateral walls of the tooth's cavity therefore the wall is subjected to a cyclical bending.

If the filling material has a higher value of the Poisson ratio, then a smaller amount of a given mastication force becomes a horizontal load which is, in fact, a periodic load exerted on the tooth's walls. Consequently the tooth filled has a better reliability.

2. B ASICS ON THE ESPI/SHEAROGRAPHY EXPERIMENTAL METHOD

The authors, in their experimental investigations, used an ESPI (**E**lectronic **S**peckle **P**attern **I**nterferometry)/Shearography System (ISI-Sys GmbH, Kassel, Germany). This system allows a high-accuracy evaluation (with some *nanometres* resolution) of the displacements field. In this sense, the so-called *reference plate method* was applied. Figure 1 shows the principle of the optical set-up for ESPI/Shearography systems [2]. Two or four special conceived laser diodes (which will represent coherent light sources) illuminate the object and consequently the interference phenomenon will be produced.

The tested object presents a diffusely scattering surface and thus each surface point is in fact a secondary light source, which reflects (in every direction) the incoming light. Consequently a part of the incident light will arrive to the camera, where, in the focal plane of the camera, one obtains the image of the object. In this sense the light passes through the Michelson interferometer consisting of a divisor cube and the two mirrors (*shear mirror* and *phase shift mirror*).

By tilting the shear mirror with a small angle $\alpha_s / 2$ two points on the surface of the object (P_1 and P_2) will be superposed in a single point P of the focal plane.

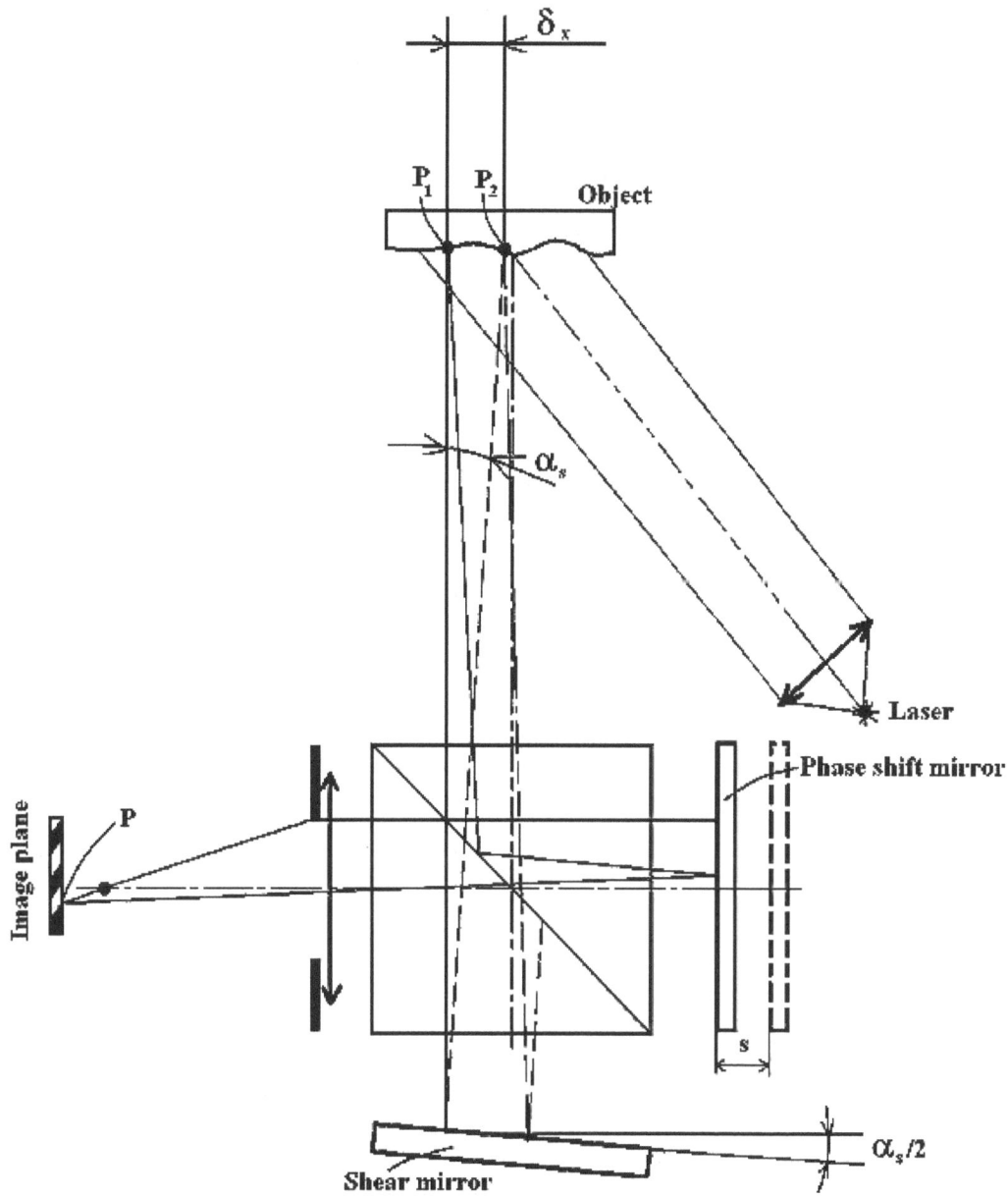

Fig. 1
The ESPI/Shearography system's optical set-up [2]

Therefore the corresponding light beam paths interfere on the focal plane of the camera (in the image plane). The corresponding $\delta_x = P_1 - P_2$ *shear amount* of the mentioned *shear angle* α_s can be set in various axial directions, depending, of course, also on the laser diodes spatial positioning.

Using the second mirror (*phase shift mirror*) of the Michelson interferometer, one can actively create light path changes for one of the partial beams. Consequently, by using multiple intensity measurements with actively altered light path changes for the partial beam, we can determine *the relative phase length for the respective pixel* (in the image plane of the camera) as regards the interference phenomenon. The path length changes may be used in establishing useful conclusions which, for example, are produced /caused by the deformation of the object surface.

In practical application of ESPI/Shearography, the relative phase changes between two (static) states of the tested object are determined.

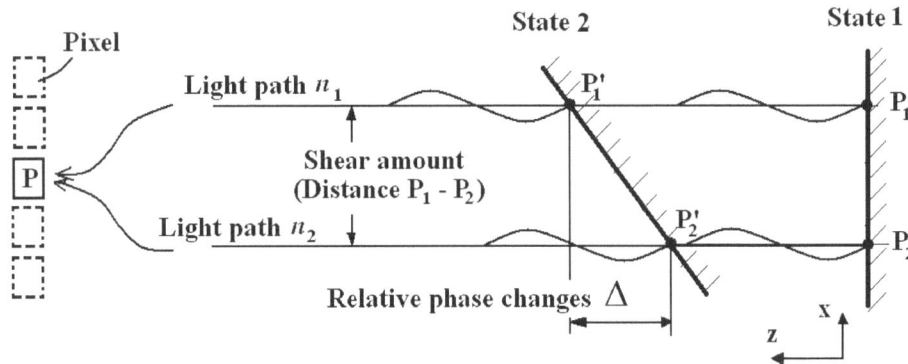

Fig. 2 Simplified correlation between
(a) the shear amount δ_x, the relative phase change Δ and
(b) the state's change of the object due to deformation [2]

Based on reference [2], Figure 2 shows the simplified correlation between (a) *the shear amount δ_x*, *the relative phase changes Δ* and (b) the state change of the object due to deformation (translation or/and rotation). This figure/scheme doesn't include the practical realisation of the radiation overlapping via interferometer, as is mentioned in [2].

For practical applications overlapping of the surface of the body to be investigated is realised by the use of an un-deformed surface (*reference object*). For a relatively small overlapping (shear amount) we have *Shearography*, for a large one: *ESPI (with the Reference Plate Method)*.

Consequently it becomes possible to determine and evaluate the displacement components and some deformations (both in 2D and in 3D) on the surface of the object tested.

3. EXPERIMENTAL RESULTS

The authors of the present paper conceived an original loading device and testing bench. To determine the Young modulus of the distinct dental filling materials small cylindrical samples were made and then we subjected them to uni-axial compression.

Figure 3 show the experimental set-up and the sizes of the small specimen, respectively. The small reference plate *1* (with a width of 5 *mms*) is appropriately (place / set beside) put beside the specimen tested (*2*). The laser diodes *3* and *4* assure a good and equal illumination on the surface observed. They are fixed onto the high-stiffened polycarbonate rods (*7, 8*). The Michelson Interferometer *5* and the four-Mega pixel CCD cameras *6* are disposed in normal direction to the object. The distances in Figure 3 are given in *mms*. A comparison of the image that belongs to the small and unloaded plate to that of the specimen makes possible a good and accurate in-plane strain analysis.

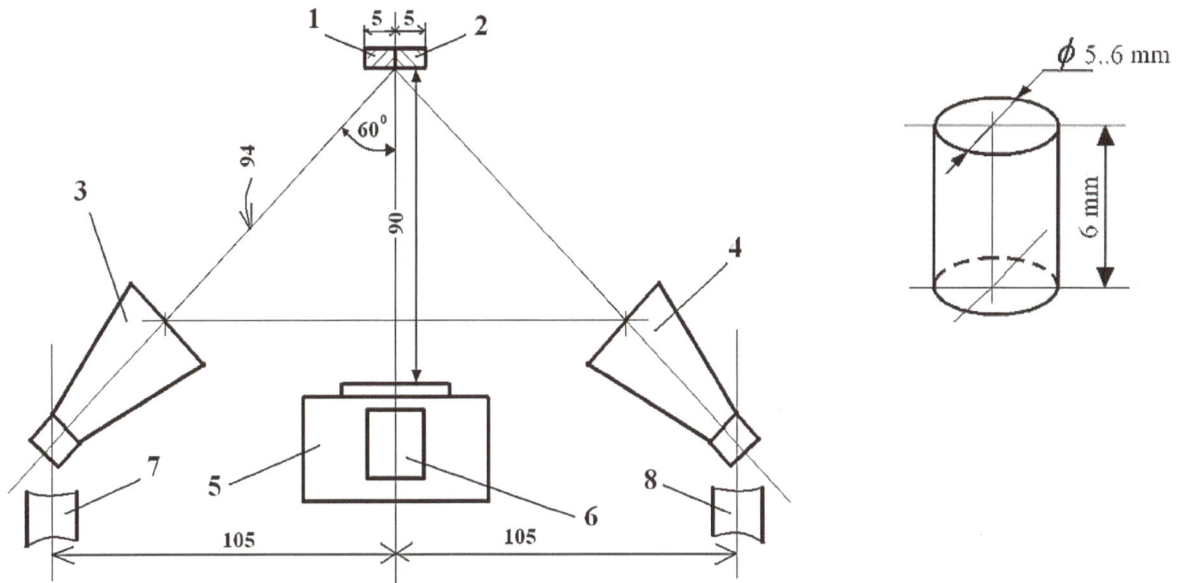

Fig.3 Experimental set-up and dimensions for the specimen tested

Were tested two types of filling materials:

- *composites* (TE-Econom, Carisma, Extra Fil) and
- *glass-ionomers* (Fuji Gp IX, Ketach Molar, Ionofil +) .

One has to remark that an adequate number of specimens were manufactured in order to ensure a reliable statistical evaluation.

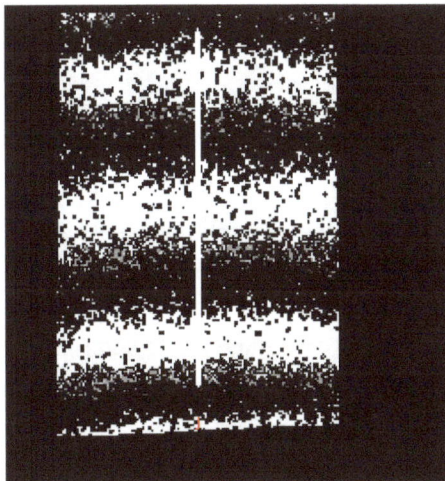

Fig. 4
Cylindrical specimen #1, Filtered data

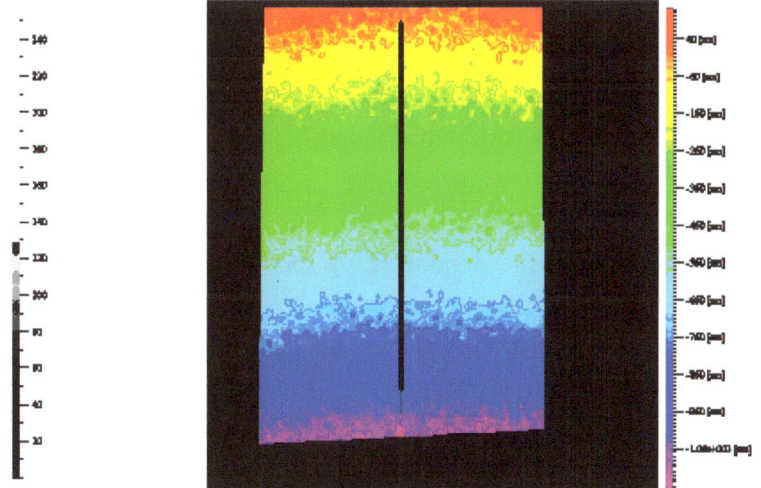

Fig. 5
Cylindrical specimen #1, Evaluated data

In order to illustrate the most important steps of the displacement's field evaluation, Figures 4 and 5 show, for a cylinder-shape specimen, two of the most important steps in this sense: the filtering and the final evaluation along a central vertical line (having a known length ℓ_0), which represents the x-axis of the specimen.

Using the values of the displacements on a selected line (e.g. a central/median line of the specimen, but not only there!) one can determine the elongation $\Delta\ell$ and then the corresponding axial strain ε_x. With the knowledge of the load applied and the cross sectional area of the specimen one can determine the axial stress σ_x. Consequently one can draw the stress-strain curve $\sigma_x - \varepsilon_x$ from where the sought Young modulus can be determined without difficulties. Figure 6 illustrates the accuracy of the strain-evaluation procedure.

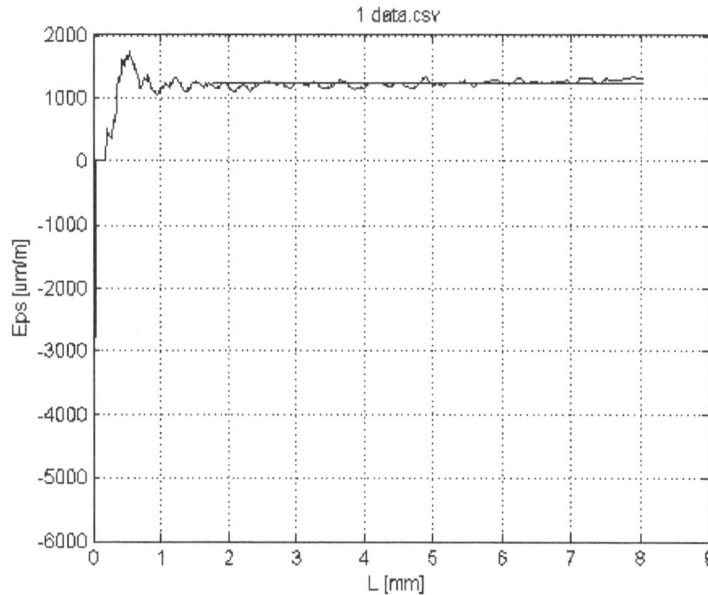

Fig. 6 Axial strains for a given force

The selected initial length $\ell_0 = 8\,mm$ was divided in a very high number of pixels and starting from the 100th pixel (!), corresponding to approximately $0.4\,mm$, it became possible an adequate calculus, presented by the zigzag – graph. Taking a mean value (represented by a straight line) of the zigzag graph (starting from the mentioned 100th pixel), finally was obtained the strain $\varepsilon_x = 1240.1\ \mu m/m$. It should be mentioned that this value belongs to a fixed value of the applied compressive load (force). This procedure was repeated for each load so that one could draw the final stress-strain curve $\sigma_x - \varepsilon_x$ and could determine the Young modulus as well. Figure 7 shows the Young moduli for the distinct filling materials which were tested.

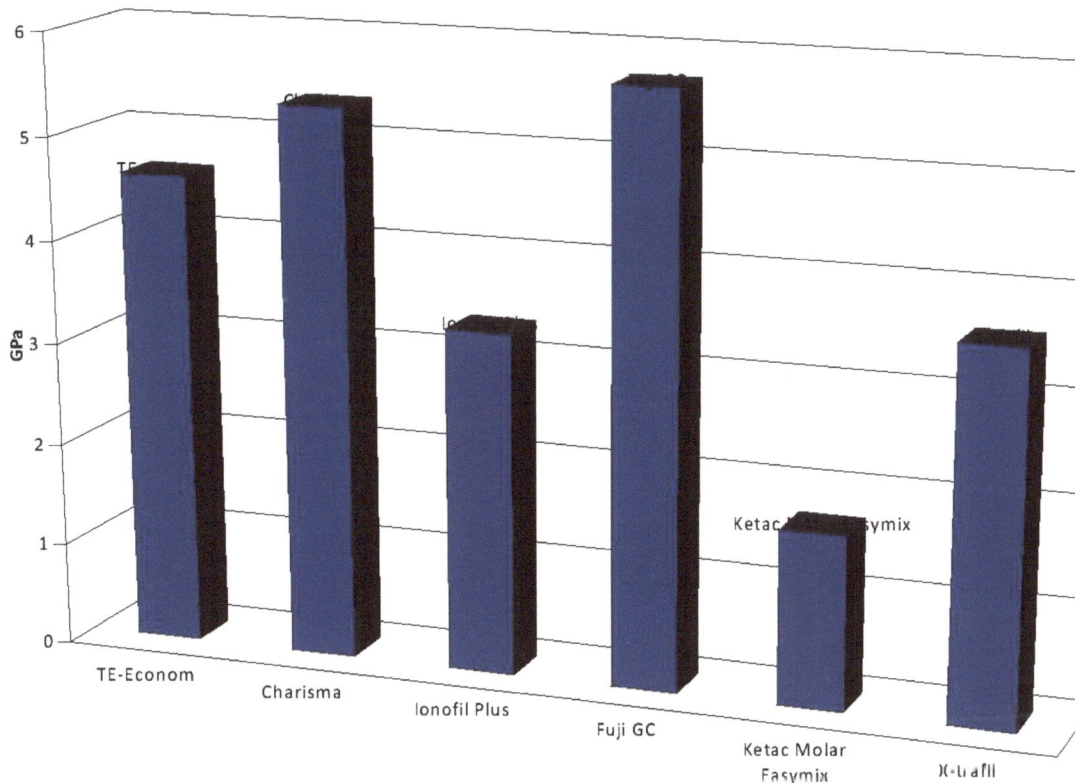

Fig.7 Young moduli for the distinct filling materials

Based on these differences in the mechanical characteristics one can evaluate and decide what filling material should be preferred or used.

REFERENCES

[1] Sutton, A. M., Orteu, J. J., Schreier, W. H., *Image Correlation for Shape, Motion and Deformation Measurements,* Springer Verlag, 2010

[2] *** *User Manual and Service Instructions SE3 Shearography/ESPI-System*, ISI-Sys GmbH, Kassel, Germany, 2009.

[3] Száva, D., Gálfi, B., Száva, I., Orbán, Piroska, *Research and Academic Exploration about Optimal Attitudes of Dental Filling Materials*, Proceedings of the 10th YSESM Symposium (Youth Symposium on Experimental Solid Mechanics), Chemnitz University of Technology, Germany, 25th-28th May, 2011, pp. 107-108, ISBN: 978-3-941003-34-7.

[4] Kobayashi, A.S. Handbook of Experimental Mechanics. Prentice Hall, Upper Saddle River, New Jersey, 1987.

[5] Jones, R., C. Wykes. Holographic and Speckle Interferometry, A Discussion of the Theory, Practice and Application of the Techniques. Cambridge University Press, Cambridge, 1983.

DESIGN OPTIMIZATION OF COMPOSITE AEROSPACE STRUCTURES

C. Casavola, L. Lamberti, C. Pappalettere
Dipartimento di Meccanica, Management e Matematica
Politecnico di Bari, Viale Japigia 182 - 70126, Bari, Italy
casavola@poliba.it; lamberti@poliba.it; c.pappalettere@poliba.it

1. INTRODUCTION

Thin-walled elements are widely utilized in the design of aerospace structures as they allow to obtain lightweight structures. However, while the structures so designed may be sufficient to carry the in-plane tensile loads and satisfy strength requirements, they are often prone to fail under buckling induced by compressive or shear loads. Buckling is a highly non-linear phenomenon caused by the sudden conversion of a large amount of in-plane strain energy into bending strain energy [1]. This process may be expressed in different modes. For example, global buckling is said the catastrophic collapse of the entire structure; local buckling affects a portion of the skin; stiffener buckling occurs in correspondence of stiffener segments. Other "special" buckling modes are typical of sandwich structures (wrinkling, dimpling, etc.).

In order to guarantee structural safety against buckling, reinforcing elements (stiffeners) running in the longitudinal and transverse directions are added to the panel skin to increase the stiffness of the structure. The proper selection of the stiffening configuration and materials leads to minimize structural weight thus maximizing payload. A number of trade studies on the sensitivity of structural weight of stiffened panels to parameters such as stiffener geometry, materials, and type of construction and manufacturing methods were presented in literature (see for example, [2,3]). Design concepts are usually compared and preliminarily selected to be further developed later on the basis of their structural weight (typically, the weight per unit area) which is minimized by means of optimization methods.

Structural optimization is mandatory in the design of aerospace structures. Design variables are repeatedly perturbed to satisfy non-linear constraints on displacements, stresses and critical buckling loads. Optimization methods can be divided in two main groups: Approximate Optimization Methods (AOM) and Global Optimization Methods (GOM). AOM formulate and solve a set of sub-problems where the original non-linear functions of the optimization problem are replaced by linear, quadratic or higher order approximations built including gradient information. Larger fractions of design space can be explored using multi-start approximate optimization (MSAO) where different optimization runs are performed starting from points generated randomly. The main difficulty of approximate optimization is to keep the quality of the approximation as highest as possible and always reliable in the region of design space currently being searched. Furthermore, approximate models should change during the optimization process based on the sequence in which design constraints become active.

Global optimization methods search the optimum design by generating randomly a certain number of trial designs. This is done in purpose to expand the portion of design space explored by the optimizer thus increasing the probability of finding the global optimum without being stuck in local optima. Meta-heuristic optimization methods are more efficient than purely random search as they generate trial designs based on some inspiring principle taken from physics, biology, astronomy, evolution theory, music, social sciences etc [4].

While stiffeners increase the bending stiffness of thin-walled members, they add an extra dimension of complexity to the model compared to unstiffened plates and shells. Furthermore, stiffened structures often employ a repeating stiffener pattern. The repeated (periodic) nature of the geometry allows the use of simplifying assumptions to obtain approximate analyses.

As is mentioned before, design is repeatedly perturbed in order to find the optimal configuration. Hence, optimization of aerospace structures may entail several thousands of finite element analyses including hundreds of thousands of elements and computationally expensive tasks such as nonlinear buckling analysis. Because this may be openly in contrast with constraints on time and computational resources, engineers must dispose of surrogate models able to quickly evaluate structural response under loading conditions. However, surrogate models may err on the conservative side thus leading to unnecessarily heavy designs. Furthermore, optimizers may exploit any weakness or deficiency in modeling. Finally, the level of refinement adopted in discretizing models may affect computation of sensitivities in terms of a certain amount of noise introduced in the optimization process. In view of this, the structural optimization problem must include additional constraints to keep the design search in regions where modeling assumptions on which surrogate models lay may hold true.

The design of composite stiffened panels presents even more challenges due the additional failure modes and anisotropy effects introduced by composite materials [2-3, 5]. Optimizing ply orientations in preliminary design is not recommended for a number of reasons. Composite laminates have failure modes that are difficult or expensive to model and analyze during design optimization. Furthermore, preliminary design optimizations are generally performed by considering a small number of load cases. Optimizing ply orientations using a small set of design load cases and simple analyses can produce laminates that will fail under neglected failure modes or loading conditions.

In addition to sizing optimization for fixed laminate thickness and stacking sequence, individual ply thicknesses can be optimized keeping the same stacking sequence. These optimizations require two iterations. In the first iteration, the ply thickness and size variables such as stiffener dimensions and spacing are optimized as continuous variables. The optimum ply thickness from the continuous optimization is rounded to the nearest integer multiple of pre-preg ply thickness. The panel design variables, excluding ply thicknesses are once again optimized to ensure that the ply thickness rounding did not result in suboptimal or infeasible designs.

Composite stiffened shells hence permit optimization on more design variables because the individual ply thickness can be changed in addition to the stiffener sizes and spacing. Increasing the number of optimization variables (design freedom) allows to design lighter structures but also increases the imperfection sensitivity of the optimum designs and makes them less robust.

Composite designs may be significantly affected by thermal considerations. The optimized design results from a complex combination between structural safety requirements, imperfection sensitivity and the layout and materials of thermal insulation/protection systems. For examples, sandwich shells may be not very efficient in terms of weight if face-sheets must be designed thick enough to avoid permeation of liquid propellants. Using non-symmetric layup constructions permits to overcome this limitation but introduces issues on structural stability.

This paper will review the most important aspects involved in the design optimization of composite stiffened structures for aerospace applications. It will be shown how structural weight may be affected not only by the number of design variables but also by the level of accuracy of the surrogate models employed in the structural analysis. An example of trade study carried out on composite stiffened panels utilized for second generation reusable launch vehicles is included in the discussion. A comparison between deterministic optimization and probabilistic optimization is presented as well.

Finally, a general approach to structural design of composite aerospace structures is discussed. This approach is based on a multi-level architecture. In Level 1, a series of detailed finite element analyses of the entire structure or segments/parts serve to establish exact correlations between numerical models and experimental tests. This information is included in Level 2 which is based on the use of approximate analysis methods corrected by response surfaces depending on design variables. In Level 3, software tools developed in Level 2 serve to optimize the structure ensuring that the constraints put on structural response reproduce reliably the overall behaviour of the structure. In Level 4, optimized designs found in Level 3 must be verified by means of the detailed finite element model built in Level 1. Sensitivity of optimized design configurations with respect to design variables (for example, in terms of changes in critical buckling loads for small design perturbations in the neighbourhood the optimum design) is evaluated. Sub-optimal designs could even be chosen if they were judged more affordable in terms of manufacturing processes.

BASIC REFERENCES

[1] D. Bushnell *Computerized Buckling Analysis of Shells*. Kluver Academic Publishers: Dordrecht (The Netherlands), 1992.

[2] L. Lamberti, S. Venkataraman, R.T. Haftka, F.F. Johnson. Preliminary design optimization of stiffened panels using approximate analysis models. *International Journal for Numerical Methods in Engineering* 57 (2003) 1351-1380

[3] L. Lamberti, S. Venkataraman, R.T. Haftka, F.F. Johnson. Challenges in comparing numerical solutions for optimum weights of stiffened shells. *AIAA Journal of Spacecrafts and Rockets* 40 (2003) 183-192.

[4] A.H. Gandomi, X.S. Yang, S. Talatahari, A.H. Alavi (Eds.) *Metaheuristic Applications in Structures and Infrastructures*, Elsevier (The Netherlands), 2012.

[5] C. Casavola, L. Lamberti, C. Pappalettere. Experimental testing, analysis and optimization of aerospace structures and components. In: *Advances in Engineering Research* (V.M. Petrova, Ed.), Vol. 1, Chapter 14, pp. 437-462. Nova Science Publishers, New York (USA), 2012.

FEM MODELLING OF AN AUTOMOTIVE DOOR TRIM PANEL MADE OF LIGNOCELULOZIC COMPOSITES IN CASE OF A DOOR SLAM SIMULATION

O.M. Terciu[1], I. Curtu[1], C. Cerbu[1]

[1] Transilvania University of Brasov, Romania

ovidiu-mihai.terciu@unitbv.ro; curtui@unitbv.ro; cerbu@unitbv.ro

Abstract: The paper presents the finite element analysis of an automotive door trim panel made of lignocelluloses composites in order to determine the stresses and displacements in case of a door slam simulation. In research was investigated a new lignocelluloses composites made of polymers reinforced with woven fabrics of natural fibres and wood particles. The mechanical properties of this material were determined experimentally and used as input data in FEM simulations. The FEM results emphasized that new material improves component stiffness compare with classical materials used.

Keywords: lignocelluloses composites, mechanical properties, automotive, door slam, door trim panel, FEM

1. INTRODUCTION

By reducing vehicle weight will reduce the amount of energy required for movement, which has major implications on the level of environmental pollution. An obvious way to this is to extend the use of lightweight composite materials with multiple functions in the body structure components without using one material for each of the requirements.

Current trends in the field of composite materials, due to environmental and economic considerations, focus on achieving low environmental impact materials and possible usage of waste resulting from other industrial processes. In this class is enclosed wood waste as sawdust resulting from cutting wood [6].

Usage of materials reinforced with natural fibres for vehicles construction and not only has become a challenge and a subject of great interest due to the great volume and variety of renewable resources and biodegradable materials that influence production cost of materials.

Composite materials made of natural fibres and polymer matrix provides synergistic properties, improving their strength and durability, thus recommending the use of matrices with high resistance to aggressive environmental factors [3, 6]. These materials are suitable for achieving automotive interior components, where in addition to their low weight have also high rigidity and good thermal and sound insulation. The most important vehicle elements include car door panels.

2. MATERIALS AND METHODS

Door panel materials used in FEM analysis were the classic ones, such as plastic polypropylene (PP) and also some new materials were used. The new lignocellulosic composite material is a laminate having two layers made of polyester resin reinforced with plain weave fabric of flax fibres (FUP) and one middle layer made of polyester resin reinforced with wood sawdust of oak (OUP). In order to determine the main mechanical properties of new natural fibre reinforced composite material, each layer of material has been tested to tensile stresses [1, 5]. Tensile test is known to be the most important and commonly used static test due to the procedure's simplicity on obtaining the strength and stiffness characteristics [2, 4]. Mechanical characteristics of the new material were needed to simulate the behaviour of parts made of these materials by finite element method (FEM).

Elastic properties of composite layers and the polypropylene properties are shown in Table 1. Direction of the weft yarn fabric reinforced lamina corresponds to the x direction of the part.

Table 1: Materials properties

Materials	E_1 [MPa]	E_2 [MPa]	υ_{12}	G_{12} [MPa]	G_{13} [MPa]	G_{23} [MPa]	Density, ρ [Kg/m³]
FUP	4711	2787	0.35	1800	1800	1800	1187
OUP	3041	3041	0.37	-	-	-	1077
PP	1300	1300	0.35	-	-	-	906

In Figure 1 are presented images with left front door trim panel designed in Catia V5 software.

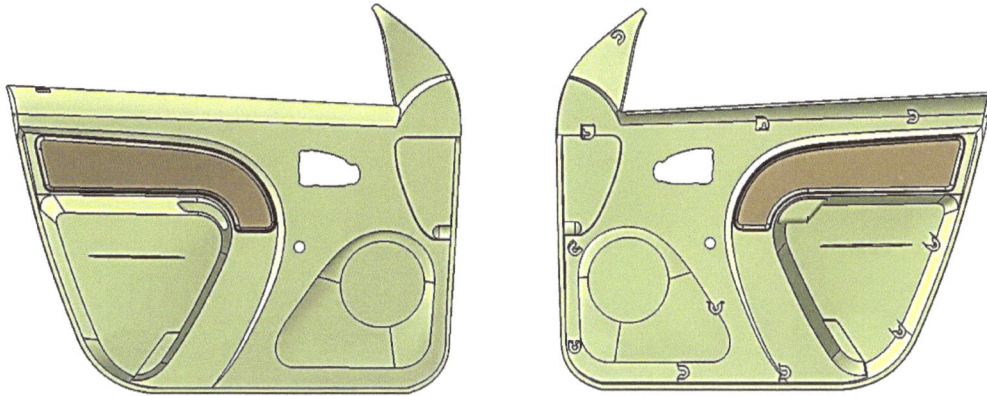

Figure 1: Door trim panel designed in Catia V5 software

Structure of the new lignocellulosic material is presented in figura2.c. According to the literature the main common mechanical stress, which affects the life of the door panels, is the shock produced by closing the door. In this case the door trim panel is loaded with an acceleration field resulting from inertia forces due to its own weight (Fig. 2.a and 2.b).

Meshing was done using shell elements with 6 graders of freedom, with dimensions of about 5 mm. As much as possible quad elements were chosen, but in areas with pronounced 3d curves triangular elements were chosen. Figure 3 presentet the door trim panel discretized in finite elements.

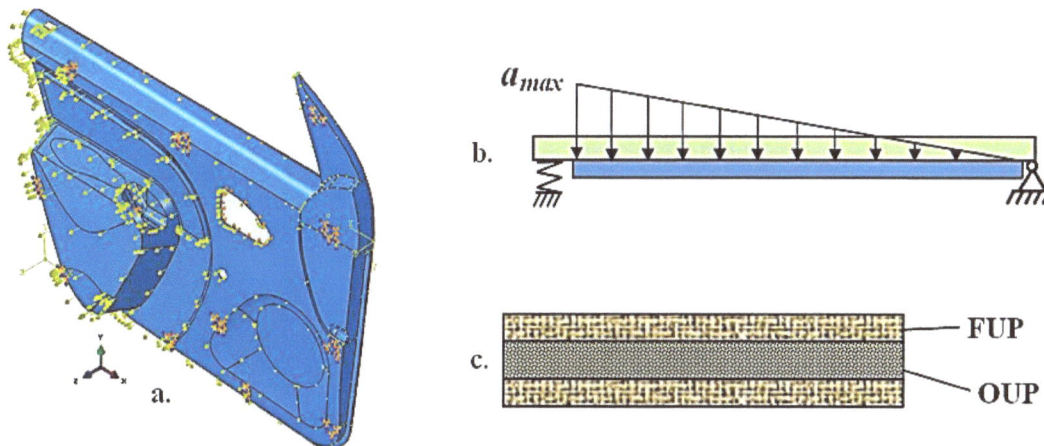

Figure 2: Acceleration distribution on door trim panel and composite material structure

Door panel has several areas where constraints are applied as follows:
- on clips systems panel mounting metal structure areas, shiftings were blocked on all 3 directions (U1, U2 and U3);
- upper part rests on a metal door structure, thus blocking shiftings on direction 2 (U2) (Fig. 6.7, b);

• in screw mounting areas shiftings were blocked on all 3 directions (U1, U2 and U3) and rotations on two directions (Fig. 4.b);

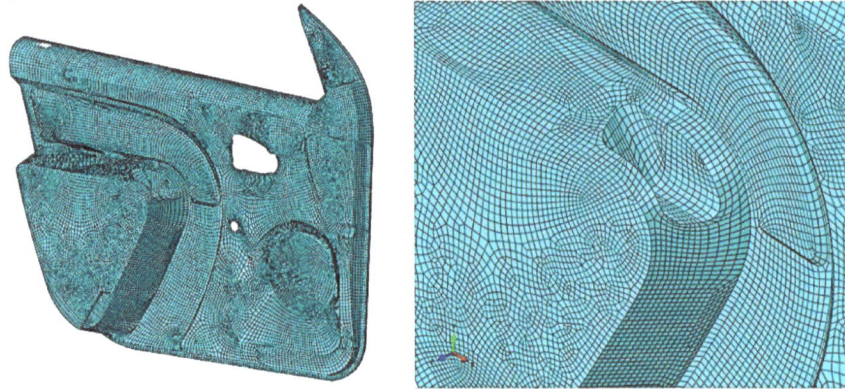

Figure 3: Discretized door trim panel

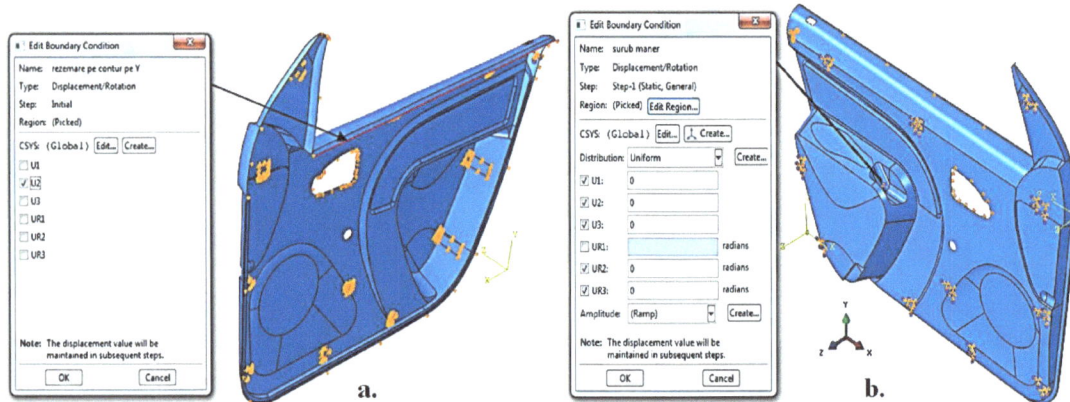

Figure 4: Setting constraints for clips fixing areas and bearing on metal structure

Figure 5 presents the equivalent stress distribution on the middle surface of the layer 1 and displacements distribution on the axis z (w) and position of the maximum values.

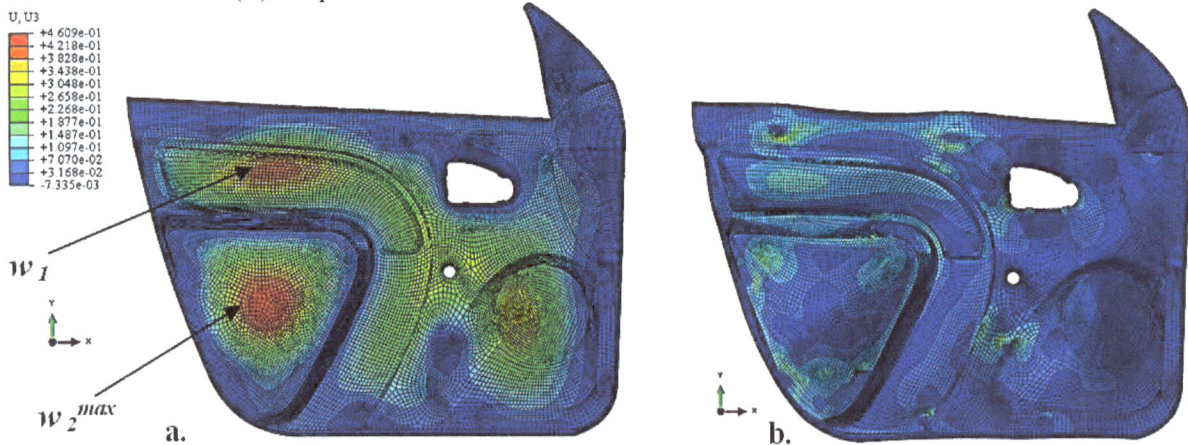

Figure 5: Displacements distribution on the z axis (w) and tensions distribution of composite laminated panel trim door

3. RESULTS AND DISCUSSION

Comparing the results obtained by finite element modelling for the two types of structures, in Figure 7 can be seen that for an acceleration of the impact of 350 m/s^2, displacements of the two points in the lignocellulosic composite moulded panel are lower by 43% than the plastic moulded panel with classic material like polypropylene. This values displacements decrease is due to greater rigidity of lignocellulose material and smaller mass of the panel, given its reduced thickness. Since when slamming door, door panel is loaded with inertial forces due to own weight, results that it's mass influence the load degree.

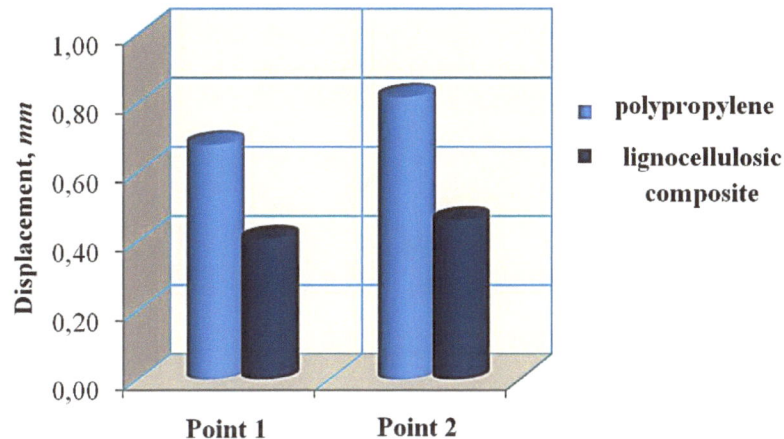

Figure 6: Displacements values in the two points of the door panel for the two materials studied at an acceleration of 350 m/s^2

4. CONCLUSION

Modelling parts with complex geometry involves the use of specialized software to achieve virtual model.

From equivalent stress distribution of analysed component can be seen that these areas have high levels on fixing areas to metal structure.

Displacements of lignocellulose materials door trim panel obtained by FEM are smaller by 43% than those of polypropylene panel.

The small values of displacements resulting for lignocellulosic composite component are due to the high rigidity and also low weight material layer given its lower thickness of the panel, this decreasing from 1.81kg to 1.49kg.

ACKNOWLEDGEMENT

This paper is supported by the Sectoral Operational Programme Human Resources Development (SOP HRD), financed from the European Social Fund and by the Romanian Government under the contract number POSDRU/88/1.5/S/59321

REFERENCES

[1]   Cerbu, C., Curtu, I., Constantinescu, D. M., Miron, M. C., "Aspects concerning the transversal   contraction in the case of some composite materials reinforced with glass fabric", MATERIALE   PLASTICE,   vol.   48,   No.4, 2011, ISSN: 0025-5289;

[2]   Cerbu, C., Curtu, I., Ciofoaia, V., Rosca I. C., Hanganu, L. C., "Effects of the wood species on the mechanical characteristics in case of some E-glass fibres/wood flour/polyester composite materials", MPLAAM, vol. 47, no. 1, pp. 109-114, 2010, ISSN 0025/5289;

[3]   Curtu, I., Stanciu, M.D., Motoc, D.L., "Diagnosis of dynamic behaviour of lignocelluloses composite plates", in Proceedings of 14th European Conference on Composite Materials, Budapest, 2010, p. 75, ISBN   978-963-313-008-7;

[4]   Terciu, O.M., Curtu, I., "New Hybrid Lignocellulosic Composite made of Epoxy Resin Reinforced with   Flax Fibres and Wood Sawdust ", REVISTA MATERIALE PLASTICE, vol. 49, No. 2, pp. 114-117,   2012,   ISSN 0025/5289;

[5]   Terciu, O.M., Curtu, I., Teodorescu-Draghicescu, H., "Effect of wood particle size on tensile strength   in case of polymeric composites", in Proceedings of the 8th International Conference of DAAAM Baltic, INDUSTRIAL ENGINEERING, Tallinn, Estonia, 2012, ISBN 978-9949-23-265-9;

[6]   Vezzoli, C., Manzini, E., Design for enviromental sustainability, Verlag, London, Springer, 2008, ISBN 978-1-84800-162-6;

A NONLINEAR MECHANICAL MODEL FOR HETEROGENEOUS CURVED BEAMS

György Szeidl[1], László Kiss[2]

[1,2] Department of Mechanics, University of Miskolc, Miskolc, HUNGARY

gyorgy.szeidl@uni-miskolc.hu, mechkiss@uni-miskolc.hu

Abstract: *This paper is devoted to the stability problem of curved beams provided that the beam is made of heterogeneous material. It is assumed that (a) the radius of curvature is constant and (b) the Young modulus and Poisson number depend on the cross sectional coordinates only. We have the following objectives: (1) derivation of an appropriate model provided that the beam is not a shallow one, (2) determination of the critical load assuming that the beam is subjected to a constant radial load at the crown point, (3) comparison of the results obtained with solutions valid for shallow arches.*

Keywords: *curved beams, heterogeneous material, stability problem, critical load*

1. INTRODUCTION

The research into the mechanical behavior of curved beams began in the 19th century [1]. As regards stability problems a number of papers could be cited. It might be worth mentioning book [2] which is a collection of the earlier results. As regards the behavior of shallow arches we should cite papers [3], [4], [5], [6] and book [7] by Dym. The results achieved by Dym have been more accurate in the paper series [8], [9], [10], [11] written by Y. L. Pi, M. A. Bradford and their co-authors in Australia.

The present paper is organized into five sections. Section 2 presents the fundamental relations. Section 3 is devoted to the derivation of the governing equations. The results are presented in Sections 4 and 5. The last section contains the conclusions. Because the length of this paper is limited, the derivations are in general shortened or omitted. Instead, we have laid the emphasis on the line of thought and the results.

We shall assume that the curved beam is made of nonhomogeneous, isotropic, linearly elastic material. Within this framework the elastic parameters can be varied arbitrarily over the beam cross section but they are independent of the coordinate perpendicular to the cross section.

2. FUNDAMENTAL RELATIONS

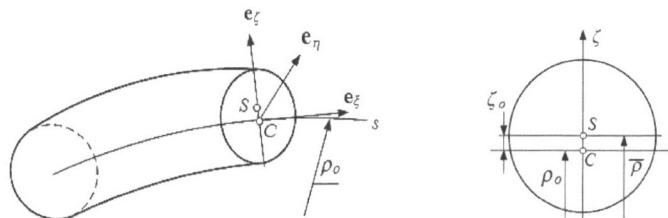

Figure 1

Some fundamental relations are gathered here on the base of lecture [12] presented at the COMEC conference in Braşov. Figure 1 shows the curvilinear coordinate system ($\xi = s, \eta, \zeta$). The cross section is symmetric with respect to the axis ζ. S is the point in which the axis ζ intersects the neutral surface in case of pure bending. The Young modulus and the Poisson number are assumed to satisfy the conditions $E(\eta,\zeta) = E(-\eta,\zeta)$; $\nu(\eta,\zeta) = \nu(-\eta,\zeta)$. The E-weighted centerline (or centerline for short) is chosen in such a way that

$$Q_{e\eta} = \int_A E(\eta,\zeta)\zeta \, dA = 0, \tag{1}$$

where $Q_{e\eta}$ is the E-weighted first moment of the cross section with respect to the axis η. Coordinate line $\xi = s$ coincides with the E-weighted centerline. The local base is formed by the unit vectors $\mathbf{e}_\xi$ (tangent to the centerline with radius ρ_o), $\mathbf{e}_\zeta$ (normal to the centerline) and $\mathbf{e}_\eta$ (perpendicular to the plane of the centerline).

For the prebuckling state the displacement vector of an arbitrary point on the cross section takes the form

$$\mathbf{u} = \mathbf{u}_o + \psi_{o\eta}\zeta\mathbf{e}_\xi = w_o\mathbf{e}_\zeta + (u_o + \psi_{o\eta}\zeta)\mathbf{e}_\xi, \qquad \psi_{o\eta} = \frac{u_o}{\rho_o} - \frac{dw_o}{ds}, \tag{2}$$

where $\mathbf{u}_o = u_o\mathbf{e}_\xi + w_o\mathbf{e}_\zeta$ and $\boldsymbol{\psi} = \psi_{o\eta}\mathbf{e}_\eta$ are the displacement and rotation on the centerline. It can be shown that the axial strain is

$$\varepsilon_\xi = \frac{\rho_o}{\rho_o + \zeta}\left(\varepsilon_{o\xi} + \zeta\kappa_o\right) + \frac{1}{2}\psi_{o\eta}^2 \tag{3a}$$

in which

$$\varepsilon_{o\xi} = \frac{du_o}{ds} + \frac{w_o}{\rho_o}, \qquad \frac{d\psi_{o\eta}}{ds} = \kappa_o = -\frac{d}{ds}\left(\frac{dw_o}{ds} - \frac{u_o}{\rho_o}\right) \quad \text{and} \quad \varepsilon_m = \varepsilon_{o\xi} + \frac{1}{2}\psi_{o\eta}^2. \tag{3b}$$

Here $\varepsilon_{o\xi}$ and ε_m are the linear and nonlinear axial strain on the centerline while κ_o is the change of curvature. We shall assume that $|\sigma_\xi| \gg |\sigma_\eta|, |\sigma_\zeta|$, consequently $\sigma_\xi = E(\eta,\zeta)\varepsilon_\xi$. For our later considerations we shall introduce the following quantities

$$A_{eR} = \int_A \frac{\rho_o}{\rho_o + \zeta}E(\eta,\zeta)dA \simeq \int_A E(\eta,\zeta)dA = A_e, \quad I_{eR} = \int_A \frac{\rho_o}{\rho_o + \zeta}E(\eta,\zeta)\zeta^2 dA \simeq \int_A \zeta^2 E(\eta,\zeta)dA = I_{e\eta},$$

$$Q_{eR} = \int_A \frac{\rho_o}{\rho_o + \zeta}E(\eta,\zeta)\zeta dA \simeq \frac{1}{\rho_o}\int_A \zeta^2 E(\eta,\zeta)dA = -\frac{I_{e\eta}}{\rho_o} \tag{4}$$

referred to as the E-weighted and reduced area, moment of inertia and first moment. These can be given with a good accuracy in terms of the E-weighted area A_e and moment of inertia $I_{e\eta}$. In what follows when writing derivatives we shall use the transformation

$$\frac{d\ldots}{ds} = \frac{1}{\rho_o}\frac{d\ldots}{d\varphi} = (\ldots)^{(1)}. \tag{5}$$

Making use of the Hook law we have the bending moment:

$$M = \int_A E\varepsilon_\xi\zeta dA = \underbrace{\int_A E\frac{\rho_o\zeta}{\rho_o+\zeta}dA}_{Q_{eR}\simeq\frac{I_{e\eta}}{\rho_o}}\varepsilon_{o\xi} + \underbrace{\int_A E\frac{\rho_o\zeta^2}{\rho_o+\zeta}dA}_{I_{eR}\simeq I_{e\eta}}\kappa_o + \underbrace{\int_A E\zeta dA\frac{1}{2}\psi_{o\eta}^2 dA}_{s_e=0} = -\frac{I_{e\eta}}{\rho_o^2}\left(w_o^{(2)} + w_o\right), \tag{6}$$

and the axial force

$$N = \int_A E\varepsilon_\xi dA = A_e\left[\left(\frac{du_o}{ds} + \frac{w_o}{\rho_o} + \frac{1}{2}\psi_{o\eta}^2\right) + \frac{I_{e\eta}}{\rho_o A_e}\frac{d}{ds}\left(\frac{dw_o}{ds} - \frac{u_o}{\rho_o}\right)\right] = \frac{I_{e\eta}}{\rho_o^2}\left(\frac{A_e\rho_o^2}{I_{e\eta}} - 1\right)\varepsilon_m - \frac{M}{\rho_o} \approx A_e\varepsilon_{o\xi} - \frac{M}{\rho_o}. \tag{7}$$

Consequently

$$N + \frac{M}{\rho_o} = \frac{I_{e\eta}}{\rho_o^2}\left(\frac{A_e\rho_o^2}{I_{e\eta}} - 1\right)\varepsilon_m \approx \frac{I_{e\eta}}{\rho_o^2}\frac{A_e\rho_o^2}{I_{e\eta}}\varepsilon_m \approx \frac{I_{e\eta}}{\rho_o^2}\mu\varepsilon_m \approx A_e\varepsilon_m \approx A_e\varepsilon_{o\xi} = \frac{A_e}{\rho_o}\left(u_o^{(1)} + w_o\right). \tag{8}$$

For the postbuckling state

$$u_o^* = u_o + u_{ob}, \quad w_o^* = w_o + w_{ob}, \quad \varepsilon_\xi^* = \varepsilon_\xi + \varepsilon_{\xi b}, \quad \psi_{o\eta}^* = \psi_{o\eta} + \psi_{o\eta b}, \quad N^* = N + N_b, \quad M^* = M + M_b, \tag{9}$$

are the displacements, strains and inner forces where the subscript $_b$ identifies the increments. It can be checked by applying the kinematic equations (3) that

$$\psi_{o\eta b} = \frac{u_{ob}}{\rho_o} - \frac{dw_{ob}}{ds}, \quad \varepsilon_{\xi b} = \frac{\rho_o}{\rho_o + \zeta}\left(\varepsilon_{o\xi b} + \zeta\kappa_{ob}\right) + \psi_{o\eta}\psi_{o\eta b} + \frac{1}{2}\psi_{o\eta b}^2. \tag{10}$$

It can also be shown by following the line of thought that has resulted in equations (6), (7) and (8)

$$M_b = -I_{e\eta}\left(\frac{d^2 w_{ob}}{ds^2} + \frac{w_{ob}}{\rho_o^2}\right), \quad N_b = \frac{I_{e\eta}}{\rho_o^2}\mu\varepsilon_{mb} + \frac{I_{e\eta}}{\rho_o^3}\left(w_{ob}^{(2)} + w_{ob}\right) \quad and$$

$$N_b + \frac{M_b}{\rho_o} \approx A_e\left(\varepsilon_{o\xi b} + \psi_{o\eta}\psi_{o\eta b} + \frac{1}{2}\psi_{o\eta b}^2\right) = A_e\varepsilon_{mb}, \quad where \quad \varepsilon_{mb} = \varepsilon_{\xi b}(\zeta = 0). \tag{11}$$

3. GOVERNING EQUATIONS

Our aim is to determine the critical load for a simply supported heterogeneous curved beam which is subjected to a dead force P_ζ at the point. Figure 2. shows the details. However when deriving the equilibrium equations for the sake of generality we shall assume that the constant $k_\gamma \neq 0$. It is a further assumption that the beam is subjected to distributed radial and tangential forces denoted by f_n and f_t (they are not shown in the figure).

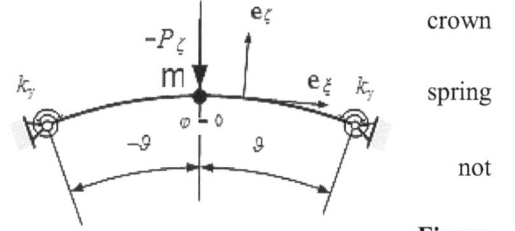

For the prebuckling state the principle of virtual work (PVW)

Figure 2.

$$\int_V \sigma_\xi\delta\varepsilon_\xi\,dV = -P_\zeta\,\delta w_o\big|_{\varphi=0} - k_t\psi_{o\eta}\delta\psi_{o\eta}\big|_{-\vartheta} - k_t\psi_{o\eta}\delta\psi_{o\eta}\big|_\vartheta + \int_{\mathcal{L}}\left(f_n\delta w_o + f_t\delta u_o\right)ds \tag{12}$$

results in the equilibrium equations (details of the formal transformations are omitted)

$$\frac{dN}{ds} + \frac{1}{\rho_o}\left[\frac{dM}{ds} - \left(N + \frac{M}{\rho_o}\right)\psi_{o\eta}\right] + f_t = 0, \quad \frac{d}{ds}\left[\frac{dM}{ds} - \left(N + \frac{M}{\rho_o}\right)\psi_{o\eta}\right] - \frac{N}{\rho_o} + f_n = 0 \tag{13}$$

which should be associated with appropriately chosen dynamical and/or geometrical boundary conditions

$$N\big|_{s(\pm\vartheta)} = 0, \quad \left[\frac{dM}{ds} - \left(N + \frac{M}{\rho_o}\right)\psi_{o\eta}\right]_{s(\pm\vartheta)} = 0, \quad \left(M \pm k_t\psi_{o\eta b}\right)\big|_{s(\pm\vartheta)} = 0, \quad u_o\big|_{s(\pm\vartheta)} = w_o\big|_{s(\pm\vartheta)} = \psi_{o\eta b}\big|_{s(\pm\vartheta)} = 0. \tag{14}$$

In addition a dynamical discontinuity condition should be satisfied at the crown point

$$\left[\frac{dM}{ds} - \left(N + \frac{M}{\rho_o}\right)\psi_{o\eta}\right]_{s(0+\varepsilon)} - \left[\frac{dM}{ds} - \left(N + \frac{M}{\rho_o}\right)\psi_{o\eta}\right]_{s(0-\varepsilon)} - P_\zeta = 0 \tag{15}$$

where ε is an infinitely small but positive quantity. We remark that each dynamical boundary and discontinuity condition follows from the PVW. For the prebuckling state we shall assume in the sequel that $f_n = f_t = 0$. Let

$$U_o = u_o/\rho_o, \quad W_o = w_o/\rho_o \tag{16}$$

be dimensionless displacement components. If we eliminate the intermediate variables by substituting (6), (7) and (8) into equilibrium equation $(13)_1$ we get

$$\frac{d}{ds}A_e\varepsilon_m = 0 \quad \rightarrow \quad \varepsilon_m = U_o^{(1)} + W_o + \frac{1}{2}\psi_{o\eta}^2 = constant, \quad or \quad \varepsilon_m \simeq \varepsilon_{o\xi} = U_o^{(1)} + W_o = constant. \tag{17}$$

If we neglect the quadratic term we can manipulate the term $\rho_o\varepsilon_m\left(1 + \psi_{o\eta}^{(1)}\right)$ in the following manner

$$\rho_o\varepsilon_m\left(1 + \psi_{o\eta}^{(1)}\right) \approx \rho_o\varepsilon_m\left(1 + \frac{1}{\rho_o}\left(\rho_o\varepsilon_m - w_o - w_o^{(2)}\right)\right) = \rho_o\varepsilon_m(1 + \varepsilon_m) - \varepsilon_m\left(w_o^{(2)} + w_o\right) \approx \rho_o\varepsilon_m - \varepsilon_m\left(w_o^{(2)} + w_o\right). \tag{18}$$

Making use of the above result and eliminating the intermediate variables in equation $(13)_2$ we have

$$W_o^{(4)} + \left(2 - \mu\varepsilon_m\right)W_o^{(2)} + \left(1 - \mu\varepsilon_m\right)W_o = -\mu\varepsilon_m \quad or \quad W_o^{(4)} + \left(1 + \chi^2\right)W_o^{(2)} + \chi^2 W_o = -\mu\varepsilon_m, \quad \chi^2 = 1 - \mu\varepsilon_m. \tag{19}$$

If we drop the terms underlined this equation coincides with the equation used for investigating the prebuckling state of a shallow arch – see for instance [13].

For the postbuckling case

$$\int_V \sigma_\xi^*\delta\varepsilon_\xi^*\,dV = -P_\zeta^*\,\delta w_o^*\big|_{\varphi=0} - P_\xi^*\,\delta u_o^*\big|_{\varphi=0} - m\ddot{w}_o^*\delta w_o^*\big|_{\varphi=0} - m\ddot{u}_o^*\delta u_o^*\big|_{\varphi=0} - k_t\psi_{o\eta}^*\delta\psi_{o\eta}^*\big|_{\pm\vartheta} + \int_{\mathcal{L}}\left(f_n^*\delta w_o^* + f_t^*\delta u_o^*\right)ds \tag{20}$$

is the principle of virtual work where P_ζ^*, P_ξ^* are the forces exerted at the crown point to which a mass m is attached (it is assumed that the stability loss is such a dynamical process for which the mass distribution on the centerline is neglected). Then the principle of virtual work – details are omitted here – yields the equilibrium equations

$$\frac{dN_b}{ds} - \frac{1}{\rho_o}\left(N + \frac{M}{\rho_o}\right)\psi_{o\eta b} + \frac{1}{\rho_o}\left[\frac{dM_b}{ds} - \left(N_b + \frac{M_b}{\rho_o}\right)\psi_{o\eta b}\right] + f_{tb} = 0, \tag{21}$$

$$\frac{d^2 M_b}{ds^2} - \frac{N_b}{\rho_o} - \frac{d}{ds}\left[\left(N + N_b + \frac{M + M_b}{\rho_o}\right)\psi_{o\eta b} + \left(N_b + \frac{M_b}{\rho_o}\right)\psi_{o\eta}\right] + f_{nb} = 0, \tag{22}$$

the discontinuty conditions at the crown point

$$\left[\frac{dM_b}{ds} - \left(N + N_b + \frac{M + M_b}{\rho_o}\right)\psi_{o\eta b} - \left(N_b + \frac{M_b}{\rho_o}\right)\psi_{o\eta}\right]_{s(0-\varepsilon)} - \left[\frac{dM_b}{ds} - \left(N + N_b + \frac{M + M_b}{\rho_o}\right)\psi_{o\eta b} - \right.$$

$$\left. - \left(N_b + \frac{M_b}{\rho_o}\right)\psi_{o\eta}\right]_{s(0+\varepsilon)} + \mathrm{m}\frac{d^2 w_{ob}}{dt^2}\bigg|_{s=0} = 0, \quad N_b\big|_{s(0-\varepsilon)} - N_b\big|_{s(0+\varepsilon)} + P_{\zeta b} + \mathrm{m}\frac{d^2 u_{ob}}{dt^2}\bigg|_{s=0} = 0 \tag{23}$$

and the dynamical boundary conditions

$$N_b\big|_{s(\pm\vartheta)} = 0, \quad \left(M_b \pm k_t\psi_{o\eta b}\right)\big|_{s(\pm\vartheta)} = 0; \quad \left[\frac{dM_b}{ds} - \left(N + N_b + \frac{M + M_b}{\rho_o}\right)\psi_{o\eta b} - \left(N_b + \frac{M_b}{\rho_o}\right)\psi_{o\eta}\right]_{s(\pm\vartheta)} = 0. \tag{24}$$

Depending on what the supports are, geometrical conditions should be satisfied instead of some dynamical boundary and discontinuity conditions:

$$u_{ob}\big|_{s(0-\varepsilon)} = u_{ob}\big|_{s(0+\varepsilon)}, \quad w_{ob}\big|_{s(0-\varepsilon)} = w_{ob}\big|_{s(0+\varepsilon)}, \quad \psi_{o\eta b}\big|_{s(0-\varepsilon)} = \psi_{o\eta b}\big|_{s(0+\varepsilon)}, u_{ob}\big|_{s(\pm\vartheta)} = w_{ob}\big|_{s(\pm\vartheta)} = \psi_{o\eta b}\big|_{s(\pm\vartheta)} = 0. \tag{25}$$

In the sequel it is assumed that $f_{tb} = f_{nb} = P_{\zeta b} = P_{\xi b} = \mathrm{m} = 0$. Upon substitution of equations (11) into equilibrium equation (22)$_1$ in which we have dropped the quadratic terms that belong to the postbuckling state we have

$$\frac{d}{ds}A_e\varepsilon_{mb} - \frac{1}{\rho_o}A_e\varepsilon_m\psi_{o\eta b} = 0, \quad \text{or} \quad \frac{d}{d\varphi}\varepsilon_{mb} - \varepsilon_m\psi_{o\eta b} = 0. \tag{26}$$

Since the quadratic term $\varepsilon_m\psi_{o\eta b}$ can also be neglected we obtain

$$\frac{d}{ds}A_e\varepsilon_{mb} = 0, \quad \text{or} \quad \varepsilon_{mb} = \text{constant}. \tag{27}$$

Equilibrium equation (22)$_2$ can also be manipulated into an appropriate form:

$$\frac{d^2 M_b}{ds^2} - \frac{N_b}{\rho_o} - \frac{d}{ds}\left[A_e\varepsilon_m\psi_{o\eta b} + A_e\varepsilon_{mb}\psi_{o\eta}\right] = \frac{d^2 M_b}{ds^2} - \frac{N_b}{\rho_o} - \underbrace{\frac{A_e\rho_o^2}{I_{e\eta}}}_{\mu}\frac{I_{e\eta}}{\rho_o^2}\frac{d}{ds}\left[\varepsilon_m\psi_{o\eta b} + \varepsilon_{mb}\psi_{o\eta}\right] = 0.$$

After substituting equations (11) and assuming that $\varepsilon_{mb} = \text{constant}$ we arrive at equation

$$w_{ob}^{(4)} + 2w_{ob}^{(2)} + w_{ob} + \rho_o\mu\varepsilon_m\psi_{o\eta b}^{(1)} + \rho_o\mu\varepsilon_{mb}\left(1 + \psi_{o\eta}^{(1)}\right) = 0. \tag{28}$$

Making use of (18) ($1 \gg |\varepsilon_m|$ by assumption) and taking into account

$$\rho_o\varepsilon_{mb} \approx u_{ob}^{(1)} + w_{ob} + \left(\psi_{o\eta}\psi_{o\eta b} + \frac{1}{2}\psi_{o\eta b}^2\right)\rho_o, \quad u_{ob}^{(1)} = \rho_o\varepsilon_{mb} - w_{ob} - \left(\psi_{o\eta}\psi_{o\eta b} + \frac{1}{2}\psi_{o\eta b}^2\right)\rho_o$$

by which we have

$$\psi_{o\eta b}^{(1)} = \frac{1}{\rho_o}\left(u_{ob}^{(1)} - w_{ob}^{(2)}\right) = \frac{1}{\rho_o}\left[\rho_o\varepsilon_{mb} - \left(\psi_{o\eta}\psi_{o\eta b} + \frac{1}{2}\psi_{o\eta b}^2\right)\rho_o - w_{ob} - w_{ob}^{(2)}\right] = \frac{1}{\rho_o}\left(\rho_o\varepsilon_{mb} - w_{ob} - w_o^{(2)}\right),$$

finally we obtain from (18) that

$$w_{ob}^{(4)} + 2w_{ob}^{(2)} + w_{ob} + \mu\varepsilon_{mb}\rho_o\left(1 + \varepsilon_m\right) - \mu\varepsilon_m\left(w_{ob} + w_{ob}^{(2)}\right) = -\mu\varepsilon_{mb}\left(w_o^{(2)} + w_o\right)$$

or – $\left(1 + \varepsilon_m\right) \approx 1$ – by switching over to dimensionless variables that

$$W_{ob}^{(4)} + \left(2 - \mu\varepsilon_m\right)W_{ob}^{(2)} + \left(1 - \mu\varepsilon_m\right)W_{ob} = \mu\varepsilon_{mb} - \mu\varepsilon_{mb}\left(W_o^{(2)} + W_o\right),$$

$$W_{ob}^{(4)} + \left(1 + \chi^2\right)W_{ob}^{(2)} + \chi^2 W_{ob} = \mu\varepsilon_{mb} - \mu\varepsilon_{mb}\left(W_o^{(2)} + W_o\right), \quad \chi^2 = 1 - \mu\varepsilon_m. \tag{29}$$

If we drop the terms underlined this equation coincides with the equation used for investigating the postbuckling state of a shallow arch – see for instance [13].

4. SOLUTIONS

Taking into account that for a simply supported beam ($k_\gamma = 0$) solution to differential equation (DE) (19) in the interval $\varphi \in [0; \vartheta]$ should satisfy the boundary conditions

Table 1.

Simply supported beam	
Crown point	Right support
$\psi_{o\eta}(\varphi)\big|_{\varphi=0+\varepsilon} = 0$	$W_o(\varphi)\big|_{\varphi=\vartheta} = 0$
$-\dfrac{dM(\varphi)}{ds}\bigg|_{\varphi=0+\varepsilon} + \dfrac{P_\zeta}{2} = 0$	$M(\varphi)\big|_{\varphi=\vartheta} = 0$

we get for the integration constants $A_1, ..., A_4$ in the general solution

$$W_o = \frac{\chi^2 - 1}{\chi^2} + A_1 \cos\varphi + A_2 \sin\varphi - \frac{A_3}{\chi^2} \cos\chi\varphi - \frac{A_4}{\chi^2} \sin\chi\varphi \tag{30}$$

that

$$A_1 = -\frac{1}{\chi^2 - 1}\left(\frac{\chi^2 - 1}{\cos\vartheta} + \frac{\sin\vartheta}{\vartheta\cos\vartheta}\frac{\mathcal{P}}{\vartheta}\right), \quad A_2 - \frac{1}{\chi^2 - 1}\frac{\mathcal{P}}{\vartheta},$$

$$A_3 = -\frac{1}{\chi^2 - 1}\left(\frac{\chi^2 - 1}{\cos\kappa\vartheta} + \chi\frac{\sin\chi\vartheta}{\cos\chi\vartheta}\frac{\mathcal{P}}{\vartheta}\right), \quad A_4 = \frac{\kappa}{\chi^2 - 1}\frac{\mathcal{P}}{\vartheta}, \qquad \mathcal{P} = \frac{P_\zeta}{2}\frac{\rho_o^2\vartheta}{I_{e\eta}} \tag{31}$$

where $\mathcal{P}$ is a dimensionless load. Within the framework of the linear theory solution of $U_o^{(1)} + W_o = \varepsilon_m$ provides U_o in the form

$$U_o = D_1 + \underbrace{\frac{(1-\chi^2)}{\mu}\left(1 + \frac{\mu}{\chi^2}\right)}_{D_2}\varphi - A_1 \sin\varphi + A_2 \cos\varphi \cos\chi\varphi \tag{32}$$

in which D_1 and D_2 are constants – D_1 is obtained from the condition $\psi_{o\eta}\big|_{\varphi=0}$. With the knowledge of U_o

$$\psi_{o\eta} = U_o - W_o^{(1)} = D_1 + D_2\varphi + D_3 \sin\chi\varphi + D_4 \cos\chi\varphi =$$

$$= (D_{21}\varphi + D_{31}\sin\kappa\varphi) + (D_{12} + D_{22}\varphi + D_{32}\sin\kappa\varphi + D_{42}\cos\kappa\varphi)\frac{\mathcal{P}}{\vartheta} \tag{33}$$

is the rotation. Here

$$D_1 = -D_4 = -\frac{1}{\chi}\left(1 - \frac{1}{\chi^2}\right)A_4 = -\frac{1}{\chi^2}\frac{\mathcal{P}}{\vartheta}, \qquad D_3 = -\frac{1}{\chi}\left(1 - \frac{1}{\chi^2}\right)A_3. \tag{34}$$

Observe that the constants $D_1, ..., D_4$ are resolved into two parts: D_{i1} does not contain D_{i2} ($i = 1, 2, 3, 4$) contains $\mathcal{P}$ and $D_{11} = D_{41} = 0$. Within the framework of the linear theory

$$\varepsilon_m\vartheta = \int_0^\vartheta (U_o^{(1)}(\varphi) + W_o(\varphi))d\varphi = \int_0^\vartheta W_o(\varphi)d\varphi \tag{35}$$

is the axial strain. After performing the integration on the right side we obtain an equation for the critical load $\mathcal{P}$

$$\varepsilon_m\vartheta = I_{ow} + I_{1w}\frac{\mathcal{P}}{\vartheta} \quad \rightarrow \quad \mathcal{P} = \frac{1}{I_{1w}}(\varepsilon_m\vartheta - I_{ow})\vartheta \tag{36}$$

in which

$$I_{0w} = \frac{\chi^2 - 1}{\chi^2} \vartheta - \frac{\sin \vartheta}{\cos \vartheta} + \frac{\sin \chi \vartheta}{\chi^3 \cos \chi \vartheta}, \qquad I_{1w} = \frac{1}{(\chi^2 - 1)} \left(1 - \frac{1}{\chi^2} - \frac{1}{\cos \vartheta} + \frac{1}{\chi^2 \cos \chi \vartheta} \right). \tag{37}$$

If we know the critical value of the axial strain ε_m then equation (36) makes possible to determine the dimensionless critical load $\mathcal{P}$. Observe that the effect of heterogeneity is implied in the formulation and solutions via the parameter $\mu = \rho_o{}^2 A_e / I_{e\eta}$ (or which is the same via χ). Therefore everything is valid for a constant Young modulus, i.e., for homogeneous beams as well.

Within the framework of the nonlinear theory equation

$$\varepsilon_m \vartheta = \int_0^\vartheta \left(U_o^{(1)}(\varphi) + W_o(\varphi) + \frac{1}{2} \psi_{o\eta}^2 \right) d\varphi = \int_0^\vartheta \left(W_o(\varphi) + \frac{1}{2} \psi_{o\eta}^2(\varphi) \right) d\varphi \tag{38}$$

is our point of departure – $\psi_{o\eta}$ is substituted from the linear solution (33). After performing the integration we obtain a quadratic equation for the critical load $\mathcal{P}$

$$\varepsilon_m \vartheta = \left(I_{0w} + I_{0\psi} \right) + \left(I_{1w} + I_{1\psi} \right) \frac{\mathcal{P}}{\vartheta} + I_{2\psi} \left(\frac{\mathcal{P}}{\vartheta} \right)^2 \tag{39a}$$

in which

$$I_{0\psi} = \frac{1}{12\chi^2} \left(2\chi^2 \vartheta^3 D_{21}^2 + 3 D_{31}^2 \chi^2 \vartheta + 12 D_{21} D_{31} \sin \chi \vartheta - 12 D_{21} D_{31} \chi \vartheta \cos \chi \vartheta - 3 D_{31}^2 \chi \cos \chi \vartheta \sin \chi \vartheta \right), \tag{39b}$$

$$I_{1\psi} = \frac{1}{6\chi^2} \left(3 \chi^2 \vartheta^2 D_{21} D_{12} + 2 \chi^2 \vartheta^3 D_{21} D_{22} - 6 D_{21} D_{42} + 6 \chi D_{31} D_{12} + 3 \chi D_{31} D_{42} + \right.$$

$$+ 6 \left(D_{31} D_{22} + \chi \vartheta D_{21} D_{42} + D_{21} D_{32} \right) \sin \chi \vartheta + 6 \left(D_{21} D_{42} - \chi D_{31} D_{12} - \chi \vartheta D_{31} D_{22} - \chi \vartheta D_{21} D_{32} \right) \cos \chi \vartheta - \tag{39c}$$

$$\left. - 3 D_{31} \left(\chi D_{32} \cos \chi \vartheta \sin \chi \vartheta - \chi^2 \vartheta D_{32} + D_{42} \chi \cos^2 \chi \vartheta \right) \right),$$

$$I_{2\psi} = \frac{1}{12\chi^2} \left(3 \chi^2 \vartheta \left(D_{32}^2 + D_{42}^2 + 2 D_{12}^2 \right) - 12 D_{42} D_{22} + 12 \chi D_{32} D_{12} + 3 \chi \left(D_{42}^2 - D_{32}^2 \right) \cos \chi \vartheta \sin \chi \vartheta \right.$$

$$+ 6 \chi D_{32} D_{42} + 6 \chi^2 \vartheta^2 D_{12} D_{22} + 2 \chi^2 \vartheta^3 D_{22}^2 + 12 \left(D_{42} D_{22} - \chi \vartheta D_{32} D_{22} - \chi D_{32} D_{12} \right) \cos \chi \vartheta + \tag{39d}$$

$$\left. + 12 \left(\chi D_{42} D_{12} + D_{32} D_{22} + \chi \vartheta D_{42} D_{22} \right) \sin \chi \vartheta - 6 \chi D_{32} D_{42} \cos^2 \chi \vartheta \right).$$

As regards the postbuckling state we have to solve DE (29) by taking into the fact that $\varepsilon_{mb} = 0$ for asymmetric buckling. Consequently we should find the solution of equation

$$W_{ob}^{(4)} + \left(1 + \chi^2 \right) W_{ob}^{(2)} + \chi^2 W_{ob} = 0 \tag{40}$$

which is associated with the boundary conditions

Table 2.

Simply supported beam	
Left support	Right support
$W_{ob}(\varphi)\big|_{\varphi=-\vartheta} = 0$	$W_{ob}(\varphi)\big|_{\varphi=\vartheta} = 0$
$W_{ob}^{(2)}(\varphi)\big|_{\varphi=-\vartheta} = 0$	$W_{ob}^{(2)}(\varphi)\big|_{\varphi=\vartheta} = 0$

With the general solution

$$W_{ob} = B_1 \cos \varphi - B_2 \sin \varphi - B_3 \sin \chi \varphi + B_4 \cos \chi \varphi \tag{41}$$

the boundary conditions result in a homogeneous system of linear equations for the integration constants $B_1, \ldots, B_4$. Its determinant, which should vanish, assumes the form

$$D = \left(1 + \chi \right)^2 \left(1 - \chi \right)^2 \cos \vartheta \sin \vartheta \cos \chi \vartheta \sin \chi \vartheta = 0. \tag{42}$$

Hence

$$\chi = 1, \quad \chi = -1, \quad \chi = \frac{\pi}{\vartheta}, \quad \text{or} \quad \chi = \frac{\pi}{2\vartheta}. \tag{43}$$

Recalling equation $\chi^2 = 1 - \mu\varepsilon_m$ we can come to the following conclusions: (a) if $\chi = 1$ then $\varepsilon_m = 0$, (b) if $\chi = -1$ then $\varepsilon_m > 0$, (c) if $\chi = \pi / \vartheta$ or (d) $\chi = \pi / 2\vartheta$ then

$$\varepsilon_m = -\frac{1}{\mu}\left(\chi^2 - 1\right) = -\frac{1}{\mu}\left[\left(\frac{g_{11(\vartheta)}}{\vartheta}\right)^2 - 1\right] \qquad \text{where} \qquad g_{11(\vartheta)} = \begin{cases} \pi & \text{for case (c)} \\ \pi / 2 & \text{for case (d)} \end{cases}, \qquad (44)$$

We remark that cases (a) and (b) have no physical sense. It can be checked that the eigenfunction for case (c) is asymmetric for case (d) is symmetric.

5. COMPUTATIONAL RESULTS

We remark that paper [13] by Bradford and his co-authors provides the following formula for the critical axial strain assuming homogeneous material and asymmetric buckling:

$$\varepsilon_m = -\frac{1}{\mu}\left(\frac{\pi}{\vartheta}\right)^2. \qquad (45)$$

This equation coincides with (44) if ϑ is small. However the greater is ϑ the greater is the difference between the two solutions. For the sake of a comparison we have determined the membrane stress for the two solutions as a function of 2ϑ if the Young modulus $E = 10^5$ MPa. Figure 3 shows the results for various μ values.

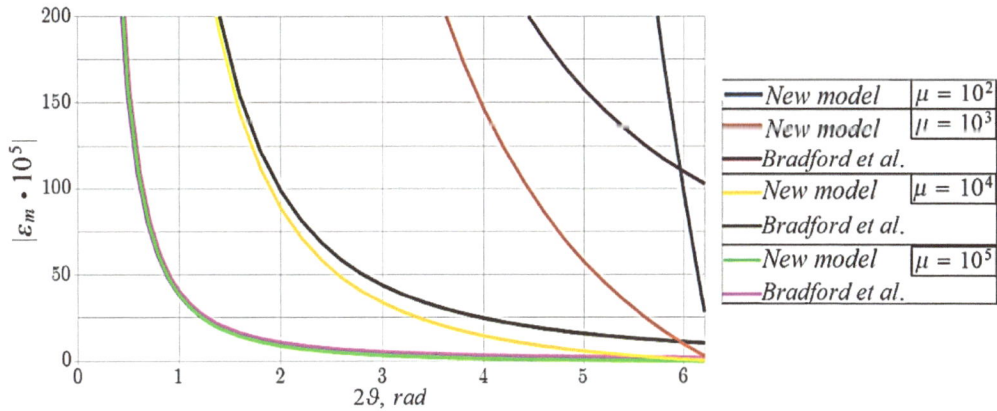

Figure 3.

It is clear from Figure 3 that the membrane stress is increasing very fast as $\vartheta \to 0$. Recalling that we have assumed elastic behaviour we can conclude that the results should be used with care since there is an increasing probability of plastic deformation for small values of ϑ.

As regards the critical load it is worth introducing the parameter

$$\lambda = 2\ell_e^2 / \left(4\rho_o\sqrt{\frac{I_{e\eta}}{A_e}}\right), \qquad \ell_e = 2\rho_o\vartheta \qquad (46)$$

and plotting $\mathcal{P}$ against λ and ϑ. Figure 4 shows the dimensionless critical load $\mathcal{P}$ as a function of λ.

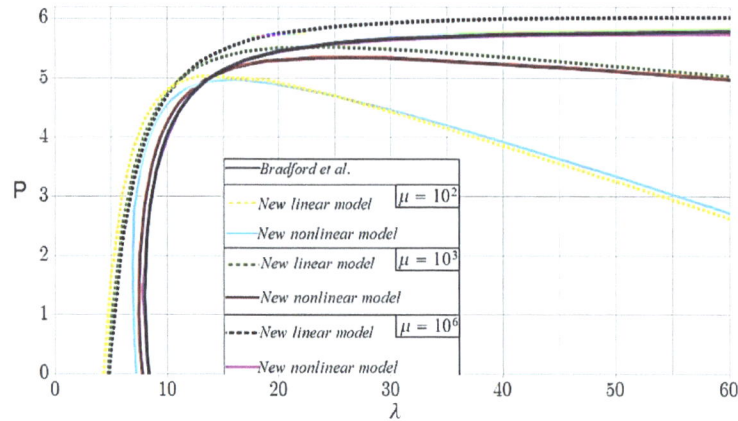

Figure 4.

It follows from the computational results that

• Within the framework of linear theory the results for $\mathcal{P}$ are practically independent of μ if $\lambda \in [4.4, 10.7]$. In addition there is no asymmetric buckling if $\lambda < 4.4$.

• Within the framework of nonlinear theory the results for $\mathcal{P}$ published in [13] for shallow arches – see the dark blue parabola[1] – are independent of μ. In contrast with this our results – see the light blue, red and violet parabolas – (i) depend on μ, (ii) are (greater)[less] than those for shallow arches if ($\lambda < 13.2$) [$\lambda > 13.2$]. It is also worth mentioning that the least value of λ, for which asymmetric buckling is possible, also depends on μ. Its minimum is about 6.9. Observe that the greater is μ the closer are our results to those valid for shallow arches.

We are making only one comment to Figure 5 which shows the dimensionless load as a function of ϑ. The linear and nonlinear models provide practically the same results if $\vartheta > \pi / 2$. However they differ significantly form those – see the black curve – published in [13] without reasoning. This issue deserves further investigations.

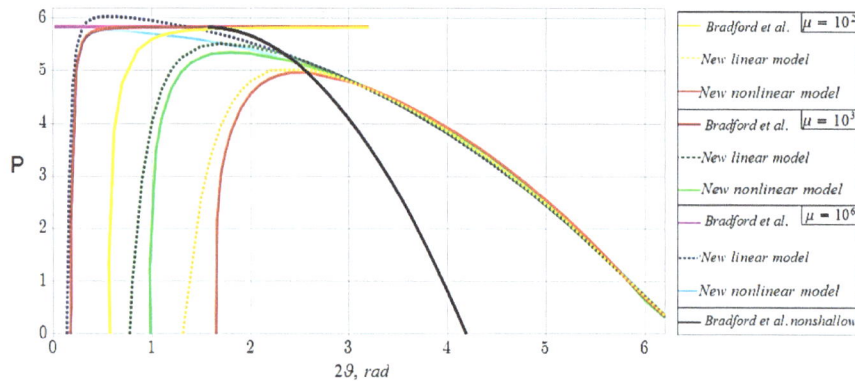

Figure 5.

6. CONCLUDING REMARKS

• We have established the governing equations for the stability problem of curved beams made of heterogeneous material. The equations valid for the radial displacements in the prebuckling and postbuckling states have the same structure. We have also shown how the equations valid for shallow arches follow from those we have set up.

• Assuming asymmetric stability loss we have determined the critical load both from a linear model and from a nonlinear one. The results obtained are compared to those valid for shallow arches. It has turned out that the critical membrane strain for small ϑ (or λ) can reach such a value which would result in plastic deformations.

---

[1]For a given λ always the grater value of the dimensionless $\mathcal{P}$ is the critical load.

• Further investigations should be carried out in order to clarify what happens if the stability loss is symmetric and if $\vartheta > \pi / 2$.

REFERENCES

3. A.E.H. Love. *Treatese on the mathematical theory of elasticity*. New York, Dower, 1944
4. S.P. Timoshenko and J.M. Gere. *Theory of elastic stability*. New York, McGraw-Hill, 2 edition, 1961.
5. H.L. Schreyer and E.F. Masur. Buckling of shallow arches. *J. Engng. Mech. Div*. ASCE, 92:1–17, 1966.
6. Clive L. Dym. On extensibility and the buckling of shallow and steep arches. In *Proceedings of the Sixth Southeastern Conference on Theoretical and Applied Mechanics*, Tampa Florida, March, 1972.
7. Clive L. Dym. Buckling and postbuckling behaviour of steep compressible arches. *International Journal of Solids and Structures*, 9(1):129, January 1973.
8. Clive L. Dym. Bifurcation analyses for shallow arches. *Journal of the Engineering Mechanics Division*, ASCE, 99(EM2):287, April 1973.
9. Clive L. Dym. *Stability Theory and Its Applications to Structural Mechanics*. Dover, 2002.
10. Y. L. Pi, M. A. Bradford, and F. Tin-Loi. Nonlinear analysis and buckling of elastically supported circular shallow arches. *International Journal of Solids and Structures*, 44:2401–2425, 2007.
11. M. A. Bradford, Y.-L. Pi and Francis Tin-Loi. Non-linear in-plane buckling of rotationally restrained shallow arches under a central concentrated load. *International Journal of Non-Linear Mechanics*, 43:1–17, 2008.
12. Y. L. Pi and M. A. Bradford. Non-linear in-plane postbuckling of arches with rotational end restraints under uniform radial loading. *International Journal of Non-Linear Mechanics*, 44:975–989, 2009.
13. Y. L. Pi and M. A. Bradford. Effects of prebuckling analyses on determining buckling loads of pin-ended circular arches. *Mechanics Research Communications*, 37:545–553, 2010.
14. G. Szeidl and L. Kiss. Vibrations and stability of heterogeneous curved beams. In S. Vlase, editor, *Proceedings of the 4th International Conference on Computational Mechanics and Virtual Engineering*, COMEC, 20 - 22 October 2011, Brasov, Romania.
15. B. Uy M. A. Bradford and Y-L. Pi. In-plane elastic stability of arches under a centra concentrated load. *Journal of Engineering Mechanics*, 128(7):710–719, 2002.

ELECTROMAGNETIC SHIELDING PROPERTIES DETERMINATION FOR ADVANCED COMPOSITE MATERIALS

L.E. Aciu[1], P.L. Ogrutan[1], M. Volmer[1]

[1] Transivania University of Brasov, ROMANIA, lia_aciu@unitbv.ro

Abstract: *Today, modern technologies and equipment are used in every domain of activity. Some of them use high frequency signals, especially when they are wireless connected. This is why, there is a wide preoccupation to find new materials having electromagnetic shielding properties to ensure their compatibility when functioning. This paper presents some studies in order to determine the shielding capability for advanced composite materials. Knowing the electromagnetic properties for new materials enlarge the possibilities of using them in different applications. The studies are based on experimental determination and Spice simulation*

Keywords: *electromagnetic shielding properties, advanced composite materials, simulations.*

1. INTRODUCTION

Late twentieth century is considered by many experts as age materials. These materials with properties superior to traditional materials programmable entered the top technology fields such as microelectronics, aerospace technology, nuclear technology, medical implants construction technique, but also in the automotive industry, shipbuilding, chemicals, furniture, construction materials industry, sports [1].

The spread of these materials started from the main advantages is that they show [2]:

- the possibility of "modularization" properties and obtain thus the materials with very different properties,
- have a very good value compared with the materials 'classic', the report tensile strength / specific gravity,
- have a good wear resistance (surface hardness), oxidation and corrosion,
- have a good stability while the size and shape,
- have a good capacity for shock absorption, vibration and noise
- -carbon composite materials - carbon or ceramic can be used at high temperatures, up to 2200^0C.

The major advantage, essential properties of composites is the possibility modulation and thus to obtain a wide range of materials whose use can be extended to almost all technical fields.

Making composite materials has become the basis for many technical and economic considerations, among which [1]:

- need for materials with special properties unattainable with traditional materials,
- need to enhance security and reliability in operation of various construction and facilities,
- the need to reduce consumption of scarce materials, precious or precious-consumption,
- can reduce labor and shortening the manufacturing.

2. SHIELDING EFFECTIVENESS DETERMINATION WITH SPICE SIMULATION

The proposed approach consists in a Spice model using transmission line model to simulate the attenuation introduced by a material characterized by the macroscopic parameters ε, μ, σ.

The method has been validated for copper [3], the results obtained being compared with theoretical results published by White [4].

Simulation conditions require that the electromagnetic radiation source be placed at a certain distance from the shield. This simple method enables to obtain a quick shielding effectiveness evaluation for new materials only by knowing their macroscopic properties ε, μ, σ.

The capability of a shield can be expressed using Shielding Effectiveness, that can be computed by the relation (1) [5].

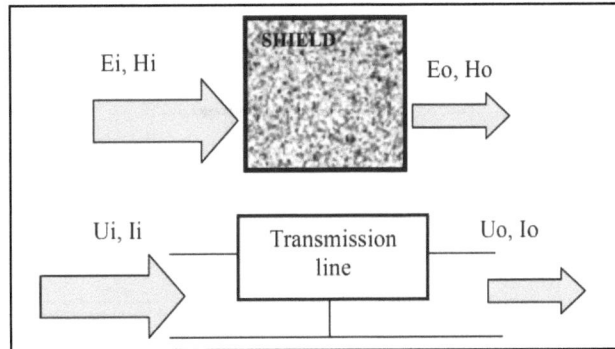

Figure 1 Loss less Transmission line model showing material properties

$$SE_{dB} = 20 lg \frac{U_i}{U_0} \tag{1}$$

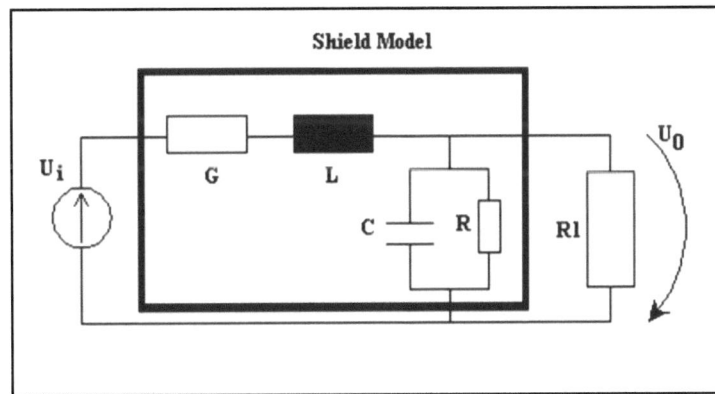

Figure 2 SPICE Model using the transmission line model for the studied material

Layered composite materials have the main advantage the economic one and qualitative reasons, because their use is by saving important quantities of expensive materials or deficient, improving at the same time, the qualities of products and increasing the duration of their operation in conditions of high performance [6].

Composite material structure, reveals itself in the fabrication, electrical characteristics of layers containing components, i.e. electric conductivity σ, the electric permittivity ε and the magnetic permeability μ is represented in Figure 3.

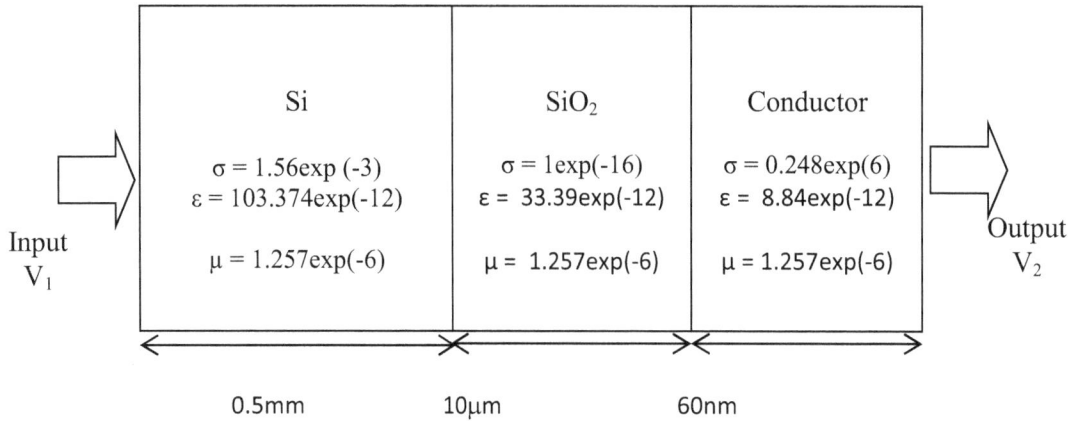

Figure 3: About the composite material construction

For simulation have been used the following model, Figure 4, where T1 is the line transmission model for Si material, T2 is the line transmission model for Si O2 and T3 is the line transmission model for the last conductive layer in the analyzed composite material.

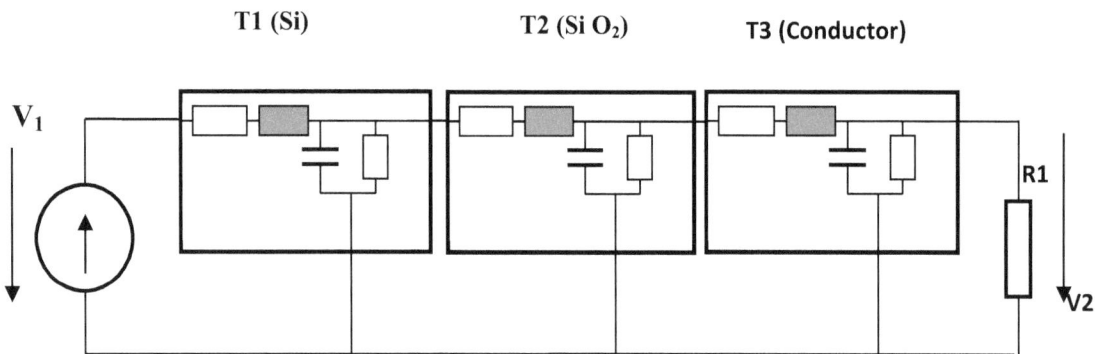

Figure 4: Layered composite material transmission line model

2.1. Spice Simulation results

After the simulation made in the frequency range of 1GHz - 500GHz, Figure 5, it can be seen that till 1GHz the electromagnetic waves attenuation is low, for all frequencies.

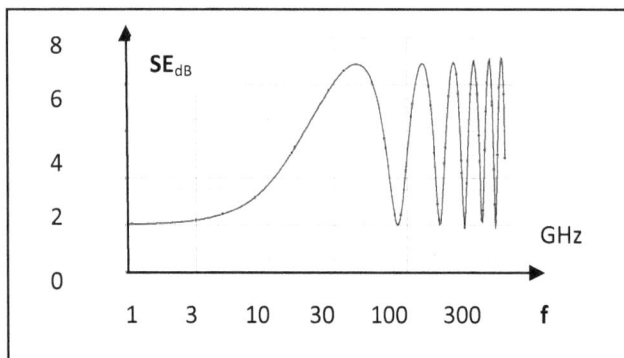

Figure 5: The material attenuation in the frequency range of 1GHz - 500GHz

The results obtained by simulations have compared with experimental determinations at 10GHz. As can be seen in Figure 5, the composite material capability to attenuate the electromagnetic waves, expressed by the shielding effectiveness, is around 2dB and was computed by relation (1) after simulations.

Were made also simulations taking into account only the conductive layer of the composite material.

The simulation results are in Figure 6.

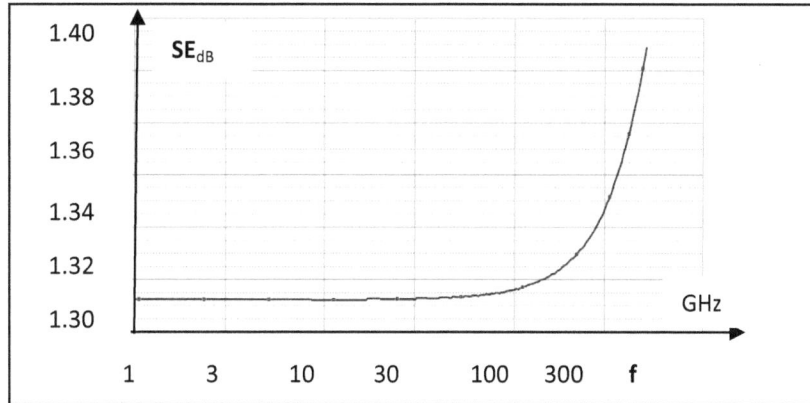

Figure 6: Simulation results taking into account only the conductive layer

Using the transmission line mode it can be studied with Spice Simulation another possible structure for the composite material in order to be used as an electromagnetic shield. Figure 7 shows the characteristic for the shielding effectiveness if the conductive layer thickness would be 60μm instead of 60nm.

It can be seen that even at low frequencies the material has the capability to be used as a shield. At high frequencies the shielding effectiveness is only a little bit higher. The characteristic in the Figure 7 is for a frequency range of 1Hz - 500GHz

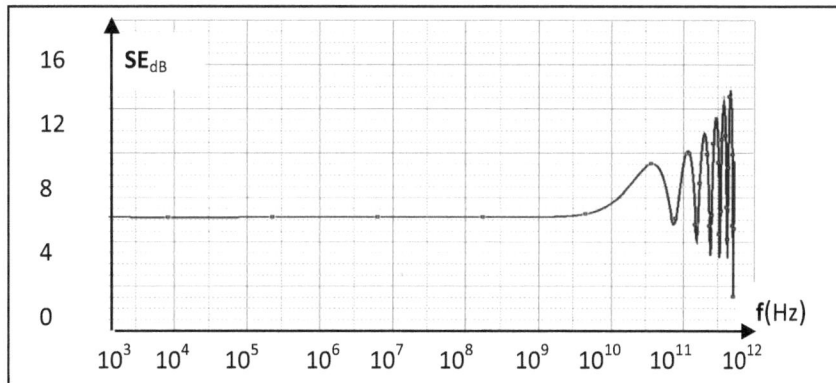

Figure 7: Simulation results taking into account a conductive layer of 60μm

3. SHIELDING EFFECTIVENESS DETERMINATION WITH SIMULINK SIMULATION

Based on the transmission line mode Simulink Simulations have been made for the composite material, Figure 8.

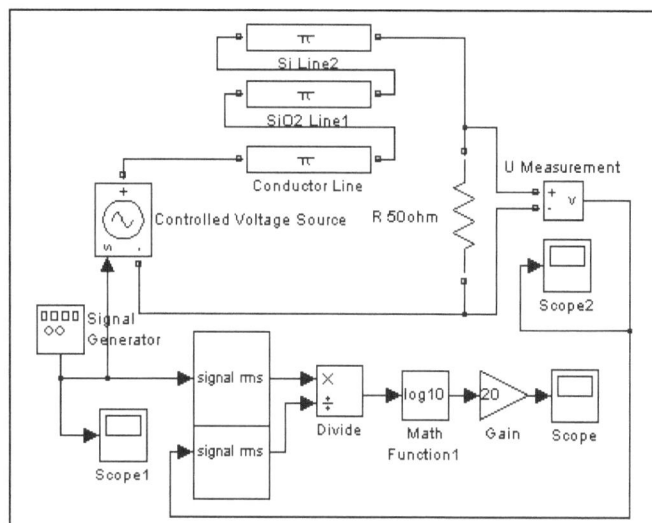

Figure 8: Simulink Simulation Model for advanced composite material

The input signal is applied to a sine 10GHz (same frequency with which the experimental system works).

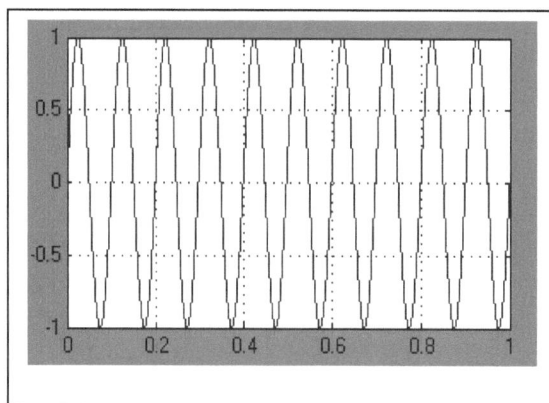

a

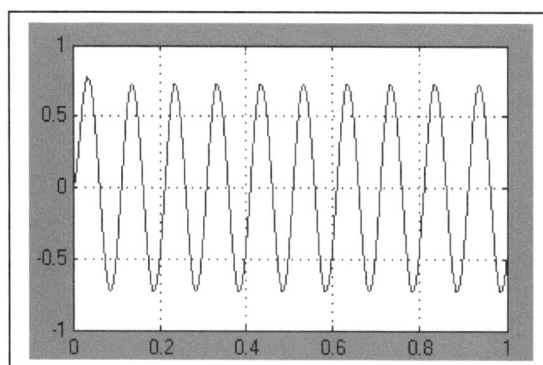

b

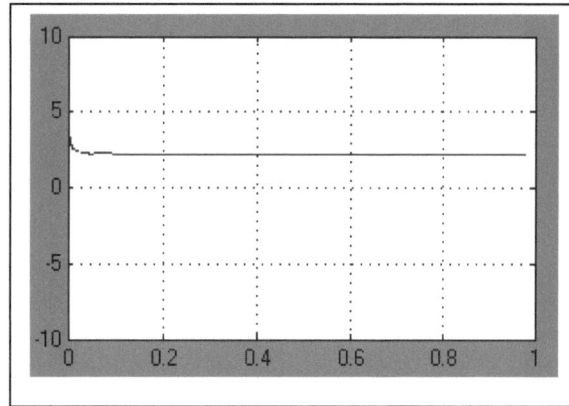

c

Figure 9 Simulink Simulation Results

The input signal is shown in Figure 9a, the output signal variation (attenuated), Figure 9b and the shielding effectiveness, Figure 9c), calculated with relation (1).

By comparing the simulation results obtained with different programs it can be seen that the attenuation introduced by the composite material at a frequency of 10GHz are in the range of (1.5dB-3dB).

To verify the correctness, simulation for the conductive layer have been made. The constant attenuation (excepting the first moments) obtained, as in Figure 10, according to the theory, shows the accuracy of the simulations

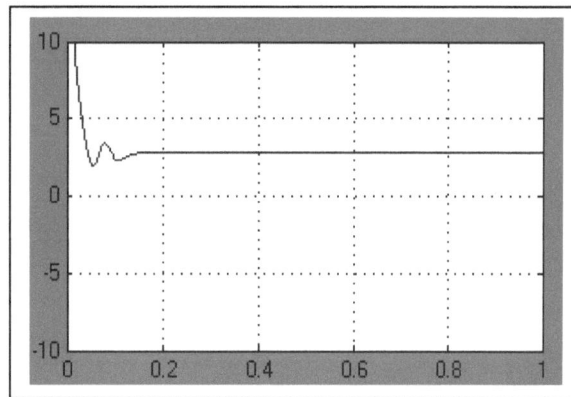

Figure 10 Simulink Simulation Results for conductive layer

Spice simulation programs and Simulink (in the Matlab R2007b) warns of the possibility of rolling over a larger calculation errors due to very low values of line capacity and inductivity.

4. EXPERIMENTAL RESULTS

For the composite material attenuation measuring in the field of radiofrequencies, was used the substitution method, known to be of high precision. Measuring scheme is given in Figure 11. For experimental determinations was used a wave guide system with an Gunn oscillator at a frequency of 10 GHz, as stated above. The result obtained through the mediation of several determinations is around $SE_{dB} = 3dB$. Comparing the value obtained with the simulations at the same frequency SE_{dB}= (1.5dB-3dB, it reveals a good precision, which means that the simulations made for the composite material are correctly made.

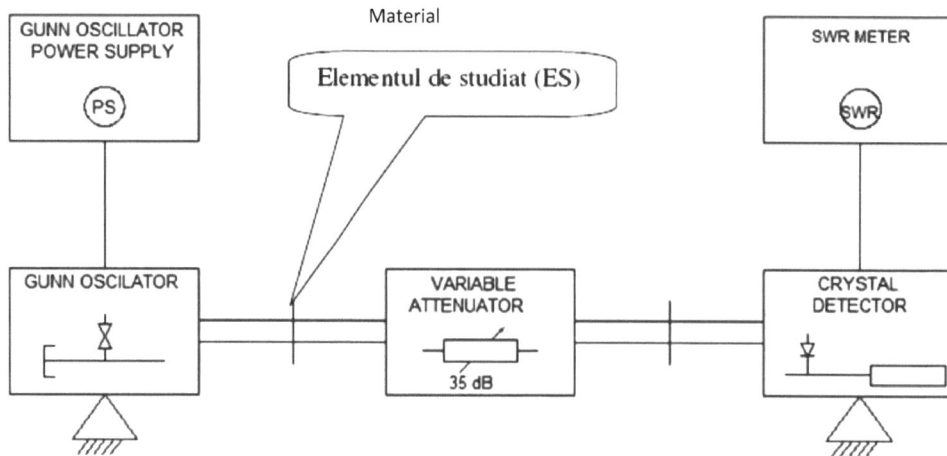

Figure 11 The block model used for experimental determinations

5. CONCLUSION

This paper proposes the use of a method of determining the shielding capacity of the electromagnetic waves composite material. Considering their many uses, knowledge of electrical characteristics of attenuation of electromagnetic waves is important.

The future can book big surprises on the border areas where composite materials will be used, and the current trend of expanding mobile communications and data transmission will be connected with them.

In addition, the simple method of simulation presented in this paper offers to the manufacturer the possibility to pre-choose the dimensions and the combination of materials in order to get superior performance of the composite material.

REFERENCES

1.   http://www.scribd.com/doc/81689631/Curs-Materiale-Compozite
2.   http://www.resist.pub.ro/Cursuri_master/AVRM/11.pdf
3.   Aciu L.E., Ogrutan P., Nicolae G., Bouriot B., New SEdB Measurement Method for Conductive Materials, PRZEGLĄD ELEKTROTECHNICZNY (Electrical Review), ISSN 0033-2097, R. 86 NR 3/2010.
4.   White D.R.J., Electromagnetic Shielding Materials and Properties, Don White Consultants, Inc., 1980.
5.   Schelkunoff A., Electromagnetic Waves, D. Van Nostrand Company, Inc., 1943.
6.   mihaelabucur.blogspot.ro/2009/05/materiale-compozite-i.html
7.   Badic M., Marinescu M.J., SE_{dB} determination for non-conductive electromagnetic absorbers, Proceedings of IEEE International Symposium on EMC, Santa Clara, California, USA, August 2004, pp. 557 - 561.

MOE AND MOR OF SOLID WOOD PANEL MADE OF TIMBER PROVIDED BY THIN (Dmax=160 mm) QUERCUS PETRAEA SPP. MATT. LIEBL. TREE

A.M. Olărescu[1], M. Cionca[1]
[1] Transilvania University of Braşov, Faculty of Wood Engineering, Braşov, ROMANIA,
a.olarescu@unitbv.ro , marinacionca@unitbv.ro

Abstract: Capitalization of secondary wooden resources represents a research priority in field of wood engineering and it's goal is to reduce human pressure on the environment. In this paper is presented the research about modulus of elasticity in bending (MOE) and of bending strength (MOR) of solid wood panels, with longitudinal texture. The solid wood panels are made of timber cut from thin (d_{max}=160 mm) sessile oak (Quercus petraea spp. Matt. Liebl.) trees, which resulted by thinning operation of forestry. Two types of joining were investigated and results show that these panels can be used successfully in furniture production.

Keywords: sustainable developments, secondary wooden resources, solid wood panels, MOE, MOR

1. INTRODUCTION

Sustainable forest management is meant to ensure forest goods and services purchased in response to immediate requirements and at the same time ensuring their continued availability and contribution to long-term development. On the 8th of July 1999, the European Council Regulation about measures to promote the conservation and management of forests came into force and refers to the potential recovery of wood secondary resources approximated to 9-10%, which opens subjects of research for a better insight to this matter.

In the capitalization of secondary wood resources technological, environmental and economic constraints that are imposed by these sorts must be kept in mind [7]. Until now, capitalization of the secondary wood resources was materialized in inferior products like fire wood, poles for scaffolding and formwork, beams for traditional constructions, pulp and crafts. Research conducted on particleboard and fibreboard which included secondary wood resources demonstrated that they have inferior mechanical qualities caused by the proportion of juvenile wood [1, 4, 6, 10, 11, 15, 16, 17, 18, 20].

The sessile oak (*Quercus petraea spp. Matt. Liebl.*) is the most important native oak species in Romania representing about 10,5 % of the total forests, the thin trees from thinning operations represent an important secondary resource that must be given a superior use. The results on its physical and mechanical properties have indicated similar strengths for wood from the thin trees compared to mature trees, but a slightly greater dimensional instability, especially in the radial direction, which compensates with a lower anisotropy coefficient [12, 13]. Similar strengths of wood from thin sessile oak trees compared to mature wood recommend it for value added capitalization providing its dimensional variation is controlled.

The objective of this paper is to study the possibility of incorporating wood provided by thin sessile oak trees (*Quercus petraea spp. (Matt.) Liebl*) into panels and their mechanical properties: MOR and MOE.

2. MATERIALS AND METHOD

Twenty *Quercus petraea Liebl.* trees with maximum diameters of 160 mm, resulted from thinning operations were taken from a forest warehouse in Stroeşti – Argeş, southern Romania. Located 45° 8' 0" North, 24° 47' 0" East with an annual rate of precipitation of 600 – 700 mm, and annual medium temperature between 8 – 9.5 °C, the area is characterized by De Martonne aridity index of 35 – 40; the Thorntwaite moisture index between 0 – 20 and Koncek humidity index 0 – 60 [2, 19].

The test area stretched on 15.1 ha, had a South-West exposure on a 15° slope and 500 – 560 m altitude. The type of forestry zone was 5152 (*Hill with Quercus petraea Bm, soil brown podzolic*) and the forestry type was 5221 (*Quercus – Fagus with Carex pilosa*). The litter was thin, with *Carex pilosa* as the characteristic flora type. The present composition of the stand is: 7 sessile oak (*Quercus petraea spp. Matt. Liebl*), 2 beech (*Fagus sylvatica L.*) and 1 hornbeam (*Carpinus betulus L.*). The stand of sessile oak trees was characterised by: medium age 45 years; medium diameter 18 cm; medium height 16 m; stand density 0.63; annual growth 5.1 m³/ha; standing volume 133 m³/ ha.

The twenty test trees were harvested and for every tree a harvest record has been made. This record contained information about position on the plot area, tree diameter, bole height, level of the crown and number of logs. The first cut was made at 30 centimetres from the ground and the last at the insertion point of the crown. Logs were cut with 2 m increments.

Then, logs were sawn into 50 mm thick timber pieces, which were kiln-dried from the initial moisture content (app.75%) to 12% final MC, according to the drying schedule presented in Table 1. Drying was followed by conditioning for one week at 20°C and 55% RH to equilibrate the internal stresses and the moisture content distribution.

Table 1: Drying schedule for 50 mm thick sessile oak timber (kiln type SEBA 17700)

Phase	Wood moisture content, [%]	Temperature, [°C]	Drying gradient	Equilibrium moisture content, [%]
Initial heating	75	31	-	18
Conditioning	85	31	-	18
Actual drying	35	29	2.1	15.0
Actual drying	26	33	2.2	11.8
Actual drying	20	38	2.4	8.3
Actual drying	12	50	2.4	3.3
Conditioning	12	52	-	7.0
Cooling	12	30	-	-

Two type of joints were investigated: edge to edge joint and finger joint.

The technological process of panels manufacture supposes: *1.* Cutting strips of dry timber. Slicing and splitting runs on Festool CS 70 Precisio universal circular saw, then straightening and planing to thickness of the combined machine Holzmann PT 260. Resulted strips have a section of 40 x 20 mm. *2.* Sort and preparation of strips to form the panel. This preparation includes trimming and marking strips in descending order of their length. *3.* Milling finger on Festool CMS + OF 1010 and finger joint cutter Festool HW S8 D34/NL 32 (just for finger joint panels). *4.* Apply adhesive to the strips edge. The adhesive used for gluing prisms is a polyurethane adhesive Jowapur 687:40. Apply adhesive can be done manually with trowel or brush, or mechanical gluing machine drum. Specific consumption of polyurethane adhesive type Jowapur 687:40 is 181 g/m^2. *5.* Panel assembly by frontal and parallel clamping. Specific pressure is 0.02 N/mm^2, pressing time is 2 h at temperature of 20 ± 2 °C and relative humidity of air φ = 60 ± 5%. *6.* Panel conditioning for 8 hours at a temperature of 20 ± 2 ° C and relative humidity of air φ = 60 ± 5%. *7.* Panel calibration. *8.* Panel sizing.

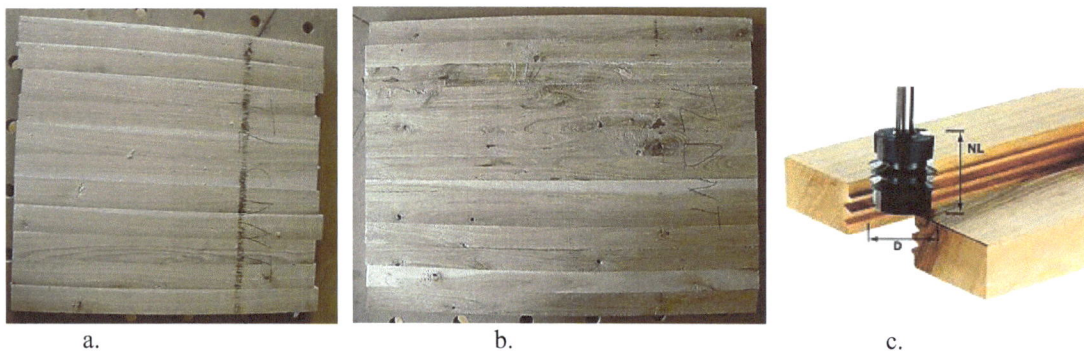

a.   b.   c.

Figure 1: a. Panel edge to edges assembly; b. panel groove assembly; c. finger joint cutter Festool HW S8 D34/NL 32.

MOE and MOR was determinate according EN 310: 1993 *Wood-based panels — Determination of modulus of elasticity in bending and of bending strength.*

The modulus of elasticity in bending and bending strength are determined by applying a load to the centre of a test piece supported at two points. The modulus of elasticity is calculated by using the slope of the linear region of the load-deflection curve; the value calculated is the apparent modulus, not the true modulus, because the test method includes shear as well as bending. The bending strength of each test piece is calculated by determining the ratio of the bending moment M, at the maximum load F_{max}, to the moment of its full cross section.

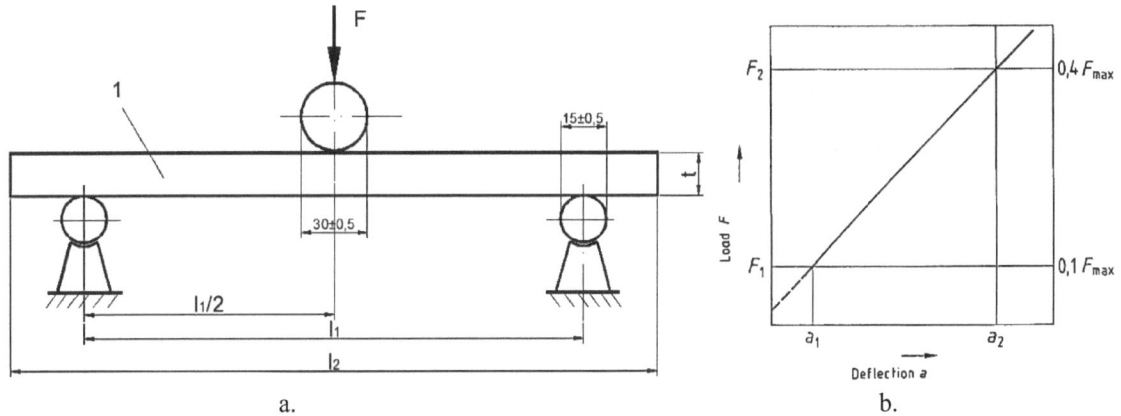

a.

b.

Figure 2: a. Arrangement of bending apparatus: 1 - test piece, 2 – load, t – thickness of test piece, $l_1 = 20$ t, $l_2 = 20t + 50$ mm; b. load – deflection curve within the range of elastic deformation (according EN 310: 1993).

The modulus of elasticity Em (in N/mm2), of each test piece, is calculated from the formula 1, where: l_1 is the distance between the centres of the supports, in millimetres; b is the width of the test piece, in millimetres; t is the thickness of the test piece, in millimetres; $F2 – F1$ is the increment of load on the straight line portion of the load-deflection curve, (Figure 2 b) in N. $F1$ shall be approximately 10 % and $F2$ shall be approximately 40 % of the maximum load; $a2 – a1$ is the increment of deflection at the mid-length of the test piece (corresponding to $F2 – F1$). The bending strength fm (in N/mm2), of each test piece, is calculated from the formula 2, where $Fmax$ is the maximum load, in newton.

$$E_m = \frac{l_1^3 (F_2 - F_1)}{4bt^3 (a_2 - a_1)} [N/mm^2] \tag{1}$$

$$\sigma_i = \frac{3F_{max} l_1}{2bt^2} [N/mm^2] \tag{2}$$

Sampling and cutting of the test pieces was according to EN 326-1 *Solid wood panels requirements*. From each panel six test pieces were cut. The test pieces were rectangular with shape and dimensions shown in figure 3. The test pieces have a symmetrical cross-sectional area with the assembly plan in the middle of the test piece (figure 3). The test pieces were conditioned to a constant mass in an atmosphere with a relative humidity of (65 ± 5) % and a temperature of (20 ± 2) °C.

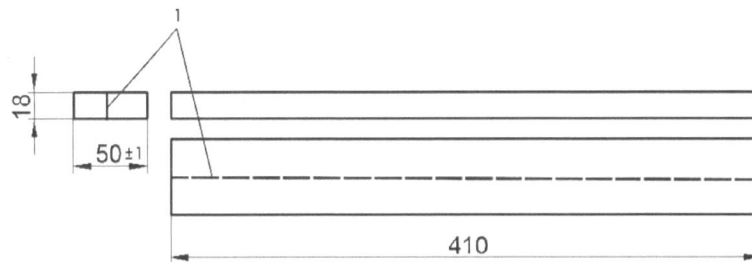

Figure 3: Shape and dimensions of test pieces: 1 – assembly line.

The test was developed in the Laboratory of Research and Testing of Wooden Products, according to European Regulations, accredited by RENAR (Romanian Accreditation Association) since 09.06.2008, in accordance with the certificate No. LI 664. An IMAL Model IBX600 universal tests machine were used. The load was applied at a constant rate of cross-head movement throughout the test. The rate of loading was adjusted so that the maximum load is reached within (60 ± 30) s. Measure the deflection in the middle of the test piece (below the loading head) to an accuracy of 0,1 mm and plot these values against the corresponding loads measured to an accuracy of 1 % of the measured value. Record the maximum load was to an accuracy of 1 % of the measured value.

The experimental data was statistically processed according to ISO 2602–2:1981, by calculating the statistical mean $\overline{x}$, the standard deviation s and the lower limit of the confidence interval $(\overline{x} - t_n s, \overline{x} + t_n s)$, which eliminates eventual errors. For a 95% confidence level, t_n is calculated according to (3). The lower limit of the confidence interval is calculated with (4) and is an important indicator for the comparison with the admissible value of a certain parameter.

$$t_n = \frac{t_{0,95}}{\sqrt{n}}$$

(3)

$$L_{5\%}^q = \overline{x} - t_n s$$

(4)

where

n - is the number of samples;

$t_{0.95}$ – value of Student distribution with n+1 degrees of freedom at 95 % confidence level. Its values are specified in a table in ISO 2602 – 2: 1981;

$L_{5\%}^q$ - is the lower limit of the confidence interval.

3. RESULTS AND DISCUSSIONS

The tests results are shown in tables 2 and 3.

Table 2: MOE and MOR for the edge to edge joint panels

Sample	Length [mm]	Width [mm]	Thickness [mm]	Weight [g]	Surface weight [Kg/m²]	Density [Kg/m³]	Maximum load [N]	MOR [N/mm²]	MOE [N/mm²]
I.T1	410.00	50.35	18.64	312.84	15.15	813.00	15.15	151.19	16632.72
I.T2	410.00	50.38	18.68	304.09	14.72	788.10	14.72	140.32	15486.21
I.T3	410.00	50.42	18.72	296.02	14.32	764.94	14.32	151.34	16115.49
I.T4	410.00	50.66	18.78	298.48	14.37	765.19	14.37	156.04	16928.73
I.T5	410.00	50.61	18.92	312.72	15.07	796.55	15.07	133.65	15483.14
I.T6	410.00	50.34	18.61	297.05	14.39	773.37	14.39	161.25	16860.40
Average	-	-	-	-	14.67	783.53	14.67	148.96	16251.12
Standard deviation [%]	-	-	-	-	0.37	19.21	0.37	10.21	658.68
$L_{5\%}^5$	-	-	-	-	**14.3663**	**767.7207**	**14.3663**	**140.5658**	**15709.0259**

The MOR was 140.565 N/mm² for edge to edge joint panels and 129.06 for finger joint panels. Also the MOE was 15709.025 for edge to edge joint panels and 14485.123 for finger joint panels. These values are higher than MOE (10635,85 N/mm²) and MOR (97,91 N/mm²) for sessile oak species solid wood with density 790 Kg /m³ [12]. This behaviour of wood composite materials was reported also in literature [5] .

MOR are comparable with laminated densified wood reported by [5]: for a density 1300±100 Kg/m³, MOR values are 180 N/mm² (type A), 130 N/mm² (type B) and 100 N/mm² (type C); for density 800±100 Kg/m³, MOR values are 100 N/mm² (type A), 100 N/mm² (type B) and 80 N/mm² (type C). Therefore the resistance/density ratio is 0,183 for edge to edge joint panels and 0,164 for finger joint panels compared with 0,076 - 0,138 for laminated densified wood. This fact indicates a greater resistance at lower density for these panels compared to laminated densified wood.

The finger joint test pieces have a long area of failure and edge to edge joint test pieces have a short area of failure. Also the specific noise which appeared at failure was almost similar with solid wood for the edge to edge joint.

For the finger joint panels, the failure type indicate a uniform panels behaviour even if MOR is less with 8,1% than edge to edge panels. This kind of joint brings to increase of dimensional stability of panels without significantly affecting mechanical resistance. The lower resistance of finger joint panels compared with edge to edge panels was reported in literature [3].

Table 3: MOE and MOR for the finger joint panels

Sample	Length [mm]	Width [mm]	Thickness [mm]	Weight [g]	Surface weight [Kg/m²]	Density [Kg/m³]	Maximum load [N]	MOR [N/mm²]	MOE [N/mm²]
I.T1	410.00	50.71	18.01	292.54	14.07	781.26	4208.00	138.15	16064.71
I.T2	410.00	49.93	18.01	313.48	15.31	850.26	4153.00	138.47	15595.86
I.T3	410.00	50.09	18.02	308.69	15.03	834.13	4814.00	159.82	17072.20
I.T4	410.00	49.07	18.06	290.69	14.45	800.04	3362.00	113.43	12791.97
I.T5	410.00	49.99	18.12	288.19	14.06	775.99	4698.00	154.56	16079.54
I.T6	410.00	50.50	18.23	311.12	15.03	824.26	4802.00	154.51	17308.20
Average	-	-	-	-	14.66	810.99	4339.50	143.16	15818.75
Standard deviation [%]	-	-	-	-	0.54	29.94	561.28	17.13	1620.44
$L^5\%$	-	-	-	-	14.2152	786.3464	3877.569	129.0601	14485.1232

4. CONCLUSION

Massive wood exploitation represents a major concern of the near future. Therefore, among other ways of reducing wood waste, new alternatives were thought to increase the added value of secondary wood resources, among which sessile oak (*Quercus petraea spp. Matt. Liebl.*) *thin trees,* which resulted by thinning operation of forestry.

This research confirmed the hypothesis that making panels out of thin sessile oak trees is feasible and rewarding. The results regarding MOE and MOR show that these panels can be used successfully in furniture production and the values, comparable with laminated densified wood, indicate expansion of utilisation field if their dimensional stability can to offer good results.

ACKNOWLEDGMENT

This paper is supported by the Sectorial Operational Programme Human Resources Development (SOPHRD), financed from the European Social Fund and by the Romanian Government under the project number POSDRU/89/1.5/S/59323.

REFERENCES

8. Baillères H., Chanson B., Fournier-Djimbil M., Plantaciones de´arboles de calidad y de r´apido crecimiento de productos forest ales en los tr´opicos, In: IUFROWP S5(3):01–04, Tema 12, August 26– 31, 1996.

9. Beldie Al. C., Rarities and floristic endemism in Arges county (In Romanian), In: Inginerul Alexandru C. Beldie, AGIR, Bucureşti, p. 112 – 117, 2010.

10. Boieriu C., Panouri composite lignocelulozice: panouri din lamele de lemn masiv, Editura Universităţii Transilvania din Braşov, 2007.

11. Cloutier A., Ananias R. A., Ballerini A., Pecho R., Effect of radiata pine juvenile wood on the physical and mechanical properties of oriented strandboard, In: Holz Roh Werkst, Vol. 65, pp. 157–162, 2007.

12. Curtu I., Ghelmeziu N., Mecanica lemnului şi materialelor pe bază de lemn, Editura Tehnică, Bucureşti, 1984.

13. Dimitri L., Bismarck C.V., Bottcher P., Schulze J.C., Production and use of poplar small-wood for particleboard manufacture, In: Holzzucht 35 (1/2), pp. 1-7, 1981.

14. Efthymiou P.N., Main technological, ecological and economic constraints in utilizing small-sized wood, Proc. of 3rd Int. Scientific Conference Fortenchenvi, may 26 -30, Prague, pp. 18 – 25, 2008.

15. EN 310: 1993 Wood-based panels — Determination of modulus of elasticity in bending and of bending strength.

16. EN 326-1: 1994 Solid wood panels requirements.

17. Geimer R., Crist J., Structural flakeboard from short-rotation intensively cultured hybrid Populus clones, In: For Prod J, Vol. 36(6), pp. 42–48, 1980.

18. Gurău L., Cionca M., Timar C., Olărescu A.,Compression strength of branch wood as alternative eco-material to stem wood, In: Environmental Engineering and Management Journal, "Gheorghe Asachi" Technical University of Iasi, Romania, July/August, Vol.8, No.4, pp. 685-690, 2009.

19. Olărescu A. M., Bădescu L. A. M., Cionca M., Gurău L., Porojan M., MOR, MOE and resistance to compression parallel to grain along the height of thin Quercus petraea spp. (Dmax=16 cm) provided from U. P. II Stroeşti – Argeş, southern of Romania. In Proceeding of International Conference Wood Science and Engineering in the Third Millennium – ICWSE 2011, Braşov, Romania, pp. 87 – 94, 2011.

20. Olărescu, A. M., Câmpean, M., Gurău, L., Effect of Heat Treatment Upon Dimensional Stability, MOE and MOR of Thin (D$_{max}$=16 cm) Sessile Oak Wood, In Pro Ligno, Vol. 7, Nr. 3, Braşov, pp. 29 – 38, 2011.

21. ISO 2602-1981 Guide to Statistical interpretation of data. Part 2: Estimation of the mean. Confidence interval.

22. Larson P., Kretchmann D., Clark A., Isebrands J., Formation and properties of juvenile wood in southern pines: A Sinopsis, In: USDA For Prod Lab Madison, p. 42, 2001.

23. Lehmann W.F., Geimer R. L., Properties of structural particleboard from Douglas – fir forest residues, Forest Products Journal, 24 (10), pp. 17 – 25, 1974.

24. Pugel A., Price E., Hse C., Shupe T., Composites from southern pine juvenile wood. Part 3, Juvenile and mature wood furnish mixtures, In For Prod J 54(1):47–52, 2004.

25. Stefaniak J., Use of juvenile wood in production of particleboard: Properties of particleboard produced from pine branch wood, In: Prace Komisji Technol.Drewna 10, pp. 95 – 116, 1981.

26. Turcu Gh. L., The Sphagnete of „Lacul cu Ochiu" (reg. Argeş) (In Romanian), In: Comunicările Academiei Republicii Populare Române, Biologie, Tomul XI, nr. 6, p. 665 – 673, 1961.

27. Zobel B. J., Sprague J. R., Juvenile wood in forest trees. Springer-Verlag, Berlin, 1998.

EVALUATION OF COATING ADHESION TO MELAMINATED PARTICLEBOARD SURFACES RECOVERED FROM OLD FURNITURE

A. Deák [1], M. Cionca [2], M.C. Timar [3]

[1] Transylvania University of Braşov, Faculty of Wood Engineering, Braşov, ROMANIA,
andrea.deak@unitbv.ro, marinacionca@unitbv.ro, cristinatimar@unitbv.ro

Abstract: *Wood and wooden based material recycling as part of wood engineering is an important scientific research field. In this paper are presented the results from adhesion tests of different coatings on the melaminated surfaces of particleboard recovered from old furniture. The tests were made according to ISO 2409 : 2007 Paints and varnishes - Cross-cut test. Results show that 77 % of the tested coatings have good adhesion to the substrate (class 1 or 0) regardless to the nature of the substrate (totally pickled by sanding, partially pickled, unpickled).*
Keywords: *wood recycling, coatings, adhesion*

1. INTRODUCTION

Although signals of environmentalism, of sustainable development and of returning to green products and service appear everywhere, today we are still in an era of consumerism. These signals are sometimes strong, sometimes weak, but always isolated in spite of national or international legislation providing sustainable development and ecology. Ecology now days is seen as a fashion and some go so far as considering it "a new religion"[3].

Over the years environmental philosophy has evolved from green design to eco-design and is moving towards sustainable design [1]. Eco-design's goal is to reduce environmental impact at every stage of the product's life cycle: raw material extraction, manufacturing, distribution, use and end of life [1].

Recycling represents the process of reusing old materials and developing new products, reducing, as a result, the emission of greenhouse gases, energy consumption required for the extraction of raw materials, or waste disposal. Recycling is actually the basic process of eco-design strategy for extending the life of materials [2]. Wood recycling may take numerous forms, from using wood waste as compost, mulch, wood brick fuel, animal bedding, particle board, MDF board, furniture, etc. It is important to recycle wood and wood based materials, if possible, with minimal energy consumption [4, 5].

In this context a solution for recycling and capitalization of the recovered melaminated particle boards should be found with minimum intervention by possibly refinishing them. For this study a 30 year old particle board was selected. The melaminated particleboard had numerous aesthetic defects such as scratches, peelings of the melamine foil and lastly, it was old fashioned. So the objectives derived from these observations include finding opaque finishing techniques for the recovered PB and methods of finishing with special effects with minimal energy consumption.

Figure 1: Surface detail.

2. MATERIALS AND METHOD

Thirty year old recovered melaminated 18 mm PB was used as substrate, medium grit (P80) and fine (P120) sandpaper was used for sanding the sample surfaces.

For performing the resistance test of paint coatings to separation from substrates when a right-angle lattice pattern is cut into the coating, penetrating through to the substrate six samples of 300x50x18 mm were cut from the recovered PB. These samples were then prepared by tree methods described in table 1 resulting tree surface types: unpickled, partially pickled, totally pickled by sanding. After all the samples had been cut and prepared their surfaces were wiped clean with methyl alcohol.

Table 1: Sample types and their preparation

Sample code	Characteristics	Used method
Type 1	Unpickled	Mechanical removal of dirt by hand sanding with 80 grit abrasive paper
Type 2	Partially pickled	Sanding by vertical belt sander Festool CMS-GM semiportabile with P120 grit abrasive paper
Type 3	Totally pickled by sanding	Sanding by vertical belt sander Festool CMS-GM semiportabile with P120 grit abrasive paper

The used paints and finishing products can be seen in table 2.

Table 2: Used paints and finishing products

No.	Type of finishing product	Brand	Colour
1	Acrylic enamel	*TESSAROL*	white
2	Acrylic lacquer	*TESSAROL-Helios*	transparent
3	Washable wall paint	Certificate, *KOBER*	white
4	Acrylic primer	made from Ecolak, *KOBER*	transparent
5	Nitrocellulose based enamel		black
6	PVAc based adhesive	*Aracet Aderoll, KOBER*	
7	Acrylic paint	*LOUVRE*	white
8	Acrylic paint	*DEKA*	metallic, azure
9	Acrylic paint	*PENTART*	pearl shine, white
10	Acrylic paint	*PENTART*	pearl shine, green
11	Acrylic paint	*PENTART*	metallic, purple

On every sample type a number of coatings were applied by brush in stripes horizontally on the samples as shown on figure 2. The coating and their combinations can be seen in table 3.

Table 3: Tested finishing products

Code	Finishing products and combinations
a	Acrylic enamel
b	Washable wall paint, 1 coating
c	Washable wall paint, 2 coatings
d	Acrylic primer + Acrylic enamel
e	Acrylic primer + Washable wall paint, 1 coating
f	Acrylic primer + Washable wall paint, 2 coatings
g	Nitrocellulose based enamel + undiluted PVAc based adhesive + Acrylic enamel (cracked finish)
h	Acrylic paint *Louvre*, white+ Acrylic paint *Deka*, metallic, azure
I	Acrylic paint *Louvre,* white

J	Acrylic paint *Pentart*, pearl shine, green
k	Acrylic paint *Pentart*, metallic, purple
l	Acrylic paint *Deka*, metallic, azure (2 coatings)
m	Acrylic paint *Pentart*, pearl shine, white (2 coatings)
n	Acrylic paint *Pentart*, pearl shine, white + Acrylic paint *Pentart*, pearl shine, green
o	Washable wall paint 2 coatings + Acrylic lacquer
p	Washable wall paint 2 coatings+ Acrylic paint *Louvre*, green + Acrylic lacquer

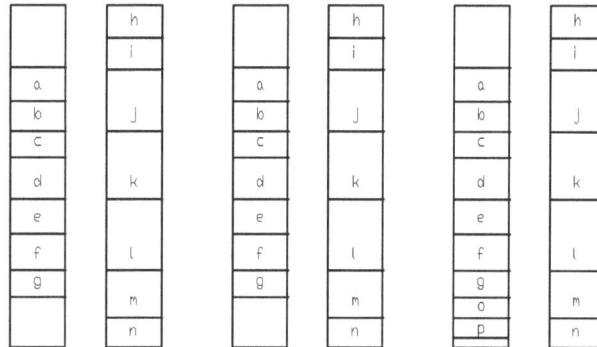

Figure 2: Distribution of the tested areas on the samples

Assessing the resistance of paint coatings to separation from substrates of the finished surfaces has been made in accordance to ISO 2409.2007(E) *Paints and varnishes Cross cut test* [6].

The test was carried out at a temperature of (23 ±2) °C and a relative humidity of (50 ± 5) %. The test panels were conditioned immediately prior to the test at a temperature of (23 ± 2) °C and a relative humidity of (50 ± 5) % for 24 h.

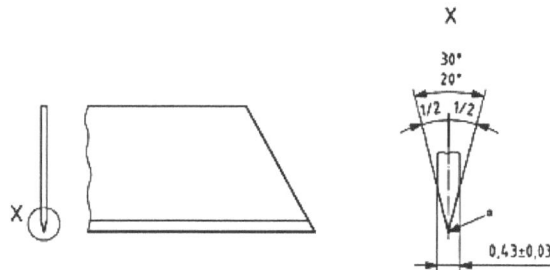

a) Single-blade cutting tool

Figure 3: Single blade cutting tool

A single-blade cutting tool was used with a 20° to 30° edge and a blade thickness of $\left(0.43^{-0.03}_{-0.03}\right)$ as shown in Figure 3. The number of cuts in each direction of the lattice pattern was six. The spacing of the cuts in each direction was equal and with 2 mm spacing as indicated by the standard for coating thicknesses from61 µm to 120 µm.

The test sample was placed on a rigid, flat surface to prevent any deformation of the sample during the test. Before the test, the cutting edge of the blade had been inspected.

The cutting was performed manually, as in the specified procedure: with cutting tool held with the blade normal to the test sample surface, with uniform pressure on the cutting tool the agreed number of cuts were made in the coating at a uniform cutting rate. All the cuts penetrated to the substrate surface. This operation was repeated, making further parallel cuts of equal number, crossing the original cuts at 90° to them so that a lattice pattern is formed.

The sample was brushed lightly with the soft brush several times backwards and several times forwards along each of the diagonals of the lattice pattern.

The center of the tape was placed over the lattice in a direction parallel to one set of cuts as shown in Figure 4, smoothening the tape into place over the area of the lattice and for a distance of at least 20 mm beyond with a finger.

To ensure good contact with the coating, the tape was rubbed firmly with a fingertip. Within 5 min of applying the tape, the tape was removed by grasping the free end and pulling it off steadily in 0,5 sto 1,0 s at an angle which is as close as possible to 60° (see Figure 4).

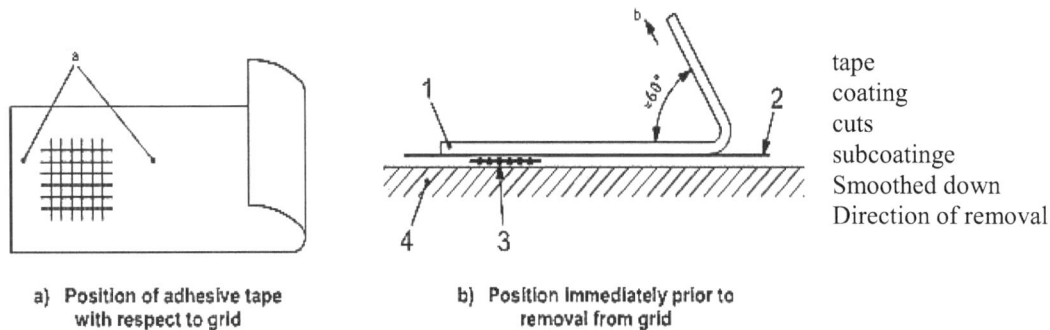

a) Position of adhesive tape with respect to grid

b) Position immediately prior to removal from grid

tape
coating
cuts
subcoatinge
Smoothed down
Direction of removal

Figure 4: Positioning of adhesive tape

The cut area of the test coating carefully examined in good lighting using normal. During the viewing process, the samples were rotated so that the viewing and lighting of the test area are not confined to one direction.

The tape was also examined in a similar manner. An additional examination was performed on magnified photographs made by a photo camera Olympus SP-560UZ 18x optical zoom set to super macro mode. The last mentioned examination method was proven to be more conclusive and exact (figure 5).

The test area was classified in accordance with Table 1, by comparison with the illustrations. In Table 4, a six-step classification is given.

Table 4: Classification of test results

Classification	Description	Appearance of surface of cross-cut **area** from which flaking has occurred (Example for six parallel cuts)
0	The edges of the cuts are completely smooth; none of the squares of the lattice is detached.	-
1	Detachment of small flakes of the coating at the intersections of the cuts. A cross-cut area not greater than 5 % is affected.	
2	The coating has flaked along the edges and/or at the intersections of the cuts. A cross-cut area greater than 5 %, but not greater than 15 %, is affected.	
3	The coating has flaked along the edges of the cuts partly or wholly in large ribbons, and/or it has flaked partly or wholly on different parts of the squares. A cross-cut area greater than 15 %, but not greater than 35 %, is affected.	
4	The coating has flaked along the edges of the cuts in large ribbons and/or some squares have detached partly or wholly. A cross-cut area greater than 35 %, but not greater than 65 %, is affected.	
5	Any degree of flaking that cannot even be classified by classification 4.	-

3. RESULTS AND DISCUSSIONS

Figure 5: Close-ups of the cut lattices

In the case of acrylic enamel (coating type a) and acrylic primer with washable wall paint in two coatings (coating type f) the adhesion increases with sanding of the substrate.

In case of washable wall paint 1 coating (sample b), acrylic primer with acrylic enamel (sample d) and all the acrylic paints (coating type c, e, h, i, j, k, l, m, n,) the adhesion is better if the tested substrate is totally pickled. For these samples both partially and totally pickled samples had similarly good results.

The adhesion of the washable wall paint in two coatings and the cracked finish doesn't depend on the type of substrate surface (coating type g) because it has the same class on both unpickled and partially pickled samples. The cracked finish on the unpickled sample was classified in class 3, so the test was repeated on the same coating a few centimetres further with the same result. On a closer observation it was obvious that the adhesion between the first coating of nitrocellulose based enamel and the second layer of undiluted PVAc based adhesive failed and caused this lower classification. This can be an expected phenomenon because the two coating are hardly compatible but precisely this incompatibility results the specific cracked surfaces. But the adhesion between the substrate and first coat of nitrocellulose based enamel was very good so this kind of coating technique can be applied on totally pickled melaminated PB with special attention to the quality and the applying method of the second and the layer of undiluted PVAc based adhesive.

The resulted quality classes are presented in table 5.

Table 5: Results of the cross cut test

COATINGS	TYPE 1	TYPE 2	TYPE 3
a	3	2	0
b	1	1	0
c	1	1	1
d	1	1	0
e	2	0	1
f	2	1	0
g	1	1	3
h	2	2	1
I	2	2	1
J	1	1	0
k	1	2	0
l	1	1	0
m	1	1	0
n	1	1	0
o	-	-	0
p	-	-	1
Average	**1,428571**	**1,214286**	**0,5**

As for the sanding method applied the fallowing can be observed:

- For the unpickled samples (type 1) the resistance of paint coatings to separation from substrates of the finished surfaces qualifies in class 1 for the majority of the coatings, excepting coating type e, f, h, i (which are class 2) and a (which is class 3).
- For partially pickled samples (type 2) the majority of the applied coatings are in quality class 1, meaning that the resistance to separation of coatings is good. Coating type e (Acrylic primer + Washable wall paint, 1 coating) has a high resistance and has a classification of 0. As for coating types a, h, i, k the resistance is lower, class 2.
- As expected, for totally pickled samples (type 3) the majority of the coatings have a veri good resistance to separation from substrate and qualify in class 0. Coating types c, e, h and i have a slightly lower resistance, quality class 1 but are acceptable.

The recommendations for different types of melaminated PB surfaces and applicable coatings are presented in table 6, which also shows that 77% of the tested coatings have a good resistance to separation from substrate (class 1 and 0) regardless of the sanding method (pickled or not).

Table 6: Recommendations for applicable coatings on different types of melaminated PB surfaces

Sample types	Finishing products and their combinations													
	a	b	c	d	e	f	g	h	i	j	k	l	m	n
Type 1	○	●	●	●	◖	◖	●	◖	◖	●	●	●	●	●
Type 2	◖	●	●	●	●	●	●	◖	◖	●	◖	●	●	●
Type 3	●	●	●	●	●	●	●	●	●	●	●	●	●	-
●	Highly recommended – quality class 0 and 1													
◖	Recommended only if the surface is protected from wear and tear – quality class 2													
○	Not recommended – quality class 3-5													

4. CONCLUSIONS

Melamine surfaces of the recovered particleboard can be finished with opaque coatings without sanding them until they are totally pickled, provided that the surface is not damaged in any way. If the melamine surface is damaged to an extent greater than 50%, sanding is recommended until a totally pickled and smooth surface is obtained. In this case one should always put in balance the energy and labor consumption with the results and always apply the least energy consumer method.

REFERENCES

[1]   Bhamra, T., Lofthouse, V., *Design for sustainability – A practical approach*. Hampshire: Gower Publishing Limited, England, 2007.
[2]   Deák A.,*Reciclarea mobilierului*. Studiu de caz: canapeaua „My NorWay". În: Şoptana, Anul 6, nr.1(19), p.18-20.,   2011.
[3]   Olărescu A. M., *Eco-design de produs*. Vol.I. Braşov: Editura Universităţii Transilvania din Braşov, 2010.
[4]   Papanek, V., *The green imperative. Natural design for the real world*. London: Thames & Hudson Press, 1995
[5   ]Vezzoli, C., Manzini, E., *Design for environmental sustainability*. Springer – Verlag, London, 2008
[6   ] ***ISO 2409:2007(E) *Paints and varnishes- Cross-cut test*.

SEM STUDY ON FRACTURE BEHAVIOR FPR PA 6 COMPOSITES USED ON AUTOMOTIVE PARTS

A.D. Botezatu*, M.I. Ursu, O.E. Gradin, M.C. Rusu

Department of Materials Engineering, Renault Technologie Roumanie,

*domnica-angelica.botezatu@renault.com

Abstract: This work presents a study on failure behavior for PA 6 matrices reinforced with different type of glass charge. SEM analysis of the fracture surfaces from reinforced polymers revealed differences between fracture behaviors due to reinforcement's type, taking into account the contact surface between the polymeric matrices and glass charge. PA 6 samples reinforced with glass fibers and glass spheres were analyzed. For polymeric matrix reinforced with glass spheres it was observed a fracture behavior represented by ductile faces. Matrices reinforced with long glass fiber showed a fragile fracture behavior due to the plasticization effect of the polymer. A scanning electronic microscope (SEM) analysis was used to visualize the damage process on the crack faces.

Keywords: PA6 composite, SEM analysis, glass fibers reinforced polymer, fracture behavior

1. INTRODUCTION

Polymer composites represent an important category of materials used in automotive industry due to their advantages compared to traditional materials: improved physical - mechanical properties and non-expensive processing. One of most significant class of polymeric composites is represented by the fiber reinforced polyamides composites, which are able to offer strength and toughness comparable or better than traditional metallic materials. Having a lower density, specific strength and modulus these fibers reinforced polyamides could be considered good substitutes for metals in many weight-critical applications [1-3].

As it is already known, polyamides represents polymers with good mechanical properties, high strain and wear strength [4-5]. There are many studies in literature concerning the glass fibers reinforced polyamides which showed that the good mechanical properties are given by the matrix and glass fibers properties, combined with the load transfer capacity at the interface matrix - glass fibers [6].

Glass fibers mainly used for glass-reinforced plastics are obtained by fusion spinning of alumino-borosilicate glass having in its composition less than 1 wt% alkali oxides [7]. The incorporation of short and long glass fibers in polyamide 6 matrix increased elastic modulus, work of fracture and flexural strength [4-7].

A large number of auto vehicle parts (dashboard elements, break pedals, fans, handles etc.) based on glass fibers reinforced PA 6 give a significant weight reduction of cars, resulting a higher performances and lower fuel consumptions. During the service life of the cars, the vehicle parts are submitted to impact, cyclic loads or vibrations (low or high intensity). The influence of these factors could lead to their failure in service and their replacement, consequently.

Various parameters affect the fracture behavior of glass fibers reinforced PA6 composites, including the impact velocity, the specimen geometry, the impactor size, impact energy, clamping mode, the matrix properties, and the reinforcement geometry. There are numerous possible modes of damage in polymeric composites: matrix deformation, micro-cracking, interfacial debonding, lamina splitting, delamination, fiber breakage and fiber pull-out. In some of the cases when the damage cannot be visually detected, damaged components will continue to fulfill their function unless the damage obviously affects the safety of the structure or aesthetic aspect [8-10]. Taking into account these aspects, clearly results why it is very important to predict and to understand the failure behavior in the damaged component during its service life.

The objective of this paper is to investigate the fracture behavior of two types of glass fibers reinforced PA 6 (30¨% GF and 20%GS – 10%GF). Morphology and fractography were done using scanning electronic microscopy (SEM).

2. RESULTS AND DISCUSSION

A large number of studies have indicated that plastic deformation of fiber-reinforced polyamides is governed by the nature of fibers, its percentage, size, shape and orientation in the polymer matrix [11,12]. Simultaneously, the fragile or ductile behavior of the fiber reinforced polymer composites is conditioned by the thickness of the material subjected to stress (Figure 1). Thus, a thick material will limit the detachment phenomena due to a high level of shear shrinkage into the constraint area, favoring a fragile fracture of the material.

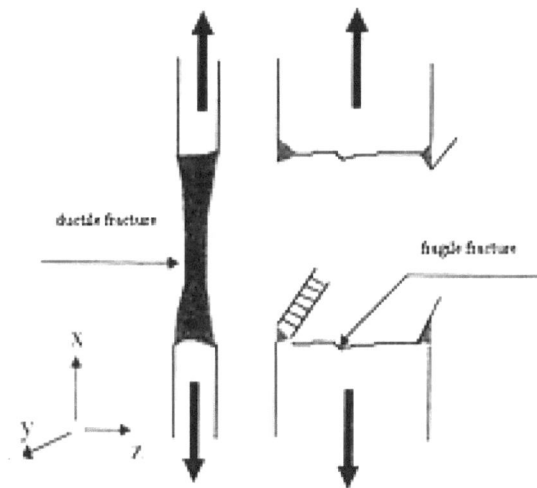

Figure 1: The influence of material thickness on the fracture behavior

The investigations of the fracture behavior presented by this paper were carried out on different vehicle parts based on polyamide 6 composite reinforced with glass fibers and glass spheres. The fracture behavior was investigated by SEM analysis and it was focused on the influence of the type of glass charge.

Representative SEM micrographs of the surface from a fractured 30% glass fibers reinforced polyamide 6 composite are presented in Figure 2. Short glass fibers are intimately mixed in the matrix and are distinguishable. The fibers are preferentially aligned in flow direction for injection molded part. Both the micrographs show that most of the glass fibers are pulled out (see the pull out fibers and pull out holes). The fibers surface seems to be smooth and without any presence of residual matrix indicating a poor adhesion between glass fibers and PA 6 matrix, most probably due to the non-treatment of the fibers. Thus, the fracture takes place at the interface matrix – fiber. These characteristics clearly show a brutal fragile fracture for the 30% GF reinforced PA 6 composite.

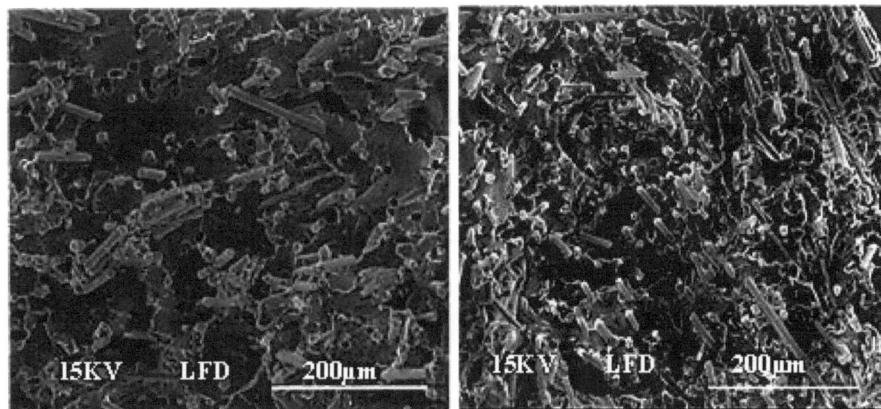

Figure 2: Scanning electron micrographs of brutal fragile fracture for 30% GF reinforced PA 6

The glass fibers take up greater stress in the composite, reducing the matrix stress, which is too small to promote matrix-governed deformation such as rubber cavitations and matrix shear yielding. In contrast, the micrographs from Figure 3 show ductile fracture behavior further reduced in the ligament region of the matrix. Some plastic deformation can be observed in the matrix ligaments. This kind of fracture behavior mostly appears in the propagating area of the crack. After a ductile behavior presented in the propagating area, the crack is propagated into the rest of the polymer matrix following a brutal fragile behavior.

Figure 4 represents scanning electron micrographs of a fracture behavior of 20% glass spheres and 10% glass fibers reinforced polyamide 6 composite. Analysis of these cracked surfaces puts in evidence a fragile fracture behavior, poorly revealed on this type of material. Glass fibers and glass spheres are mixed in the matrix and are clearly distinguishable.

The fracture behavior is accompanied with noticeable de-bonding between the matrix and the reinforcing fibers. However, the adhesion between glass spheres and PA 6 matrix appears to be stronger. This observation is indicative of a strong interface with good stress transfer from the polymer matrix to the glass spheres, due to a higher contact between the glass spheres and the PA 6 matrix.

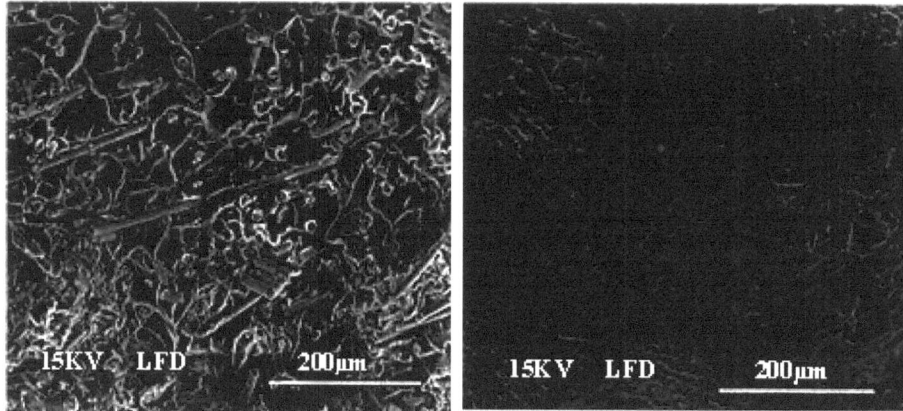

Figure 3: Scanning electron micrographs of ductile fracture for 30% GF reinforced PA 6

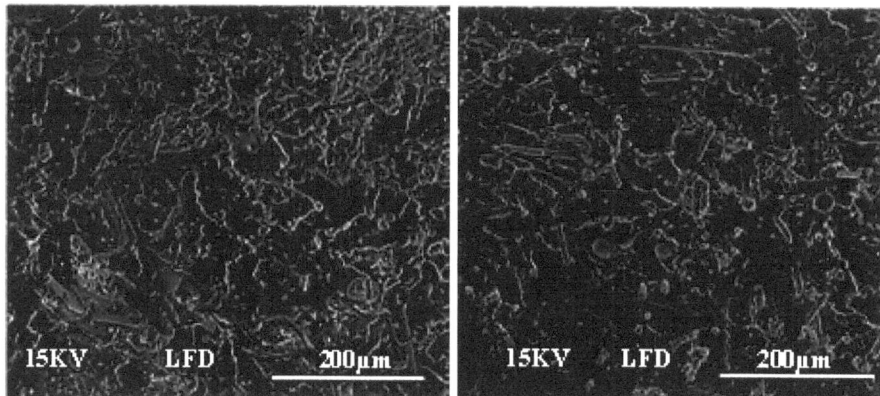

Figure 4: Scanning electron micrographs of a fragile fracture for 20% GS – 10% GF reinforced PA 6

Figure 5 presents a ductile fracture for the 20% glass spheres and 10% glass fibers reinforced polyamide 6 composite. Ductile fracture is preferred in most applications and is characterized by an extensive plastic deformation ahead of crack (necking). In these cases, the crack resists further extension unless applied stress is increased. The ductile behavior of fracture is accompanied by slow propagation and important energy absorption before fracture.

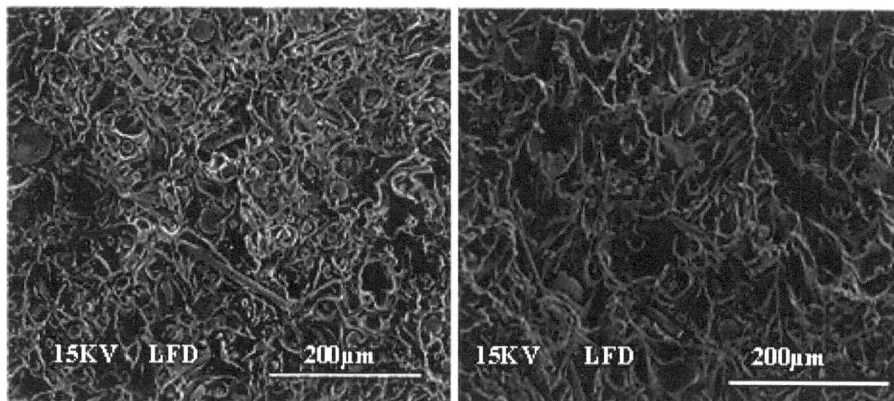

Figure 5: Scanning electron micrographs of a ductile fracture for 20% GS – 10% GF reinforced PA 6

In general, three main steps are distinguished for the ductile fracture behavior: formation of micro-voids, growth of these voids and formation of the shear bands. At microscopic scale ductile fracture surfaces are represented by rough and irregular areas, consist in micro-voids and dimples. The ductile fracture behavior is a slow process giving a large time for the problem to be corrected. Also, due to the high plastic deformation, more strain energy is needed to cause ductile fracture.

3. CONCLUSION

In this study, the effect of glass charge on fracture behavior of the PA 6 composites was investigated using scanning electron microscopy (SEM). Differences have been observed in the fracture behavior of glass fibers reinforced PA 6 and glass spheres reinforced PA 6.

The crack faces were observed into various micrographs illustrating the type fracture for the analyzed materials: micro-cracks, matrix cracks, fracture at the interface polymer matrix – glass charge and fibers crack. For the 30% glass fibers reinforced polyamide 6 were observed micro-cracks accompanied by de-bonding polymer matrix – glass fibers.

For the 20% glass spheres and 10% glass fibers reinforced polyamide 6 composite were observed a good adhesion between glass spheres and polymer matrix.

In conclusion, the 30% GF reinforced PA 6 composite presents a brutal fragile fracture behavior, while a ductile fracture behavior is corresponding to the 20% glass spheres and 10% glass fibers reinforced polyamide 6.

REFERENCES

[1]   Reis P.N.B., Ferreira J.M., Richardson M.O.W., Effect of the Surface Preparation on PP Reinforced Glass Fiber Adhesive Lap Joints Strength, J. Thermoplast. Compos., 25(3), 2012.

[2]   Cosmi F., Bernasconi A., Elasticity of Short Fibre Reinforced Polyamide Morphological and   Numerical Analysis of Fibre Orientation Effects, Mat. Eng., 17(2), 2010.

[3]   Yuanjian T., Isaac D.H., Combined Impact and Fatigue of Glass Fiber Reinforced Composites,   Composites. Part B: Eng., 32, 2008.

[4]   Mouhmid B., Imad A., Abdelaziz M.N., Benmedakhene S., Mechanical Behavior of a Glass Fiber Reinforced Polyamide, 17eme Congres Francais de Mecanique, 2005.

[5]   Thomason J.L., The Influence of Fiber Properties of the Performance of Glass-fibre-reinforced   Polyamide 66, Composites science and technology, 59, 1999.

[6]   Surampadi N.L., Ramisetti N.K., Misra R.D.K., On Scratch Deformation of Glass Fiber Reinforced Nylon 66, Mat. Sci. Eng. A, 456, 2007.

[7]   Pedroso A.G., Mei L.H.I., Agnelli J.A.M., Rosa D.S., Properties That Characterize the Propagation of Cracks of Recycled Glass Fiber Reinforced Polyamide 6, Polymer Testing, 18, 1999.

[8]   Thomason J.L., The Influence of Fiber Length, Diameter and Concentration on the Modulus of Glass   Fiber Reinforced Polyamide 6.6, Composites: Part A, 39, 2008.

[9]   Hartmann J., Moosbrugger E., Buter A., Variable Amplitude Loading with Components Made of Short   Fiber Reinforced Polyamide 6.6, Eng. Procedia, 10, 2011.

[10]   Levay I., Lenkey G.B., Toth L., Major Z., The Effect of the Testing Conditions on the Fracture   Mechanics Characteristics of Short Glass Fiber Reinforced Polyamide, 133, 2003.

[11]   De Monte M., Moosbrugger E., Jaschek K., Quaresimin M., Multiaxial Fatigue of a Short Glass Fiber Reinforced Polyamide 6.6 – Fatigue and Fracture Behavior, Int. J. Fatigue, 32, 2010.

[12]   Ferreno D., Carrascal I., Ruiz E., Casado J.A., Characterisation by Means of a Finite Element Model of   the Influence of Moisture Content on the Mechanical and Fracture Properties of the Polyamide 6   Reinforced with Short Glass Fiber, Polymer Testing, 30, 2011.

DYNAMIC TESTS ON ASPHALT MIXTURES WITH POLYPROPYLENE FIBERS

Carmen Răcănel[1], Adrian Burlacu[1]

[1] Technical University of Civil Engineering, Bucharest, Romania,
carmen.racanel@yahoo.com, adrian_burlacu@yahoo.com

Abstract: *The asphalt mixtures with polypropylene fibers are a SMA (stone mastic asphalt) asphalt mixtures – a blend with crushed stone skeleton, with voids filled with mortar with a higher bitumen percent. This concept is based on achieving high stability and durability. In Romania this kind of mixtures is noted by MASF. The polypropylene fibers are used as additions in asphalt mixture, like other fibers (e.g. cellulose fibers), 0.3 – 0.5 % in content. The aim of this paper is to compare, by laboratory studies, the contributions of polypropylene fiber and cellulose fiber on behavior of asphalt mixtures to creep and fatigue damages. The laboratory tests taken into account are Marshall test, cyclic compression test, indirect tension test, four point bending test and wheel tracking test. The study was carried out in Roads Laboratory of Technical University of Civil Engineering of Bucharest and the results are presented in tables and graphs.*
Keywords: *asphalt mixture, polypropylene fiber, stiffness, creep, fatigue*

1. INTRODUCTION

The asphalt mixtures with polypropylene fibers are a SMA (stone mastic asphalt) asphalt mixtures – a blend with crushed stone skeleton, with voids filled with mortar with a higher bitumen percent. This concept is based on achieving high stability and durability. In Romania this kind of mixtures is noted by MASF (asphalt mixture stabilized with fibers).

This type of asphalt mixture is successfully used in wearing courses, both abroad and in our country, for heavy traffic roads and airport pavements.

The results for MASF asphalt mixture are a good stability at high temperatures, a good flexibility at low temperatures, a good wearing resistance provided by high quality components, an increased adhesion, a good roughness and abrasion resistance.

The MASF differs from the asphalt concrete mainly by a high amount of chipping and bitumen-rich asphalt mastic. Comparing the grading curves of the BA asphalt concrete, BAR rough asphalt concrete and asphalt mixture stabilized with fibers, we can notice the difference in the aggregate composition of the three types of asphalt mixture (figure 1). Initially, the fibers were introduced into the asphalt mixture to allow the increment of bitumen amount in the recipe, without incurring the phenomenon of flowing. Further research demonstrated that the fibers in the recipe fulfill also other functions, such as enhancement of asphalt grout flexibility and creation of a three-dimensional structure, which allows the interaction of all compounds.

In order to obtain the same effect into the mixture, with different fibers, it turns out that these fibers have a different efficiency, according to their nature (e.g. polypropylene fibers, cellulose fibers, wood fibers, mineral fibers, glass fibers).

Cellulose fibers look like grains or grey fibers, sometimes impregnated with bitumen, and function as stabilizers for the asphalt mixtures, due to the spatial distribution of fibers. They exhibit thermal stability and are not polluting products or harmful to health.

The polypropylene fibers are used as additions in asphalt mixture, like other fibers, 0.3 – 0.5 % in content. They are yellow-black-gray synthetic polypropylene fibers, having physical characteristics in Table 1. These fibers have the following advantages: insensitivity to moisture, stabilizing effect, possibility of automatic proportioning and productivity growth, as they do not require dry mixing time.

The manufacturing and placing of the asphalt mixtures containing added polypropylene fibers are practically achieved by the same equipment and machinery, like the common asphalt mixtures. Figure 2 presents the polypropylene and cellulose fibers.

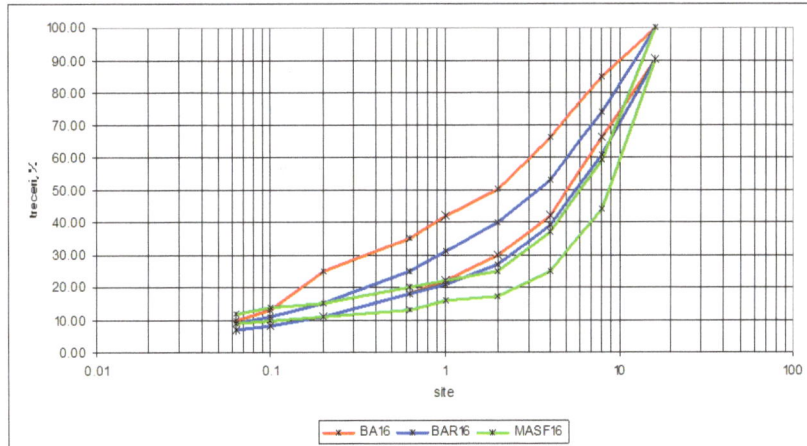

Figure 1: Grading curves for a BA16 asphalt concrete, BAR16 rough asphalt concrete and MASF16 asphalt mixture stabilized with fibers, according to SR 174 Romanian standard

Figure 2: Polypropylene fibers (left) and cellulose fibers (right)

Table 1: Physical characteristics of polypropylene fibers

Physical characteristics	Imposed limits according to the Technical Specification Polypropylene / Aramid
Specific density, g/cm^3	0,91/1,45
Tensile strength, MPa	483
Length , mm	19
Resistance to acids/alkalis	Inert/good
Melting point, ^{o}C	157

The aim of paper is to compare by laboratory studies, the contributions of polypropylene fiber and cellulose fiber on behavior of asphalt mixtures to creep and fatigue damages. The laboratory tests taken into account are Marshall test, cyclic compression test, indirect tension test, four point bending test, wheel tracking test. The study was carried out in Roads Laboratory of Technical University of Civil Engineering of Bucharest and the results are presented in table and graphs.

65

2. USED MATERIALS AND RECIPE

The two types of asphalt mixtures for thin surfacing considered in this study stand on SR 174 Romanian standard and are denoted as "MASF16m-T" (Topcel-fiber asphalt mixture) and "MASF16m-P" (polypropylene-fiber asphalt mixture). The materials used for the asphalt mixture (aggregate, fiber and bitumen), as well as the recipe, are presented in Table 2.

Table 2: The used asphalt mixtures materials and the recipes of the used asphalt mixtures

Asphalt Mixtures	Source /Type and %	Crushed Rock				Filler	Fiber by mix	Bitumen by mix
		0-4	4-8	8-16	16-25			
MASF16m-T	Source /Type	Turcoaia				Limestone Holcim	Topcel	25/55-65 PMB
	%	13	25	45	-	11	0.3	5.7
MASF16m-P	Source /Type	Turcoaia				Limestone Holcim	Poly-propylene	25/55-65 PMB
	%	13	25	45	-	11	0.3	5.7

3. SAMPLES PREPARATION

The manufacturing of the asphalt mixture samples meets the requirements of European standards appropriate for each type of laboratory test presented in the next chapter:
- cylindrical samples, measuring ϕ=100 mm and h≈63mm, have been carried out, using the Marshall hammer with 75 blows per side, for the Marshall test, IT-CY test and cyclic compression test;
- prismatic samples, measuring L=405 mm, l=50 mm and h=50 mm, have been carried out for 4PB-PR test, by cutting the roller-compacted slabs;
- slabs measuring 30.5x30.5x5 cm have been compacted with the roller, for the wheel tracking test.

4. TYPE OF TESTS AND TESTING CONDITIONS

In order to achieve the aim of this paper, the following tests have been carried out in the "Roads Laboratory" at the Faculty of Railways, Roads and Bridges (Technical University of Civil Engineering Bucharest):
- Marshall test on cylindrical samples, according to SR EN 12697-34, at a temperature of 60°C, resulting the values of the Marshall stability and flow;
- indirect tension test on IT-CY cylindrical samples, according to SR EN 12697-26 C Annex, a temperature of 15°C and 20°C, resulting the stiffness of the asphalt mixture;
- four point bending test on 4PB-PR prismatic samples, according to SR EN 12697-26 B Annex and SR EN 12697-24 D Annex, at a temperature of 20°C and 30°C, resulting the stiffness and fatigue resistance of the asphalt mixture;
- cyclic compression test on cylindrical samples, according to SR EN 12697-25, for tri-axial compression at a temperature of 50°C, resulting the resistance to permanent deformations;
- wheel tracking test on plates, according to SR EN 12697-22, small equipment, B outdoor method, at a temperature of 60°C, resulting the resistance to permanent deformations.
The test conditions are those found in the SR EN 13108-20 European standard, as well as in the SR174 Romanian standard.

5. EXPERIMENTAL RESULTS

The experimental results obtained on "MASF16m-T" and "MASF16m-P" asphalt mixture samples, through laboratory tests, are ranged in different categories according to SR EN 13108-1, being comparatively described in the Figures 3-10, as follows:

- in the Figures 3 and 4 – Marshall stability, S [kN] at 60°C, flow index, I [mm] at 60°C, S/I [kN/mm] ratio at 60°C, bulk density [kg/m³];
- in the Figure 5 - stiffness, S [MPa] by IT-CY indirect tension at 15°C and 20°C and 4PB-PR four point blending stiffness at 20°C;
- in the Figures 6 and 7 – creep rate, f_c and calculated permanent stain ε_{1000} and ε_{10000} of tri-axial compression test;
- in the Figures 8 and 9 – rut depth for 10^4 load cycles (%) and the "Wheel Tracking" slope (wheel-tracking speed) [mm/10^3 load cycle] in the wheel tracking test;
- in the Figure 10 - initial strain corresponding with a fatigue life of 10^6 cycles ε_6 from the 4PB-PR fatigue test for T = 30°C.

6. CONCLUSIONS

The conclusions drawn from this study are the following:
- The fiber asphalt mixtures have a good behavior regarding the main degradations occurring on a road during the service period: permanent deformations and fatigue cracking;
- Marshall test provides comparative results between the two types of fibers used in the asphalt mixture: for an almost equal bulk density obtained on Marshall cylinders, there are the same categories, according to SR EN 13108-1 for stability, flow index and ratio stability/flow index;
- The 4PB-PR test for determining the modulus of stiffness leads to different categories, both for the maximum and for the minimum values, the Topcel-fiber asphalt mixture presenting higher values;
- The IT-CY test for determining the stiffness modulus leads to the same maximum and minimum categories at a temperature of 15°C; when the temperature rises (20°C), the stiffness modulus of the MASF16-P asphalt mixture decreases more comparing to that of the MASF16-T mixture;
- The triaxial compression test leads to the same types of values for the creep rate that characterizes the resistance to permanent deformations; however there are permanent strains at 1000 and 10000 load cycles, 12-15% bigger for the MASF16-P asphalt mixture, which indicates a lower resistance to permanent deformations, comparing to the MASF6-T asphalt mixture;
- The wheel tracking test leads to the same categories for the rut depth and the wheel tracking slope, hence there is a similar behavior under this type of degradation;
- The 4PB-PR fatigue test indicates a fatigue strength characterized by the specific strain estimated to be reached after 10^6 load cycles, similar to both types of fibers;
- According to the SR 174 Romanian standard, the both asphalt mixture recipes meet the requirements for the performed tests;
- Considering the obtained categories of the studied tests according to SR EN 13108-1, it can be said that the polypropylene fiber leads to good results if used into an asphalt mixture stabilized with fibers for the thin surfacing.

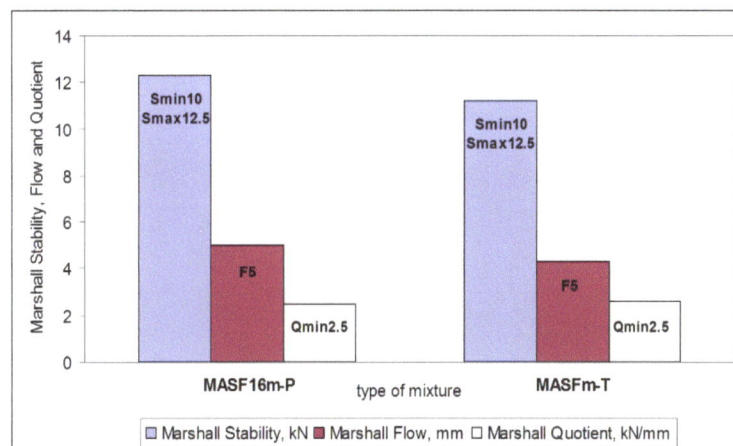

Figure 3: Marshall stability, flow index and the S/I ratio [kN/mm] at 60°C for the studied asphalt mixtures

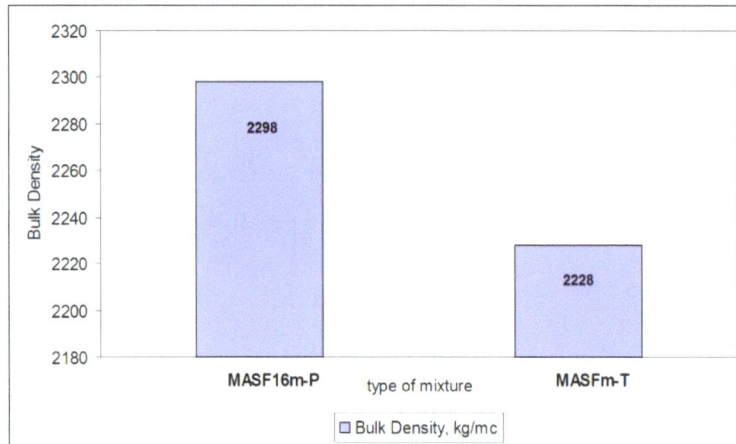

Figure 4: Bulk density for the studied asphalt mixtures

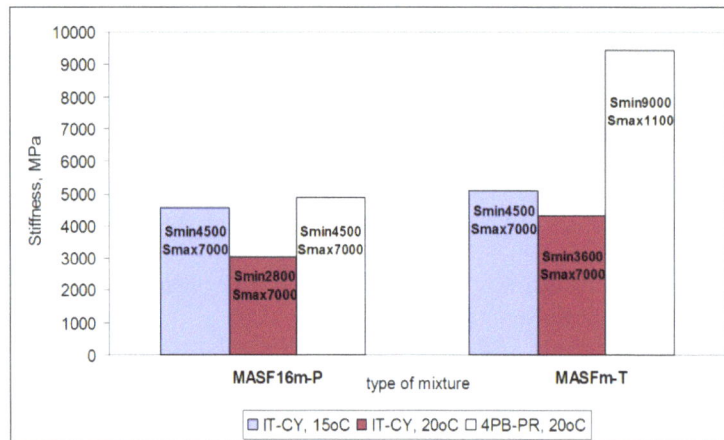

Figure 5: Stiffness by IT-CY indirect tension at 15°C and 20°C and four point blending stiffness at 20°C for the studied asphalt mixtures

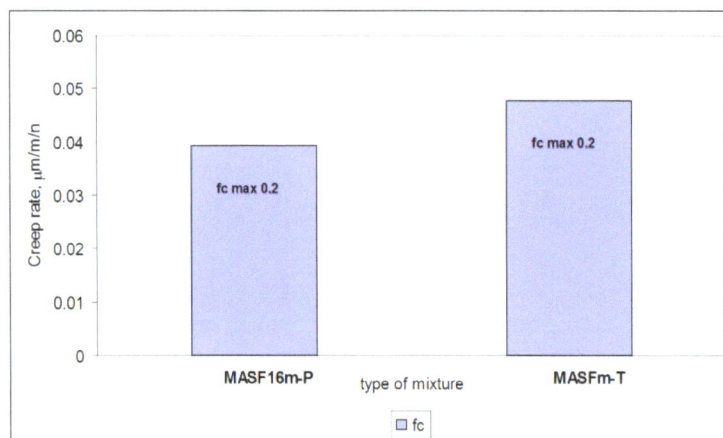

Figure 6: Creep rate calculated through the triaxial compression test for the studied asphalt mixtures

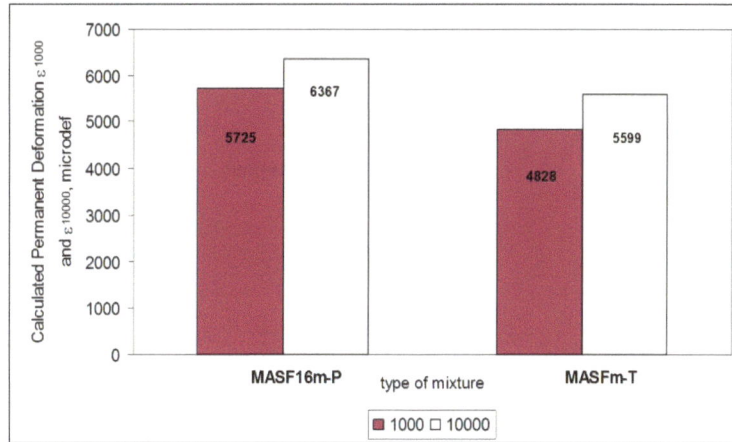

Figure 7: Permanent deformation calculated through the triaxial compression test for the studied asphalt mixtures

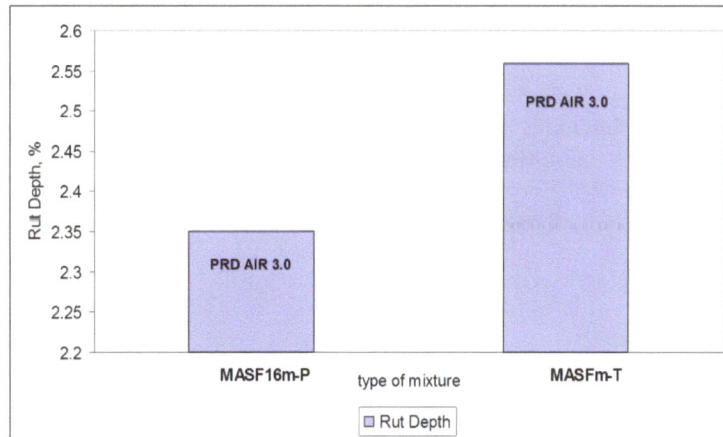

Figure 8: Rut depth for 10^4 load cycles from the wheel tracking test for the studied asphalt mixtures

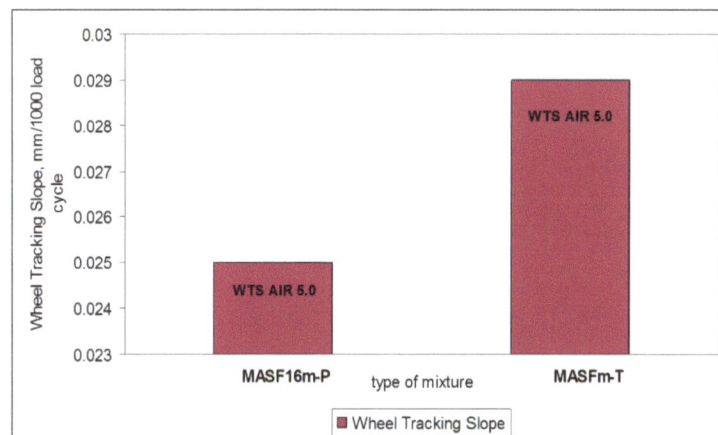

Figure 9: "Wheel Tracking" slope from the wheel tracking test for the studied asphalt mixtures

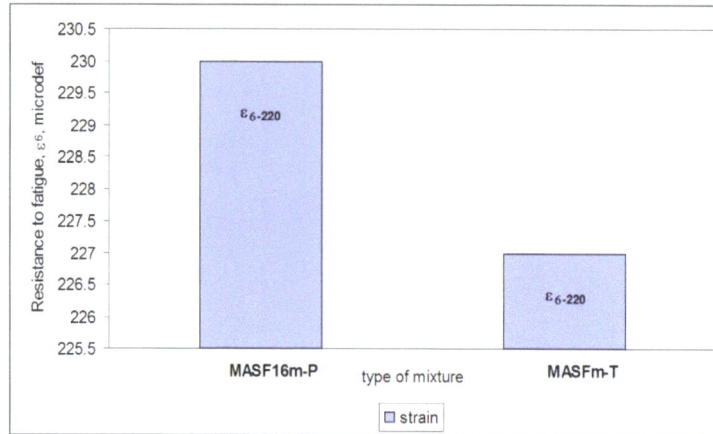

Figure 10: Initial strain corresponding with a fatigue life of 10^6 cycles from the 4PB-PR fatigue test for T = 30°C for the studied asphalt mixtures

REFERENCES

28. R.C. RĂDUCANU: "Utilizarea fibrelor de polipropilenă în mixturi asfaltice. Studiul comportării la oboseală", Lucrare de Disertație – îndrumător conf.dr.ing. C. Răcănel, Universitatea Tehnică de Construcții București, Facultatea de Căi Ferate, Drumuri și Poduri, Bucuresti, martie 20012.
29. E.L. VOICU: "Utilizarea fibrelor de polipropilenă în mixturi asfaltice. Studiul modulului de rigiditate", Lucrare de Disertație – îndrumător conf.dr.ing. C. Răcănel, Universitatea Tehnică de Construcții București, Facultatea de Căi Ferate, Drumuri și Poduri, Bucuresti, martie 20012.

A SHORT REVIEW OF RECENT RESEARCH ON THE MECHANICS OF FRACTURE AND FAILURE IN COMPOSITE MATERIALS

Catalin Iulian Pruncu

Dipartimento di Meccanica, Management e Matematica
Politecnico di Bari, Viale Japigia, 182 - 70126 Bari, Italy, c.pruncu@poliba.it

Abstract: *Composite materials have excellent properties in terms of fracture toughness and low weight. These materials are used for various industrial applications in the field of wind turbine, naval, aircraft and aerospace engineering. However, composite materials usually exhibit far more complex failure mechanisms than traditional metallic alloys. These failure mechanisms involve, for example, matrix deformation, fracture of fibers, interfacial debonding and crack deflection. Extensive studies have been documented in literature using theoretical models, numerical analysis and experimental methods. The article will review the most recent trends of the research on mechanical behavior of composites materials outlining also directions for future investigations.*

Keywords: *Composite material; Fracture; Delamination; Fiber–matrix interface; Interlaminar debonding; Micro-cracking.*

1. INTRODUCTION.

Market requirements pushed people in industry to gradually improve products quality by performing a careful selection of materials and manufacturing technologies. An important example of this philosophy is represented by the use of composite materials. Automotive, biomedical, naval and aerospace industries received a great impulse from the use of composites that allowed specific stiffness of manufactured composites to be significantly improved [1].

Composite materials are comprised of a polymeric matrix including some reinforcement such as fibers. The heterogeneous nature of composite materials pushed designers towards searching the best combination of phases that allows mechanical properties of the material to be optimized. The growing interest in the use of composite materials is confirmed by the blooming of technical papers published on the mechanical behavior of these materials (see, for example, the study carried out by Gibson [2]). Data relative to the US market indicate that the request for composites is expected to grow annually by 10.3% through 2013 [3]. Figure 1, taken from Beck [4], demonstrates how the utilization of composite materials is growing exponentially in strategic fields such as, for example, the aircraft industry.

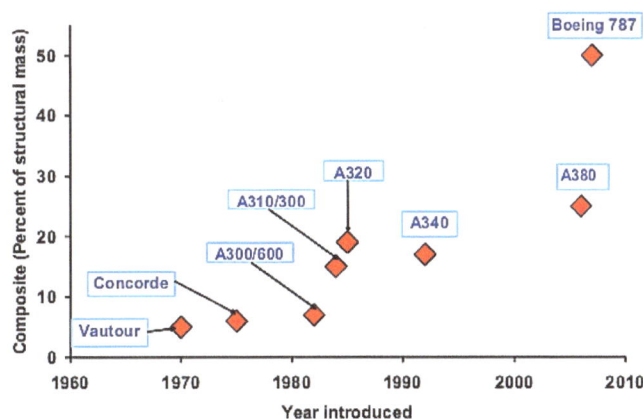

Figure1. Trend of use of composite materials in aircraft industry since 1970 (taken from [4])

However, composite materials still are less diffused than one might expect in view of their excellent mechanical properties. This is due to the limitations put by the heterogeneous nature of these materials. Material heterogeneity manifests itself in different ways and may cause a variety of failure modes during service life. For example, these problems are prominent if the composite matrix has a high degree of porosity (Figure 2) or inclusions [5]. Figure 3 shows that interface debonding will occur as the crack-tip approaches a fiber [6].

Defects may became more severe under the action of applied loads and cause changes in material microstructure that affect at different extents the mechanical behavior of the composite structure. In general, the final effect is to reduce mechanical properties and introduce various types of damage. In view of this, the paper will review the most important failure mechanisms in composite materials highlighted in technical literature.

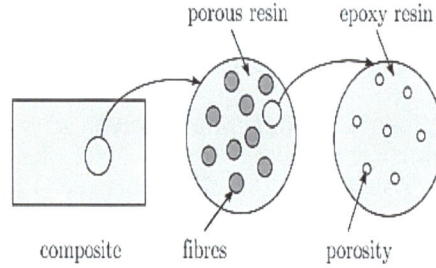

Figure 2. Porosity defect in a composite material [5].

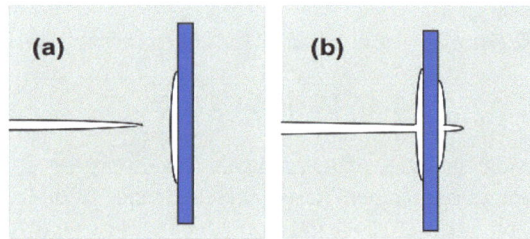

Figure 3. a) Interface debonding starts from left side of a fiber; b) matrix cracking occurs on the right [6].

2. OVERVIEW ON DAMAGE BEHAVIOR OF COMPOSITE MATERIALS

2.1. Analytical and numerical methods

The first approach to the analysis of the mechanical behavior of a material is represented by theoretical/analytical models. These models may describe some specific damage process (for example, crack initiation or/and propagation), Progressive failure analysis ranging from the most simple load-unload cycling to complex simulations involving linear or non-linear behavior can be done as well. Ribeiroa *et al.* [7] examined the damage process of the composite matrix by defining the strain energy density parameter written in terms of effective stresses:

$$E_d = \frac{1}{2}\left[\frac{\langle\sigma_{22}^2\rangle_+}{E_{22_0}(1-d_2)} + \frac{\langle\sigma_{22}^2\rangle_-}{E_{22_0}} + \frac{\langle\tau_{12}^2\rangle_+}{G_{12_0}(1-d_6)}\right] \qquad (1)$$

where σ_{22} and τ_{12}, respectively, are the stress in the direction transverse to fibers and the shear stress; E_{22_0} and G_{12_0}, respectively, are the initial values of the elastic modulus in the direction transverse to fibers and shear modulus; d_2 and d_6 are damage parameters related to σ_{22} and τ_{12}, respectively.

The mechanisms of interlaminar fracture were analyzed by Krause *et al.* [8]. Fibre–matrix interface failure between plies and multiple small delaminations were modeled with Linear Elastic Fracture Mechanics (LEFM). Crack propagation in mode I from pre-existing defects was computed using the critical energy release rate. For that purpose, two main approaches can be followed: "area" methods expressed by Eq. (2) and "compliance" methods expressed by Eq. (3).

$$G_{IC} = \frac{1}{b(a_{i+1}-a_i)}\int_{P_s} P(\delta)d\delta \qquad (2)$$

In the above equation, δ is the displacement of the cantilever arms loading the specimen and equals the crack opening; b is the width of the tested specimen; a_i and a_{i+1} are the crack lengths before and after crack progress; P_s is the area swept by the loading curve $P(\delta)$.

$$G_{IC} = \frac{P_C^2}{2b} \frac{dC}{da} \qquad (3)$$

In Eq. (3), $C=\delta/P$ is the compliance evaluated at each point of the load–displacement curve; δ is the relative displacement between two adjacent points of the load path (i.e. corresponding to crack lengths a_i and a_{i+1}, respectively); P_c is the critical load at which fracture propagates.

For mixed mode loading, Zhang et al. [9] proposed the strain energy release rate parameter (SERR), normalized with respect to the fatigue delamination resistance (G_c) (Eq. 4).

$$G_C = \int_0^{\delta^*} \sigma_b(\delta) d\delta + G_0 \qquad (4)$$

where: σ_b and δ, respectively, are the bridging stress and the crack opening displacement (COD); G_0 is the critical value of the SERR parameter in presence of initial delamination; δ^* is the COD at the pre-crack tip.

The above mentioned approach was utilized to determine the delamination growth rates and threshold of composite laminates subject to mixed I/II mode fatigue loading.

Another technique for predicting the damage of fiber-matrix composite structures is the progressive failure analysis. Different models were developed by Fan et al. [10] and Liu et al. [11]. The most important approach followed to detect the micromechanical damage of composite pressure vessels is the modified Mises failure criterion:

$$\left(\frac{\sigma_{eq}}{\sigma_{eq}^{cr}}\right)^2 + \frac{J_1}{J_1^{cr}} = 1 \qquad (5)$$

where: $\sigma_{eq} = \sqrt{\frac{3}{2} S:S}$ is the Mises effective stress, $J_1 = tr(\sigma)$ is the first stress invariant; $S = \sigma - tr(\sigma)/3I_2$ is the deviatory stress tensor; I_2 is the second-order unit tensor; $\sigma_{eq}^{cr} = \sqrt{T_m C_m}$ and $J_1^{cr} = C_m T_m / (C_m - T_m)$ are critical values.

Sun et al. [12] applied the progressive failure analysis approach to assess mechanical behavior of fiber-reinforced composites. They mentioned two conditions for micromechanics-based failure: one refers to the fiber failure state, Eq. (6); the other refers to matrix (inter-fiber) failure, Eq. (7).

$$\frac{\sigma_{f1}^2}{T_f C_f} + \left(\frac{1}{T_f} - \frac{1}{C_f}\right)\sigma_{f1} = 1 \qquad (6)$$

$$\frac{\sigma_{VM}^2}{T_m C_m} + \left(\frac{1}{T_m} - \frac{1}{C_m}\right)I_1 = 1 \qquad (7)$$

In the above equations, T_f, C_f, T_m and C_m are the tensile and compressive strengths of fibers and matrix, respectively (all these entities are calculated from ply strengths); σ_{f1} is the micro-longitudinal stress in the fibre; σ_{VM}, I_1 and I_2, respectively, are the von Mises stress, the first and second stresses invariant computed for the composite matrix;

$$\begin{cases} \sigma_{VM} = \sqrt{I_1^2 - 3I_2} \\ I_1 = \sigma_1 + \sigma_2 + \sigma_3 \\ I_2 = \sigma_1\sigma_2 + \sigma_2\sigma_3 + \sigma_3\sigma_1 - \left(\tau_{12}^2 + \tau_{23}^2 + \tau_{31}^2\right) \end{cases} \qquad (8)$$

In order to define the damage model, Sun et al. [12] used a non-iterative element-failure method criterion based on the equivalent stress:

$$\sigma_{eq} = \frac{(\beta_m - 1)I_1 + \sqrt{(\beta_m - 1)^2 I_1^2 + 4\beta_m \sigma_{VM}^2}}{2\beta_m} \qquad (9)$$

where $\beta_m = C_m/T_m$.

Pemberton et al. [13] developed another model that predicts the fracture energy of ceramic–matrix composites. In this approach, fracture is caused by pull-out and/or plastic deformation of fibres bridging the crack plane. The contribution of pull-out to the fracture energy is expressed as:

$$G_{cpo} = \left(\frac{f}{2\pi R^2} \right) \pi R x_0^2 \tau_{i*} = \frac{f s_{po}^2 R \tau_{i*}}{2} \qquad (10)$$

where the fiber protrusion aspect ratio, s_{po}, is equal to x_{0po}/R (see Figure 4b) or, equivalently, can be expressed by the ratio of δ_{po} (average length of fiber protruding beyond the crack plane), to R.

A model involving both plastic deformation and fiber rupture should assume that the interfacial debonding is an extension of crack plane (Figure 4c). The resistance to fracture debonding can be expressed in terms of energy as:

$$G_{cfd} = 2x_{ofd} \left(\frac{f}{2\pi R^2} \right) W_{fd} \pi R^2 = x_{ofd} f W_{fd} \qquad (11)$$

where: $2x_{ofd}$ is the initial length of fiber; U_{fd} and W_{fd} are the deformation work for the fiber expressed per fiber unit length (J/m) or unit volume (J/m^3), respectively. The fracture process is schematized in Figure 4.

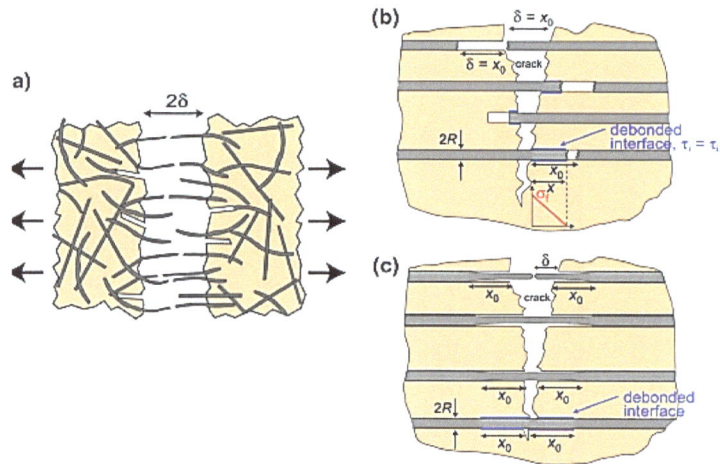

Figure 4. Representation of the fracture process [13]: a) general fracture geometry;
b) debonding at matrix-fiber interface causing fracture and then frictional pull-out;
c) fibers undergoing debonding, plastic deformation and then fracture.

Patrícioa and Mattheij [14] developed an incremental algorithm to predict the evolution path of pre-existing cracks in composite materials. They utilized the maximum circumferential tensile stress criterion: crack growth occurs when the maximum of $K_{\theta\theta}(\theta)$ reaches the value of the critical stress intensity factor K_{IC}:

$$\max_{\theta} K_{\theta\theta}(\theta) = K_{IC} \qquad (12)$$

The angle defining the direction of propagation of the crack was computed with Eq. (13) once that the propagation mechanism was defined (see Figure 5). Figure 6 shows how the crack will propagate between composite layers.

$$\theta_p^{(K)} = 2\arctan\left(\frac{K_I - \sqrt{K_I^2 + 8K_{II}^2}}{4K_{II}} \right) \qquad (13)$$

where K_I and K_{II}, are the stress intensity factors for modes I and II, respectively.

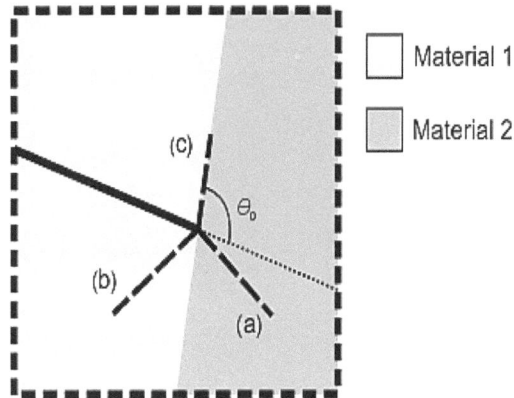

Figure 5. a) Penetrating crack; b) reflected crack; c) crack deviated along the interface [14].

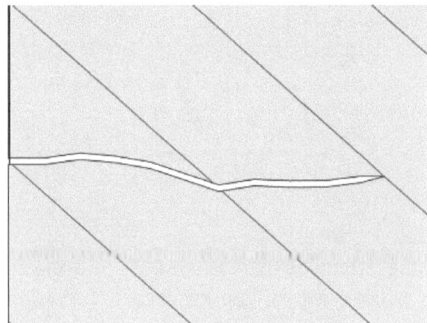

Figure 6. Crack propagating through layers [14].

2.2. Experimental methods

Analytical and numerical models of fracture must be corroborated by experimental evidence. Researchers focused their attention on nondestructive methods. In particular, we should mention: optical microscopy, scanning electron microscopy (SEM), nano-indentation, Eddy currents, Dye penetrant inspection (DPI) (also called liquid penetrant inspection (LPI)), ultrasonic tests, magnetic particle methods and acoustic emission.

Canal *et al*. [15] carried out nano-indentation tests on matrix pockets. The tested specimens are fiber-reinforced composites applied to notched beams. The interface strength was measured by means of push-out tests done on thin slices of material. Specimens were inspected with a SEM microscope. Figure 7 shows the typical pattern obtained in the experiments including a number of decohesed interfaces connected by matrix ligaments.

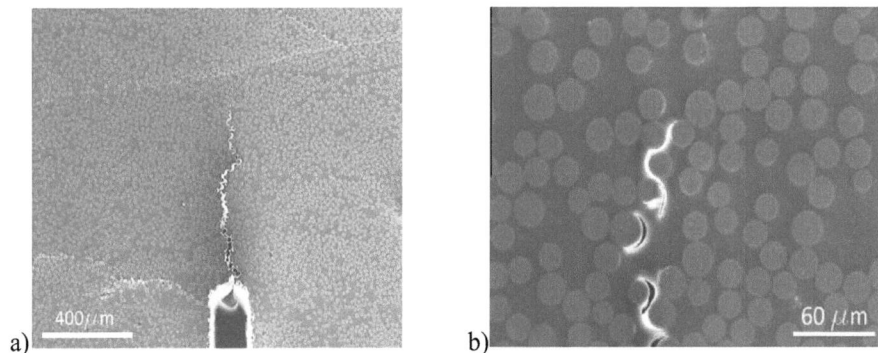

Figure 7. Damage processes by interface decohesion and matrix failure in front of the notch tip [15].

Okoli and Abdul-Latif [16] considered the pullout of fibers to depend on the matrix-fiber bond strength and load transfer mechanisms. The progress of damage depend on the interaction between the increasing in loading rate and the interfacial bond strength. As fibers are pulled out, matrix debonding occurs and produces cracking and disintegration of matrix (Figure 8).

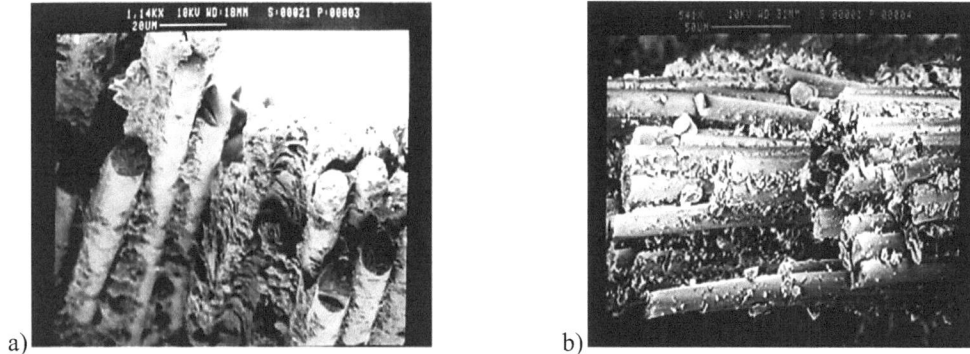

Figure 8. a) Tufnol 10G/40 laminate showing fibre-matrix delamination and debonding (×1140); b) Tufnol 10G/40 laminate showing brittle failure with fiber breaking (×541) [16]

Abdizadeh and Baghchesara [17] observed that fracture is controlled by the inter-dendritic cracking of the matrix. In addition, a number of dimples were observed on the fractured surfaces of all samples. This may be due to void nucleation and subsequent coalescence during the fracture process.

Sabirov and Kolednik [18] analyzed the effect of matrix strength and the mechanism of void initiation. Maximum principal stresses generated in the matrix structure at the moment of void initiation are not constant, but show a dependency on the yield strength of the composite material. Local changes in material properties depend on the void initiation, but may also be caused by debonding rather than matrix fracture.

Amato de Campos et al. [19] used a conventional optical microscope combined with the extended depth-of-field 3-D reconstruction method to detect intense fiber decohesion in warp and interlaminar fracture mechanisms in a plain- weave carbon/epoxy composite. Furthermore, during tests, resistance to delamination failure was measured with high-speed video photography by recording the specimen displacement and crack length history [20].

Xu et al. [21] carried out seawater tests over a period of about 29 months. The level of degradation of matrix and fibers caused by the exposure to seawater was around 10%. Several failure mechanisms including interlaminar debonding, microcracking and sublaminate buckling were detected.

3. CONCLUSIONS

This paper presented a short review of state-of-the-art failure models and experimental investigations on the fatigue and fracture behavior of composite materials and structures recently published in literature. It appears that the fracture process starts with void nucleation and is followed by crack growth, matrix debonding, delamination and fiber breaking. Special cares should be taken to keep under control the rate of reduction of mechanical properties which may occur during life service. This requirement brings a number of issues relatively to experimental testing and numerical modeling. In fact, either experimental techniques for real time monitoring of damage evolution should be available or multi-scale analyses of damage mechanisms should be performed. These multi-scale models must be corroborated by experimental evidence and may be multi-fidelity models entailing different level of complexity in the same analytical/numerical framework. Parametric studies may help analysts to tune multi-scale models as well as to design the experimental set up best suited for the particular problem dealt with.

4. ACKNOWLEDGEMENTS

The support given by Professor Carmine Pappalettere, Professor Luciano Lamberti and Professor Caterina Casavola is gratefully acknowledged.

REFERENCES

[1]. Automotive Plastics & Composites. *Worldwide Markets and Trends to 2007*. Elsevier Science Ltd. (1999).

[2]. R.F. Gibson. A review of recent research on mechanics of multifunctional composite materials and structures. *Composite Structures* 92 (2010) 2793–2810.

[3]. *http://www.reinforcedplastics.com/view/13565/composites-2011-looking-to-the-us-industry-s-future/*

[4]. A.J. Beck, A. Hodzic, C. Soutis, C.W. Wilson. *Influence of Implementation of Composite Materials in Civil Aircraft Industry on Reduction of Environmental Pollution and Greenhouse Effect.* University of Sheffield (UK) (2008).

[5]. H. Welemane, H. Dehmous,. Reliability analysis and micromechanics: A coupled approach for composite. *International Journal of Mechanical Sciences* 53 (2011) 935–945.

[6]. R. Zhou, Z. Li, J. Sun. Crack deflection and interface debonding in composite materials elucidated by the configuration force theory. *Composites: Part B* 42 (2011) 1999–2003.

[7]. M.L. Ribeiroa, V. Tita, D. Vandepitte. A new damage model for composite laminates. *Composite Structures* 94 (2012) 635–642.

[8]. T. Krause, K. Tushtev, D. Koch, G. Grathwohl. Interlaminar Mode I crack growth energy release rate of carbon/carbon composites. *Engineering Fracture Mechanics* (2012) (In press).

[9]. J. Zhang, L. Peng, L. Zhao, B. Fei. Fatigue delamination growth rates and thresholds of composite laminates under mixed mode loading. *International Journal of Fatigue* 40 (2012) 7–15.

[10]. X. L. Fan, T. J. Wang, Q. Sun. Damage evolution of sandwich composite structure using a progressive failure analysis methodology. *Procedia Engineering* 10 (2011) 530–535.

[11]. P. F. Liu, J. K. Chu, S. J. Hou, J. Y. Zheng. Micromechanical damage modeling and multiscale progressive failure analysis of composite pressure vessel. *Computational Materials Science* 60 (2012) 137–148.

[12]. X. S. Sun, V. B. C. Tan, T. E. Tay. Micromechanics-based progressive failure analysis of fibre-reinforced composites with non-iterative element-failure method. *Computers and Structures* 89 (2011) 1103–1116.

[13]. S. R. Pemberton, E. K. Oberg, J. Dean, D. Tsarouchas, A. E. Markaki, L. Marston, T. W. Clyne. The fracture energy of metal fibre reinforced ceramic composites (MFCs). *Composites Science and Technology* 71 (2011) 266–275.

[14]. R.M.M. Mattheij, M. Patrícioa. Crack paths in composite materials. *Engineering Fracture Mechanics* 77 (2010) 2251–2262.

[15]. L.P. Canal, C. González, J. Segurado, J. LLorca. Intraply fracture of fiber-reinforced composites: microscopic mechanisms and modeling. *Composites Science and Technology* 72 (2012) 1223–1232.

[16]. O. Okoli, A. Abdul-Latif. Failure in laminate composites: overview of an attempt at prediction. *Composites. Part A* 33 (2002) 315–321.

[17]. H. Abdizadeh, M.A. Baghchesara. Investigation on mechanical properties and fracture behavior of A356 aluminum alloy based ZrO_2 particle reinforced metal-matrix composites. *Ceramics International* (2012) (In Press).

[18]. Sabirov, O. Kolednik. Local and global measures of the fracture toughness of metal matrix composites. *Materials Science and Engineering* A527 (2010) 3100–3110.

[19]. K. Amato de Campos, J. Augusto, T.A. Pereira, L.R. de Oliveira Hein. 3-D reconstruction by extended depth-of-field in failure analysis – Case study II: Fractal analysis of interlaminar fracture in carbon/epoxy composites. *Engineering Failure Analysis* 25 (2012) 271–279.

[20]. B.R.K. Blackman, A. J. Kinloch, F.S. Rodriguez-Sanchez, W.S. Teo. The fracture behaviour of adhesively-bonded composite joints: Effects of rate of test and mode of loading. *International Journal of Solids and Structures* 49 (2012) 1434–1452.

[21]. L. Roy Xu, A. Krishnan, H. Ning, U. Vaidya. A seawater tank approach to evaluate the dynamic failure and durability of E-glass/vinyl ester marine composites. *Composites: Part B* 43 (2012) 2480–2486.

RESEARCH REGARDING THE INFLUENCE OF THE REINFORCEMENT DEGREE UPON THE TENSILE STRENGTH OF SOME COMPOSITE MATERIALS

Claudiu FLOREA, Horaţiu IANCĂU, Liana HANCU, Melinte SIMON

Technical University, Cluj-Napoca, ROMANIA, florea_claudiu_81@yahoo.com
Technical University, Cluj-Napoca, ROMANIA, Iancau.Horatiu@tcm.utcluj.ro
Technical University, Cluj-Napoca, ROMANIA, Liana.Hancu@tcm.utcluj.ro
Technical University, Cluj-Napoca, ROMANIA, melinte.simon@yahoo.com

Abstract: *In this paper there is presented a study on the mechanical characteristics of some polymeric composite structures with different degrees of reinforcement. The composite structures were produced by Resin Transfer Moulding process (RTM). Based on the results of the tensile tests, comparative studies were made in order to determine the influence of reinforcement degree upon the tensile strength.*

Keywords: *composite materials, Resin Transfer Moulding, tensile strength.*

1. INTRODUCTION

Composite materials are not an entirely new concept. Examples are found in nature, such as wood, which is made of cellulose fibers linked to lignin or bone, comprising the periphery of compact tissue. Another example is the inside bone marrow, the whole being embedded in a fibro-elastic membrane, [IAN 03].

As a general definition, composite materials are a mixture of two or more different components whose properties complement each other, and the result is a material with superior properties to those specific to each component, [ŞOM 00].

Composite's manufacturing technologies are numerous and the basic concept of each one is different from the others. The choice of the proper technology depends on the following factors: the geometrical shape of the part or product, the structure of the material, part's dimensions, dimensional accuracy and quality of the parts, production batch, mechanical stresses, part's destination, and so on [IAN 03].

To highlight the influence of the reinforcement degree upon the tensile strength, composite plates were made by RTM procedure. This is a process characterized by low pressure in a closed mold where a mixed resin and catalyst are injected. The mold is usually containing a fiber pack or a fiber preform. After the resin is cured, the mold can be opened and the finished component removed.

A wide range of resin systems can be used including: polyester, vinylester, epoxy, phenolic and methyl methacrylates, combined with pigments and fillers including aluminum trihydrates and calcium carbonates, if required. The fiber pack can be glass, carbon, aramid, or a combination of these.

Figure 1 shows an RTM installation.

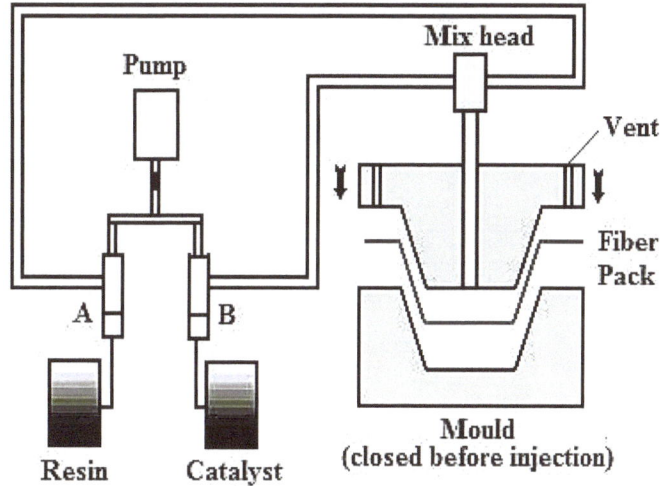

Figure 1 Resin Transfer Molding installation [FLO 11a]

2. EQUIPMENT FOR COMPOSITE PLATES MANUFACTURING

2.1. RTM installation

For the manufacture of composite structures, fiberglass mat, *Unifilo U 813 300 g/m²*, made of continuous fibers without binder to facilitate the flow, was used as a reinforcement material. It was impregnated with a special polyester resin for RTM process, *Norsodyne I 20282 I*.

Composite plates were made using an experimental installation (fig. 2), located in *The Research Laboratory of Materials and Competitive Manufacturing Parts* from Technical University of Cluj-Napoca.

Figure 2 Experimental installation of RTM

2.2. Specimens

For tests were made 7 composite structures with different reinforcement degree. They were coded according to Table 1. The specimens cut from these plates were labeled with the same code as the plate.

Table 1 Coding the composite plates

Nr.	Specimen code	Number of layers	Reinforcement degree [%]
1	P1	5	37
2	P2	6	42
3	P3	7	47
4	P4	8	52
5	P5	9	56
6	P6	10	61

The reinforcement degree M_f of composite plates was calculated using the formula:

$$M_f = \frac{m_f}{m_c} \times 100 \quad [\%] \tag{1}$$

where m_f - fiber weight and m_c - total weight of the composite material.

3. EXPERIMENTS AND RESULTS

Determination of tensile strength through tensile testing is the most general and important resistance test. The method consists in applying a load along the main axis of the specimen, with a constant speed until failure or until the elongation reaches a predetermined value. For the tensile tests, 7 specimens were taken from each plate. They were cut from the composite plates to the desired dimensions on *Water Jet Cutting Machine* OMAX Jet Machining Center 2626 (fig 3), located in *Unconventional Technologies Research Laboratory and Competitive Manufacturing* from the Technical University of Cluj-Napoca.

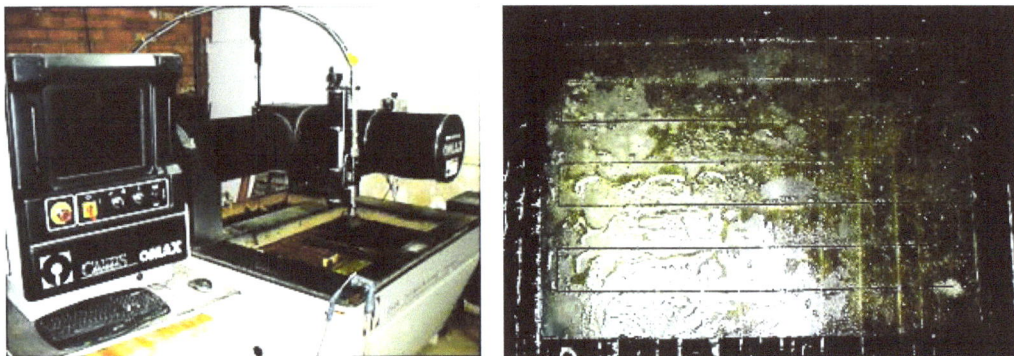

Figure 3 Water jet Cutting Machine OMAX Jet Machining Center 2626

Specimens dimensions and test data are in accordance with ISO 527-4 3 [ISO 97]. Specimens dimensions shown in figure 4 are: L = 250 mm, h = 25 mm, b = 2 – 3,2 mm.
To prevent the crushing or the breaking of the specimen because of the pressing from the machine jaws, end tabs were glued using a slow curing adhesive, Bison Epoxy Universal.
End tabs were made from a laminate composed of fiber reinforced epoxy resin.

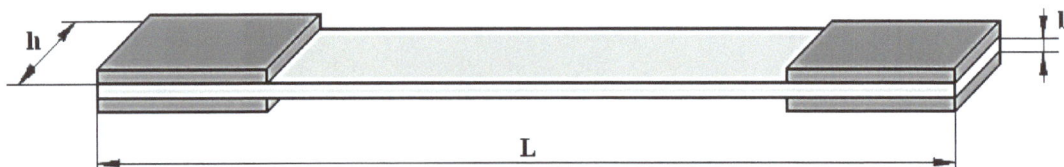

Figure 4 Tensile test specimen

For the tensile test was used an universal testing machine type Zwick/Roell Z150 (fig. 5), located in *The Research Centre on Sheet Metal Forming Technology CERTETA* from the Technical University of Cluj-Napoca.

Figure 5 Tensile test machine Zwick/Roell Z150

The tensile test machine is equipped with a computer control system with the ability to draw the diagrams of the variation of the force during traction and to calculate, based on section area specimens, the breaking strength, the elasticity modulus and the maximum strength. For fixing, longitudinal axis of specimens must be aligned to the test machine columns.

Speed test was 2 mm/min., and the tensile strength (σ_r) was calculated using the relationship:

$$\sigma_r = \frac{F_{max}}{A} [MPa] \tag{2}$$

where F_{max} - the maximum force required to break the specimens, [N] and A - initial cross-sectional area, [mm^2].

Tensile test were made at a temperature of 20 -22 ° C.

The tensile test results, the maximum breaking strength and the elasticity modulus for composite structures made by RTM process are presented in table 2. The specimens, subjected to tensile strength, are shown in Figure 6.

Figure 6 Specimens subjected to tensile strength

Table 2. Mechanical characteristics obtained after tensile test

Nr. Crt.	Specimen code		Specimens size		Section area [mm²]	Longitudinal modulus E [MPa]	Maximum force [N]	Tensile strength [MPa]	Average tensile strength [MPa]
			Width [mm]	Thickness [mm]					
1	P1	P11	25.4	2.4	60.9	4259	4530	74.3	
2		P12	25.3	2.4	60.7	4078	4310	71	72.3
3		P13	25.3	2.4	60.7	4217	4360	71.8	
4	P2	P21	25.3	2.5	63.2	4897	5770	91.2	
5		P22	25.3	2.6	65.7	4235	5200	79.1	85.9
6		P23	25.3	2.6	65.7	4684	5750	87.5	
7	P3	P31	25.3	2.9	73.3	4007	6780	92.4	
8		P32	25.4	3.1	78.7	4079	7450	94.6	95.1
9		P33	25.3	2.9	73.3	4202	7220	98.4	
10	P4	P41	25.3	2.8	70.8	4489	7930	112	
11		P42	25.3	3	75.9	3479	7960	104.8	107.1
12		P43	25.4	3.4	86.3	3510	9030	104.6	
13	P5	P51	25.4	3.5	88.9	2947	9570	107.6	
14		P52	25.4	3.4	86.3	2260	9090	105.3	106.8
15		P53	25.4	3.8	96.5	3365	10400	107.7	
16	P6	P61	25.4	3.6	91.4	1968	9880	108.1	
17		P62	25.4	4.2	106.6	2294	10830	101.6	106.1
18		P63	25.4	4.2	106.6	2635	11600	108.8	

Variation of the tensile strength according to the reinforcement degree is shown in Figure 7.

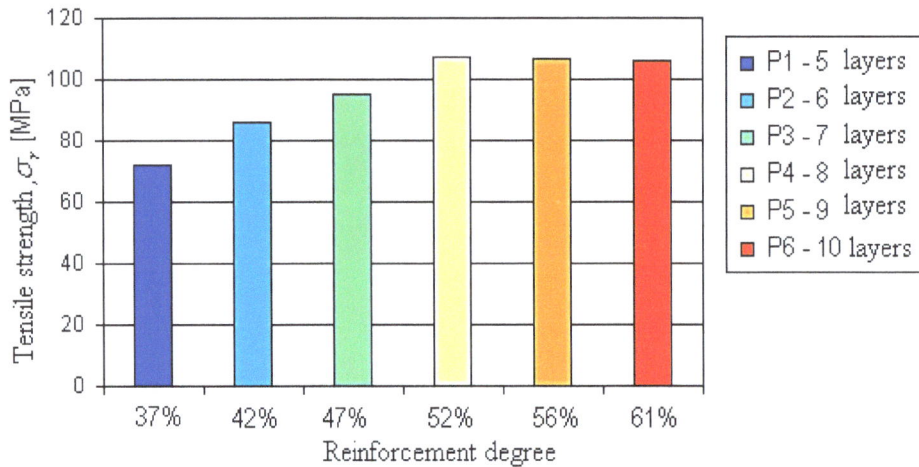

Figure 7 Variation of the tensile strength according to the reinforcement degree

4. CONCLUSIONS

Considering the results of the tensile test it is obvious that increasing the reinforcement degree, it also increases the tensile strength of the composite material. It appears to be an optimum on a degree of reinforcement for the composites studied by us, at 52%, corresponding to a composite with 8 layers.

It is noted that if exceed this degree of reinforcement, the tensile strength e begins to decrease. The explanation comes from the fact that the manufacturing process using RTM procedure takes place in a closed mold. That's why if the reinforcement degree is increased by increasing the number of layers, the quantity of the resin from the mold becomes insufficient to transmit filaments efforts. In this way they are not bound by the matrix, thus producing composite depreciation.

ACKNOWLEDGMENT

This paper was supported by the project "Doctoral studies in engineering sciences for developing the knowledge based society-SIDOC" contract no. POSDRU/88/1.5/S/60078, project co-funded from European Social Fund through Sectorial Operational Program Human Resources 2007-2013.

REFERENCES

[1]. Barrau, J.J., Laroze, S., *Calcul des structures en materiaux composites*, Tome 4, Ecole nationale superieure de l'aeronautique et de l'espace, Toulouse, 1987;

[2]. Berthelot, J.M., *Matériaux composites. Comportement mécanique et analyse des structures.*, 3e édition, Editions TEC & DOC, Paris, 1999, ISBN 2-7430-0349-9;

[3]. Decolon, C., *Structures composites. calcul des plaques et des poutres multicouches*, Hermès Science Publications, Paris, 2000, ISBN 2-74620-114-3;

[4]. Florea C., Iancău, H., Popescu A., Simon, M., and Bere, P., *Consideration regarding the manufacturing of the composite structures using the RTM process,* 3rd International Conference "Advanced Composite Materials Engineering " COMAT, ISSN 1844 – 9336, pag. 92 – 97, Brasov, 27- 29 October, 2010;

[5]. Florea C., Iancău, H., *Consideration regarding the influence of manufacturing process of the composite structures on their mechanical behaviour,* The 10th International Conference "Modern Tehnologies in Manufacturing" Cluj-Napoca, Romania, 6-8 Octombrie, 2011.

[6]. Hadăr, A., *Structuri din compozite stratificate. Metode, algoritmi şi programe de calcul*, Ed. Academiei Române: AGIR, Bucureşti, 2002, ISBN 973-8466-25-3;

[7]. Iancău, H.,Nemes, O., Materiale compozite. Concepŭie si fabricaŭie., Ed. Mediamira, Cluj-Napoca, 2003, ISBN 973-9357-24-5;

[8]. Manolea, Gh.: *Bazele cercetării creative,* Editura AGIR, Bucureşti, 2006.

[9]. Şomotecan M., Compozite. Calcul de rezistenţă., Ed. U.T. Pres, Cluj-Napoca, 2000, ISBN 973-9471-28-5;

[10]. Tero, M. Materiale compozite realizate prin procedeul RTM. Cluj-Napoca, Editura Napoca Star,2005.

ELECTRICAL RESISTANCE CHANGE IN CFRP COMPOSITES WITH DIFFERENT FIBRE VOLUME FRACTION UNDER TENSILE LOADING

I. Dumitrascu[1], P.D. Barsanescu[1*], [1]V. Goanta[1]

[1]"Gheorghe Asachi" Technical University of Iasi, Faculty of Mechanical,
Department of I.M.M.R, Iasi, Romania, *e-mail: paulbarsanescu@yahoo.com

Abstract: Carbon fibers have high electric conductivity, while the matrix surrounding them is an insulator. Carbon fibers can be represented in materials as some resistors. If a fiber breaks, the current cannot pass through this fiber. Thus the electrical resistance material will increase because the current will have to find another route. In this study, the electrical resistance change method is used. Two type of cross-ply laminate composites is used in this study: carbon fiber reinforced polypropylene and epoxy matrix. First, the composites production and set-up used is presented. Then, tensile tests on the laminate are performed. The influence of fiber volume fraction on the measurements is examined. It may be concluded that the volume fraction of fibers has a some influence on the resistance change. If the fibers volume fraction is higher, the contact between fiber is greater and the resistance is lower, but during the test, its value increases with applied load.

Keywords: Composites, CFRP, Carbon Fiber, Electrical resistance change

1. INTRODUCTION

Carbon fiber reinforced polymer-matrix composites (CFRP) have become in recent years the basic material used in many engineering areas where high mechanical properties (i.e. stiffness, strength, etc.) and light weight is necessary. In general, CFRP is formed intro laminated structures made from unidirectional prepreg or woven fabric sheet, positioned in different stacking sequences. To detect internal damage like matrix cracks or delamination is very difficult because this damage is not visible from the outside of structure.

Several NDT techniques (e.g. ultrasound, radiography, thermography, acoustic emission, etc.) have been developed for structural health monitoring and detected of damage in composites structure. However, these techniques require expensive equipment, extensive human involvement and often, inspected structure must be switched off. Fiber optic sensor or piezoelectric sensor is a "sensitive" materials used for health monitoring of composites structures [1-4]. One disadvantage of using them it is their tendency to compromise the structural integrity of the host components. Electrical resistance change method (ERCM) is of great interest as an in-service structural health monitoring system for CFRP [5-11]. This method is employed to detect the internal damages of CFRP like matrix cracking [5,6], delamination [7,8,9] or fiber breakages [10,11]. The electrical resistance change method does not require expensive instruments. Since, the method adopts a reinforcement carbon fiber itself as a sensor, and it does not require additional sensors except for electrodes to make contact with carbon fibers in the CFRP structures. Therefore this method is called "self-sensing method", and the basic principle is that the damages due to carbon fiber breakage or delamination in the laminate will cause an increase in electrical resistance, resulting in voltage change in the damaged region. To measure resistance, a current is introduced in the materials through the electrodes and the voltage is measured. In literature, there are two measurement techniques [5-17]: (i) the two-probe technique, which means that the voltage is measured on the same electrodes where the current is introduced; and (ii) the four-probe technique, where the first two contacts are needed for the current injection and the voltage is measured with the other two. The first technique is mostly used in tests in which current is introduced by bonding electrodes on opposite edges or on the surface of the sample. They provide a better contact with carbon fibers in the edges of the samples [17]. Most research on the use of electric resistance change method was performed on unidirectional composites, and this is because current flows more easily in the direction of fibers and measured resistance is much higher [12-17]. In transvers to the fiber and in the specimen thickness direction, the current will flow with more difficulty because the fiber is isolated of the resin and are coated with agents for interface strength improvement. In this case the transverse and thickness conduction can only be achieved via fiber to fiber contact from many longitudinal paths. This means that transverse resistivity will also be a function of fiber volume fraction as well as the number and area of contact points [7, 18].

In multidirectional laminates the present of different orientations of fibers in the lamina can complicate the conduction process of current. Thus, in this investigation the effect of differing fiber orientations and the fiber volume fraction on the electrical resistance change in multidirectional laminates will be investigated.

2. MATERIALS, EQUIPEMENT AND METHOD

2.1. Raw Materials

Two types of unidirectional CFRP prepregs were used:
- Unidirectional prepreg carbon fiber/polypropylene (CF/PP), namely CarbostampTM UD tape, produced by Soficar (France), with a nominal ply thickness of 0.25 mm. The carbon fibers used were T700 (made of Toray), with the following tensile properties: modulus of 230 GPa, strength to failure of 4900 MPa, density 1.80 g/cm3 and strain to failure of 2.0 % and filament diameter 7 µm.
- Unidirectional prepreg carbon fibre/epoxy (CF/Epoxy) type Q-1112 product by TohoTenax Europe GmbH, with a nominal ply thickness of 0.15 mm. The carbon fibres is STS 40 24K, modulus of 230 GPa, strength of 4240 MPa, density 1.77 g/cm^3 and strain to failure of 1.75 % and filament diameter 7 µm.

2.2. Production of plate laminates composites

The stacking sequence used for production of the plates laminate composites were $[(0,90),\overline{0}]_s$ for CF/PP and $[(0,90)_2,\overline{0}]_s$ for CF/Epoxy. The carbon fiber/polypropylene cross-ply laminates were fabricated with the hot pressing machine Zenith 2 (Pinette Emidecau Industries). A mold was designed for 25 mm width samples. The slender channels of the mold provided the individual samples, avoiding cutting the samples from the molded plates. This was a countermeasure against the fact that sticky thermoplastic polymer, especially PP, makes it impossible to cut the samples with a water-cooled diamond saw. Therefore, 5 layers were cut from the prepregs into the length and width as needed for the objective samples. Thickness of the coupons was controlled by spacers at the both edges of the slender channels between upper and lower mold. The cut plies were stacked in designated stacking sequences and filled into the mold. Then, the mold was placed into the hot-pressing machine and immediately heated up. When reaching the designated temperature of 165 °C, the mold is pressed under 3 bar for 10 min. Then, the mold was cooled down to temperature of curing polymer (50 °C) and after that, removed from the machine.

To produce the carbon fiber/epoxy cross-ply laminates CF/Epoxy, the prepregs were cut into sheets with dimensions 300 mm by 300 mm. Nine such prepregs were stacked one on top of the other and cured at 90°C for 60 min followed by a post curing step at 130°C for 90 min. The laminates were produced in an autoclave at vacuum of -0.68 to -0.70 bar.

The fiber volume fraction of composites is 48% for cross-ply CF/PP and 71% for cross-ply CF/Epoxy, and was determined by matrix digestion method, described in ASTM D3171-99 [19], using sulfuric acid/hydrogen peroxide.

The cross-ply CF/PP samples have not required cutting from composite plate, because there are already resulting to size recommended in ASTM D3039 [20], from the manufacturing process. The cross-ply CF/Epoxy samples were cut from the composite plate to size recommended in ASTM D3039, with a water-cooled diamond saw machine.

The size of samples were 260 mm length, 25 mm width and around 1.3 mm thickness, see Figure 1. Tabs made of glass fiber woven/epoxy prepregs stacked in $[\pm45°]_{2s}$ direction, with 2 mm thickness, were cut to size 40 mm length and 25 mm width (without taper) and bonded to the sand-blasted surface of the samples, using Araldite 2011TM epoxy glue.

2.3. Equipment

The tensile test were performed on a testing machine Instron 8801 with a load cell of 100 kN and hydraulic grips. Tensile tests were done according to ASTM D3039 [20]. The test was displacement controlled with a speed of 1 mm/min. For measuring the electrical resistance, a multimeter Agilent 34401 A was used. The measured data of the instrument were recording to the computer via data cable type RS-232. Using the Agilent IO Control program, we could control time (of 1 sec.) in which the multimeter to perform two measurements.

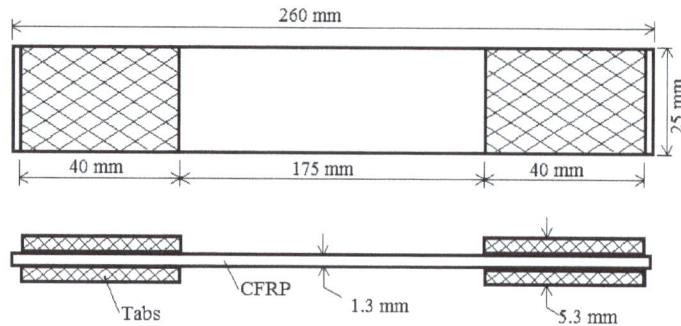

Figure 1: Dimension of the tensile samples used and bonded tabs

2.4. Method

These tests were performed to confirm the electrical resistance change during tensile loading in the fiber direction. In order, to measure the resistance, a DC current of 10 mA is introduced in the samples and the voltage is measured using data acquisition system. To eliminate any electrical contact between samples and the Instron machine, the tabs made GFRP composites was used. In this study, the two-probe technique is used. To provide a good contact between electrodes and carbon fibers, the edges of the samples was polished. After polished, the electrodes were attached to the edges of the samples through silver paste. To protect the electrode, they were covered with epoxy resin. In Figure 2., is represented an example of the sample with bonded electrodes on the edges used for this study.

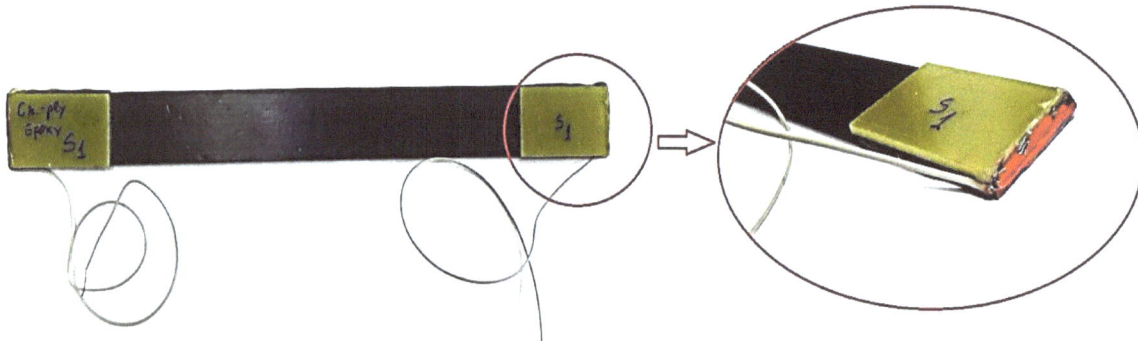

Figure 2: Sample prepared with electrodes for electrical resistance measurement

3. RESULTS AND DISCUTION

Electrical resistance values vs strain and stress-strain diagrams are show for cross-ply CF/PP in Firure 3. and for cross-ply CF/Epoxy in Figure 4. Before starting the test, initial electrical resistance R_0 of each sample tested, after she was clamping the test machine, was recorded. So, for cross-ply CF/PP, R_0=5.466 Ω and the failure stress of specimen was 941 MPa and for cross-ply CF/Epoxy, R_0 =0,820 Ω and the failure stress of specimen was 1223 MPa. As show in these figures, for both materials, the electrical resistance increase with increase of applied tensile strain.   The   lower values of initial resistance obtained for cross-ply CF/Epoxy, is a consequence of the high volume fiber which provides a larger area the current flowing. Electrical resistance values measured for cross-ply CF/PP, remain relatively close in most of the test and after ε =1.6 %, the values increase suddenly. This suggests that in the samples the damage like delamination or fibers break growth and the end of the failure of sample is close. Not the same thing is observed for cross-ply CF/Epoxy. Even if the electrical resistance values recorded are much lower, they grow steadily during of the test. Same results were obtained by I De Baere [21] but for materials reinforced with fabrics. Contrast to cross-ply CF/PP, sudden increase of resistance values is observed for these materials close to the end of the test, which means that the damages growth and the end of the failure of sample is close.

In Figure 5 and Figure 6 the evolution of electrical resistance change function of the strain for the two types of materials is shows. Can observe that for cross-ply CF/ PP the evolution of electrical resistance change can be divided in two separate phases: a "steady state phase", where the increase of $\Delta R/R_0$ are small and fluctuate to positive to negative values, and an "end of life" phase, where the growth suddenly increases (from $\varepsilon = 1.6$ %, and onward). Pentru cross-ply CF/Epoxy, as show in Figure 6, the electrical resistance change increases whit the increase of strain. This increase, compared to cross-ply CF/PP, is almost linearly, which means the piesoresistivity, which represent electrical resistance change of material, is positive and constant.

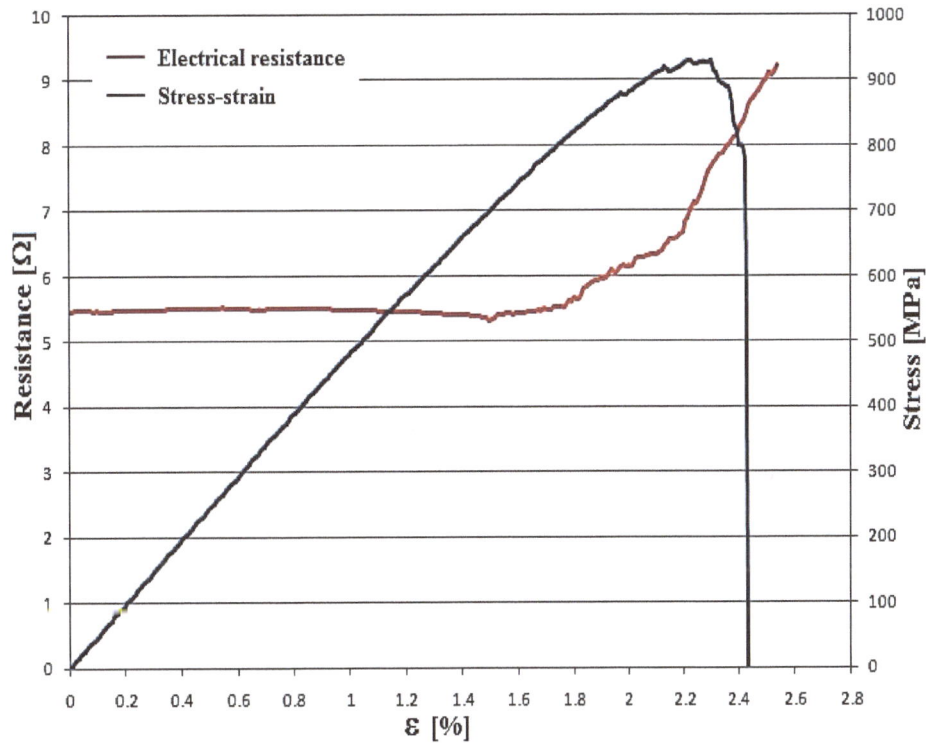

Figure 3: The evolution of resistance vs. strain (red) and stress-strain diagram (blue) for cross-ply CF/PP

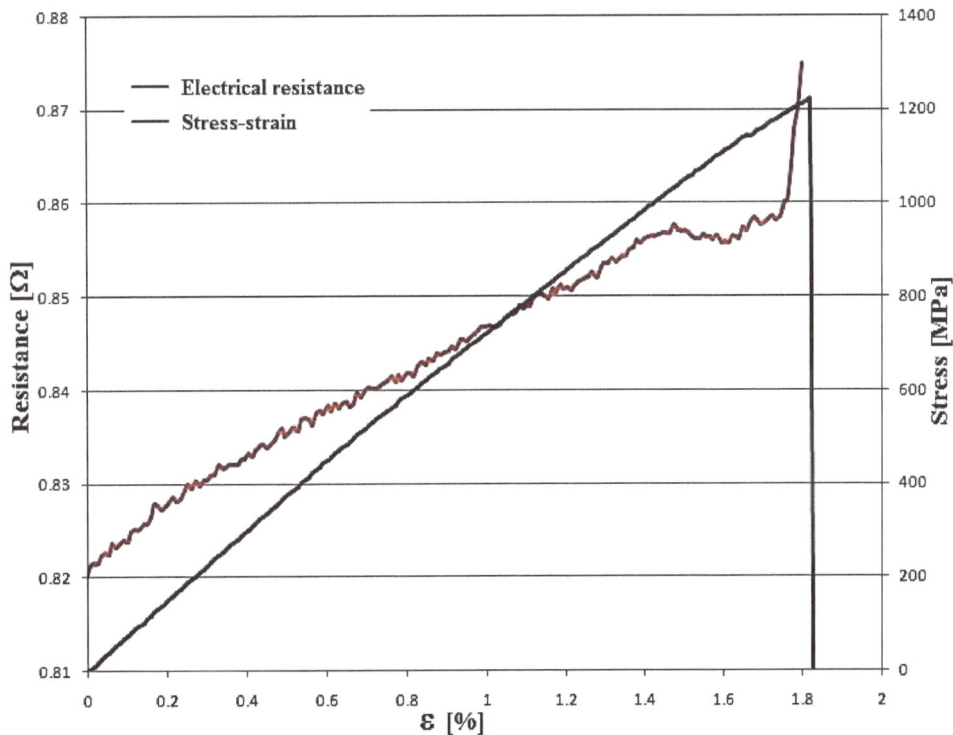

Figure 4: The evolution of resistance vs. strain (red) and stress-strain diagram (blue) for cross-ply CF/Epoxy

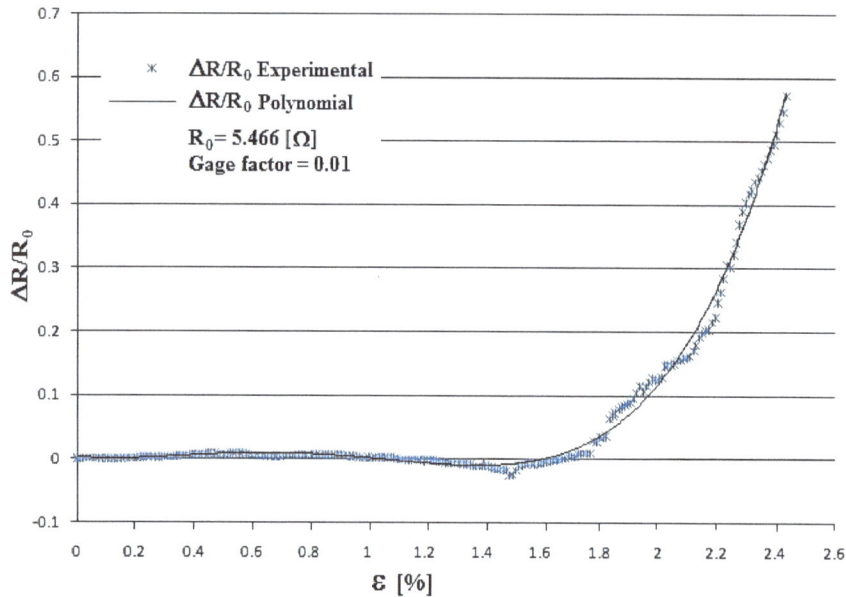

Figure 5: The evolution of electrical resistance change vs. strain for cross-ply CF/PP

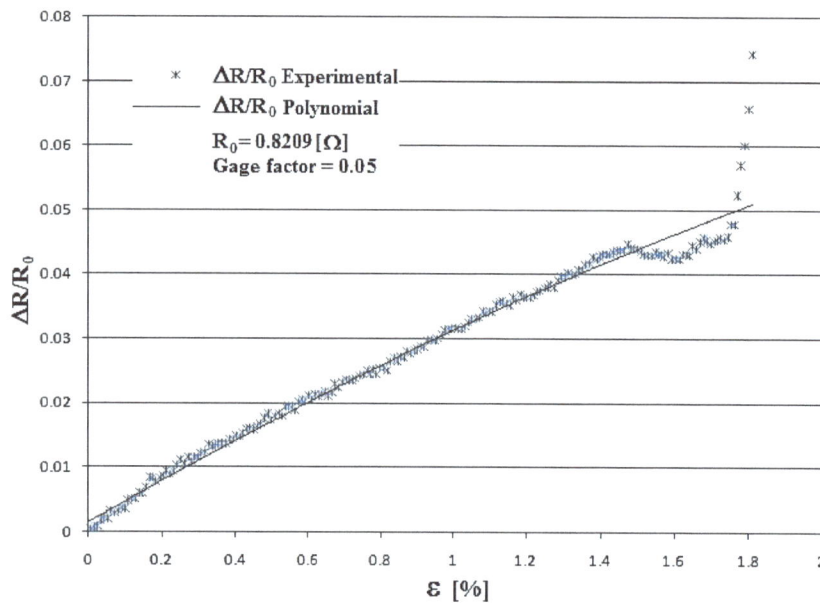

Figure 6: The evolution of electrical resistance change vs. strain for cross-ply CF/Epoxy

4. CONCLUSIONS

Based on the results obtained in this study, the following conclusions can be drawn:
- The electrical resistance in both types of materials with different fiber volume fraction increases with applied load. For CFRP composites with lower fiber volume (cross-ply CF/PP), this increase is insignificant till to the value of 1.6% of strain, but, after that, the electrical resistance increases pronounced. Not the same thing was observed for cross-ply CF/Epoxy, where electrical resistance increases steadily with increasing of strain.

- The electrical resistance change in cross-ply CF/PP in the first part of the tensile are small and fluctuate to positive to negative values which meant that in inside of materials no major changes of the circuits through which the current flows, occur.
- Constant increase of change of electrical resistance in cross-ply CF/Epoxy composite, where the volume of fiber is higher, represents the constant interruption of contacts between fibers, leading to the number reducing of circuits in which current flows.
- Increased resistance to the end of the test as a result of damage (delamination or fiber breakage), suggests that this method can be used successfully for predicting the end of life of carbon-reinforced composites.

ACKNOWLEDGMENT

This paper was realized with the support of EURODOC "Doctoral Scholarships for research performance at European level" project, financed by the European Social Found and Romanian Government.

REFERENCES

[1] De Waele W., Degrieck J., De Baets P., Moerman W. and Taerwe L., Feasibility of integrated optical fibre sensors for condition monitoring of composite structures—part II: combination of Bragg-sensors and acoustic emission detection, Insight 45:542-546, 2003.

[2] Kuang K.S.C., Kenny R., Whelan M.P., Cantwell W.J., Chalker P.R., Residual strain measurement and impact response of optical fiber Bragg grating sensors in fiber metal laminates, Smart Materials and Structures 10(2):338–46, 2001.

[3] Imai S., Tokuyama M., Hirose S., Burger G.J., Lammerink T.S.J., Fluitman J.H.J., Thin film piezoelectric impact sensor array fabricated on a Si slider for measuring head-disk interaction, IEEE Trans Magn 31(6.1):3009–11, 1995

[4] Haywood J., Coverley P.T., Staszewski W.I., Worden K., An automatic impact monitor for a composite panel employing smart sensor technology, Smart Materials and Structures 14(1):265–71, 2005.

[5] Omagari K., Todoroki A., Shimamura Y., Kobayashi H., Detection of matrix craking of CFRP using electrical resistance changes, Key Engineering Materials 297-300, pp 2096-2101, 2005.

[6] Todoroki A., Tanaka M., Shimamura Y., Kobayashi H., Effects with a matrix crack on monitoring by electrical resistance method, Advance Composite Materials., Vol. 13, No. 2, pp. 107–120, 2004.

[7] Wang X., Chung D.D.L., Sensing delamination in a corbon fiber polymer-matrix composites during fatigue by electrical resistance measurement, Polymer Composites 18(6), pp.692-700, 1997.

[8] Ueda M., Todoroki A., Shimamura Y., Kobayashi H., Monitoring delamination of laminated CFRP using the electric potential change method (application of normalization method and the effect of the shape of a delamination crack), Advance Composites Materials 13(3–4):311–24, 2004.

[9] Todoroki A., Tanaka Y., Shimamura Y., Delamination monitoring of graphite/epoxy laminated composite plate of electric resistance change method, Composites Science and Technolgy 62(9):1151–60, 2002.

[10]   Kaddoour A.S., Al-Salehi F.A., Al-Hassani S.T.S., Electrical resistance measurement technique for detecting failure in CFRP materials at high strain rate, Composites Science and Technolgy 51(3):377–85, 1994.

[11]   Schueler R., Joshi S.P., Schulte K., Damage detection in CFRP by electrical conductance mapping, Composites Science and Technolgy 61(6):921–30, 2001

[12]   Gordon D.A., Wang S., Chung D.D.L., Piezoresistivity in unidirectional continuous carbon fibre polymer matrix composites: single lamina composite vs 2 lamina composite, Composite Interfaces, 11:95–103, 2004.

[13]   Xia Z., Okabe T., Park J. B., Curtin W. A., Takeda N., Quantitative damage detection in CFRP composites: coupled mechanical and electrical models, Composites Science and Technology 63 :1411–22, 2003.

[14]   Todoroki A, Samejima Y, Hirano Y and Matsuzaki R. Piezoresistivity of unidirectional carbon/epoxy composites for multiaxial loading, Composites Science and Technology 69: 1841–1846, 2009.

[15]   Todoroki A, Yoshida J., Electrical resistance change of unidirectional CFRP due to applied load, JSME International Journal Serie A, vol.47, nr.3, pp.357-364, 2004.

[16]   Park J.B., Okabe T., Takeda N., New concept for modeling the electromechanical behavior of unidirectional carbon-fiber-reinforced plastic under tensile loading, Smart Materials and Structures 12:105–14, 2003.

[17]   Angelidis N., Wei C.Y., Irving P.E., The electrical resistance response of continuous carbon fibre composite laminates to mechanical strain, Composites Part A, 35:1135–47, 2004.

[18]   Todoroki A., Tanaka M., Shimamura Y., Measurement of orthotropic electric conductance of CFRP laminates and analysis of the effect on delamination monitoring with an electric resistance change method, Composites Science and Tehnology 62(5):619-28, 2001.

[19]   ASTM Standard ASTM D3171 - 11, "Standard Test Methods for Constituent Content of Composite Materials", ASTM International, West Conshohocken, PA, 2011.

[20]   ASTM Standard D 3039/D3039M-08, " Standard Test Method for Tensile Properties of Polymer Matrix Composite Materials", ASTM International, West Conshohocken, PA, 2008.

[21]   De Baere I., Van Paepegem W., Degrieck J., The use of rivets for electrical resistance measurement on carbon fiber-reinforced thermoplastics, Smart Material and Structures 16, pp.1821–1828, 2007.

INORGANIC NANOPARTICLES IN POLYMER MATRIX COMPOSITES

A. Matei[1,2], L. Dumitrescu[2], I. Cernica[1], Vasilica Schiopu[1], I. Manciulea[2]

[1]National Institute for Research and Development in Microtehnologies IMT-Bucharest, Romania, alina.matei@imt.ro
[2]Transilvania University of Brasov, Research Centre Renewable Energy Systems and Recycling, Romania

Abstract: Polymer – inorganic nanoparticles composites present an interesting approach, because by combining the attractive functionality of both components result new materials with synergistically improved properties. The objective of this paper was to examine the structure and properties of a new type of composite materials based on thermoplastic polymers and inorganic nanoparticles.
It is very difficult for inorganic nanoparticles to disperse in the polymer matrix through conventional mixing, because the nanoparticles have high surface energy and have tendency to agglomerate during mixing. So, surface of inorganic nanoparticles was modified with a capping agent, who it was used to improve the interface between the organic and inorganic phases. Chemical structure, particle size distribution and surface morphology of the obtained composites were characterized using Fourier infrared spectra (FTIR) and scanning electronic microscopy (SEM).
Keywords: nanoparticles, composites, polymer matrix.

1. INTRODUCTION

For several years, incorporation of different inorganic nanoparticles in polymer matrix is the most interesting approach to synthesize nanocomposites. These materials are of growing interest due to improve properties by combining of both inorganic nanoparticles and organic molecules, and because extensive potential applications in a wide range of industrial fields, such as textile, paints, magnetic fluids, high-quality paper coatings to catalysis, microelectronics and biotechnology [1, 2].

According to the studies, the nanoparticles can be synthesized from a variety of materials with controllable sizes, shapes and composition (semiconductors, oxides, carbon-based materials, nanoparticles, etc.), which can be embedded in various polymeric matrix depending of the potential applications. [3].

The main problem in obtaining nanocomposites is the prevention of particles agglomeration, by modification of the surface of the nanoparticles, in order to improve the compatibility between the inorganic particles and the polymer matrix [4, 5].

In the present work, ZnO nanoparticles were used as inorganic nanoparticles and ZnO-PMMA nanocomposites were obtained by dispersing the ZnO nanoparticles in polymethyl methacrylate (PMMA).

Among inorganic nanoparticles, zinc oxide (ZnO) is one of the most researched, because it can be synthesized in a wide range of particle sizes, shapes, and due to its physical and chemical properties, such as chemical stability, low dielectric constant, high catalysis activity, effective antibacterial and bactericide, intensive ultraviolet and infrared absorption [6].

Polymethyl methacrylate (PMMA) is selected as matrix because is known to be suitable medium to disperse inorganic nanoparticles to improve the dispersion stability. It is also an important polymer with excellent transparency and processabil due to its properties and extensive applications [7, 8].

The aim of the present work was to synthesize ZnO nanoparticles by modifying the surface of particles with capping agent, then dispersing in polymer matrix to obtain ZnO-PMMA nanocomposites. It is known that interactions between the nanoparticles and matrix play an important role in determining the quality and properties of the nanocomposites.

The ZnO nanoparticles and ZnO- PMMA nanocomposites were characterized by evaluating their phase analysis through FTIR spectroscopy, and their shape size and distribution in polymer matrix by SEM.

2. EXPERIMENTAL

2.1. Synthesis of ZnO nanoparticles

The ZnO were prepared by adding an amount of $Zn(CH_3COO)_2 \cdot 2H_2O$ to ethanol (C_2H_5OH) at room temperature and stirred on a magnetic stirrer for 1h. Then, a small amount of glacial acetic acid [CH_3COOH] was added to the mixture and stirred at 60°C for 2h. The addition of NH_3 changed the pH values of the sol from 5 (acidic condition) to 12 (alkaline condition). The resulting white gel was stirred for another 1h. The sample was left alone for 2 days to allow the sol–gel process to finish.

After that, the samples were heat treated at different temperatures 250°C, 350°C, 450°C, 550°C, for 3h with a heating rate of 10°C/min.

2.2. Surface modification of ZnO nanoparticles

Firstly, the surface of ZnO nanoparticles was modified with capping agent (elaidic acid, trans isomer of the oleic acid ($C_{18}H_{34}O_2$).
The capping agent plays a significant role in improvement the compatibility between inorganic nanoparticles and organic matrix. Also, nanoparticles have the tendency to agglomerate, which makes them lose their specific characteristics.

In our experiments, elaidic acid was solubilized in ethanol solution and an amount of ZnO was introduced into the solution and stirred for a good dispersion. The solution was filtered and the precipitate was washed with the mixture of solvent and deionized water. The precipitate was dried and kept in a vacuum desiccator.
Secondly, ZnO nanoparticles were dispersed in polymethyl methacrylate (PMMA) matrix under ultrasonic stirring condition at room temperature until a stable suspension was obtained.

2.3. Characterization methods

The chemical compositions of the synthesized ZnO nanoparticles and ZnO–PMMA composites were analyzed by Fourier Transform Infrared (FTIR) spectrometry at room temperature using a Bruker Tensor 27 spectrometer, in the spectral range of 4000 - 400cm^{-1}. The spectral resolution was 4cm^{-1} and 64 scans were averaged and KBr pellet technique was used. The discs were prepared by compressing a mixture, formatted from the samples powder and KBr, at pressure of 10 tons for 5 min. in a hydraulic press, and then the discs were scanned to obtain FTIR spectra.

The morphology and the distribution of the ZnO particles in PMMA matrix were studied by Scanning Electron Microscopy (SEM - VEGA II LMU).

3. RESULTS

3.1. Phase analysis of the samples

Figure 1 shows the spectra of zinc oxide sample obtained in the first part of our study. Using FTIR spectrometry we studied the sintering process evolution at different temperatures of the ZnO powder. Spectrum from figure 1(a) is mainly characterized of bands attributable to organic groups. At 250°C it is not noticeable any characteristic band to carboxylic group, which indicates decomposition of acetate, used as raw material. Further increase of the temperature showed major spectral changes (figure 1 (b) and (c)). With the disappearance of the absorption bands characteristic to organic bonds can be seen a better definition of the absorption band attributed to Zn-O bond, and at the end (figure 1 (d)) we can speak of a single Zn-O bond characteristic absorption band that can be observed at 420cm^{-1}.

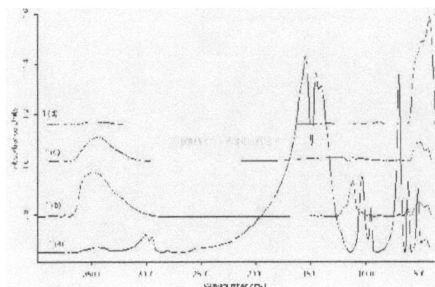

Figure 1: FTIR spectra of ZnO sample sintered at: (a) 250°C, (b) 350°C, (c) 450°C, (d) 550°C

FTIR spectrum of pure elaidic acid (Figure 2 (a)) is characterized by the bands attributed to stretching vibration of C=O mode (1711 cm^{-1}) and C-H (2923 and 2854 cm^{-1}). Figure 2(b) presents the FTIR spectrum of ZnO modified with elaidic acid. FTIR spectra of this sample show a typical Zn-O absorption band at 480cm^{-1}, as well as the absorption bands of elaidic acid onto the ZnO nanoparticles surface. Formation of a complex between elaidic acid and ZnO is emphasized by comparing the spectra of 2(a) and 2(b). It can be observed the complete disappearance of the band at 1711cm^{-1} and the appearance of a new broad band at about 1550 cm^{-1} attributed to stretching vibration of COO-.

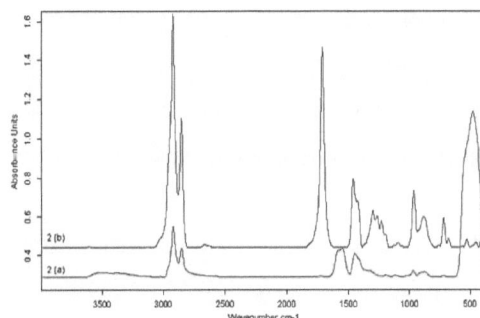

Figure 2: FTIR spectra of (a) elaidic acid and (b) elaidic acid - ZnO samples

FTIR spectra of the sample of pure polymethyl methacrylate (PMMA) (Figure 3(a)) is characterized by absorption bands appearing in the range 1940-1730cm^{-1} which can be assigned to the vibration mode of C=O bond and bands in the range 3000 - 2900cm^{-1} that can be attributed to the stretching mode of C-H bond. The bands in the range 1445-1370cm^{-1} can also be attributed to the vibration mode of C-H bond and the band centered at 1271cm^{-1} can be attributed to the group C-H deformation and the last peak at 736cm^{-1} corresponds to the vibration out of plane of C-H bond. The band at 1151cm^{-1} can be attributed to the group C-O from the ester, the band centered at 1236cm^{-1} can be associated with vibration mode of the group O-CH$_3$. The absorption band centered at 902cm^{-1} is correlated to C-C bond vibration mode. FTIR spectra of the samples of ZnO - PMMA (Figure 3(b) didn't show any spectral band characteristic to metal-oxygen bond of zinc oxide. Finally, the disappearance of the band Zn-O indicates total embedding of the ZnO particles in the polymer matrix of PMMA.s.

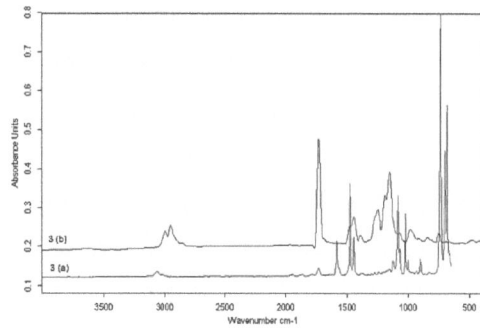

Figure 3: FTIR spectra of (a) polymethyl methacrylate (PMMA) and (b) ZnO-PMMA samples

3.2. Morphological characterizations of the samples

Scanning electron micrographs (SEM) of the ZnO particles, elaidic acid modified ZnO nanoparticles and ZnO –PMMA composites are showed in Figure 4 (a) – (c), respectively.

In Figure 4 (a), we can easily observe that the particles are granular and have a slight tendency of agglomeration. The size of ZnO nanoparticles is in the range from 20 nm to 100 nm.

It is observed in Fig. 4 (b) that elaidic acid modified ZnO nanoparticles consists of particles irregular in shape and presents a good distribution and homogeneity.

In Figure 4 (c), the ZnO nanoparticles are seen as white regions in the darkness PMMA matrix. Also, it is observed that the nanoparticles are non-agglomerated and well dispersed in PMMA matrix.

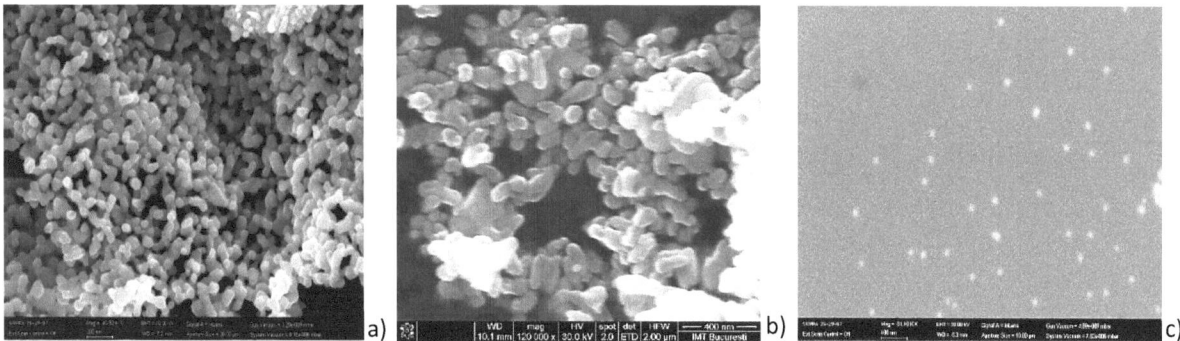

Figure 4: SEM images of the a) ZnO nanoparticles; b) elaidic acid –ZnO; c) ZnO - PMMA

4. CONCLUSION

The ZnO nanoparticles were synthesized by sol-gel method. The nanoparticles were modified with the capping agent (elaidic acid) and then dispersed in organic matrix of PMMA.

FTIR spectra show that 550°C is the optimal sintering temperature for obtaining of ZnO nanopowder. Formation of a complex between elaidic acid and ZnO was emphasized by the disappearance of C=O absorption band and the appearance of a new band attributed to COO- group from COO-Zn. Based on the FTIR spectra, elaidic acid covers the surface of the ZnO particles and is totally embedded in the matrix of PMMA.

From the results obtained, it was concluded that the surface modification of ZnO nanoparticles improves the dispersability in different matrices and also reduces the tendency of agglomeration of the nanoparticles.

ACKNOWLEDGEMENTS

The authors kindly thank to Dr. Adrian Dinescu for the SEM characterization.

REFERENCES

[1]   Jeon I. Y., Baek J.B., "Nanocomposites Derived from Polymers and Inorganic Nanoparticles",   Materials, 3, 3654-3674, 2010.

[2]   Li S., Lin M. M., Toprak M. S., Kim D. K., Muhammed M., "Nanocomposites of polymer and   inorganic nanoparticles for optical and magnetic applications", Nano Reviews. 2010

[3]   Hanemann T., Szabó D.V., "Polymer-Nanoparticle Composites: From Synthesis to Modern   Applications". Materials, 3, 3468-3517, 2010.

[4]   Honga R., Pan T., Qian J., Li H., "Synthesis and surface modification of ZnO nanoparticles", Chemical   Engineering Journal 119, 71–81, 2006.

[5]   Zou H., Wu S., Shen J., "Polymer/Silica Nanocomposites: Preparation, Characterization, Properties, and   Applications", Chem. Rev. 108 (9), 3893–3957, 2008.

[6]   Wang Z. L., "Zinc oxide nanostructures: growth, properties and applications", Journal of Physics: Condensed Matter, Vol. 16, R829-R858, 2004.

[7]   Anzlovar A., Zorica C.O., Zigon M., "Nanocomposites with nano to sub micrometer size zinc oxide is   an effective UV absorber", Polimeri, 29, 84-87, 2008.

[8]   Demir, M.M., Memesa M., Castignolles P., Wegner G., "PMMA/ ZnO nanocomposites prepared by in   situ bulk polymerization", Macromol. Rapid Commun., 27, 763-770, 2006.

DETERMINATION OF MECHANICAL PROPERTIES FOR IMPACT AND BENDING A BUMPER SHOCK ABSORBER MADE OF STEEL COMPARED TO BUMPER SHOCK ABSORBER MADE FROM A NEW COMPOSITE MATERIAL FOR AUTOMOTIVE INDUSTRY - PART 1

A. Calienciug[1], G.N. Radu[2]
[1] "Transilvania" University of Brasov, Brasov, ROMANIA, adrian.calienciug@unitbv.ro
[2] "Transilvania" University of Brasov, Brasov, ROMANIA, rngh@unitbv.ro

Abstract: *A current issue is related to increasing levels of environmental pollution and global warming caused by industrialization especially excessive burning of fossil fuels or liquid. This is due in part to motor vehicle fuel in the combustion process they eliminate environmentally harmful emissions. By reducing the weight of vehicles will be allowed to reduce the amount of energy needed to produce mechanical work. The purpose of this study is to design and create a new lightweight composite material used in the structure of automotive bumper shock absorber. This item is made of composite material and bumper shock absorber type was compared with the current bumper shock absorber made from steel. New laminated composite type must meet the standards of today, have minimized weight and impact absorption capacity at low speed.*
Keywords: *absorption, weight, stratified, bumper car design.*

1. INTRODUCTION

In the present context, excessive industrialization hardly degradable materials is that researchers create new materials that meet the requirements of present but that is biodegradable and recyclable. Thus by reducing the weight of vehicles using composites in automotive safety bumper structure, reduce resource consumption and wear occurred due to vehicle weight. But bumper shock absorber has an important role in reducing injuries to pedestrians in collision bumpers with them. Most accidents are low speed frontal, such as the first element in contact with: other vehicles, pedestrians, items on or off road, is car bumper. Car bumper consists of several elements that make up its structure as we can see in Figure.1.:

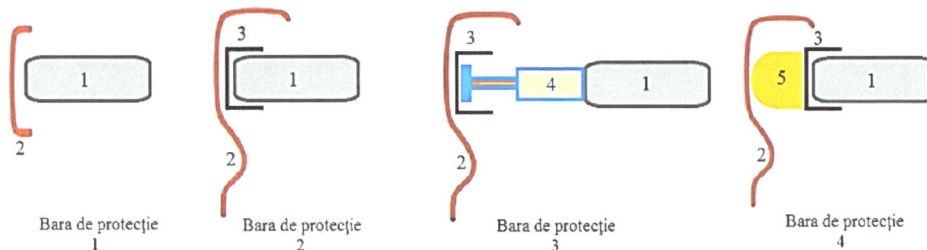

Figure 1 Car bumper structure section [1,2]

Mechanical test methods for case study namely layered polymeric composite materials consist of car bumper shock absorber made of steel bending test in comparison with car bumper shock absorber made of composites.

2. WORKING METHOD

In this study, to get a bumper for a car was used as bumper mold present existence on cars made of steel. Thus the overlap of 6 layers of textile fabric warp yarns to reinforce polyester resin Polylite 440-M880 and adding foam cork tiles in Figure 3 was obtained by manual or molding in Figure 2 bumper shock absorber in Figure 5.

From an economic perspective was used the manual method of molding the polymer immersed in resin with hardener overlapping cork and foam.

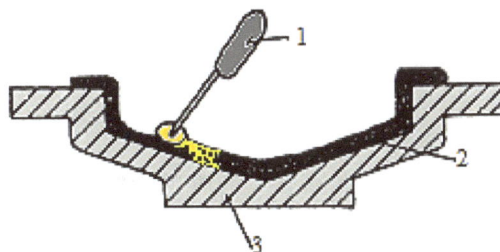

Figure 2 Lay-up method [4]

1.roll;2.layer (resin-impregnated reinforcement material);3.mold open

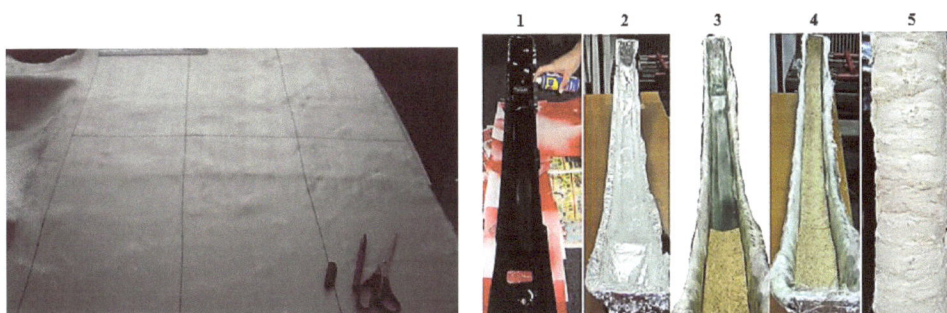

Figure 3 Lay-up method for obtaining composite structure for bumper shock absorber [1]

The temperature is the most important factor in this process because the polymer is plastify and pressure die cast him in a certain period of time. After the time-out, the finished product can be removed from the mold as can be seen in Figure 4:

Figure 4 Bumper shock absorber made of composite materials removed from the mold [1]

Figure 5 Bumper shock absorber made of composite materials [1]

3. EQUIPMENT USED FOR BENDINDING TEST FOR BUMPER SHOCK ABSORBERS

In Figure 6 is presented the bending testing machine and in Figure 7 analyzer designed to transmit data to the computer, the data obtained from the bending test:

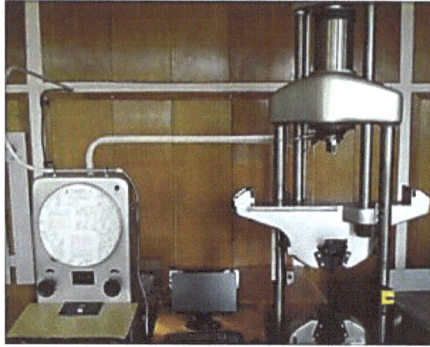

Figure 6 Bending testing machine [1]   **Figure 7** Actuator signal reception
and transmission of registered [1]

Here are some details as follows:
    Manufacturer: JIANIN-CHINA;
    Type: WE-60 with measuring range: 5...550kN; res=0,001kN

4. BENDING TESTS OF BUMPER SHOCK ABSORBER FOR AUTOMOTIVE

For three-point bending test using test machine described above were used as car bumper shock absorber made of steel black painted and composite car bumper shock absorber yellow colored as can be seen in Figure 8

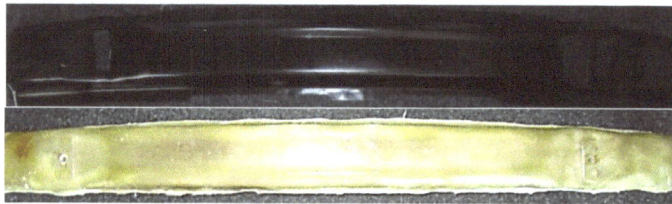

Figure 8 Steel and Composite car bumpers shock absorbers [1]

Were weighed bumpers car shock absorber made of steel in Figure 9 and in Figure 10 were weighed both of them and the mass difference obtaining will be the weight of composite bumper car shock absorber. For steel car bumper shock absorber 4kg mass is fitting and sheet thickness is 0.2 mm. For composite car bumper shock absorber the mass is 1.3 kg;

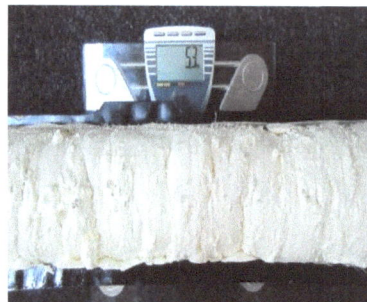

Figure 9 Weight of steel car bumper   **Figure 10** Weight of both of them [1]
shock absorber [1]

Knowing the drive test data and the items to be tested, the testing was determined as follows:
the car bumper shock absorber made of steel like in Figure 11, and composite car bumper like in Figure 12, were mounted on the bending test machine:

Figure 11 Bending test
for steel car bumper shock absorber [1]

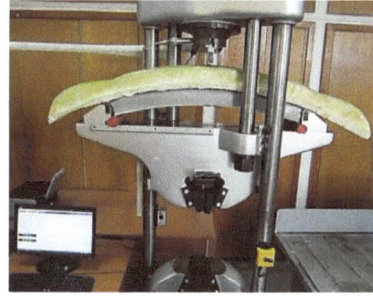

Figure 12 Bending test
for composite car bumper shock absorber [1]

Figure 13 Bending detail
for steel car bumper shock absorber [1]

Figure 14 Bending detail
for composite car bumper shock absorber [1]

5. RESULTS

The bending tests where you can see details of test and behavior of car bumper shock absorber respectively in Figures 13. and in Figure 14. We can see actuator by receiving and transmitting data to the system were obtained for fitting the force-displacement graphs in Figure 15. and for composite car bumper shock absorber in Figure 16.:

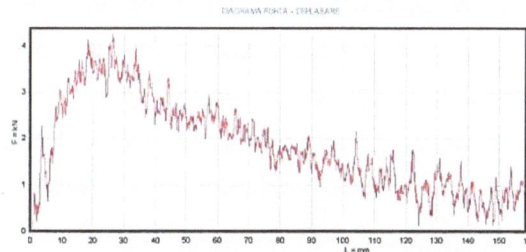

Figure 15 Force-displacement graphic
stell car bumper shock absorber [1]

UNIVERSITATEA TRANSILVANIA BRASOV
LABORATORUL DE ÎNCERCARI MECANICE - REZISTENTA MATERIALELOR

RAPORT DE ÎNCERCARE
Nr. 42   16/07/2012 12:39:55

BENEFICAR:
METODA DE ÎNCERCARE   Compresiune   REPER: Rm: 0.091 [N/mm²]
MATERIAL: SARJA: Fmax: 2.957 [kN]
TIPUL EPRUVETEI :   Plata   VOLUM ESANTION: Modulul de   0 [N/mm²]
DIAMETRUL EPRUVETEI (mm):   3240   VOLUM CONTROLAT: elasticitate [E]
SECTIUNEA EPRUVETEI (mm²): 32400   REFERENTIAL: SR EN 10002 - 1/ 2002   Limita de curgere   0 [N/mm²]
                                                                                     Rp0.2
ECHIPAMENT UTILIZAT: Masina de incercare la tractiune, compresiune si incovoiere
                          Prod. JINAN - CHINA, Tip: WE - 60   SN: 088 / 1979.4   Alungirea relativa   150.971 [%]

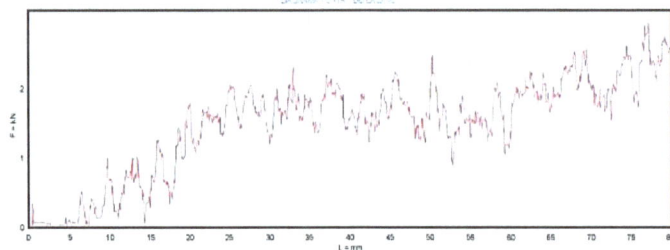

Figure 16 Force-displacement graphic
composite car bumper shock absorber [1]

In Table 1 were displayed peaks resulting from force-displacement graphs for steel shock absorber bumper car and separate composite shock absorber bumper car made from a new composite material:

Table 1: Values obtained from flexural reinforcement and self-bumper bar [1]

Values	Car bumper shock absorber made of steel	Car bumper shock absorber made of composite materials
Fmax [kN]	4.258	2.957
Rm [N/mm²]	1.314	0.091

6. CONCLUSIONS

By reducing the weight of the bumper using biodegradable and recyclable materials such as cork can make the claim that depreciation costs of producing this type of bar can be achieved by reducing the vehicle weight, reducing mechanical work and thus lower consumption.

ACKNOWLEDGEMENT

This papers supported by the Sectorial Operational Programme Human Resources Development (SOP HRD), financed from the European Social Fund and by the Romanian Government under the contract number POSDRU/88/1.5/S/59321

REFERENCES

[1] Calienciug A., Cercetări privind bara paraşoc pentru autovehicule, confecţionată din materiale compozite   noi, Teza de doctorat , Universitatea Transilvania din Brasov, Braşov, 2012.

[2] Davoodi M.M., S.M. Sapuan, A. Aidy, N.A. Abu Osman, A.A. Oshkour, W.A.B. Wan Abas, Development process of new bumper beam for passenger car: A review, Elsevier, Materials and Design 40, pp. 304-313, www.elsevier.com, 2012.

[3] Dogaru F., Mecanica compozitelor laminate, Editura universităţii Transilvania, ISBN 978-973-598-251-5, 2008.

[4] Preda M., G., Şontea S., Procedee de elaborare a pieselor pentru automobile din materiale compozite stratificate, in The 8th International Conference University Constantin Brâncuşi, Târgu Jiu, 2002.

THE INFLUENCE OF THE TRIBOLOGICAL TESTING CONDITIONS ON THE WEAR RATE FOR SOME TITANIUM PLATES PROCESSED BY SPARK PLASMA SINTERING ROUTE

Adrian Olei[1], Gabriel Benga[1], Iulian Stefan[1]

[1] University of Craiova, Faculty of Mechanics, Department of IMST, Drobeta Turnu Severin, ROMANIA
adrian_olei@yahoo.com, gabrielbenga@yahoo.com, stefan_iuly@yahoo.com

Abstract: *The paper presents the experimental results obtained during the tri-biological tests developed on Ti plates processed by SPS route. The material used in the current research is micrometric Ti powder (<150μm). The Ti plates are processed by SPS at (1000-1100) C and (10-20) min. dwell time. The wear rate and wear tracks' morphology are studied vs. the sintering route parameters as well as the wear testing conditions (stainless steel and sapphire testing ball, normal load=2N, sliding speed=1cm/s).*

Keywords: *powder metallurgy, titanium, sparks plasma sintering, tribology*

1. INTRODUCTION

Modern medicine has made advances by improving the biomedical implants as the population ages and needs implants. Titanium and its alloys, introduced in the early 1950s, have been used as materials for biomedical implants, due to their proved biocompatibility and physical-chemical and mechanical properties related to these kind of applications [1-3]: replacement of hard tissues in devices such as artificial hip joints, total knee replacement and dental implants.

One of the main mechanical performances requested for these applications is the wear strength. Experimental data on Ti wear strength provide information about different processing technologies for Ti implants able to perform low wear rates. For example, Masmoudi et al. [4] used the powder metallurgy (PM) technology to obtain TiO_2 passive films at the Ti and Ti6Al4V surface. These films are lubricating and, by consequence, provide low wear rates about 1.7-$2.3 \times 10^{-4} mm^3/Nm$ range.

Other researchers [5] have applied the passivation technology on Ti-5Al-4V and Ti-6Al-3,5Fe alloys in order to obtain a multi-film structure. The results were confirmed by the mechanical point of view that is the model is realized by two films deposition (one interior which acts as a barrier and has a high compacticity and one exterior, porous), and thus the technology facilitates the mineral ions incorporation and improves the wear resistance.

In this paper, the spark plasma sintering route is analyzed by the influence on the wear strength point of view.

2. MATERIALS AND EXPERIMENTAL PROCEDURE

The material used in the research is micrometric Ti powder (<150μm). The Ti powder is water atomized and the particles are irregular-shaped. Figure 1 presents a SEM image of the Ti micrometric powder used in the research.

SPS technology is a new PM route to obtain high performance PM materials due to the advantages offered by lower sintering temperatures and dwell times as compared to the classic sintering parameters (see figure 2).

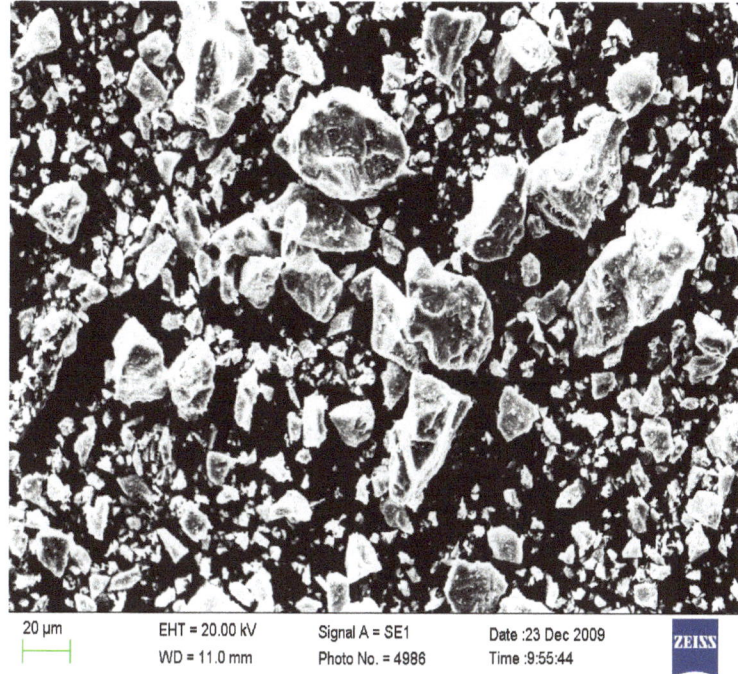

Figure 1: SEM image of the Ti powder particles

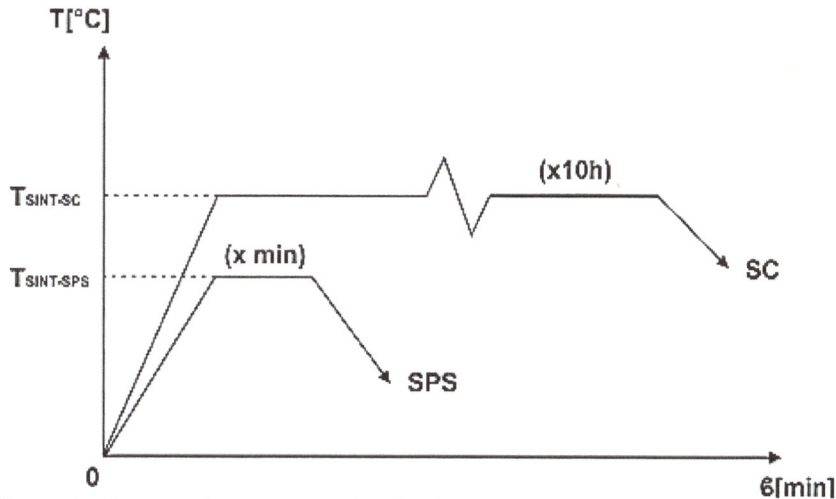

Figure 2: Comparative representation for the SPS with classical sintering parameters

For this research the SPS treatment has been developed in vacuum at 1000^0C, respectively 1100^0C, the dwell times for each temperature being 10 min. and 20 min. The heating rate was for 10^0C/min. and the punches load was 7kN.

Table 1: The parameters of the SPS process

Sample number	Sintering route	Sintering parameters	
		Temperature	Maintaining time
1		1000	20
2	SPS	1000	10
3		1100	20
4		1100	10

The wear tests have been performed on a TRB 01-02541 tribometer (CSM Instruments SA, Switzerland – figure 3), with linear reciprocating module, equipped by InstrumX software. The parameters determined by the tribometer are the friction coefficient, μ, using ball on disc friction couple, and the wear rate. The friction parameters are: sliding linear velocity =

1cm/s; room temperature (RT) = 23^0C, room humidity = 30%; the normal load is 2N. These parameters were chosen according to the literature [4, 6] in order to compare the experimental results obtained by this research.

Figure 3: CSM Instruments tribometer

3. RESULTS AND DISCUSSIONS

Table 2 presents the results obtained during the wear tests.

Table 2: Results obtained for the Ti samples sintered through SPS

Sintering technology	T_1/τ_2 [°C/min]	Friction coefficient, μ		Wear rate, ω [mm³/Nm]		Worn track depth, h [µm]	
		Steel	Sapphire	Steel	Sapphire	Steel	Sapphire
SPS	1000/10	0.49	0.49	$2.2 \cdot 10^{-4}$	$5.759 \cdot 10^{-4}$	3.17	5.07
	1000/20	0.52	0.43	$2.764 \cdot 10^{-4}$	$1.709 \cdot 10^{-3}$	3.18	5.36
	1100/10	0.57	0.74	$1.019 \cdot 10^{-4}$	$3.022 \cdot 10^{-4}$	2.6	2.98
	1100/20	0.52	0.27	$3.603 \cdot 10^{-4}$	$2.814 \cdot 10^{-4}$	4.12	3.47

The lowest wear rate is obtained for the Ti sample processed at (1100°/10'), tested with the steel ball for which the wear track has a 2.6 µm depth. The morphology of the worn track for this sample and the coefficient of friction are presented in figures 4 and 5.

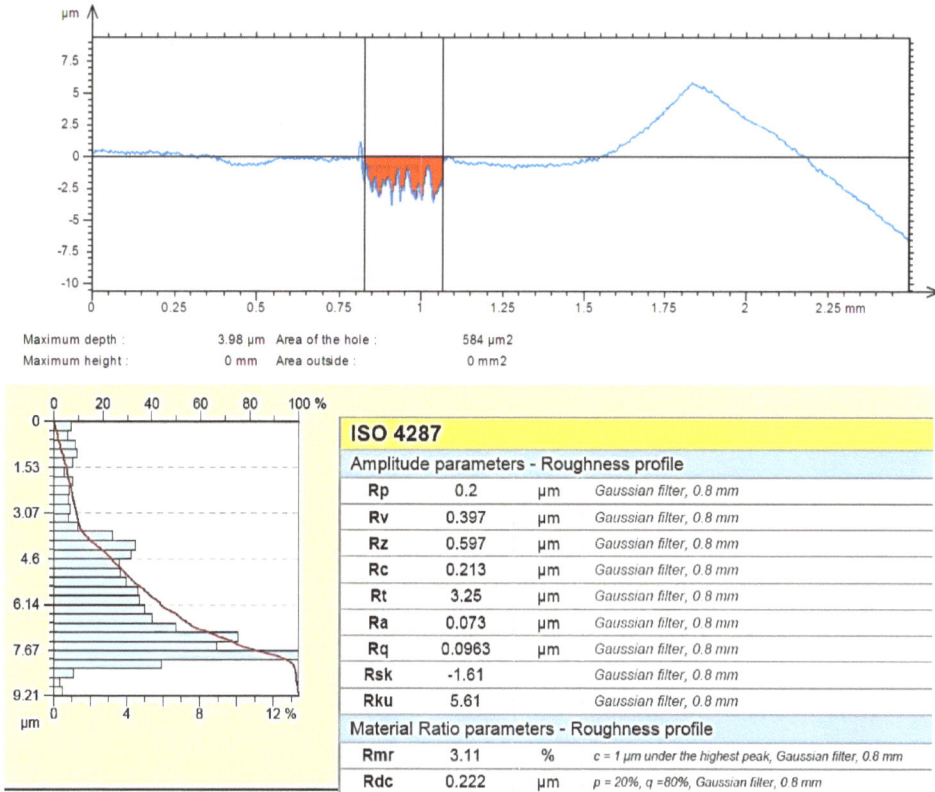

Figure 4: The morphology of the worn track for the sample processed at 1100°/10'

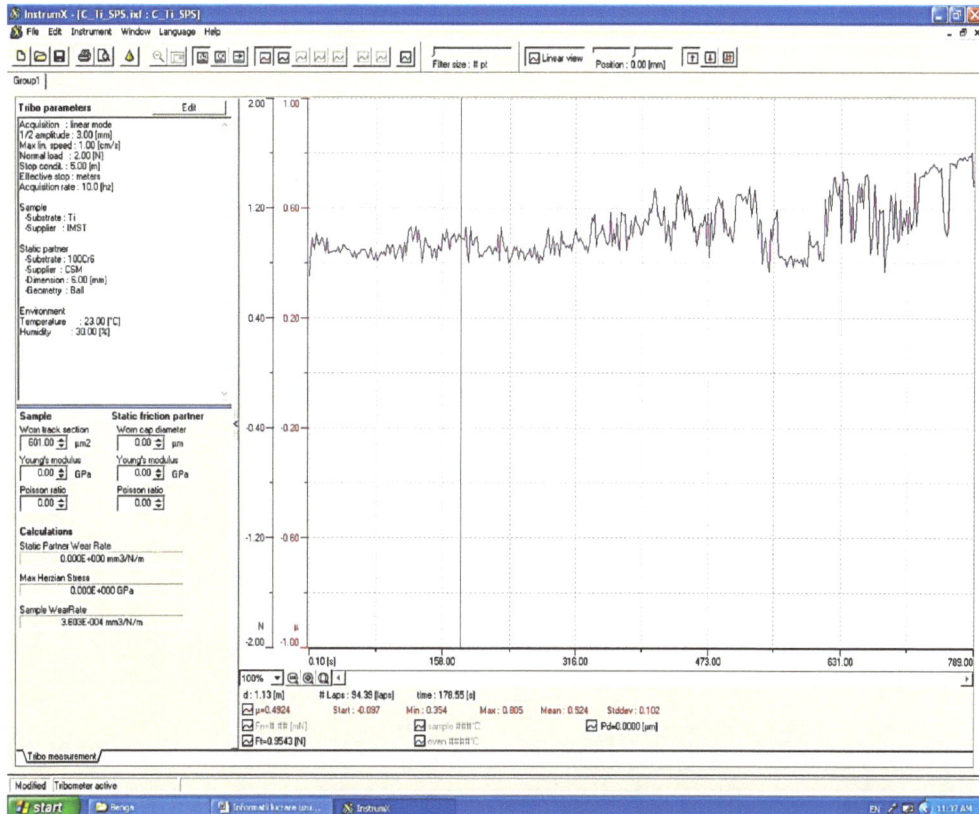

Figure 5: Variation of the friction coefficient tested with the steel ball for the sample processed at 1100°/10'

The highest wear rate is obtained for the Ti sample processed at 1000°/20 min., tested with sapphire ball. The sample has a 5.36 μm wear track depth. The morphology of the worn track for this sample and the coefficient of friction are presented in figures 6 and 7.

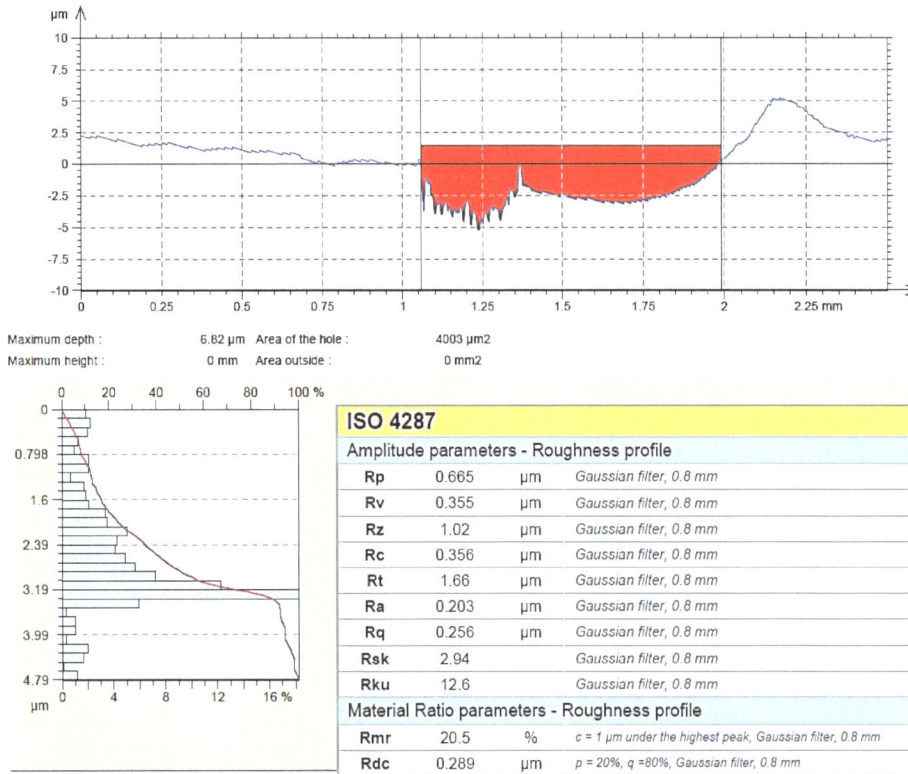

Maximum depth :	6.82 μm	Area of the hole :	4003 μm2
Maximum height :	0 mm	Area outside :	0 mm2

ISO 4287			
Amplitude parameters - Roughness profile			
Rp	0.665	μm	Gaussian filter, 0.8 mm
Rv	0.355	μm	Gaussian filter, 0.8 mm
Rz	1.02	μm	Gaussian filter, 0.8 mm
Rc	0.356	μm	Gaussian filter, 0.8 mm
Rt	1.66	μm	Gaussian filter, 0.8 mm
Ra	0.203	μm	Gaussian filter, 0.8 mm
Rq	0.256	μm	Gaussian filter, 0.8 mm
Rsk	2.94		Gaussian filter, 0.8 mm
Rku	12.6		Gaussian filter, 0.8 mm
Material Ratio parameters - Roughness profile			
Rmr	20.5	%	c = 1 μm under the highest peak, Gaussian filter, 0.8 mm
Rdc	0.289	μm	p = 20%, q =80%, Gaussian filter, 0.8 mm

Figure 6: The morphology of the worn track for the sample processed at 1000°/20'

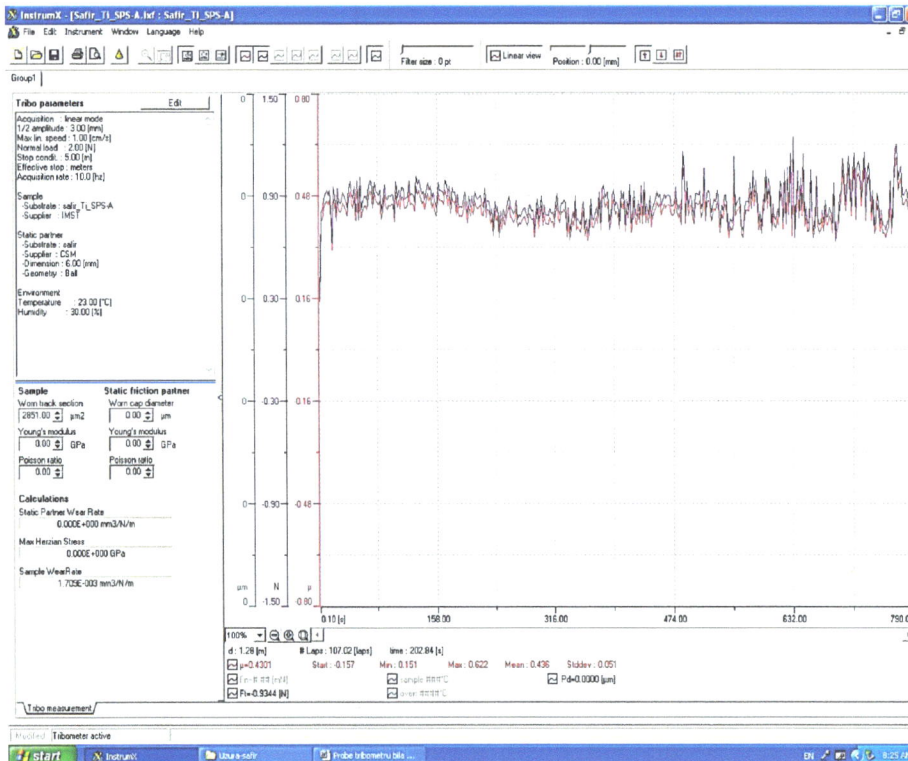

Figure 7: Variation of the friction coefficient for the sample processed at 1000°/20'

4. CONCLUSIONS

The most important issue regarding the wear behavior of PM Ti implants is the influence of the wear debris on the human health. These debris could be metallic (from the steel ball) or ceramic (from the sapphire ball) and can remain in the human body along the patient life and may cause/or not the hard/soft tissue inflammation nearby the implant place. In conditions of the testing with the steel ball, the lowest coefficient of friction is obtained for the sample processed at 1100°/10 min, while the highest coefficient of friction is obtained for the 1100°/20 min.

In conditions of the testing with the sapphire ball, the lowest coefficient of friction is obtained for the sample processed at 1000°/10 min, while the highest coefficient of friction is obtained for the 1000°/20 min.

REFERENCES

30. Zioupos, P., Recent developments in the study of solid biomaterials and bone: "fracture" and "pre-fracture" toughness, Materials Science and Engineering, C, 6, 33-40, 1998.
31. Phelps, J. et al., Microstructural heterogeneity and the fracture toughness of bone, Journal of Biomedical Research, 51, 735-741, 2000.
32. Moore, W. R., Graves, S. E., Bain, G. I., Synthetic bonne graft substitutes, Journal of Surgery, 71, 354-361, 2001.
33. M. Masmoudi et al., Friction and wear behaviour of Ti and Ti6Al4V following nitric acid passivation, Applied Surface Science, 2006, 253, 2237-2243.
34. Marza Rosca et al., Microsoc. Microanalisys, 12 (supp. 2), 2006.
35. M. Wang et al., Journal of Materials Science: Materials in Medicine, 2002, 13, 607-611.

FATIGUE CRACKS AND MICROSTRUCTURAL CONSTITUENTS

C.S. Bit[1], T. Bolfa[1]

[1] *Transilvania* University, Braşov, ROMANIA, e-mail: cbit@unitbv.ro, t.bolfa@unitbv.ro

Abstract: *This paper is concentrated on some microstructural issues concerning the mechanical interactions between fatigue cracks and microstructural constituents of an aluminum alloy subjected to fatigue cycles. A diffusion process involving different manganese compounds developed within the immediate area of the fatigue cracks has been also revealed.*
Keywords: *crack, fatigue, microstructural constituent*

1. INTRODUCTION

The present-day fatigue research is concentrated on materials cracks behavior, considering that such cracks are present to some degree in all mechanical structures. They may exist as basic defects in the constituent materials – assimilated to material deficiencies in the form of pre-existing flaws – or they may be induced in a certain engineering structure during the service life. In Fig. 1a a fatigue crack surface in Al-alloy has been represented. The whole life time of a certain structure subjected to fatigue cycles (or to static loads as well) depends upon the way in which material cracks do propagate until the final failure. In Fig. 1 and Fig. 2 two fatigue crack surfaces have been

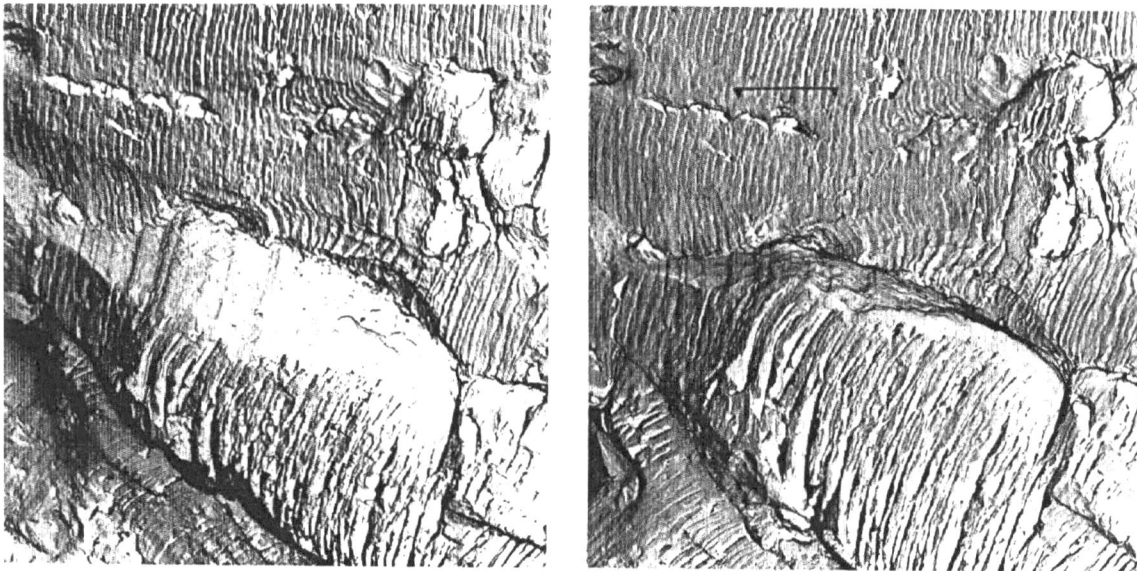

Figure 1a: A fatigue crack surface in Al-alloy

represented. It is to be noted the direct influence of the fatigue crack propagation on the material grains structures. This is why the fatigue phenomenon gives a great importance to the interaction between the fatigue cracks and the microstructural constituents of the investigated alloy. A direct and an important consequence of such a position is that Miner's linear criterion of damage mechanics loses its validity.

This paper has been focused on the mechanical interaction between fatigue cracks and microstructural constituents of an aluminum alloy (6061 T651) subjected to fatigue cycles.

2. PHYSICAL ASPECTS OF THE MECHANICAL INTERACTION BETWEEN FATIGUE CRACKS AND MECHANICAL STRUCTURE

The analysis of mechanical interaction between the microstructural components of the aluminum alloy Al 6061 T 651 and the fatigue cracks has been done within an original fatigue testing program, using specialized specimens. The specimens material used for the experimental investigations – aluminum alloy 6061 T651 was in form of rolled plates with initial crack (Fig. 3).

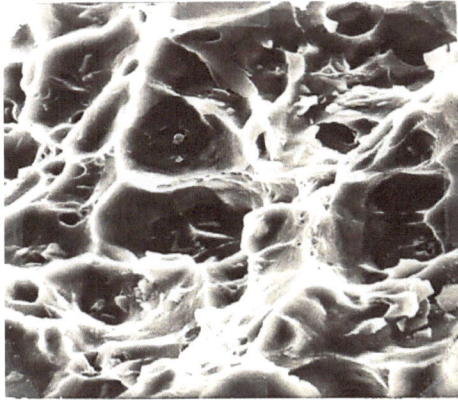

Figure 1: Fatigue crack surface (x3000)

Figure 2: Fatigue crack surface (x2000)

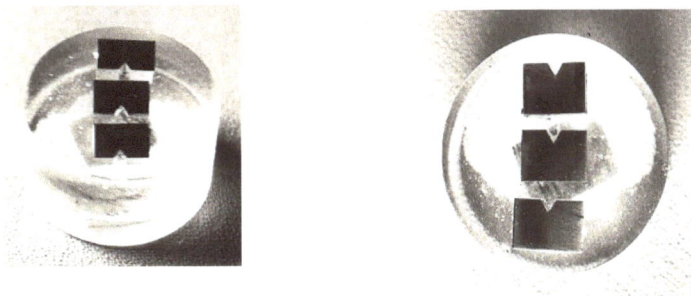

Figure 3: Specimens used within investigations

Figure 4: **Microstructure of Al 6061 T651 (x300)**

Figure 5: **Microstructure of Al 6061 T651 (x300)**

In Fig. 4 and Fig. 5 the structures of the considered aluminum alloy, before subjected to fatigue cycles have been represented. In Fig. 6 different types of fatigue cracks interacting with the microstructural constituents of the investigated aluminum alloy have been presented. For a small number of fatigue cycles the cracks produced may be very fine and short (~0.04-0.07mm) - propagating along a single direction (Fig.6 a, b) or may be short and rough, with a bifurcation at end (Fig.6c). One may also notice that in Fig. 6a the crack stopped in its interaction with an above presented chemical compound. There are also cracks which are fine but long (0.1-0.3 mm) that usually propagate at the level of inter-granular area, following the grains borders and passing through the fine precipitated compounds or rounding the components.

a

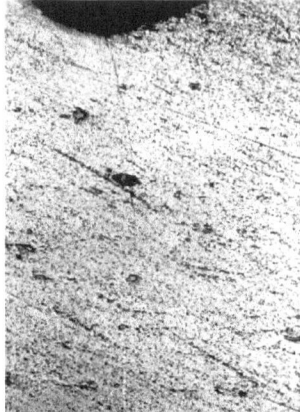

b

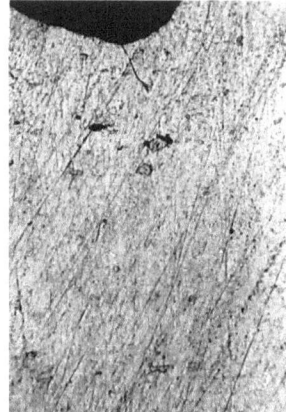

d

e

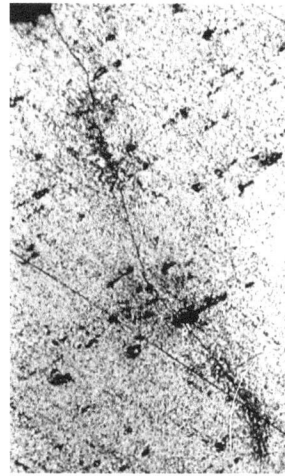

f.

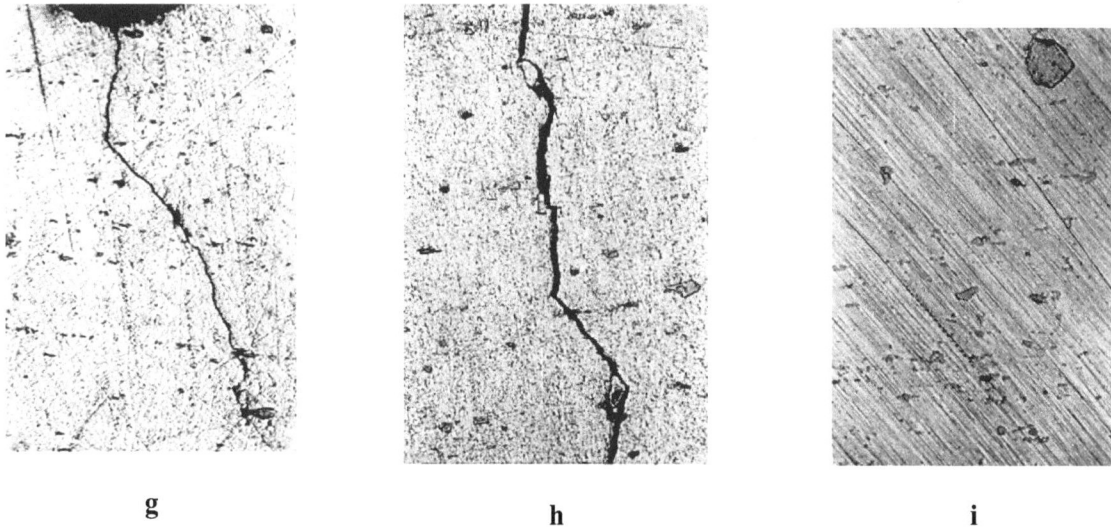

g   h   i

Figure 6: Mechanical interaction between fatigue cracks and microstructural constituents (continued)

In Fig. 6 d, e, f it is to be noticed that the crack propagated along the direction with an increased chemical inhomogeneity. For a large number of fatigue cycles the cracks are long, thick and linearly propagated (Fig.6 g, h, l) or present short bifurcations when meeting a structural constituent. In Fig. 6 h, i, j, k one could observe that the crack does round the chemical compound in form of Chinese letter – identified as being $Fe_3Si_2Al_{12}$ and shown in Fig. 4i. At the same time, different manganese compounds have been observed at the level of the investigated fatigue surfaces that could be the result of a diffusion process involving different one-dimensional faults (interstitial atoms, foreign atoms, second-phase particles etc.) with very important consequences for the fatigue crack propagation and short crack growth (Fig. 6f).

3. CONCLUSIONS

All the above presented aspects concerning the interaction between fatigue cracks and microstructural constituents of the investigated alloy may represent the physical base in creating a mathematical model for fatigue cracks propagation analysis.

REFERENCES

[1]   Broek D.: Elementary engineering fracture mechanics, Martinus Nijhoff Publishers, 1982, London.
[2]   Cioclov D.: Rezistenta si fiabilitate la solicitari variabile, Editura Facla, 1995, Timisoara.
[3]   Bit C.: Elementary strength of materials, Risoprint Publisher, 2005, Cluj-Napoca, Romania.
[4]   Bit C.: Puncte de vedere asupra oboselii mecanice, Editura Universităţii Transilvania, 2001, Brasov, Romania.

EFFECTS OF DIFFERENT REINFORCING PHASE IN CORDIERITE MATRIX COMPOSITES

C. Atănăsoaei[1], F. C. Oliveira[2]

[1]Technical University "Gheorghe Asachi"of Iasi, ROMANIA, catanasoaei@tuiasi.ro
[2]Laboratório Nacional de Energia e Geologia, Lisbon, PORTUGAL, fernando.oliveira@lneg.pt

Abstract: *This paper proposes a comparison between the effects of reinforcing cordierite matrix with three different types of reinforcing materials: ZrO_2, TiC and SiC. All samples were made following the same processing steps. The technological process adopted was intended to be easily repeated on an industrial scale and therefore we tried to use methods and equipments that are easily accessible. The volume fraction of the reinforcing phase was set at 15% for all three types of composites. Shaping the samples was done by uniaxial pressing at 40MPa. Sintering was performed for one hour at 1375°C in air - for ZrO_2 addition, and in vacuum – for the reinforcement with TiC and SiC. Current work focuses on the effects on porosity and elastic properties.*

Keywords: *cordierite, particulate reinforcement,*

1. INTRODUCTION

Cordierite is a ceramic material which due to its properties is found in many technical applications, especially in the manufacture of structural materials, refractories, electrical and thermal insulation etc. However, the use of cordierite is limited by its poor mechanical properties. To overcome this obstacle the best solution found is mixing cordierite with powders made from a diverse variety of materials such as diamond [1], mullite ($3Al2O_3 \cdot 2SiO_2$ and $2Al_2O_3 \cdot SiO_2$) [2], Si_3N_4 [3], ZrO_2 [4], SiC etc. Of the reinforcement materials listed above, probably the most common example is ZrO_2 because it allows the use of two strengthening mechanisms: residual stresses caused by the thermal expansion mismatch and phase transformation mechanism (transformation toughening). However, is very difficult to obtain significant strengthening results by normal processing technology, mainly due to the natural tendency of ZrO_2 to react with SiO_2 to form zircon (Zr_2SO_3). This compound, as has been observed in several studies, allows poor improvement of mechanical performance of the composite [5].

The present study is intended to follow the global trends of developing advanced materials using techniques that are simple an easy to be reproduced at an industrial scale. Thus we propose a comparative study between the effects of reinforcing cordierite with different compounds: ZrO_2, TiC and SiC.

TiC reinforcements are commonly found in a variety of matrices (Al_2O_3, Si_3N_4, SiC, TiB_2, $MoSi_2$ and so on). However, at the date on which this paper was written we did not found any studies focused on using TiC as reinforcement in a cordierite matrix. The situation is similar to that of SiC additions: they are rarely used in cordierite matrices, mainly as fibers reinforcement (whiskers).

The concerning of the present paper are the effects of the additions over the porosity and Young's modulus.

2. SAMPLE'S PREPARATION

The whole process of sample's preparation was conducted at National Laboratory of Energy and Geology - LNEG (Lisbon, Portugal). The raw materials used were: Cordierite - BALCO SPA (Italy), ZrO_2 + 3% mol.Y_2O_3 (UNITEC CERAMICS LIMITED - England), TiC (Alfa Aesar GmbH & Co.Co.KG - Germany) and SiC (Arendal Smelteverk AS - Norway). Powder characterization consisted in measurements of the particle size (laser diffractometer - Cillas 1064) and density (pycnometer AccuPyc 1330). The results are presented in Table 1.

Table 1: Average particle size of the starting powders

Starting powder	cordierite	ZrO_2	TiC	SiC
Average diameter [μm]	7.11	1.21	2.03	3.31
Density [g/cm^3]	2.63	5.40	4.89	3.24

After consulting much of the existing publications concerning cordierite composites, the addition of ZrO_2 was set at 15% vol. This value has been adopted for other types of composites in order to make direct comparisons between the three types of additions,

The step of mixing the starting powders was made by ball milling for an hour at a speed of 1400rpm. Mixing medium consisted of distilled water and ZrO_2 balls (for the addition with ZrO_2) and Al_2O_3 (for the additions with TiC and SiC). Particle size measurement was repeated after mixing raw materials and we obtained average values between 3.03 and 5.2μm.

The shaping of samples was done by uniaxial pressing using a pressure of 40 MPa. In order to maintain a low level of contamination we didn't used lubricants. Samples were uniaxial pressed in two geometries: cylindrical (Φ = 30 mm, h = 3.3 mm) and rectangular (50 x 4.5 x 3.3 mm). In order to have reference samples, besides composite green-bodies we have also pressed a batch of pure cordierite samples. The cordierite and cordierite+ZrO_2 samples were sintered air in an electrically heated furnace at a temperature of 1375°C for one hour. In order to avoid oxidation of TiC and SiC, the cordierite+TiC and cordierite +SiC samples were sintered in vacuum using the same temperature and duration.

Irrespective of the sintering atmosphere used, both types of samples had deviations from the original geometry (bending) that can be attributed to density gradients introduced during uniaxial pressing. For this reason it was necessary to include a mechanical machining step to correct the sample geometry.

In the case of vacuum sintering was found that residual vapors emanated from the furnace residual substances formed precipitates in the surface layers of the samples. We tried to remove the contaminated layer by mechanical machining, but this process could not be completed for all samples because it was quite difficult to determine the boundary between contaminated and uncontaminated layer (both of them had the same shade of color).

3. EVALUATION OF APPARENT POROSITY AND YOUNG'S MODULUS

Given the fact that the samples placed near the oven walls had variations of porosity greater than those placed in the center, for further evaluations were used only samples placed in the center of the sintering chamber.

Porosity evaluation was made using a technique based on Archimedes principle, which is described in detail in ASTM B962 [7]. According to it, density and apparent porosity can be calculated from values obtained by weighing the dry sample in dry, soaked and immersed state. In this study, wetting the samples was done through boiling for one hour in distilled water. Weighing samples was performed using a hydrostatic balance which allows a resolution of 10^{-4} [g]. Each sample was evaluated twice and we couldn't notice significant differences between the obtained values.

Determination of Young's modulus. It was performed using the method of mechanical resonance, using the regulations found in ASTM C 1198 [8]. All samples analyzed were previously mechanically machined in order to reduce geometric deviations. Before testing, the samples were dehumidified by drying for one hour at a temperature of 85°C. The evaluation was repeated at least twice for each sample and we found that the difference between the results is negligible (below 0.1%).

4. RESULTS AND DISCUSSIONS

As shown in figure 3, the conditions adopted for sintering in air allowed reaching of apparent porosities of less than 1%. There were no means of determining the closed porosity, but using SEM analysis (figure 1) one can conclude that the addition of ZrO_2 allows a closed porosities lower than the base material (pure cordierite). The causes are multiple: increased surface contact between particles (ZrO_2 powders are much finer than those of cordierite), the appearance of a small amount of vitreous phase due to presence of Y_2O_3 etc. Consequently, the difference in porosity of the composite material confirms the efficiency and ZrO_2 as sintering additive

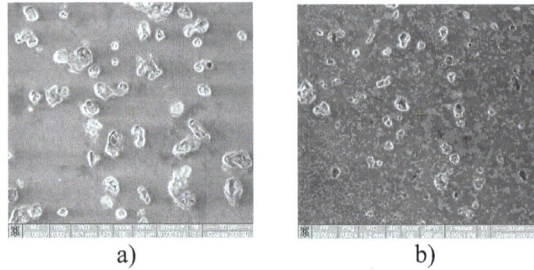

Figure 1: SEM microstructure analysis: a) cordierite b) cordierite+ZrO_2

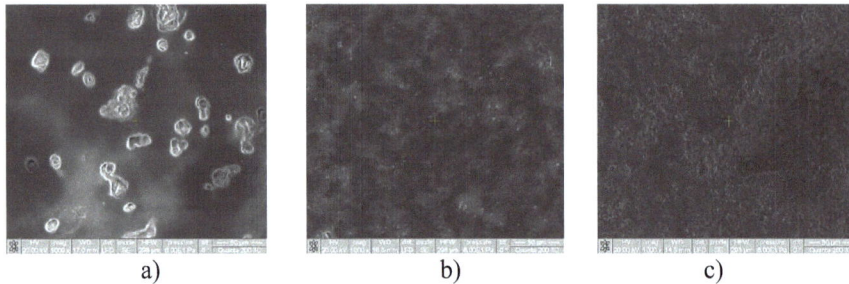

Figure 2: SEM microstructure analysis: a) cordierite (vacuum), b) cordierite + TiC and c) cordierite + SiC

For a given material, sample geometry has a negligible influence on apparent porosity values of samples sintered in air. In contrast, in samples sintered in vacuum was found that bars (rectangular shaped samples) shows porosities higher than those found in the discs (cylindrical shaped samples). This difference is due to the fact that removal of the contaminated layers of the bar samples could not be performed with the same efficiency as in the case of disc samples.

As shown in figure 2, of all the samples sintered in vacuum only cordierite samples could be completely decontaminated. For the TiC and SiC composites was found that contaminating vapors formed precipitates deep in the sample's body (figure 2.b and 2.c), so that the complete removal of contaminated layer was impossible. Infiltration of contaminants throughout the sample volume is probably the main cause for the fact that composites sintered in vacuum shows high values of apparent porosity.

In this paper are discussed only the results obtained from the samples. The reasons behind this choice are:
- in the case of sintering in air, the results do not depend on sample's geometry;
- in the case of sintering in vacuum, the most efficient decontamination was obtained for the disc samples and thus. This means that the results obtained on the disc samples are closest to reality.

It is important to note that regardless of the sintering medium used, cordierite samples shows always the same range of apparent porosities. This suggests that, if we avoid the sample's contamination, composites sintered in vacuum and those sintered in air may have similar porosities.

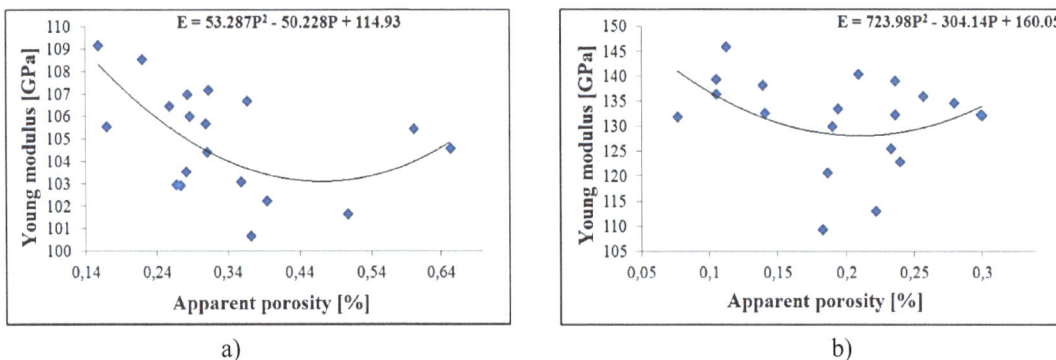

Figure 3: Young's modulus as a function of apparent porosity for air sintered samples:
a) cordierite; b) cordierite + ZrO_2

Young's modulus values obtained for samples made of cordierite (figure 3.a) are within the range of 95-110 GPa and are about 20-25% higher than those that can be found in similar works. This discrepancy can be attributed to the fact that, in general, the values found in the literature were determined on cordierite synthesized from different mixtures of powders which, besides the constituent phases (Al_2O_3, SiO_2 and MgO) presents a certain amount of impurities. In contrast, in this study the raw materials used were of high purity.

The addition of ZrO_2 leads to an increase of Young's modulus by about 25% compared with the base material (figures 3.a and 3.b). The possible reasons are multiple: modulus of ZrO_2 which is about two times higher than that of cordierite, increasing of the surface contact between particles as a result of mixing the base material with a much finer powder (15% vol. ZrO_2), accumulation of residual stresses due to the difference of thermal expansion coefficients of the composite compounds etc. The XRD analysis revealed that most of the ZrO_2 reacted to zircon, so in this case the influence of the transformation toughening mechanism can be neglected.

Unlike pure cordierite samples, those mixed with ZrO_2 are characterized by a higher dispersion values. This is probably because mixing of starting powders wasn't enough efficient in order to avoid powder agglomeration (figure 1.b).

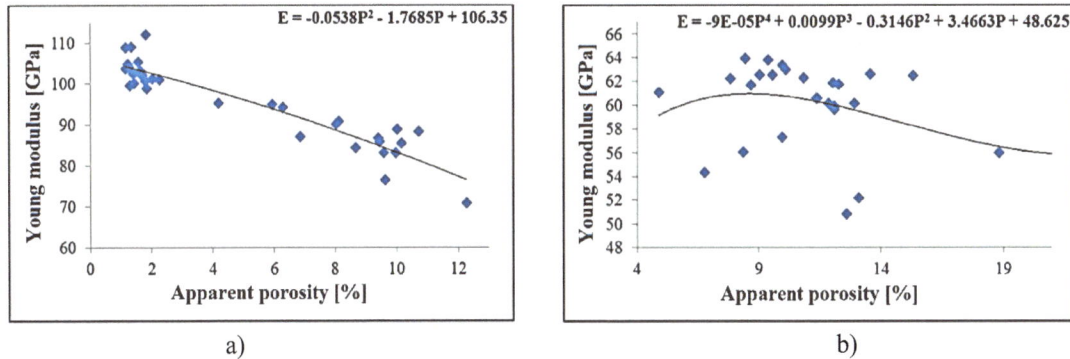

a)   b)

Figure 4: Young's modulus as a function of apparent porosity for vacuum sintered samples:
a) cordierite; b) cordierite + TiC

From the graphs shown in figures 3 and 4 it can be seen that, irrespective of the sintering medium used (air or vacuum) the Young's modulus of cordierite most often falls in about the same range (95-110 MPa).

In the case of cordierite+TiC results shows that, in comparison with pure cordierite samples, the high levels of apparent porosity are reflected in low values of Young's modulus. The value's scatter follows quite closely the apparent porosity values, which means that the porosity is predominantly of open type and the initial surface contact between the particles did not increase significantly after sintering. The most common values are found in the range of 55-65 GPa.

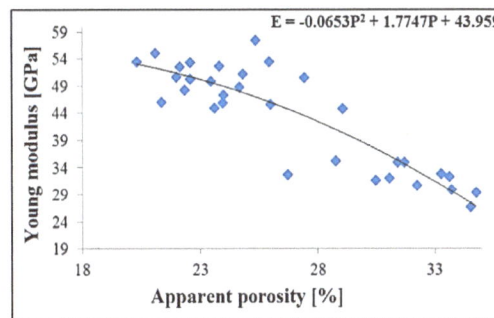

Figure 5: Young's modulus as a function of apparent porosity for air sintered samples.
The case of SiC addition

The high degree of apparent porosity that can be seen in figure 5 shows that the samples of cordierite + SiC have the highest sensitivity to the action of contaminants. In addition, major differences of measured values were observed between disc and bar samples of SiC composites. The main reason for this was the fact that this type of composites had the lowest strength, which made the decontamination machining to be particularly difficult.

It is noted that although the most common values that can be seen in figure 5 are similar to those found in figure 4.b, the difference between the apparent porosity of the two materials is significant. This suggests that in case of some technological improvements that would allow a full densification of the two types of materials, cordierite+SiC samples would have a much greater Young's modulus that that of cordierite+TiC composites.

Irrespective of the sintering medium used it is found that the samples sintered in vacuum shows a clear dependence of Young's modulus with apparent porosity. In contrast, samples that were sintered in air shows values that don't have a clear dependence with the apparent porosity, because in their case the total porosity consist mainly in closed pores, whose volume could not be taken into account.

5. CONCLUSIONS

Methods that were used in this work are nondestructive, which allowed testing of samples several times. The results showed that testing conditions allowed a good reproducibility of the results.

The sintering condition used does not avoid formation of zircon, mainly due to lower heating rates used. However the addition of ZrO_2 allows increased Young's modulus of base material by improving densification.

The addition of TiC and SiC fail to exceed the elastic performance of the base material because the contamination occurred during the sintering step greatly reduced the densification rate. However, there is a huge difference between porosity of the composites sintered in vacuum and base material and a little difference between their Young's modulus. This fact suggests that by reducing the sample's porosity, the Young's modulus of these composites could easily have superior values compared with the base material.

So, in qualitative terms, we can say that all the reinforcements used are able to increase the Young's modulus of the base material. It is found that all the values obtained are in close touch with the sample's apparent porosity. All of this remarks suggests that by improving the sintering parameters while keeping the initial phase composition, would allow the discussed composites to reach a much higher range of Young's modulus values.

In the case of TiC and SiC additions, this optimization could consists in prolonging sintering time and avoid contamination of samples during sintering (e.g. using sintering in inert atmosphere instead of vacuum).

In the care of ZrO_2 additions, much better results would be possible by using higher heating rates (>10K/min) in order to quickly overcome the thermal range in which the reaction rate of zircon formation has maximum values ($\sim$ 1200-1250 $^\circ$ C) [6].

REFERENCES

[1] Hasselman D. P. H., Kimberly Y. D., Liu J., Gauckler L. J., Ownby P. D., Thermal conductivity of a Particulate-Diamond-Reinforced Cordierite Matrix Composite. J. Am. Ceram. Soc. 77 [7] 1757-60, (1994)

[2] Ebadzadeh T., Lee W.E., Processing-microstructure-property relations in mullite-cordierite composites, Vol.18, Issue 7, pp 837-848, (1998)

[3] Zamir S. S., Jafari M., Nourbakhsh a. A., Monshi A., Formation of in situ Si3N4 composite produced by nano and micro silicon particles, Journal of Materials Science, vol-2, No.3, 64-77, (2010)

[4] Wadsworth I., Wang J. and Stevens R., Zirconia toughened cordierite, Journal of Materials Science, **25**, 3982-3989, (1990)

[5] Sun E., Kusunose T., Sekino T., Niihara K. Fabrication and Characterisation of Cordierite/Zircon Composites by Reaction Sintering: Formation Mechanism of Zircon, Journal of American Ceramic Society, Vol. 85, 1430-1434, (2002)

[6] B. C. Lim and H. M. Jang, Homogeneous Fabrication and Densification of Cordierite-Zirconia composites by a Mixed Colloidal Processing Route, Journal of the American Ceramic Society, vol.76, 182-90, (1993)

[7]ASTM B962 - 08 Standard Test Methods for Density of Compacted or Sintered Powder Metallurgy (PM) Products Using Archimedes' Principle, http://www.astm.org/Standards/B962.htm

[8] ASTM C1198 - 09 Standard Test Method for Dynamic Young's Modulus, Shear Modulus, and Poisson's Ratio for Advanced Ceramics by Sonic Resonance, http://www.astm.org/Standards/C1198.htm

DETERMINATION OF THE STRAINS DEVELOPED IN FURNITURE PARTS MADE OF GLASS FIBRES COMPOSITE MATERIALS

C. Cerbu

Transylvania University, Brasov, ROMANIA, cerbu@unitbv.ro

Abstract: *The paper shows the results experimentally obtained concerning to the strains developed in the seat-backrest part of a chair. This component was made of an hybrid composite material reinforced with both glass fabric and wood flour. The resistive tensometry method was used to measure the strains developed in the seat-backrest. Therefore, some strain gage rosettes 0/45/90 and two universal amplifier elements MX840 with eight channels were used for data acquisition. The numerical analysis shows that the greatest values of the stresses develop in the seat part of the component made of composite material. Taking into account these results, the chair was mechanically loaded just on the seat part by using the test weights. Finally, the values of strains experimentally measured were compared with the ones obtained by numerical modeling with the finite element method (FEM).*

Keywords: *composite, glass fibers, tensometry method, strains.*

1. INTRODUCTION

Nowadays, the incorporation of the recycled materials in composite materials is known as a new research direction in the field of manufacturing of the composite materials. For this purpose, there was interest for recycling of the large amount of wood waste [1, 2] obtained during the different stages in the wood processing and wood applications such as furniture industry and building constructions.

It is already known [2, 3] that cellulose fibres are recommended as fillers for plastic composites because these kinds of fibres lead to reducing of the costs concerning materials while mechanical characteristics are better than those that characterized the plastics with no reinforcing. These were the reasons for manufacturing and mechanically testing some hybrid composite materials reinforced both with glass woven fabric and with wood floor. The results of those tests were published in some previous papers [4, 5].

Then, some applications of that kind of composite were found such as the seat-backrest component of a chair (Figure 1, a). Moreover, that kind of chair could be successfully used as garden furniture because the hybrid composite material of the seat-backrest is characterized by good mechanical behaviour and stability under the actions of the aggressive environmental conditions (humidity, thermal cycles).

In a previous paper [6], it had already published some aspects concerning to the advantages of using of the composite materials reinforced with glass fibres to manufacture some components of garden chairs. Another papers [7, 8] also treated aggressively aspects regarding the environmental effects on composite materials reinforced with glass fabric and / or wood floor.

The main objective of this work consists in to experimentally analyse the chair shown in the Figure 1,a from strains state point of view. The seat-backrest component of the chair is made of a hybrid composite based on resin reinforced both with glass fibres and with wood floor.

2. MATERIALS AND WORK METHOD

The first of all, it was made the numerical analysis of the chair involved. The results showed that the greatest values of the stresses develop in the seat part of the component made of composite material. Taking into account these results, the chair was mechanically loaded just on the seat part by using the test weights whose mass is equal to 10 kg (Figure 1, a). The test weights have disk shape and the contact surface with the chair seat has circular shape whose diameter is equal to 234 mm. To improve the contact between the test weights and the surface of the chair seat, it was used a rubber piece having also a circular shape whose diameter is equal to 234 mm while its thickness is equal to 6 mm.

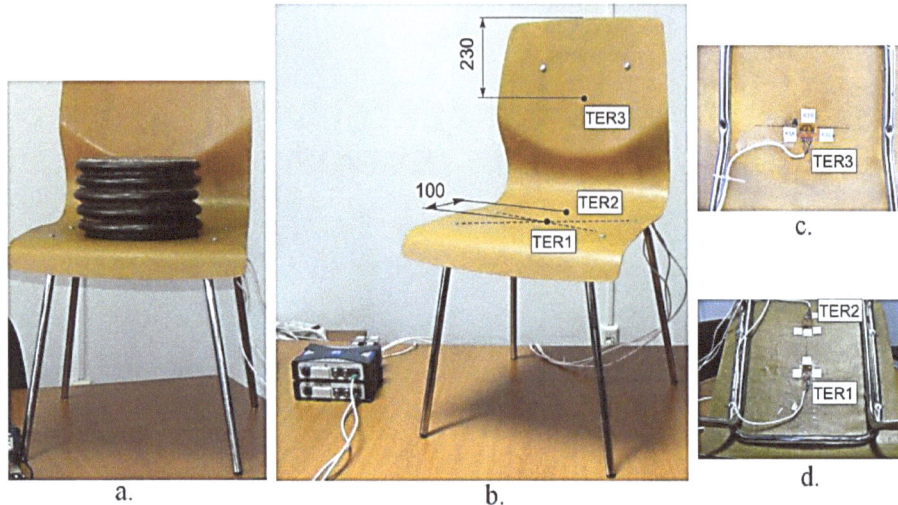

Figure 1: Mounting of the strain gauges of rosette type to measure strains developed in chair seat:
    a. The tested chair mechanically loaded;
    b. Positioning of the strain gauges on the seat-backrest part; b. TER 3 strain gauge glued on the back site of the backrest;
    c. TER1 and TER2 glued on the bottom of the chair seat.

2.1. MATERIALS

To measure the normal strains ε that develop in the seat component of the chair when this is mechanically loaded, it was used the followings:

- bonded strain gage rosettes 0/45/90, 1-RY88-6 / 350Ω (Figure 1, b si c) made by HBM having the following characteristics: electric resistance 350.0Ω ± 0.3% (at 24°C); the constants of the strain-gauge for each electric resistance $k_a = 2,3 \pm 1\%$, $k_b = 2.15 \pm 1\%$, $k_c = 2.13 \pm 1\%$ (la 24°C); transverse sensitivity $(a : 0\%, b : -0.3\%, c : 0\%)$; the length of measuring a = 6 mm;
- adapter units for working with strain-gauge ¼ or ½ bridge strain gauges with electric resistance120/350 Ω;
- two universal amplifier elements MX840 (Figure 1, a), each of these contains 8 channels for data acquisition having the following characteristics: the frequency of data acquisition up to 19.2 kHz for each channel; converter A/D of 24 bits of each channel to synchronize the parallel measuring; contains filter Bessel, Butterworth 0.01Hz - 3.2 kHz (-3dB); electric supply 10...30V; provides electric supply 5-24 V for transducers; allows the identification of the

  connected transducers TEDS; sleeve joints of type D-SUB-15HD; the range of the work temperatures $-20...+65°C$; accuracy class 0.05; compatible with transducers with strain gauges with full and half bridge deck, inductive transducers in full bridge and half bridge LVDT, voltage / current, resistance thermometers PT100 and PT1000, thermocouples, potentiometers, counter of impulses, CANbus; adapter for use with ¼ bridge strain gauges; soft Quantum X; sleeve joints for 8 gauges.
- *Easy Catman* software package for processing data: parameterization of the amplifier with automatic recognition of the type of transducer; possibility of creating virtual channels to achieve real-time the mathematical calculations for strain gauges of rosette type; setting of limits and monitoring functions; visual analysis of the data by synchronization and real-time overlay; data export in common formats of type: Excel, ASCII, DIAdem, nSoft, compatible with MX840 systems, MGCplus, MGCsplit or Spider8.

2.2. WORK METHOD

To experimentally measure of the strains developed in seat part by using the tensometry method, the following steps were covered:

- choice of the type of strain gauge - 0/90/0 rosette type, 1-RY88-6 / 350Ω;
- establishment of the piece areas where the electrical resistive strain gauges (TER) will be mounted so these places are easily accessible and free of defects or cracks;

- preparation of the surfaces where the strain gauges will be applied, step that consists in mechanical cleaning in order to bring to an adequate roughness, marking of the seating position for strain gauges, chemical cleaning of the area in order to obtain as clean surfaces and neutralize of the surface by using a solution;
- application of strain gauges on the work piece surface by bonding with special glue (Figure 1, b-d);
- wiring of the strain gauges (Figure 1, b-d);
- connection of the strain gauges to the jacks of the two amplifiers of type MX840 with 8 channels via adapters to work with strain gauges mounted in quarter (¼) bridge circuit and half (½) bridge circuit with resistance of 120/350;
- connection of the amplifiers of type MX840 to the computer;
- using of the Catman Easy software compatible with MX840 acquisition system for measuring of the deformations, step that begins with: setting of the channels (Figure 1); their initialization to 0, setting of the constants for each strain gauge and for their corresponding resistances ($k_a = 2.13 \pm 1\%$, $k_b = 2.15 \pm 1\%$, $k_c = 2.13 \pm 1\%$); setting of the transversal sensitivities $(a : 0\%, b : -0.3\%, c : 0\%)$; creating of the real-time graphics for measured quantities and for calculated quantities;
- making of the actual measurements (a sufficient number of measurements required for subsequent statistical processing of the signals) to measure specific linear deformations ε_a, ε_b, ε_c recorded by each strain gauge rosettes.

Measurements were performed for the following values of the loading force: 98.10 N, 196.2 N, 294.30 N.

Acquisition of signals from the electrical resistances of the strain gauges, was performed through acquisition board MX840 having amplification role of the signals from the channels were the resistances were connected and then, transmitting to the computer. Then, these signals can be processed either with the QuantumX software supplied with the amplifier MX840 or with Catman Easy software specialized on tensometry.

Finally, the results experimentally measured by using resistive strain gauge method were compared with the results obtained by numerical modelling.

3. RESULTS

In the Figure 2, it is shown the results of the measurements concerning to the specific normal strains ε recorded by the two strain gauges glued on the underside of the chair seat (three directions A, B, C for each strain gauge). These results were recorded during loading with three calibrated mass units: 10 kg (98.10 N); 20 kg (196.2 N); 30 kg (294.30 N).

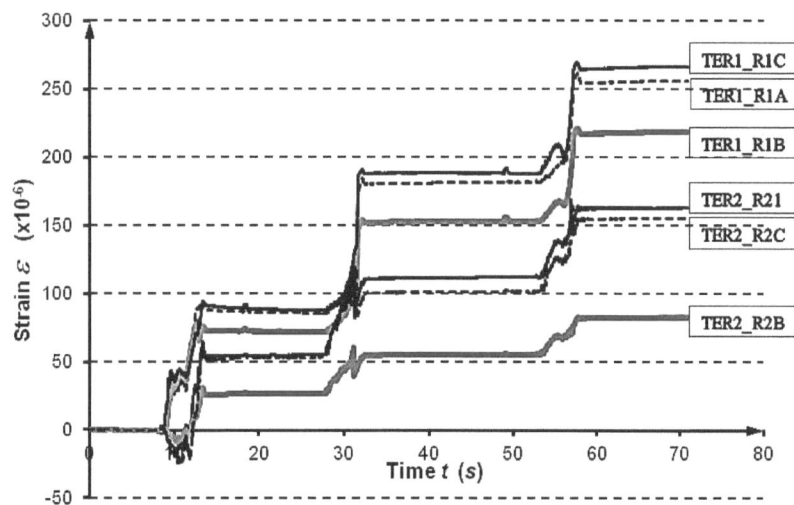

Figure 2: Time variation of the normal strains ε recorded by the both TER1 and TER2 strain gauges glued on the bottom of the chair seat

Finally, to validate the numerical model, the theoretical results were compared with the experimental ones for the cases corresponding to the loading scheme showed in the Figure 4 when the following forces were applied: 98.10 N, 196.2 N, 294.30 N.

For this purpose, Figure 3 shows the finite element model of the chair while Figure 4 presents the loading scheme used in the numerical model.

The values of the strains ε experimentally measured were compared with the ones obtained by numerical modelling for 117505 and 116967 elements (Figure 5, a) on the measurement directions corresponding to the strain gauge rosettes. In this sense, two coordinate systems were created (Figure 5, b) and the strains were displayed with respect of them: CSYS-1 coordinate system (x-axis coincides with the direction of the resistances R1B and R2B); CSYS-2 coordinate system rotated by 45° compared to the first.

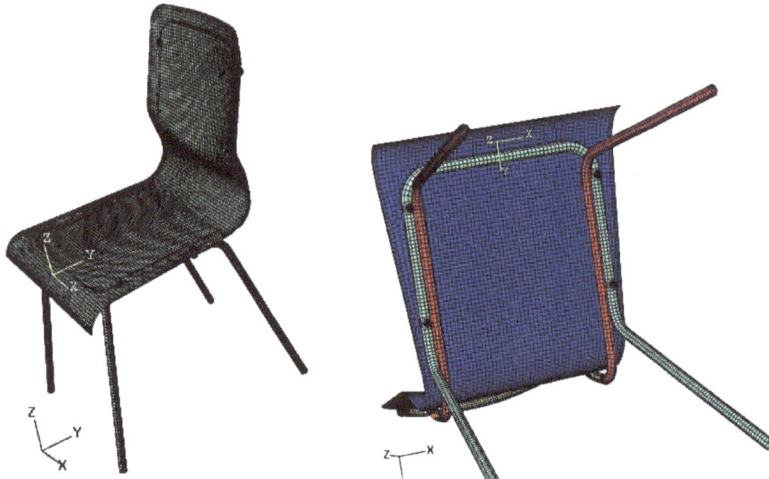

Figure 3: Finite element model of the chair

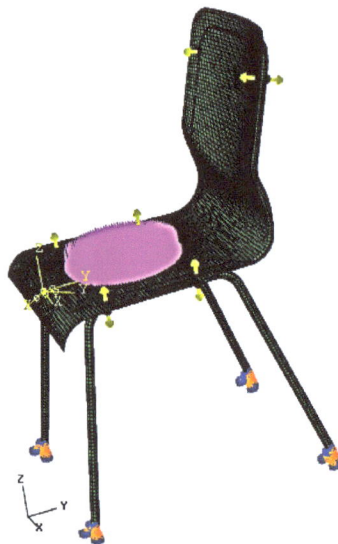

Figure 4: Loading scheme for the numerical model of the chair

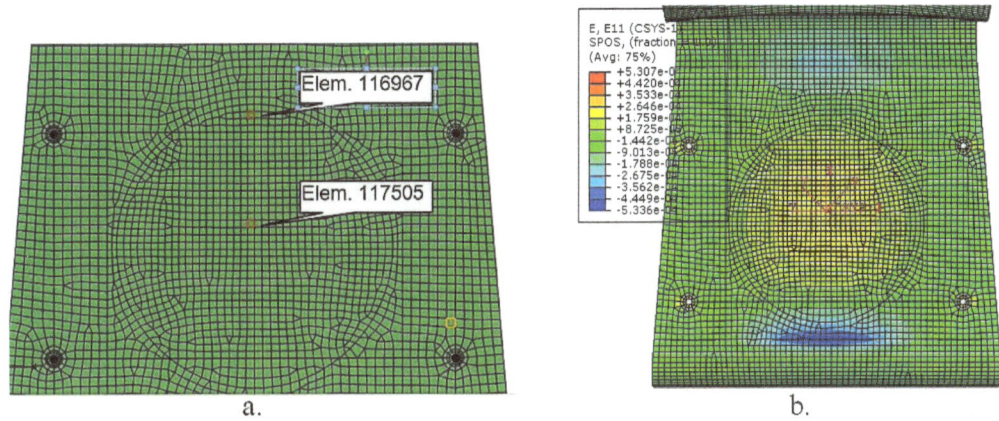

a.   b.

Figure 5: Establishing the model elements and coordinate systems used to obtain theoretical results
a.   Considered elements in the numerical model; b. CSYS-1 and CSYS-2 coordinate systems used for displaying the results in FEM analysis

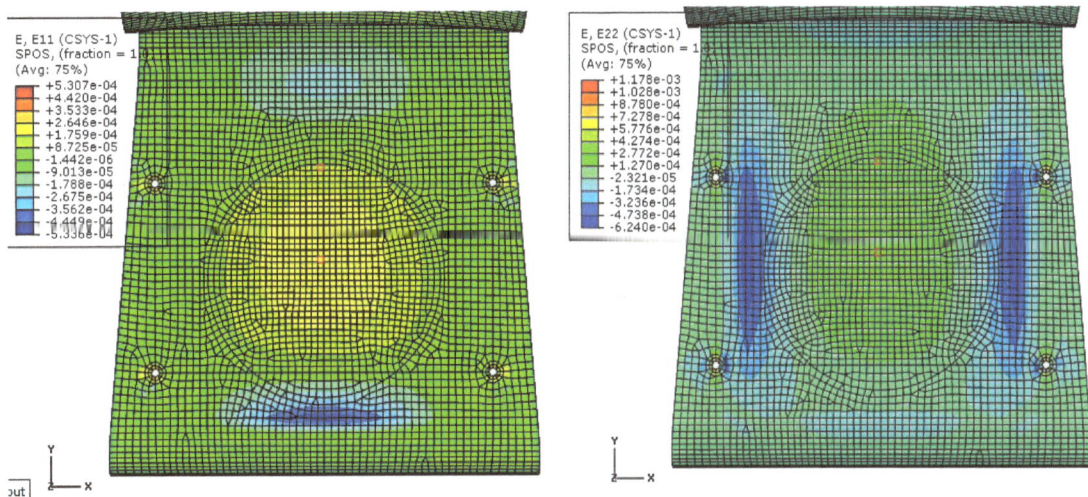

Figure 6: State of the normal strains ε relative to CSYS-1 coordinate system whose 1 axis conincides with Y axis (F=294,3 N)

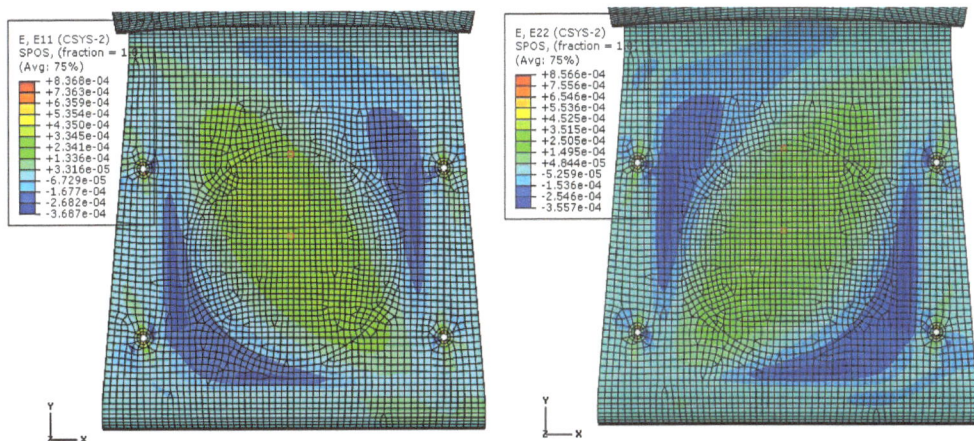

Figure 7: State of the normal strains ε relative to CSYS-2 coordinate system whose 1 axis is rotated with 45° relative to the Ox axis (F=294,3 N)

The first of all, it is shown the results obtained by finite element analysis for the case when $p = 0.004562 \text{N}/\text{mm}^2$ (corresponding to the force F = 294.3 N) was the pressure applied on the chair seat in the numerical model.

121

In this context, it is presented the distribution of specific strains ε developed on the chair seat in relation to the coordinate system CSYS-1 (Figure 6) and relative to the coordinate system CSYS-2 (Figure 7).

Tabel 1: Comparison between the theoretical results and the experimental ones

Force (N)		98.1			196.2			294.3		
Pressure p (N/mm^2)		0.00228112			0.004562			0.006843		
		ε (x 10^{-6})		Err [%]	ε (x 10^{-6})		Err [%]	ε (x 10^{-6})		Err [%]
		Theoretic	Exp.		Theoretic	Exp.		Theoretic	Exp.	
Strain gauge TER 1	R1A	97.19	86.41	11.10	194.81	181.20	6.99	292.43	255.18	12.74
	R1B	81.04	72.25	10.85	164.16	152.82	6.91	247.28	217.98	11.85
	R1C	97.19	88.56	8.87	194.81	188.43	3.28	292.43	265.50	9.21
Strain gauge TER 2	R2A	58.02	54.52	6.03	117.30	111.78	4.70	176.58	162.67	7.88
	R2B	29.01	26.52	8.58	60.35	55.22	8.50	91.68	82.34	10.18
	R2C	58.02	53.08	8.52	117.30	101.23	13.70	176.58	154.87	12.29

Both the results obtained by numerical modelling and the ones experimentally measured by using electrical resistive strain gauges, were systematized in Table 1. Finally, the errors expressed as percentage were calculated for each case. It is noted that the error values are greater for the strain gauge rosette denoted with TER2.

4. CONCLUSIONS AND DISCUSSIONS

Analysing the results shown in Table 1, it may observe that the values of the strains experimentally measured do not significantly deviate from the theoretical values obtained by analysis with the finite element method.

Measurement errors that occurred during the experimental investigation could be due to: misalignment of the strain gauge to the direction of the load application; transverse sensitivity of composite materials; improper bonding of the strain gauge rosettes, so an angular variation between 0° to 4° could lead to the increasing of the error value up to 65%.

Comparison between the theoretical results and the experimental ones (Table 1) regarding the state of strains developed in the composite part analysed (seat-backrest component of the chair), confirms the correctness of the numerical model with finite elements.

ACKNOWLEDGEMENT

This work is supported by CNCSIS –UEFISCSU, project number PNII – IDEI 733.

REFERENCES

36. Kamdem D. P., Jiang H., Cui W., Freed J., Matuana L. M., Properties of wood plastic composites made of recycled HDPE and wood flour from CCA-treated wood removed from service, Composites: Part A, vol. 35, 2004, pp. 347-355.
37. Klyosov A.A., Wood-plastic composites, Wiley Publishing, 2007, pp. 75-122.
38. Stancato A.C., Burke A.K., Beraldo A.L., Mechanism of a vegetable waste composite with polymer-modified cement (VWCPMC), Cement & Concrete Composites, ISSN 0958-9465, vol. 27, 2005, pp. 599.
39. Cerbu C., Curtu I., Ciofoaia V., Rosca I. C., Hanganu L. C., Effects of the wood species on the mechanical characteristics in case of some E-glass fibres/wood flour/polyester composite materials, in Rev. Materiale Plastice, MPLAAM 47 (1) 2010, Vol. 47. nr. 1 – martie 2010, Bucuresti Romania, ISSN 0025/5289 – ISI (CNCSIS A), pp.109-114.

40. Cerbu C., Luca-Motoc D., Solutions for Improving of the Mechanical Behaviour of the Composite Materials Filled with Wood Flour, Proceedings of The World Congress on Engineering 2010, Vol II, ISBN 978-988-18210-7-2, WCE 2010, June 30 - July 2, 2010, London, U.K; Publisher: Newswood Limited; Organization: International Association of Engineers; pp. 1097-1100.

41. Cerbu Camelia, Itu Călin, Curtu Ioan, The problem of the using of the composite materials reinforced with glass fibres to manufacturing of some components of the garden chairs, Revista ProLigno, vol. 6, Nr. 3, septembrie 2010, ISSN 1841-4737, pp. 51-60.

42. Cerbu, C., Materialele compozite şi mediul agresiv. Aplicaţii speciale, Editura Universitătii Transilvania, 2006, Brasov.

43. AdhikarY K. B., Pang S., Staiger M. P., Long-term moisture absorption and thickness swelling behaviour of recycled thermoplastics reinforced with Pinus radiata sawdust, Chemical Engineering Journal, doi: 10.1016 /j.cej.2007.11.024.

ABOUT THE MACROMECHANICAL CHARACTERISTICS OF COMPOSITE MATERIALS BY DYNAMIC IMPACT TESTS

V. Ciofoaia[1], C. Cerbu[1]

[1] Transivania University, Brasov, ROMANIA, ciofoaiav@unitbv.ro

Abstract: This paper focuses on the determination of the characteristics of composite materials reinforced with woven fabrics by using dynamic impact tests. The specimens used in our study were manufactured by reinforcing an epoxy resin with woven fabric EWR300 made of E-glass fibers. The hand lay-up technology is used to prepare the specimens with different pressures (low and high pressure) in the molding step. The composite specimens were subjected to the dynamic impact tests and the mechanical characteristics of the specimens were analyzed taking into account the different manufacturing methods used. The results obtained were compared taking into account the two kinds of reinforcements used.

Keywords: composite, manufacturing, mechanical tests, mechanical properties.

1. INTRODUCTION

Composite materials reinforced with woven fabrics combine the strength and stiffness of reinforcing fibers with the load transferring and protective properties of a polymer matrix. With a combination of low weight and excellent mechanical performance, the fiber reinforced composites have found wide use in highly demanding structural applications. Currently composite materials are used increasingly in various industry branches such as aerospace, marine, defense, road transportation, construction, energy, domestic, consumer electronics, agricultural and automotive, railroad et., mostly due to the composite materials' behavior and performance (rigidity, resistance, thermal and phonic isolation etc). Textile / woven composite materials have recently received considerable attention due to their structural advantages of high specific-strength and high specific-stiffness as well as improved resistance to impact, crash and fatigue [6,8]. Three main methods are used currently to predict the mechanical properties of woven fabrics: analytical, numerical and experimental models [2,3,4,5,6,7,9,10,11,12]. The behavior of composite material reinforced with E-glass when impacted by solid objects is the subject of significant numerical, analytical and experimental research [1,6, 9]. To assess the behavior of composite plates to concentrated loads, two types of specimens reinforced with glass fiber fabric were studied, with concentrated loads application dynamically. Specimens were obtained by two different methods, casting by using a lower pressure and a higher pressure. The present paper presents the mechanical properties of composite materials as determined when they are subjected to impact with the Charpy hammer. The purpose of this study is to develop a new fabrication procedure for manufacturing composite materials reinforced with woven.

2. EXPERIMENTAL PART

2.1 Materials

Some layers of the laminated composites are made of woven fabrics *EWR*300 / polyester *Copoly* 7233, while the others are reinforced with chopped *E*-glass fibres. A lower pressure was used to manufacture one of the specimens by using a hand lay-up technology, while higher pressure was used for the other one by using automated technology. After manufacturing the required number of specimens, a variety of tests were carried out to investigate the behavior of specimens when subjected to impact in various positions and under compression.

By using a digital microscope, sample sections photos were taken in order to analyze the structure of the composite material after loading with the Charpy hammer and to compare the two pressure manufacturing processes. Photos of the composite materials type Glass E / polyester Colpoly 7233 manufactured through

- low pressure molding (Figure1);
- high pressure molding (Figure2).

 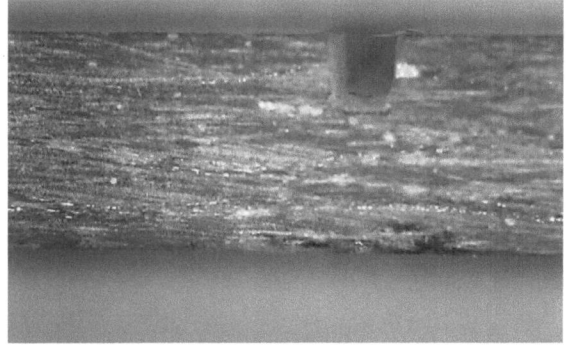

a.   b.

Figure 1. Photos of the structure of the materials in section, for the samples prepared for Charpy shock loading; composite material Glass E / polyester Colpoly 7233 (low pressure molding):

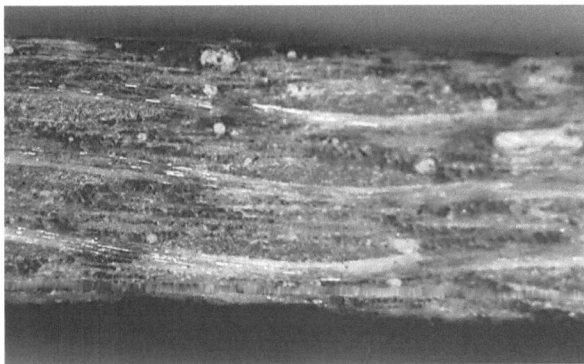 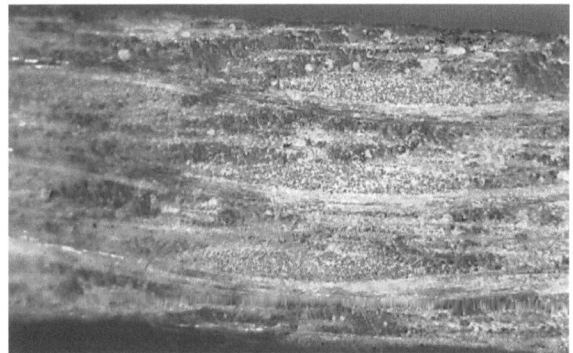

a.   b.

Figure 2. Photos of the structure of the materials in section, for the samples prepared for Charpy shock loading; composite material Glass E / polyester Colpoly 7233 (high pressure molding):

By analyzing the figures 1 and 2, a few remarks can be made:
– high pressure molding manufacturing (Figure2), leads to reinforcement layers with glass weaving to be better consolidated, more pronounced and countured on the digital microscopy photos;
– low pressure molding manufacturing (Figure1) leads to glassfibers to be harder to detect, even with a zoom factor of 200 x (Figure1, e și f).

2.2. Manufacturing technology effects by pressing on mechanical behaviour in an attempt to shock with the Charpy pendulum

The samples used for the Charpy test have rectangular shape, size 80mm x 10mm x 6mm (grosime) according to the European normatives EN ISO 179-1 (2001) for plastic reinforcement materials. The section size was recorded for eacg sample before the impact test. ; afterwards the samples were subjected to the Charpy loading. The impact is produced by raising the hammer at the height h.When the hammer is released, it follows a circular path, hitting the target sample; after impact the hammer reaches height *h'. The difference between the initial and after impact potential energy represents a measure of the energy necessary to break the sample. This energy is names the break energy and is noted U.* The results obtained in the attempt to break the Charpy pendulum were systematized in Table 1, for the two types of samples that differ in table 1 (manufactured under low, respectively high pressure).

Table 1. Results: breaking Energy U values for the sample; resilience U/A as determined through the Charpy hammer breaking challenge

No.	Composite material / Molding pressure type used for manufacturing	Cod sample	Sample size		Width left at the notch	Transversal section area at the notch	Breaking energy	Resilience
			b (mm)	h (mm)	h_n (mm)	A (mm^2)	U (J)	U/A (kJ/m^2)
1.	Glass E / polyester Colpoly 7233 / Low pressure molding	P211	9.70	4.30	3.30	32.01	2.90	90.60
		P212	9.80	4.60	3.60	35.28	3.91	110.83
		P213	9.50	4.50	3.50	33.25	3.25	97.74
		P214	10.40	4.40	3.40	35.36	3.42	96.72
		P215	10.10	4.30	3.30	33.33	3.34	100.21
		P216	9.80	4.50	3.50	34.30	3.44	100.29
		P217	9.80	4.50	3.50	34.30	3.28	95.63
		P218	10.70	4.00	3.00	32.10	3.12	97.20
		P219	10.60	4.40	3.40	36.04	3.56	98.78
		P220	9.60	3.90	2.90	27.84	2.98	107.04
							Average value	99.50
2.	Glass E / polyester Colpoly 7233 / High pressure molding	N46	11.80	3.90	2.90	34.22	4.13	120.69
		N47	10.90	4.00	3.00	32.70	4.06	124.16
		N48	11.50	4.00	3.00	34.50	4.20	121.74
		N49	10.90	3.80	2.80	30.52	3.75	122.87
		N50	11.20	3.90	2.90	32.48	3.79	116.69
		N51	11.10	3.50	2.50	27.75	3.47	125.05
		N52	10.00	3.90	2.90	29.00	3.63	125.17
		N53	10.40	3.70	2.70	28.08	3.45	122.86
		N54	10.40	3.90	2.90	30.16	3.78	125.33
		N55	11.20	3.60	2.60	29.12	3.52	120.88
							Average value	122.54

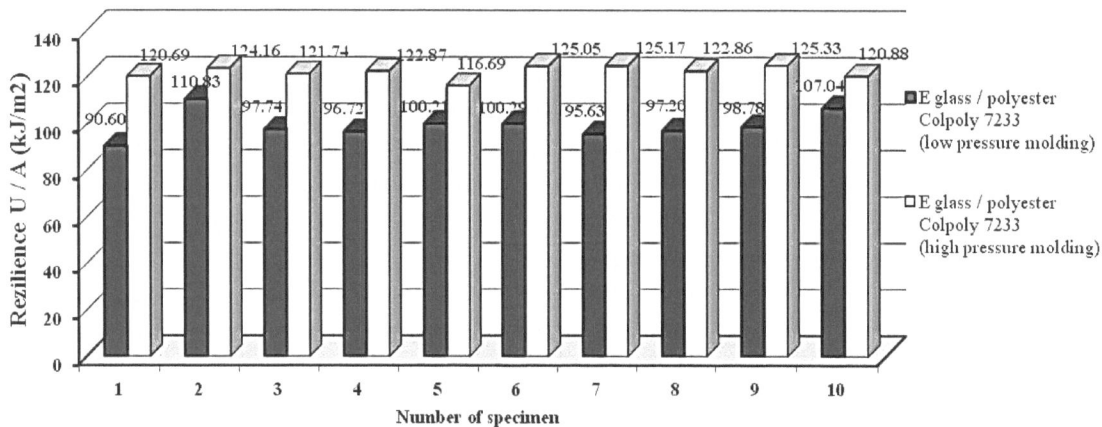

Figure 3. – Effect of the manufacturing pressure type on the resilience U / A (resistance to impact) for Glass E / polyester Colpoly 7233

3. CONCLUSION

Figure 3 presents graphically the comparative results in the last column of table 1. This figure presents the K resilience, the ratio of the rupture energy U and the transversal sectional aria A [at the notch]. This ration is higher for composite materials manufactured by using high pressure. In this case, the medial value of the K impact resistance is 122,54 kJ/m^2 (Table 1), 23,16% higher than the median value 99,50 kJ/m^2 (Table 1), value that was recorded in the case of composite materials samples manufactured under low pressure.

4. REFERENCES

[1]. Abrate,S.: Impact on composite structures, Cambrige University Press, 2001.

[2]. Cerbu A., Ciofoaia, V., Curtu I., Visan A.,The Effects of the Immersion Time on the Mechanical Behaviour   in Case of he Composite Materials Reinforced with E-glass Woven Fabrics. Materiale Plastice, 46, nr.2,   p.201-205, 2009.

[3]. Cerbu, C., Ciofoaia, V., Teodorescu-Draghicescu, H., Rosca, I.C., Water effects on the composites made of   E-gass woven fabrics. Proceeding of International Conference Advanced Composite Materials Engineering,   COMAT 2008, Brasov, 0-11 octombrie 2008, p.310-313.

[4]. Cerbu Camelia, V. Ciofoaia, Curtu I. – The effects of the manufacturing on the mechanical characteristics   of the E-glass / epoxy composites, Proceedings of The 12th International Research / Expert Conference   "Trends in the development of machinery and associated technology"- TMT2008, Istanbul (Turkey), 26-30   august, 2008, p.229-232.

[5]. Cerbu Camelia, Ciofoaia, V., Curtu I., Visan, A. The Effects of the Immersion Time on the Mechanical Behaviour in Case of the Composite Materials Reinforced with E-glass Woven Fabrics. Materiale Plastice ISSN 0025 / 5289, vol. 46, nr.2, iunie 2009, pag.201-205.

[6]. Ciofoaia, V., Modelarea şi simularea comportării la factori mecanici şi de mediu agresiuv a materialelor compozite întărite cu textile. Project ID_191, no. UEFISCU 225 / 2007

[7]. Ciofoaia, V., Cerbu, Camelia Dogaru, Fl., On the determination of the mechanical elastic constants of textile composites.COMAT 2010.

[8]. Choo V.K.S Fundamntals of composite materials.   Knowen Academic Press, Inc., Dover, Delaware USA, 1990.

[9]. Gay D., Hoa S.V., Tsai S.W., Composite Materials. Design and applications. CRC Press Washington D.C. 2003.

[10].Dogaru, Fl. Baba, M.N. – Analytical study of the CFRP laminated plates subjected to low velocity impact, Proceedings of International Conference Advanced Composite Materials Engineering „COMAT 2008", Braşov, 9-11 octombrie, 2008, p. 126-129

[11].Morozov E. V. Mechanics and analysis of fabric composites and structures AUTEX Research Journal, Vol.   4, No2, June 2004

[12].Sherburn M. Geometric and Mechanical Modelling of Textiles. The University of Nottingham. 2007.

[13].Vasiliev V.V. and Morozov E.V.. Mechanics and Analysis of Composite Materials. Elsevier Science, (2001).

MATHEMATICAL MODELS OF FORCE AND MOMENT IN MACHINING PRODUCTS MADE BY SANDWICH COMPOSITES POLYMERIC MATERIALS

C. Opran[1], M.E. Lupeanu[1], C. Bivolaru[1]

[1] POLITEHNICA University of Bucharest, Bucharest, ROMANIA,
constantin.opran@ltpc.pub.ro; mihaela.lupeanu@ltpc.pub.ro; catalina.bivolaru@ltpc.pub.ro

Abstract: *In industrial application of products made by sandwich composites polymeric materials it is important to determine mathematical models of force and torque, when machining this. Thus, it would be possible to know optimum process parameters values or to predict force and torque values, once process parameters values set. Machining, more specifically milling process, for products made by materials sandwich composites polymeric, the existing data published on this topic are very poor and, rather, quantitative ones. So, a study on products made by materials sandwich composites polymeric has been done and the results presented by this paper, mainly considering the wide industrial application of this material in Romanian industry.*

Keywords: *mathematical model, regressions, milling process, force, moment.*

1. INTRODUCTION

In the 50's the development was mainly concentrate on honeycomb materials. Honeycomb was mainly used as core material in the aircraft industry. However, it had some limitations, for example there big problems with corrosion, see figure 1. At the end of 50's and during the 60's and until this days different cellular plastics where produced, suitable as core materials [1, 2, 3].

In the beginning rather soft materials were used because of their insulation properties, for example polystyrene and polyurethane, see figure 2. Later it was possible to produced harder cellular plastics with higher densities and by that time sandwich became a very useful and flexible concept, see figure 3.

In this paper the sandwich composite polymeric products are made of fiber glass and the core is polystyrene extruded, see figure 3.

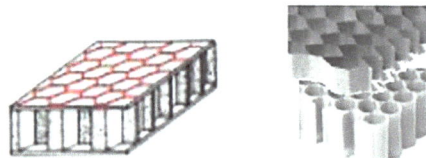

Figure 1: Example of honeycomb core

Figure 2: Example of polystyrene or polyurethane core

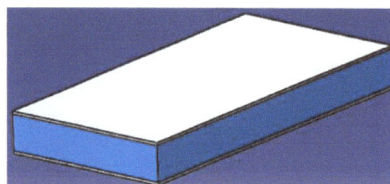

Figure 3: Example of polystyrene extruded core

The combination of two composite faces and a lightweight core allows obtaining a high flexural stiffness with a weak mass. The faces carry most of the tensile and compressive stress due to axial loading and bending whereas the core carries most of the shear stress. A priori, the weaving of the sandwich structures should constitute a non negligible value-added for the development of these structures. Every part has its specific function to make it work as a unit.

The aim is to use the material with a maximum efficiency. Two faces are placed at a distance from each other to increase the moment of inertia, and thereby the flexural rigidity, about the neutral axis of the structure [4,5,6].

2. RESEARCH APPLIED METHODS

Determining mathematical models for the relationship of parameters specific to a certain machining process, based on experimental results, involves some steps to be followed, such as: -the "definition" of both independent and dependent variables associated to this models [1]; -an appropriate experiments design type to be considered; -the regression analysis, if available; -the fitted mathematical model to be obtained.

The mathematical models for axis cutting force and torque for milling process are mentioned by most of the articles and books dealing with this problem, and are represented in relations (1) to (6).

> in milling process

$$F_x = f(v_c, v_f, a_e, a_p), \quad [N] \tag{1}$$

$$F_x = C_{Fx} \cdot v_c^{a1} \cdot v_f^{a2} \cdot a_e^{a3} \cdot a_p^{a4}, \quad [N] \tag{2}$$

$$F_y = f(v_c, v_f, a_e, a_p), \quad [N] \tag{3}$$

$$F_y = C_{Fy} \cdot v_c^{a1} \cdot v_f^{a2} \cdot a_e^{a3} \cdot a_p^{a4}, \quad [N] \tag{4}$$

$$M_z = f(v_c, v_f, a_e, a_p), \quad [Nm] \tag{5}$$

$$M_z - C_{Mz} \cdot v_c^{a1} \, v_f^{a2} \, a_e^{a3} \, a_p^{a4}, \quad [Nm] \tag{6}$$

where:

F_x, F_y – represent the axial cutting force (dependent variable);
M – the torque (dependent variable);
v_c – the cutting spead of the process [m/min] (independent variable);
v_f – the cutting feed of the process [mm/min];
a_e – axial depth of cutting process [mm] (independent variable);
a_p – redial depth of cutting process [mm] (independent variable);
a_1, a_2, a_3, a_4 – polytropic exponents;
C_F, C_M, - constants.

In fact, the aim of this paper, meaning, presenting the steps followed and, specially, the new mathematical models obtained for the axial force and torque in milling process.

For obtaining the constants and polytropic exponents' values, relations (2), (4), and (6) must be of linear type and, so, by logarithm they will "turn" into relations (7) to (9), as follows:

> in milling:

$$\lg F_x = \lg C_{Fx} \cdot \lg v_c^{a1} \cdot \lg v_f^{a2} \cdot \lg a_e^{a3} \cdot \lg a_p^{a4} \ [N] \tag{7}$$

$$\lg F_y = \lg C_{Fy} \cdot \lg v_c^{a1} \cdot \lg v_f^{a2} \cdot \lg a_e^{a3} \cdot \lg a_p^{a4} , [N] \tag{8}$$

$$\lg M_z = \lg C_{Mz} \cdot \lg v_c^{a1} \cdot \lg v_f^{a2} \cdot \lg a_e^{a3} \cdot \lg a_p^{a4} \ [Nm] \tag{9}$$

So, it can be mentioned that the first method, REGS, applied was that of solving a four / five linear equations system – as there were four / five constants to be determined (C, a_1, a_2, a_3 and a_4), five constants for the milling process. The second method applied dealt with experiments design and regression analysis – done with a special software, DOE KISS. [5] Due to limited license rights (as the authors have only the "student version"), there could only be determined regression models with three independent variables, each of them with two "levels'.

Table 1: The experiments design (Central Composite Design, CCD)

Central Composite Design (CCD)	Experiments Design		
	Run	A	B	C
	1	-1	-1	-1
	2	-1	-1	+1
	3	-1	+1	-1
	4	-1	+1	+1
	5	+1	-1	-1
	6	+1	-1	+1
	7	+1	+1	-1
	8	+1	+1	+1

So, based on the results obtained by solving relations (7) to (9), for the regression analysis there have been considered the three most "significant" variables – for axial force and torque values. The three most important variables are: v_c, v_f and a_e [9, 10]. The experiments design (Central Composite Design, CCD) is evidenced in table 1.

Regression analysis performed by the software resulted in models like the ones mentioned by relations (10) to (12):

➢ in milling:

$$Fx = a_0 + a_1 \cdot v_c + a_2 \cdot v_f + a_3 \cdot a_e + a_{12} \cdot v_c \cdot v_f + a_{13} \cdot v_c \cdot a_e + a_{23} \cdot a_e \cdot v_f + a_{123} \cdot v_c \cdot v_f \cdot a_e \text{ ,[N]} \quad (10)$$

$$Fy = a_0 + a_1 \cdot v_c + a_2 \cdot v_f + a_3 \cdot a_e + a_{12} \cdot v_c \cdot v_f + a_{13} \cdot v_c \cdot a_e + a_{23} \cdot a_e \cdot v_f + a_{123} \cdot v_c \cdot v_f \cdot a_e \text{ ,[N]} \quad (11)$$

$$Mz = a_0 + a_1 \cdot v_c + a_2 \cdot v_f + a_3 \cdot a_e + a_{12} \cdot v_c \cdot v_f + a_{13} \cdot v_c \cdot a_e + a_{23} \cdot a_e \cdot v_f + a_{123} \cdot v_c \cdot v_f \cdot a_e \text{ ,[Nm]} \quad (12)$$

The relationship of "coded" variables, and "natural" ones, z_j ($z_j = v_c$, v_f, a_e, a_p) is (13):

$$x_j = \frac{z_j - \frac{z_{min} - z_{max}}{2}}{\frac{z_{max} - z_{min}}{2}} \quad (13)$$

where:

z_{min} - is the minimum experimental value;

z_{max} - the maximum experimental value.

Figure 4: Experimental stand – force and torque study for machining milling process

Figure 5: Special dynamometric device

Equipment and materials:
a) Computer Aided Process, special software;
b) Amplifier analog-digital;

c) Special dynamometric, type 9257B, device for 6 components, Fx, Fy, Fz, Mx, My, Mz. The dynamometer has a great rigidity and consequently a hight natural frequency. Its hight resolution enables the smallest dynamic changes in large forces to be measured.

d) Studied material: Products made by materials sandwich composites products with two faces of fiber glass and the core made by polyester extruaded, some plate characteristcs are presented in table 2 and some core characteristics are presented in table 3.

Table 2: Plates characteristics

Material code	Plate dimensions, [mm]	Thickness, [mm]	Density, [kg/m^3]	Tension resistance, [MPa]	Elastic Modulus, [N/mm^2]
EC12-2400-P207	100x200	1.5/2/4	2,54x10^{-3}	3450	5000

Table 3: Core characteristics

Material code	Thickness [mm]	Density, [kg/m^3]	Compression Resistance de 10% [N/mm^2]	Conductivity, [W/m.K]
AplaXfoamBT	20	28	0.30	0.018

Experimental values obtained for the forces Fx and Fy and torque Mz measurement are shown in the tables 4, and table 5 as follows: - for the first method, REGS considered, that of solving the four / five linear equations systems – see table 4; - for the second method studied, we use just 3 variables, because the program used was DOE KISS with 3 variables and 2 levels that of experiments design (CCD) and regression analysis – see table 5.

Table 4: Experimental values obtained for the forces Fx and Fy and torque Mz

Cutting force and torque - F_x, F_y şi M_z[N]									
Factorial program P2.1				v_c [m/min]	v_f [mm/min]	a_e [mm]	a_p [mm]	F_x [N]	F_y[N]	M_z[Nm]
-1	-1	-1	-1	62.8	480	1	12	8.48	11.87	0.59
+1	-1	-1	+1	219.8	480	1	23	7.83	9.54	0.59
-1	+1	-1	+1	62,8	1680	1	23	24.65	23.31	1.68
+1	+1	-1	-1	219,8	1680	1	12	20.44	17.18	1.56
-1	-1	+1	+1	62.8	480	5	23	19.53	21.36	1.32
+1	-1	+1	-1	219.8	480	5	12	16.19	15.74	1.23
-1	+1	+1	-1	62.8	1680	5	12	51.02	38.45	3.48
+1	+1	+1	+1	219.8	1680	5	23	47.06	30.91	3.49
0	0	0	0	117,48	897,9	2,3	16.61	20.25	19.34	1.45
0	0	0	0	117,48	897,9	2,3	16.61	20.46	19.21	1.36
0	0	0	0	117,48	897,9	2,3	16.61	20.85	19.45	1.49
0	0	0	0	117,48	897,9	2,3	16.61	20.20	19.36	1.29

Table 5: Experiments design (CCD) and regression analysis

Cutting force and torque - F_x, F_y şi M_z[N]								
Factorial program P 1:2			v_c[m/min]	v_f[mm/min]	a_e[mm]	F_x	F_y	M_z
-1	-1	-1	62.8	480	1	4.95	9.73	0.37
-1	-1	1	62.8	480	5	12.95	19.75	1.54
-1	1	-1	62.8	1680	1	20.57	21.37	1.32
-1	1	1	62.8	1680	5	37.67	30.99	3.59
1	-1	-1	219.8	480	1	5.05	6.75	0.38
1	-1	1	219.8	480	5	13.95	22.98	1.98
1	1	-1	219.8	1680	1	32.91	32.85	1.97
1	1	1	219.8	1680	5	4.95	54.99	4.37

Observations: for each experience, replicates number equaled five.

3. MATHEMATICAL MODEL

A. For the first method considered, REGS, obtained experimental results were further "processed", in order to solve the four / five linear equations systems, required for models' constants and polytropic exponents values determination (C, a_1, a_2, a_3 and a_4), [5, 6].

Knowing that initial dependence relationships were exponential, – relations (7) to (9), there were obtained the mathematical models of force and torque, for machining milling process for products made by materials sandwich composites polymeric[7, 8].

So, in milling process, there are the final equations - see relations (13) to (15):

$$Fx = 0.073 \cdot v_c^{-0.107} \cdot v_f^{0.809} \cdot a_e^{0.485} \cdot a_p^{0.082} \quad ,[N] \tag{13}$$

$$Fy = 1.063 \cdot v_c^{-0.209} \cdot v_f^{0.504} \cdot a_e^{0.338} \cdot a_p^{0.067} \quad , [N] \tag{14}$$

$$M_z = 0.004 \cdot v_c^{-0.028} \cdot v_f^{0.805} \cdot a_e^{0.477} \cdot a_p^{0.056} \quad ,[Nm] \tag{15}$$

Some graphical representations of the obtained mathematical models can be noticed in figure 6a, b, c (for the milling process for REGS- Fx, Fy and Mz) and in figure 7 (for the milling process for DOE KISS).

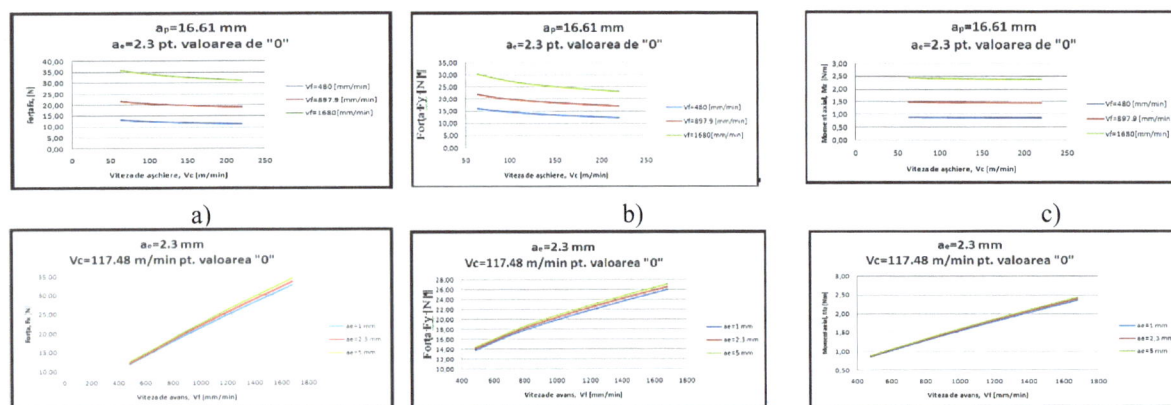

a)   b)   c)

Figure 6: Graphical representation of Fx-a, Fy-b and Mz-c with REGS models - relations (13), (14), (15).

As observation, there should be mentioned the fact that these graphs were plotted for the variables with higher influence on force and torque values.

B. One can notice the fact that relations (13), (14) and (15) were obtained by solving classical linear equations system. It means, four / five unknown parameters and, consequently, the need for four / five linear equations systems to be solved [10]. So, an "improvement" of the method to obtain mathematical models was considered to be right [9, 10].

So, for the new sets of experiments, as mentioned before, there have been considered, both the Central Composite Design of experiments and the regression analysis, performed with the special software, DOE KISS.

The only three independent variables studied were the ones that proved (by previously obtained mathematical models) to strongly influence the dependent variable (axial force or torque).. All of these are evidenced by table 6.

Table 6: Independent variables values

	Real, jz			Coded, jx		
	min		max	min		max
	Milling					
Cutting speed, v_c[m/min]	62,8		219,8	-1		+1
Cutting feed, v_f[mm/rot]	480		1680	-1		+1
Axial depth of cutting process,a_s [mm]	1		5	-1		+1

132

Examples of the DOE KISS software results, are shown in figure 7. So, as result of this method, the obtained mathematical models of force and torque, for machining milling process are the ones mentioned above – see relations (15), (16) and (17):

$$Fx = -2.857 - 0.0152 \cdot v_c + 0.01009 \cdot v_f + 1.883 \cdot a_e + 0.00005 \cdot v_c \cdot v_f + 0.00097 \cdot v_f \cdot a_e + 0.036 \cdot v_c \cdot a_e + 0.0000034 \cdot v_c \cdot v_f \cdot a_e \quad , [N] \quad (15)$$

$$Fy = 3.064 - 0.058 \cdot v_c + 0.031 \cdot v_f + 2.556 \cdot a_e + 0.000065 \cdot v_c \cdot v_f + 0.004 \cdot v_c \cdot a_e - 0.00093 \cdot a_e \cdot v_f + 0.00001 \cdot v_c \cdot v_f \cdot a_e \quad , [N] \quad (16)$$

$$, [Nm] \quad (17) \qquad Mz = 1.549 - 0.002 \cdot v_c + 0.0033 \cdot v_f + 0.207 \cdot a_e + 0.00003 \cdot v_c \cdot v_f + 0.006 \cdot v_c \cdot a_e + 0.00016 \cdot a_e \cdot v_f$$

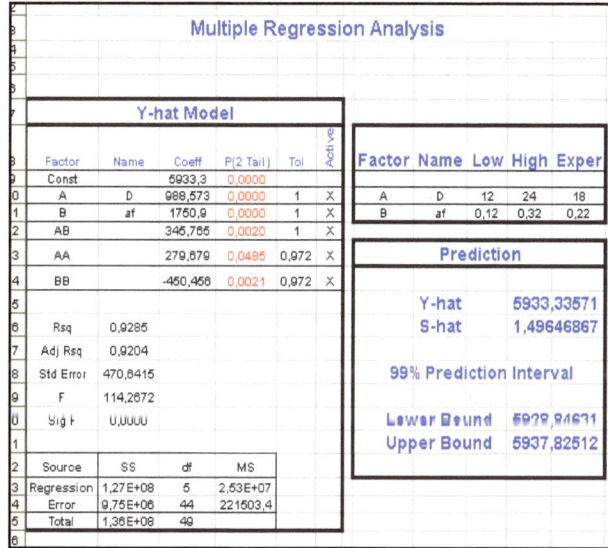

Figure 7: Examples of the DOE KISS software results

Some graphical representation for Fx , Fy and M$_z$ models - relations (15) , (16), (17) – see figure 8.

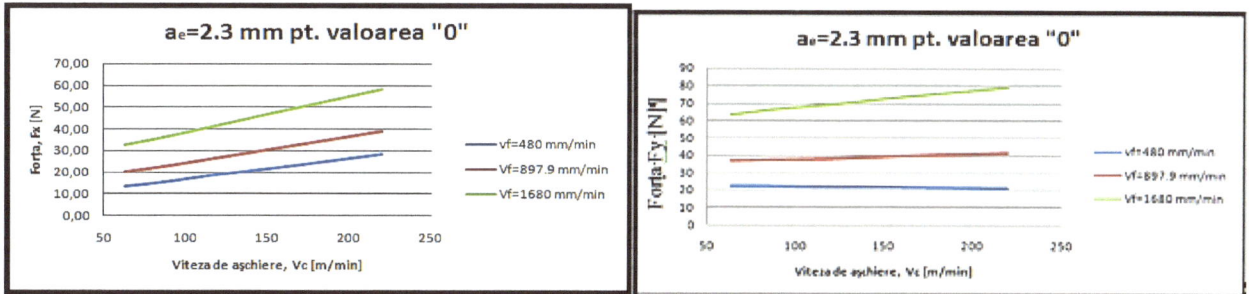

Figure 8: Graphical representation for Fx , Fy and M$_z$ models - relations (15) , (16), (17)

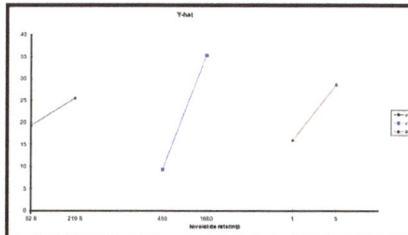

Figure 9: Examples of DOE KISS software - marginal means plots

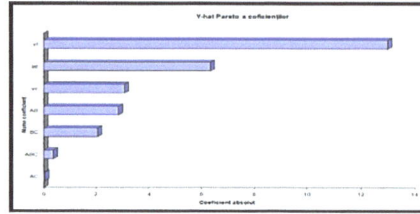

Figure 10: Examples of DOE KISS software – Pareto chart of coefficients

4. CONCLUSION

This study is about two methods REGS and DOE KISS, used for determining new adequate mathematical models of the forces Fx and Fy and torque Mz in machining, milling process of products made by materials sandwich composite polymeric. The machining procedures considered were milling process, while the material studied is one of great importance for Romanian industry.

The first method, consisted in solving the four / five linear equations systems, required for models' constants and polytrophic exponents values determination (C, a_1, a_2, a_3 and a_4).

The second method, involved design of experiments (CCD) and regression analysis performed with a special software.

There were obtained two independent variables polynomial type models and by plotting graphs, it was evidenced their influence, as well as their interaction, on the forces and torque values.

All the steps carried in order to obtain the new mathematical models can be considered as parts of a "procedure" to be followed / applied for any materials type and different machining procedures, whenever dependence relations of machining parameters are worth to be determined.

Further research involves other process parameters and materials to be studied, as well as the application of obtained models on real time control of the machining milling process.

REFERENCES

[1]   Opran C., Research concerning polymeric laminate composite materials products under impact   conditions, PROCEEDINGS CNC TEHNOLOGIES 2011, Fifth Edition, ISSN: 2068-2093, Politehnica   Press, Bucharest, May 12-13, pp.115-120, 2011.

[2]   Tsai S.W., Strength & life of composites, Editor Aeronautics & Astronautics Stanford University, SUA, 2011.

[3]   Bivolaru C., Opran C., Murar D., Milling of polymeric sandwich composites products, 22nd DAAM INTERNATIONAL, World Symposium, Uno City-Austria, Vienna, pp.1295-1296, 2011.

[4]   Hsu-Hwa C., Chih-Hsien C., Use of esponse Surface Methodology and Exponential Desirability Functions to Paper Feeder Design, WSEAS TRANSACTIONS on Applied and Theoretical Mechanics   Issue 9, Volume2, ISSN: 1991-8747, 2007.

[5]   Marcos M., Gomez- Lopez A., Batista P., Roughness based study of milled composite surfaces, 22nd DAAM INTERNATIONAL, World Symposium, Uno City-Austria, Vienna, pp.0153-0154, 2011.

[6]   Fong T. J., Composites failure criteria and estimation of associated A-and B –basis design allowable,   JEC COMPOSITES MAGAZINE, N0 71, 2012.

[7]   Camanho P., Turon A., Costa J., Structural integrity of thin –ply laminates; JEC COMPOSITES MAGAZINE, N0 71, 2012.

[8]   Bivolaru C., Opran C., Murar D., Influence of temperature on polymeric sandwich composite products during milling processing; PROCEEDINGS CNC TEHNOLOGIES 2011; Fifth Edition; ISSN: 2068-2093; Politehnica Press; Bucharest, pp. 89-93, 2011

[9]   Lupeanu M.E., Cercetări privind aplicarea analizei funcționale tehnice la dezvoltarea de noi produse realizate prin tehnologii neconvenționale de fabricare aditivă, Teză de doctorat, Coordonator: Neagu C., București, 2012.

[10]   Neagu C., Lupeanu M.E, Rennie A.E.W., Geometry optimization of tools used in milling of spherical surfaces, Proceedings of The 1st Sustainable Intelligent Manufacturing Conference, IST PRESS, ISBN   978-989-8481-03-0, Leiria, Portugalia, pg. 87-94, 2011.

CONTRIBUTIONS TO COMPUTER-AIDED EVALUATION OF MICROSTRUCTURE FOR PARTICLE REINFORCED COMPOSITES BY MEAN OF IMAGE PROCESSING

C. Locovei[1], M. Nicoară[1], A. Răduţă[1], V.A. Şerban[1]

[1] Universitatea "Politehnica" din Timişoara, Timişoara, ROMANIA, cosmin.locovei@mec.upt.ro

Abstract: Aluminum matrix composites reinforced with ceramic particles have already proved to be feasible choice when lightweight materials must provide high mechanical properties. Materials samples of composites with reinforcement levels between 0 and 20% in volume have been fabricated using standard powder metallurgy techniques. In order to improve uniformity of particle distribution and remove clustering, additional hot extrusion has been applied at different deformation ratios. Specialized software for image analysis has been used to evaluate correlation between microstructural images and ratios of hot extrusion. The image processing has included image acquisition, image processing, particle detection, particle measurement and classification in relevant granulometric fractions. Resulting histograms have determined strong correlation between particle distribution and deformations ratio and possibility to evaluate effectiveness of plastic deformation upon particle redistribution inside aluminum matrix.

Keywords: particle -reinforced aluminum, hot extrusion, particle distribution, computerized microstructure analysis.

1. INTRODUCTION

Particle reinforcement is a powder metallurgy technology that aims to improve mechanical properties of lightweight alloys, (e.g. aluminum, magnesium or titanium). As a consequence, significant increase of strength, stiffness, wear resistance and fatigue limit may be achieved. Among the fabrication techniques powder metallurgy has a special role, based on some major advantages such as structural homogeneity or possibility to embed even very small particles at reinforcement proportions up to 60%. Since fabrication costs seem to be critical for applications, some general tendencies have been observed [1],[2]:
- Fabrication techniques should be based on standard PM technologies, usually a blending-pressing-sintering route similar to classic materials, that are easy to reproduce and could be implemented with minor technological modifications;
- Both metallic particles and ceramic reinforcements have to be cheap and produced in large quantities, preferably already available on the market.
This last requirement usually produces large difference in particle size between metallic and ceramic powder that could affect material homogeneity, especially when ceramic reinforcements are added in higher proportion. Therefore conventional PM techniques will determine formation of clusters and pores, where ceramic particles are agglomerated inside metallic matrix, as seen in Figure 1. Reinforcement clustering is responsible for dramatic loss of material toughness and ductility, and for this reason secondary processing by mean of high-ratio plastic deformation; becomes necessary for improvement of particle distribution.
So far both clustering of ceramic particles during PM processing and redistribution of reinforcement have been evaluated mostly qualitative. This paper is proposing a more objective tool that could better evaluate microstructural effects of plastic deformation as secondary processing, based on image processing. The computerized processing of the images assumes that a number of mathematical operations have to be followed and logical decisions have to be made in a precise and organized manner.

2. MATERIALS AND METHODS

Samples of composite materials have been fabricated using classic blending-pressing-sintering PM technique. The powder blends consists of a mix of polyhedral-shaped metallic powders (Al - 4,5% Cu, 0,5% Si, 0,7% Mg) with particle diameters between 75 – 95 µm and polygonal silicon carbide (SiC) particles with nominal diameter size of 8,5 µm. These

SiC powders are produced at larger scale as grade F800 abrasive powder for polishing suspensions. Volume proportions of reinforcements are 5, 10, 15 and 20%, [3].

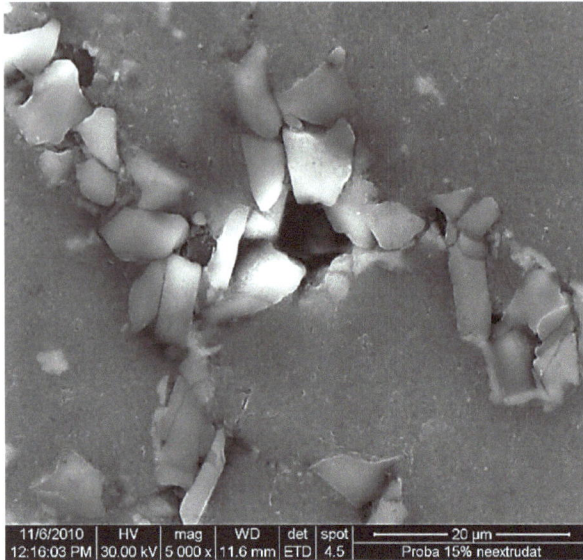

Figure 1: SEM image of clusters of ceramic particles inside metallic matrix, 15% volume of reinforcement proportion, before hot extrusion.

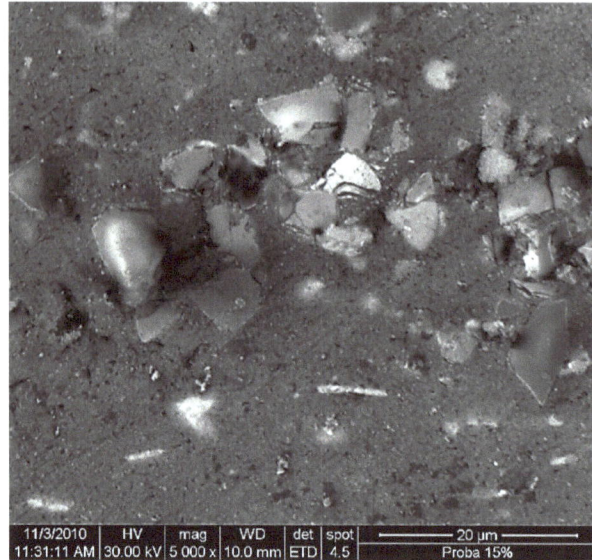

Figure2: SEM image of clusters of ceramic particles inside metallic matrix, 15% volume of reinforcement proportion, after hot extrusion, i = 2.25.

Redistribution of ceramic reinforcement has been achieved by mean of direct hot extrusion, where samples of 18 mm in diameter have been deformed at diameter values of 12, 9 and 6 mm, which correspond to deformation ratios between incident and resulting section areas ($i = d_o^2/d_1^2 = A_0/A_1$) of 2.25, 4 and 9 respectively.

In order to achieve apparent density around 95% of theoretical density, powder blends have been cold pressed up to 650 MPa. Since ceramic particles prove to have no influence on phase transformations of metallic matrix, a liquid-phase sintering has been applied at 600°C for 30 minutes in argon atmosphere.

Direct hot extrusion has required special attention, since friction between ceramic particles and tool may cause stick-and –slip deformation involving poor surface quality. Therefore a special technique has been developed using a pure aluminum pads, to reduce friction. Optimum deformation temperature has been determined to be 500°C. Remarkably, metallographic investigations determined that pores are no longer present after extrusion, even at low deformation ratio, as seen in Figure 2.

The sequence of technological operations is illustrated in Figure 3, [3] and fabrication details are presented elsewhere [1], [4], [5], [6].

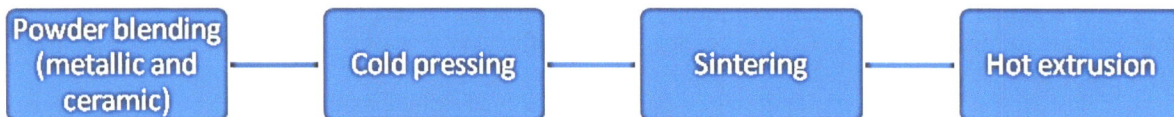

Figure3: Fabrication route of composite samples.

A proposed variant of the algorithm used to process the images have been presented previously in more detail as in Figure 4, [3].

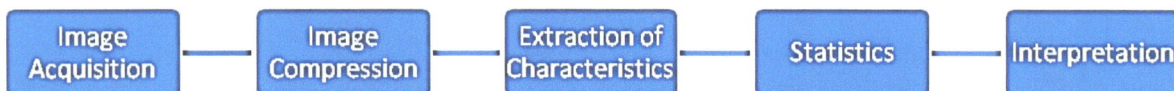

Figure 4: Algorithm used in image processing, [3].

2.1. Image acquisition.

Image acquisition assumes digitalisation (transformation of analog images into numerical ones) of the virtual image of the real objects obtained using the microscope. During the entire program of test and experiments two microscopes have been used: an optical Olympus BX51M microscope and a SEM FEI Inspect S. In order to digitize the optical images a 4.8 MP discrete sensor array on an Olympus ColorView microscope digital camera has been used. During SEM acquisition both backscattered and secondary electron detectors has been used, on 1 MP images.

2.2. Image Compression.

During all mathematical calculations and transformations only 256 gray levels, monochrome images and 2 levels black and white images has been processed. The optical microscope images are acquired in colours and therefore need to be compressed from 16.7 millions colours (24 bits images) to a 256 gray levels monochrome (8 bits) images. The colour information is stored in 24 bits RGB quantized images using 8 bits for each colour, R, G, B, including 256 levels of gray. The gray tones are obtained when R = G = B. The colour to greyscale image transformation is done by calculating a luminance value, L, for each pixel of the digital RGB colour image using a linear transformation as in Equation 1.

$$L = 0.299 \cdot R + 0.587 \cdot G + 0.114 \cdot B \tag{1}$$

The SEM images has been processed as they were, no compression needed as the depth of the SEM images is 8 bits, only.

2.3. Extraction of Characteristics.

Extraction of characteristic is the most complex step in terms of mathematical calculus. First step in extraction of characteristics is the evaluation of the background illumination. The background illumination is evaluated using morphological operations. Morphology is a broad set of image processing operations based on shapes. Morphological operations are non-linear operations that apply a structuring element to an input image in order to generate the output image. The input and output images are the same size and depth. When a morphological operation is applied to a digital image, the value of each pixel of the output image is calculated by means of comparison of the corresponding pixel in the input image with the pixels in its neighbourhood. The shape and size of the neighbourhood defines the structuring element of the morphological operation. The way the shape and size of the structuring element are assigned sets the morphological operation sensitivity to a specific shape in the input image.

2.4. Statistics.

In order to quantify the particles a connectivity test is needed. Connectivity defines which pixels form an object. In a binary image, an object consists in a group of pixels that are set to 1 and are connected to each other. The shape and dimension of an object depends on how connectivity is defined. In two dimensions, standard morphological connectivity is of two types: 4 connected and 8 connected. 4 connected type connectivity assumes that a 1 value pixel is connected to other pixels if at least one of up, down, left or right neighbor pixels are set to 1. The mask for a 4 connected type connectivity test is shown in Ecuation 2. The mask is placed centered on the test pixel.

$$4CTM = \begin{pmatrix} 0 & 1 & 0 \\ 1 & 1 & 1 \\ 0 & 1 & 0 \end{pmatrix} \tag{2}$$

8 connected type connectivity assumes that a 1 value pixel is connected to other pixels if at least one of its neighbor pixels are set to 1. The mask for an 8 connected type connectivity test is shown in Ecuation3.

$$8CTM = \begin{pmatrix} 1 & 1 & 1 \\ 1 & 1 & 1 \\ 1 & 1 & 1 \end{pmatrix} \tag{3}$$

The masks for connectivity tests can be modified to test for multiple pixel touch as a pixel is set to be a part of an object if a minimum predefined number of neighbor pixels belong to that object. In all processed images a 4 connected type with multiple pixel touch connectivity tests were applied.

All objects have been identified and tagged and the following information was processed:
- area of objects
- equivalent diameter
- position (X,Y) of the mass center.

Deeper statistics will be discussed in the results chapter of this paper.

3. RESULTS

Using commercially available image analysis software (Olympus AnalySIS Five) a complete image processing procedure has been undergone. The image processing has included particle detection, particle measurement and classification and has been done using the following procedure:
- the threshold gray values have been set to the values that enables particle separation from the background information;
    - the detection parameters has been defined as follows:
    - all particles under 1.5 µm has been considered as noise
    - all the border particles has been truncated and counted
    - all holes inside particles has been filled
    - after the detection of the particles the area of each particle has been measured
    - the classification scheme of area has been defined as in Table 1 and the area histograms have been build.
    - after the detection of the particles the position of the center of the mass of each particle has been calculated.

Table 1. Area Classification Scheme

Class ID	From [µm²]	To [µm²]
1	0	10
2	10	20
3	20	40
4	40	80
5	80	160
6	160	320
7	320	640
8	640	1280
9	1280	2560
10	2560	5120

All acquired images are of 440 x 330 µm overall dimension, in 4:3 picture format. A matrix of 100 boxes has been built, of 44 x 33 µm each. The positions of the center of the mass of each particle have been associated with one of the boxes. For each box two values have been calculated:

$N_{i,j} = $ number of particles in box (i, j), $i = \overline{1,10}; j = \overline{1,10}$.

$$A_{i,j} = \sum_k A_k \tag{5}$$

where: $A_{i,j}$ is the total area of particles in box (i, j),

particle $k \subset$ box(i, j),

A_k is the area of particle k.

138

Two set of three samples have been processed. For the first set of samples the particle to matrix ratio is 10% in volume, but the deformation ratio, i, is different 2.25, 4 and 9. The resulting images are presented in Figures 6, 7 and 9. For the second set of samples the deformation ratio is 9 and the volume of reinforcement proportion is different: 5, 10 and 15%. The resulting images are presented in Figures 8 to 10. As noticed, the 10% volume of reinforcement proportion and i=9 deformation ratio sample belongs to both sets. The average values of number of particles in a box, N, the average total area of particles, A, standard deviation of N, SD, relative standard deviation, RSD, standard deviation of A, SDA and relative standard deviation $RSDA$ are presented in Table 2.

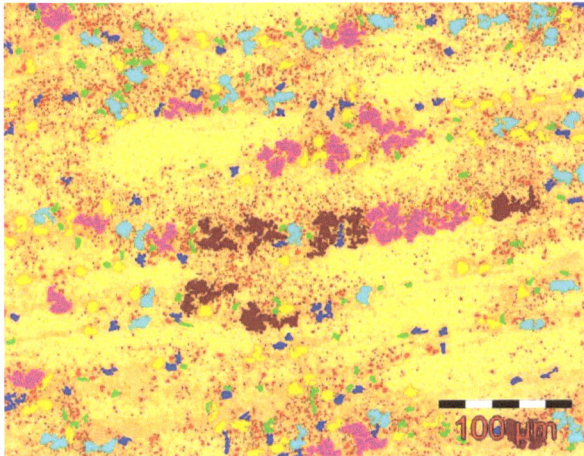

Figure 6: Distribution of area of particles, 10% volume of reinforcement proportion, i=2.25 deformation ratio

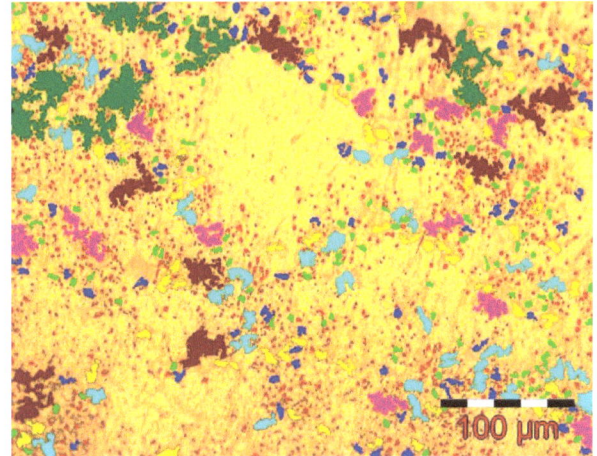

Figure 7: Distribution of area of particles, 10% volume of reinforcement proportion, i=4 deformation ratio

4. CONCLUSIONS

The computer aided method leads to an accurate evaluation of the composite samples in terms of detection of the clusters and the uniformity of the particle distribution within the matrix. In all the samples subjected to optical computer processing clustering of the particles has been observed, but the dimension of the clusters tends to decrease with the increase of the deformation ratio of the sample. The number of clusters tends to increase as the deformation ratio increases. This is a consequence of the fact that the clusters tend to brake in more, but smaller clusters when the composite material is subjected to a higher deformation state. As the deformation ratio increases the particles are distributed more uniformly within the matrix of the samples (standard deviation of $N_{i,j}$ notably decreases with the increase of deformation ratio, table 2).

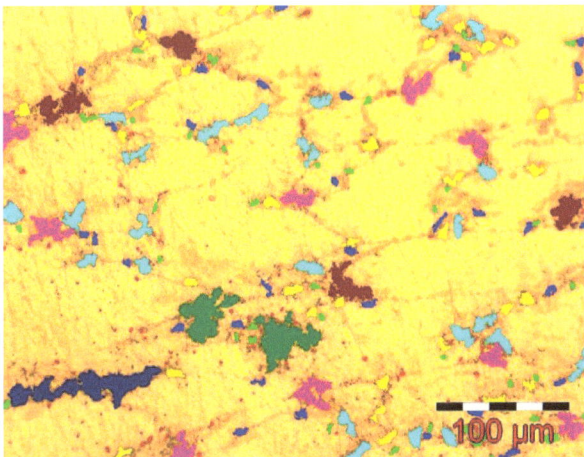

Figure 8: Distribution of area of particles, i=9 deformation ratio and 5% volume of reinforcement proportion

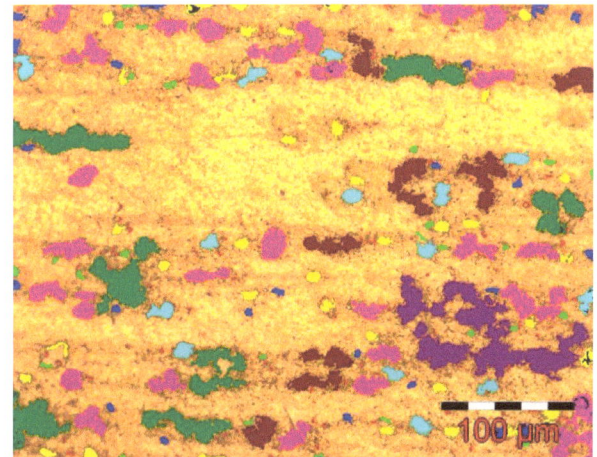

Figure 9: Distribution of area of particles, i=9 deformation ratio and 10% volume of reinforcement proportion

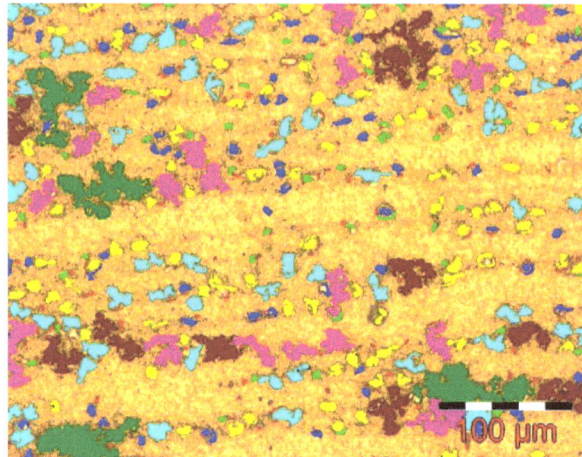

Figure 10: Distribution of area of particles, i=9 deformation ratio and 10% volume of reinforcement proportion

Table 2: Statistics for the analyzed samples

Sample	N [particles]	SD [particles]	RSD [%]	A [μm^2]	SDA [μm^2]	RSDA [%]
10%vol, i=2.25	15.34	8.60	56.06	204.22	157.33	77.04
10% vol, i=4	12.96	7.03	54.24	251.93	249.59	98.67
10% vol, i=9	3.00	2.01	67.00	250.68	421.01	167.94
i=9, 5% vol	2.44	2.19	89.92	133.77	207.72	155.28
i=9, 15% vol	4.61	2.22	48.15	289.22	264.28	91.37

A threshold value of 50% for the relative standard deviation seems to separate the non-uniform distributions from the uniform ones, as a set of 100 boxes are used to evaluate the distribution of the particles.

The tendencies observed on 10 vol.% (the reduction of the number of huge clusters with the increase of deformation factor and the increase in the uniformity of the spatial distribution of the particles with the increase of deformation factor) remain valid for the 5 and 20 vol.%. The phenomenon is more pronounced as the vol.% of the reinforcement particles of the samples increases.

ACKNOWLEDGMENT

This work was partially supported by the strategic grant POSDRU/21/1.5/G/13798, inside POSDRU Romania 2007-2013, co-financed by the European Social Fund – Investing in People.

REFERENCES

[1] Nicoara M., Raduta A., Locovei C., Serban V.-A., Contributions to Thermo-Mechanical Processing of Al-Based Particle-Reinforced Composites, Proceedings of Colloquium of European Centre For Emerging Materials And Processes Dresden, Technische Universität Dresden, December 2010, ISBN 978-3-00-   032522-9, pp. 229-253.

[2] Evans A., San Marchi C., Mortensen A. - Metal matrix composites in industry, Kluwer Academic Publisher, 2003.

[3] Nicoara M., Raduta A., Locovei C., Computerized Image Processing for Evaluation of Microstructure in Metallic Matrix Composites, Solid State Phenomena, Vol. 188, Advanced Materials and Structures IV, 2012, Trans Tech Publications, Switzerland, ISSN1012-0394, pp. 124-133

[4] Sahin Y. - Preparation and some properties of SiC particle reinforced aluminium alloy composites, Materials   & Design, Volume 24, Issue 8, December 2003.

[5] Narayanasamy R., Ramesh T., Prabhakar M. - Effect of particle size of SiC in aluminium matrix on   workability and strain hardening behaviour of P/M composite, Materials Science and Engineering: A,   Volume 504, Issues 1-2, 25 March 2009.

[6] Nicoară M., Răduță A. - The Effect of Particle Reinforcement upon Precipitation of Secondary Phases in Composites with Aluminum Matrix, Defect and Diffusion Forum, vol. "Defects and Diffusion in Metals", 2002, TRANS TECH PUBLICATIONS Ltd. Zürich – Switzerland.

DESIGN AND APPLICATION OF COMPOSITE MATERIALS IN MECHANICAL STRUCTURES

M. Růžička[1], O. Uher[2], J. Had[1], V. Kulíšek[3], P. Padovec[1]

[1] Czech Technical University in Prague, Faculty of Mechanical Engineering, Prague, CZECH REPUBLIC,
Milan.Ruzicka@fs.cvut.cz

[2] CompoTech Plus s.r.o., Družstevni 159, Sušice, CZECH REPUBLIC
Ondrej@CompoTech.com

[3] Research Center of Manufacturing Technology Czech Technical University in Prague, CZECH REPUBLIC,
V.Kulisek@rcmt.cvut.cz

Abstract: *Design of machine structures with using composite parts allows to combine different sorts of material, manufacturing technology and thereby control stiffness, strength, dynamic response, weight, as well as the costs, and achieve the products "on request". This paper describes new hybrid composites products based on combination of winding technology with 3D cellular structure and dynamic damping layers. An analytical methods and their effective use for optimization process and a multiscale homogenization on the micro and mezzo scale as well as the FE analysis of the final structure is described. Application of damping layers significantly contributes to the rate of vibration decay and dynamic damping.*

Keywords: *composite structure, dynamic response, multi-scale modeling,*

1. INTRODUCTION

During the last decade, performance composite components have been used in an increasing number of different applications, such as sport equipment, transport, marine, energy or machine building. One factor is the decreasing carbon fibre prices, which have created new demands for composite technologies. Some other aspects are going currently in application.

Composite technology enables building materials with desired properties, for example rollers or spindles with zero thermal expansion, drive shafts high resonance frequency or machine parts with high damping. Industries can now adopt better materials to produce better performing products, while maintaining efficiencies of productivity and cost. The designers and producers can change material components and manufacturing technology and thereby control stiffness, strength, dynamic response, weight, as well as the costs, and achieve an optimal product.

2. CURRENT AND NEW TECHNOLOGY

2.1. Current technology overview

There are several technologies for the production of machine parts:

- Autoclave processing is the most common method used for curing thermoset prepregs. The curing of thermoset composites involves both mechanical and chemical processes. Mechanically, pressure is applied to remove trapped air and volatiles, and to consolidate the individual plies and fibers.

- Resin Transfer Molding is a low-pressure, closed mold semi-mechanized process. The process allows fabricating simple low-performance to complex or high-performance articles in varied sizes. The fiber reinforcement, which may be pre-shaped is placed in the required arrangement in the cavity of a closed mold and a liquid resin of low viscosity is injected under pressure into the cavity, which is subsequently cured.

- Pultrusion is a continuous, automated closed-molding process that is cost effective for high volume production of constant cross section parts. Due to uniformity of cross-section, resin dispersion, fiber distribution & alignment, excellent composite structural materials can be fabricated by pultrusion. The basic process usually involves pulling of continuous fibers through a bath of resin, blended with a catalyst and then into pre-forming fixtures where the section is partially pre-shaped & excess resin is removed. It is then passed through a heated die, which determines the sectional geometry and finish of the final product. The profiles produced with this process can compete with traditional metal profiles made of steel & aluminum for strength & weight.

- Filament winding is a semi-automatic manufacturing method for making fiber reinforced composite materials by precisely laying down continuous resin impregnated roving or tows on a rotating mandrel that has the required shape. The mandrel can be cylindrical, round or of any shape that does not have a reverse curvature. The technique has the capacity to vary the winding tension, wide angle or resin content in each layer of reinforcement until the desired thickness or resin content of the composite are obtained with the required direction of strength. A large array of products can be fabricated by this technique e.g. storage tanks, pipes, pressure vessels, rocket engine cases, nose cones of missiles and other aerospace parts.

- The specifics of designing of composite structures include the fact that designers should work closely with production technology and they must discard "traditional isotropic" thinking. Czech Technical University in Prague (CTU) closely collaborates with many companies which deal with composite production. The co-author (company CompoTech Plus) has developed its own fiber laying process for structural composite tubes, which is particularly suitable for components that require high bending stiffness and stability. This is called the zero degree axial fiber laying process. Unique application allows using ultra high modulus Pitch carbon fiber to manufacture structural tubes. They can be used in almost any type of high speed industrial machine, such as milling machines, robots and printers. This pitch-based fiber offers higher bending and torsion stiffness than the usual PAN-based fiber and can be cost-effective when used in volume.

- As described in [1], a design of composite parts with ultra-high stiffness (for example of machining center spindle beams) leads to thick-walled reinforcing members characterized by the axially oriented fibers in the direction of maximal loading-flows. However, the low shear static and fatigue strengths of these unidirectional thick structures often limit their application. This is due to the low strength of the composite matrix and the multi-axial stress state. Cracks arise at several points between the fibers (thick-walled pultruded composite flanges), or delamination occur between the laminae (laminated composite plates). An increase in the shear strength is usually produced by various three dimensional (3D) laminate techniques, such as 3D braiding or 3D strengthening (transversal needling). Such technologies can improve or partially eliminate delamination or matrix cracking. However, these techniques lead to a rapid decrease in stiffness in the dominant load direction. Filament fibre winding technology combined with stamping and wrapping with using both high modulus (cell core) and high strength carbon fibers (wrap) was used to manufacture a three-dimensional cell structure.

2.2. Three dimensional cell hybrid structure

A new type of cell hybrid composite structure was developed by CompoTech Plus in research cooperation with the Czech Technical University in Prague [2]. The main application of these structures is for thick-walled or nearly solid beams with maximum bending strength (spars, wing flanges, etc.) or with high stiffness (machining center spindle beams).

A typical hybrid composite beam consists of the main supporting element (e.g. a central wound composite tube), secondary elements (a corner tubes for connection or integration of the guidelines). Thick parts of the cross section of such typical spindle beam are filled in by a sub-cell structure, as is shown in Fig 1.

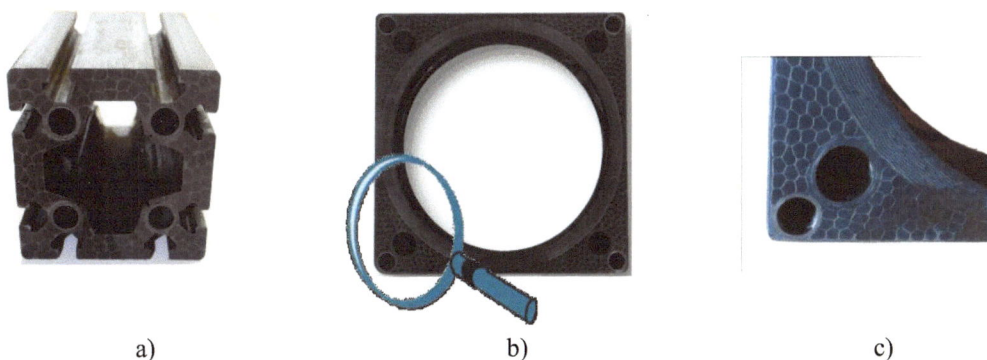

a)   b)   c)

Figure 1: A composite spindle beams cross sections with an integrated guideline formed by 3D cell structure (a) or with a corner guideline connection tubes (b), and a detail of the 3CD structure (c)

This bioinspired structure in its cross section (in the y-z plane) creates sub-cells with a volume fraction of up to 75% axial fibers, see Fig. 2. The sub-cells consist of carbon fiber tows with axial orientation (x axis). The diameter of this bundle is usually between 4 and 8 mm. In the next step, another thin layer is wound around this axially oriented core. The winding is created between 0.2 and 1 mm in thickness. The thickness can be optimized, as the orientation of the winding fibers allows, which can be made from 0 to 89 degrees. The prefabricated bundles are then put into the form, molded together and subsequently cured into the final shape.

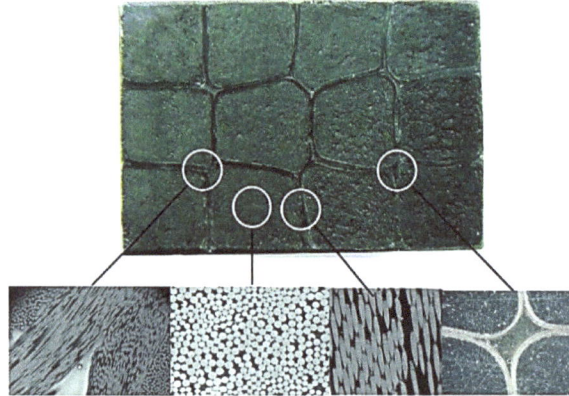

Figure 2: Details of the parts of the 3D composite structure.

2.3. Hybrid structure

As was partially showed above, the design of the more complicated final composite structure can include and conveniently combine different independently produced semi-finished products, which can be "mounted together" by final technological operations, for example stamping in the mold. It allows to include in the structure bearing layers, which transmit loads as well as technological layers to protect the surface (against impact or environment).

Special layers of cork or rubber-cork were also tested, to achieve high damping of dynamic vibration, see Fig. 3. Other possibilities are open for structural health monitoring systems (SHM). Authors successfully applied into a composite structure the integrated fiber optic sensors with Bragg grating (FBG). FGB sensors allowed a monitoring of deformation and debonding of adhesive joints, see [2].

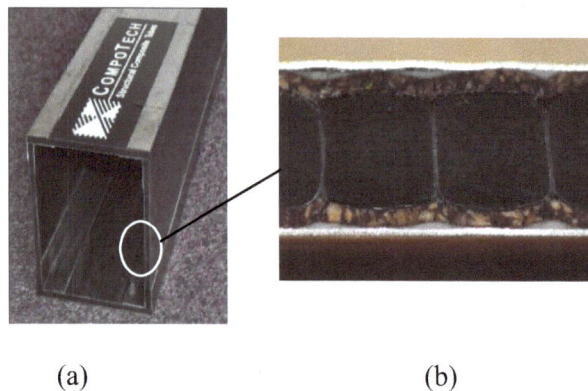

(a)   (b)

Figure 3: A thin-walled sandwich beam (a) and a detail of its wall (b) containing metal coverings, cork layers and high modulus sub-cell structure

2.4. Properties of high performance structures and their application

Design of composite hybrid structure „on request" allows to achieve high mechanical and dynamic parameters of the final product.

- For example, an axial fibre placement in the winding technology of tubes and combination of two types of carbon fibres (PAN/PITCH fibres, with positive/negative thermal expansion in longitudinal/transversal direction) enabled the production of printing cylinders, which did not change the outer diameter during a working cycle and changing of temperature.

- Why drill holes when they can be made in the process of manufacturing the beam? Authors have developed the idea of "Winding Holes" and have researched the stresses around these holes to enable secure connection to metal components using the ability of composite fibres to be placed in an advantageous way. No drilling, min. cutting fibres, fewer manufacturing steps, less waste. One example of integrated multi pin joints, which connect the bottom and other equipment with the hydraulic composite cylinder, is shown on the Fig. 4a and Fig. 4b, [3]. Another idea of a connection using the steel rod, which is inside of composite corner tube and screws, is shown in Fig 4c.

- Combination of high strength and high modulus carbon fibres and use of other material layers allows building "composite dynamic damping" hybrid structures. A high natural frequency is beneficial, but the real benefit to machine builders is the improvement in dynamic damping. Everything vibrates at its own natural frequency and every material has a damping coefficient ζ which is the rate of vibration decay. The developed composite parts have the ability to dissipate vibrational energy. Tests were done by comparing three spindle tubes: with and without a simulated tool (mass of 30 kg), see Fig. 5. Each with the same dimensions and each tested with the same load. It was shown that the graphite carbon damping spindle tube has nearly double the natural frequency, 12 x better dynamic damping and 7 x better dynamic stiffness as steel spindle. It was shown that not only carbon-graphite-damping composites have a higher stiffness and a higher natural frequency, but they also have a lower response at the points of sympathetic vibration. This reduces the response to excitation frequencies and should permit higher accuracy and / or higher cutting speed. The productivity of "removing metal" should be significantly better.

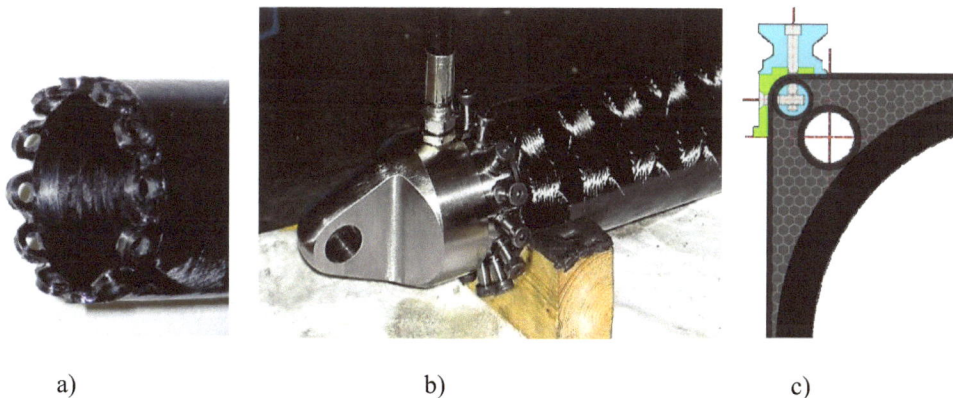

a)   b)   c)

Figure 4: Integrated loops into the composite tube (a), final hydraulic cylinder (b) and mechanical connection with using of a integrated corner tube (c)

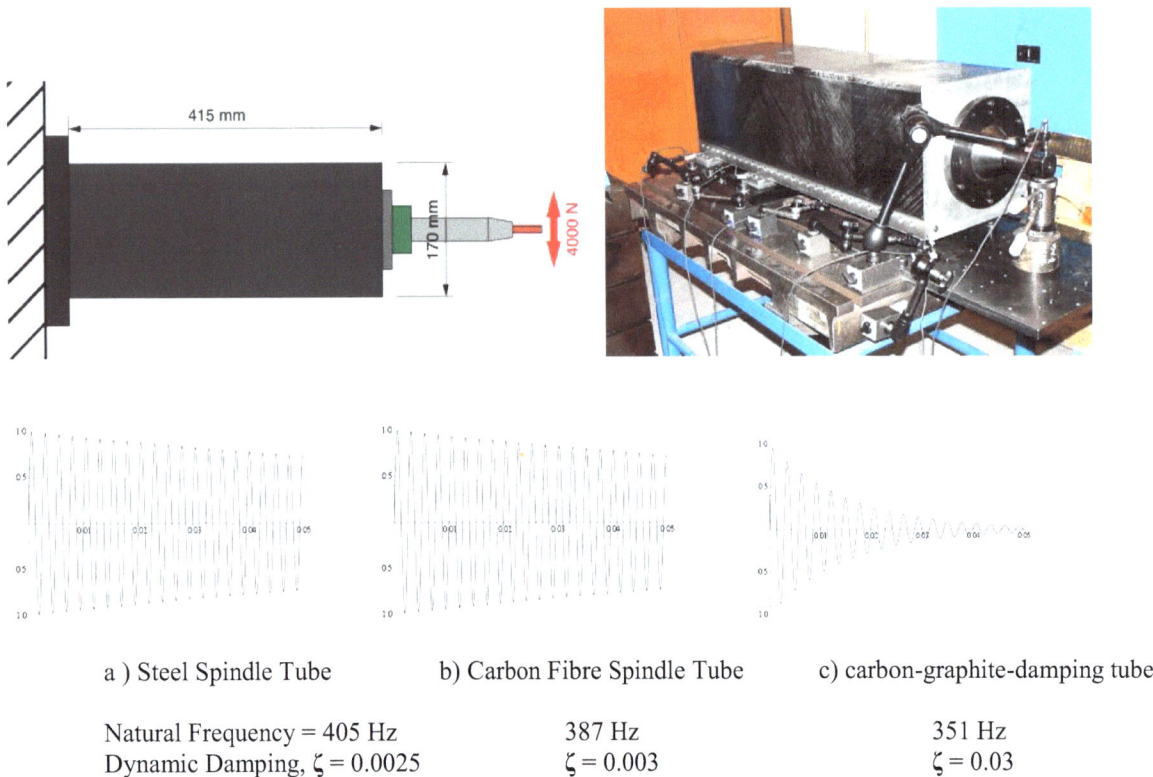

a) Steel Spindle Tube   b) Carbon Fibre Spindle Tube   c) carbon-graphite-damping tube

Natural Frequency = 405 Hz   387 Hz   351 Hz
Dynamic Damping, $\zeta = 0.0025$ $\zeta = 0.003$ $\zeta = 0.03$

Figure 5: Comparison of dynamic response of steel spindle (a), carbon spindle (b) and carbon-graphite-damping tube (c) with the end mass of 30 kg.

3. DESIGN METHODS AND SOFTWARE

3.1. Analytical solutions

The design and comparison of many variants or the optimization process required relative very fast calculation of complicated structures. For this phase of the design process, the analytical methods are more suitable, because they are more flexible for changing of stacking of layers or loading and boundary conditions. Two basic groups of software for designing and analysing fibre layed structures were developed by CompoTech in the close collaboration with the Czech Technical University in Prague.

The first group of software analyses stresses and displacements of cylindrical, fibre layed, laminated composite shells. Two approaches have been developed: The first is simple linear shell theory based on Kirhoff's conditions of shell deformation, known as Classical Laminating Theory (CLT) was modified with own method of unit forces distribution in the cross-sections of tubes. This enables different loads such as bending moment, torque, axial force or thermal load to be solved together. Because this method is based on plate theory, the advantage is a simpler and more precise two dimensional data preparation and the possibility of solving several loading states at the same time.

This method is powerful enough for most composite tube applications and has been verified by full destructive tests on test samples. The second approach is based on the theory of the cylindrically anisotropic elasticity. Using Lekhnitski's stress functions, a closed form solution for calculating stresses and displacements was found. The advantage of this approach is the exact and precise calculations of three-dimensional stresses in a fibre layered tube. It particularly enables the analysis of the stress gradients through the thickness of a laminate, which may be very substantial in the form of interlaminar stresses for thicker sections of composite tubes.

The disadvantage is a more complicated three dimensional lamina data preparation and the possibility of solving only one load state-bending moment.

The company is currently working on the solution using this method for three load states: bending, torsion and axial force.

The second group of software solves and analyses stresses and displacements of fibre layed, laminated composite shells of generally noncircular cross-sections. The principle of this model is the integration of stiffnesses of elements in a cross section. The stiffnesses is calculated as a transformation of (CLT) into a geometric three dimensional expression.

The special software was developed to calculate deflection curves and deformations of the beams or the spars. One dimensional finite element method (FEM) or method of transfer matrix (MTM) is transformed to the numerical program codes. The main advantage is that the way mathematical models are built from elements is identical to the way in which the structure is made, offering accuracy and time saving in pre-processing data entry and in transferring results directly into the production code.

3.2. FEM stress state calculation and stiffness and damage prediction

In order to analyse the details of more complicated composite products or taken into account the contacts or joints with other components, FEM software is used on CTU in Prague. In the following, a multi-scale modelling of the 3D composite structure is briefly described.

Homogenization on a micro scale. Two main different configurations of the cells can be observed in the cross-section of the 3D structural part, which is hexagonal or rectangular. The idealized geometry of the Representative Volume Elements (RVE) containing the periodic structure with both typical sub-cell shapes is shown in Fig. 6.

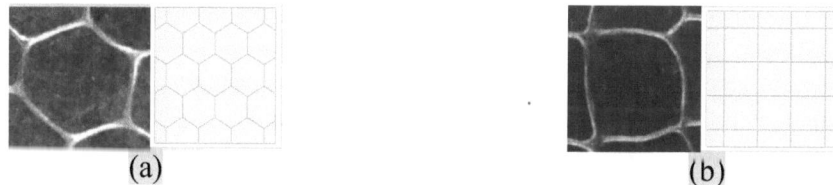

Figure 6: Hexagonal (a) and rectangular shape (b) of sub-cells and periodic RVE models

Computational homogenization on this level can be conceived analytically or with the use of FEM. For an analytical calculation of the unidirectional transversely isotropic structure, law-of-mixture and its empirical variations can be applied in the simplest simulation. A lot of authors have published a number of different more sophistical approaches, see [4], [5]. These formulas usually cannot affect the influence of the interface between fibre and matrix. A more concise estimate of the micro-scale stiffness is provided by FEM analysis with modeling of the fibre interface according [6].

Homogenization on a mezzo scale. The RVE with hexagonal and with rectangular cell shape was modeled in the FE program, according to Fig. 6. The finite element model of a periodic unit cell (PUC) consists of a core and a wrap. It is assumed that the cell contains the homogenized elastic properties of uniaxial oriented fibres tows (high modulus PITCH type fibers). The packaging wrapped around the core can be wound with various angles and thicknesses. High strength PAN type fibers are usually used for this layer. The elastic properties of a PUC calculated on a micro scale level were considered and compared with the experimental measurement of mechanical properties of the 3D structure composite specimens.

The aim of the FE mezzo-scale analysis was to obtain the matrix of homogenized elastic modules according to the Hook's relation between stress and strains tensor transformation:

$$\bar{\sigma}_{i,j} = C_{i,j,k,l} \cdot \bar{\varepsilon}_{i,j} \qquad (1)$$

Homogenization has to be implemented by the boundary conditions. The periodic boundary conditions are assumed on RVE and the displacement of the two opposite boundary surfaces were calculated according periodic condition relations published in [7].

Calculation on a macro scale. A matrix of homogenized elastic modules of PUC was applied to build the final stiffness matrix of the whole hybrid composite structure containing a 3D sub-cell technology. For example, the spindle beams of the machining center shown in Fig. 5 were calculated, manufactured and experimentally investigated. Another application is depicted in Fig. 3. This beam is composed of thin sandwiches consisting of metal coverings, cork layers and a high modulus sub-cell structure. FE calculations of such hybrid structures based on the multi-scale homogenization described above give very satisfied results, and can be used for the construction design of machines. Experimental verification of the FE calculations of static stiffness showed differences up to 10 % in deflections by bending.

4. CONCLUSION

Design of machine structures with using composite parts allows to combine different sorts of material, manufacturing technology and thereby control stiffness, strength, dynamic response, weight, as well as the costs, and achieve the product "on request". This paper described new hybrid composites products based on combination of winding technology with 3D cellular structure and dynamic damping layers. Multi-scale homogenization on a micro scale and on a mezzo scale and final FE analysis on the hybrid material composite structure leads to satisfactory design of complicated machine parts. Analytical methods can be conveniently and effective used in the first step of the design phase and for the optimization process.

Experimental investigation, as well as numerical simulations, showed that 3D cell composite structure can be more effective in transferring a shearing force than a classical unidirectional or layered structure. Application of damping layers significantly contributes to the rate of vibration decay and dynamic damping. With using of hybrid composites, parts of high stiffness and very good shear and normal strength can be designed. It was demonstrated on the prototype of the carbon/epoxy spindle for CNC machine center.

ACKNOWLEDGEMENTS

This work was supported by the Technological Agency of the Czech Republic, project number TA02010543.

REFERENCES

44. Růžička M., Had J., Kulíšek V., Uher O.: Multiscale modeling of hybrid composite structures. Key Engineering Materials, Vol. 471 - 472. (2011), p. 916.
45. Dvořák M., Růžička M., Had J., Pošvář Z.: Monitoring of 3D Composite Structures Using Fiber Optic Bragg Grating Sensors. In: Structural Health Monitoring 2011, Stanford University: Lancaster, Pennsylvania: DEStech Publications, Inc., ISBN 978-1-60595-053-2.
46. Růžička, M., Uher, O., Blahouš, K., Kulíšek, V., Dvořák, M.: Integrated High Performance Joint in Composite Vessels. In Sixteenth International Conference on Composite Materials, Japan Society for Composite Materials, Kyoto 2007, pp. 1400-1401.
47. Zeman J., Šejnoha M.: Numerical evaluation of effective elastic properties of graphite fiber tow impregnated by polymer matrix. Journal of the Mechanics and Physics of Solid 49, No. 1, (2001), p.69-90.
48. Kouznetsova V., Geers M.G.D., Brekelmans W.A.M.: Multi-scale constitutive modelling of heterogeneous materials with a gradient-enhanced computational homogenization scheme. Int. Journal for Numerical Methods in Engineering 54, No. 8, (2002), p. 1235-1260.
49. G. Steven: Int. J. for Computer-Aided Engineering and Software, Vol. 34, No.4, (2006), p. 432.
50. Barbero, E. J: Finite Element Analysis of Composite Materials. CRC Press, 2008

RESEARCH ON THE IDENTIFICATION OF HEMP COMPOSITES USING TENSILE TESTING

M. L. Scutaru[1*], C. Cofaru[1]

[1*]Transilvania University of Brasov, Brasov, Romania, e-mail: luminitascutaru@yahoo.com, ccornel@unitbv.ro

Abstract: *For use in automotive design composite components is necessary to know their mechanical properties, in order to determine the stresses that they can resist exploitation. Below is determined these properties for hemp composites using tensile tests.*

Keywords: *Mechanical properties, Hemp composites, Static tests.*

1. INTRODUCTION

Composite materials made of hemp fiber reinforced resins are frequently used materials in vehicle construction. Their use requires knowledge of the structure of their mechanical properties. There are various theoretical methods for calculating these quantities but in practice, it is found that these methods lead to values sometimes differ than actual values. For this it is necessary to experimentally determine the mechanical characteristics of these types of composite materials. Tensile testing determines these values [1], [2], [3], [4].

2. MATERIAL AND METHOD

We used specimens made according to the standards specified above to perform tensile tests. In Figure 1 are shown a series of specimens of composite materials reinforced with hemp fiber. If the break was made at one end of the grip, samples were replaced with a representative sample of reservation. For tensile test was performed a set of 8 samples

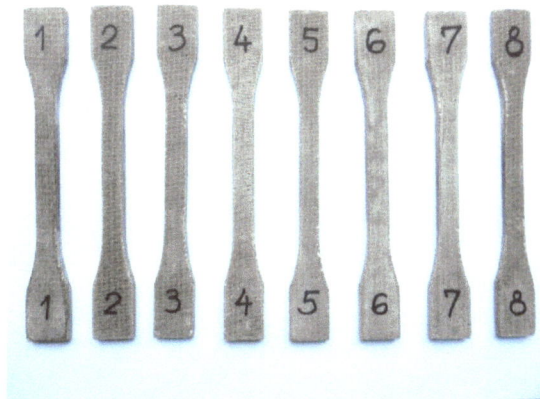

Figure 1. Composite samples of hemp fiber reinforced

Tensile testing machine tried to use existing SIM Faculty laboratory and test drive of the laboratory strength of materials. Were performed tensile tests with composite samples of hemp fiber reinforced.

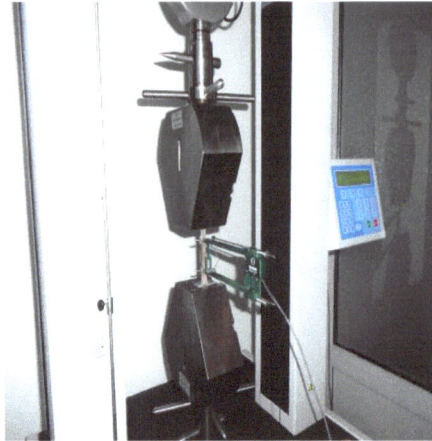

Figure 2. Tensile testing machine tried

Figure 3. Tensile testing machine tried specimen mounted between the device terminals tensile test

Following main features have been experimentally determined:
- Stiffness (N/m);
- Young's modulus (MPa);
- Load/stress/strain at maximum load;
- Load/stress/strain at maximum extension;
- Load/stress/strain at minimum load;
- Load/stress/strain at minimum extension;
- Tensile strength;
- Load/stress at break;
- Work to maximum load/extension;
- Work to minimum load/extension.

Test and specimens features are:
- Test speed: 1 mm/min;
- Number of specimens: 8;
- Preload/stress: 1.4680 kN;
- Specimens mean width: 9.8375 mm;
- Specimens mean thickness: 5.0500 mm;
- Mean cross-sectional area: 49.649 mm^2.

The materials testing machine allows determination of experimental results in electronic format by help of the NEXYGEN Plus software.

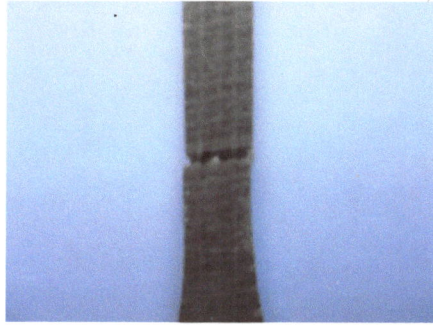

Figure 4. Resin specimens reinforced with hemp fiber, torn by traction

3. TENSILE TEST RESULTS

The maximum mechanical properties of eight composite materials made of hemp fiber reinforced resins determined in tensile tests are presented in table 1. Typical load-extension from preload distributions of eight composite materials made of hemp fiber reinforced resins is presented in figs. 5-7 and stress-strain distributions are shown in figs. 8-9.

Table 1. Specimens' maximum mechanical properties

Feature	Value
Load at maximum load (kN)	1.4680
Load at break (kN)	1.4612
Young's modulus (MPa)	10250.0
Tensile strength (MPa)	29.649
Machine extension at maximum load (mm)	0.28674
Stress at break (MPa)	29.512
Strain at break (-)	0.0057074
Machine extension at maximum extension (mm)	0.53770
Stress at minimum extension (MPa)	14.745
Strain at maximum load (-)	0.0057281
Strain at maximum extension (-)	0.010747

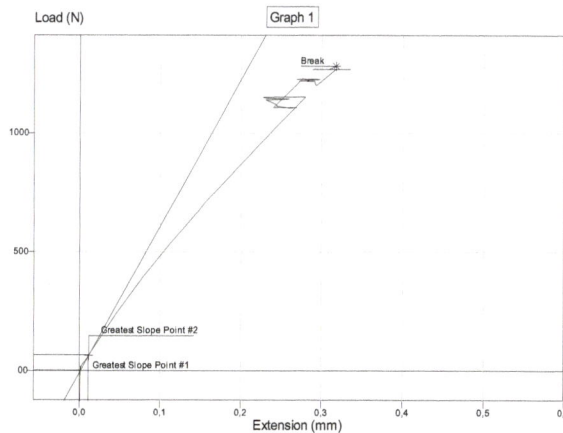

Figure 5. Load-extension from preload distribution of specimen 1

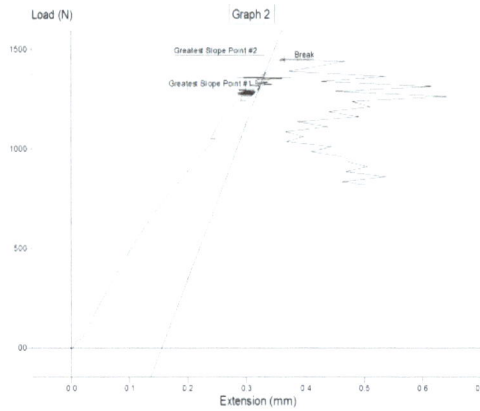

Figure 6. Load-extension from preload distribution of specimen 2

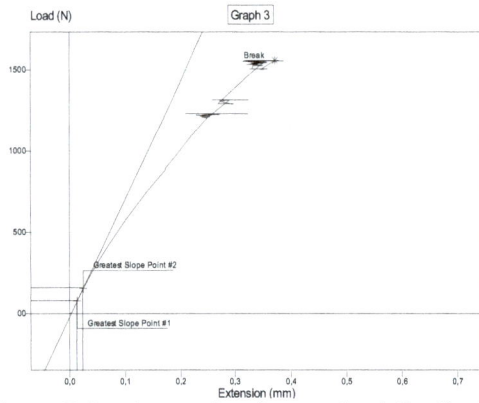

Figure 7. Load-extension from preload distribution of specimen 3

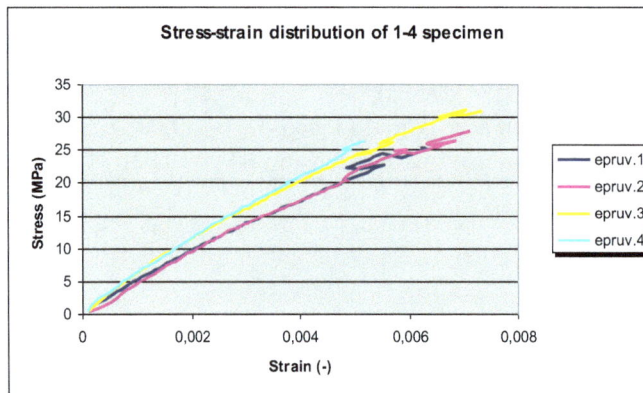

Figure 8. Stress-strain distribution of specimen 1-4

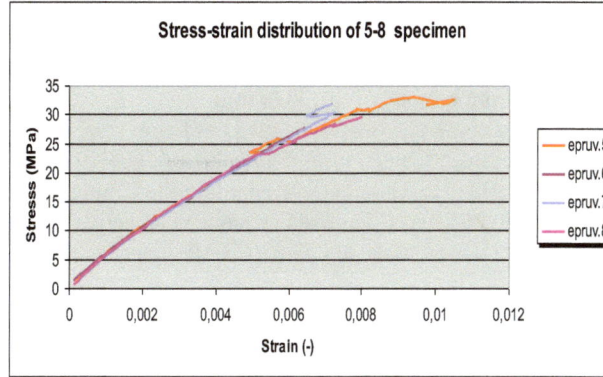

Figure 9. Stress-strain distribution of specimen 5-8

Young's modulus distributions of eight composite materials made of hemp fiber reinforced resins are presented in fig. 10 as well as a distribution between Young's modulus and tensile strength (fig. 11).

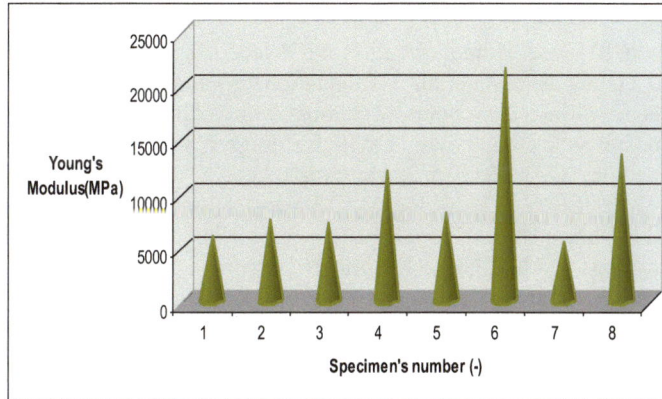

Figure 10. Young's modulus distribution of eight composite materials made of hemp fiber reinforced resins

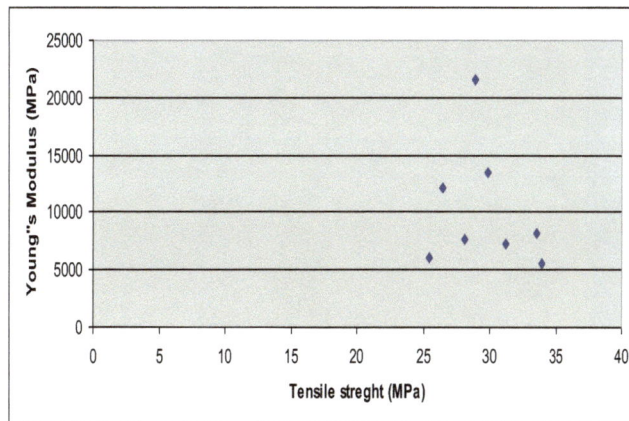

Figure 11. Young's modulus distribution versus tensile strength of eight composite materials made of hemp fiber reinforced resins

4. CONCLUSION

Regarding the load-extension in the first load-extension from preload distribution of specimen 1 presented in fig. 5, the load at break reaches a value of 1277.7 N at an extension from preload of 0.315 mm. The greatest slope points generated by the materials testing machine that give the rigidity modulus, have been determined between 0 – 500 N load and 0.0 – 0.1 mm extension from preload. In case of the second load-extension from preload distribution of specimen 2 presented in fig. 6, the load at break has been reached at 1435.5 N at an extension from preload of 0.35633 mm. The greatest slope points have been generated between 1000 – 1500 N load and 0.30 – 0.40 mm extension from preload. In the last load-extension from preload distribution of specimen 3 presented in fig. 7, the load at break has been experimentally determined at 1549.5 N for an extension from preload of 0.36779 mm.

The Young's modulus distribution of eight composite materials made of hemp fiber reinforced resins specimens presents a maximum value of 21540,85 MPa at specimen 6 and a minimum value of 5605.191 MPa at specimen 7 (fig. 10). These values have been determined using the machine extensometer. Regarding the Young's modulus distribution versus tensile strength of eight composite materials made of hemp fiber resin specimens presented in fig. 11, the eight values of Young's modulus are scattered between 25.41469 MPa and 33.88673 MPa tensile strength.

ACKNOWLEDGEMENTS

This paper is supported by the Sectoral Operational Programme Human Resources Development (SOP HRD), financed from the European Social Fund and by the Romanian Government under the contract number POSDRU/89/1.5/S/59323.

REFERENCES

[1]. H. Teodorescu-Drăghicescu, S. Vlase, Homogenization and Averaging Methods to Predict Elastic Properties of Pre-Impregnated Composite Materials, Computational Materials Science, 50, 4, Feb. (2011).

[2]. H. Teodorescu-Drăghicescu, S. Vlase, L. Scutaru, L. Serbina, M.R. Calin, Hysteresis Effect in a Three-Phase Polymer Matrix Composite Subjected to Static Cyclic Loadings, Optoelectronics and Advanced Materials – Rapid Communications (OAM-RC), 5, 3, March (2011).

[3]. S. Vlase, H. Teodorescu-Drăghicescu, D.L. Motoc, M.L. Scutaru, L. Serbina, M.R. Calin, Behavior of Multiphase Fiber-Reinforced Polymers Under Short Time Cyclic Loading, Optoelectronics and Advanced Materials – Rapid Communications (OAM-RC), 5, 4, April (2011).

[4]. H. Teodorescu-Drăghicescu, A. Stanciu, S. Vlase, L Scutaru, M.R. Calin, L. Serbina, Finite Element Method Analysis Of Some Fibre-Reinforced Composite Laminates, Optoelectronics and Advanced Materials – Rapid Communications (OAM-RC), 5, 7, July (2011)

RESEARCH ON STUDY HEMP FIBER SUBJECTED TO THREE-POINT BEND TESTS

M. L. Scutaru[1*], C. Cofaru[1], H. Teodorescu-Drăghicescu[1]

[1*]Transilvania University of Brasov, Romania, e-mail: luminitascutaru@yahoo.com, ccornel@unitbv.ro, hteodorescu@yahoo.com

Abstract: In this paper, the most important mechanical features determined in three-point bend tests of a hemp fiber composite are presented. Following main mechanical properties have been determined: stiffness, Young's modulus of bending, flexural rigidity, load/stress/strain at maximum load, load/stress/strain at maximum extension, load/stress/strain at minimum load, load/stress/strain at minimum extension, work to maximum load/extension, work to minimum load/extension and load/stress at break.

Keywords: Three-point bend, Hemp Fiber, Composite

1. INTRODUCTION

The no woven polyester mat can be used as core material in thin sandwich structures to increase their stiffness. As skins used in sandwich constructions we can commonly encounter fiber-reinforced composite materials like glass fiber-reinforced polyester resins, carbon/aramid fiber reinforced epoxies or metal sheets. Since composite materials present anisotropies, the micromechanics of these materials are more complex than the metallic ones [1]. A "polymer matrix composite" is usually a structure known also as a composite laminate formed by a number of unidirectional reinforced laminate or fabrics layers of different types. As common reinforcement types there are fiber hemp and fabrics of various specific weights. These reinforcement materials, usually with high strengths and stiffness, are embedded in matrices with lower values of these features [1].

In general, a polymer matrix composite material is formed of at least two components: resin and reinforcement. Most common composites present three-phase compounds: resin, reinforcement and filler. For such anisotropic composites to predict their elastic properties, homogenization as well as averaging methods can be used. [1].

Tensile, compression and three-point bend tests are the most common tests to determine the mechanical properties of a composite material. For three-phase composites, static cyclic tension-compression tests have been also carried out in order to determine their hysteresis. These materials have been subjected to different number of cycles as well as load limits [2].

Elastic properties like Young's module Ex, Ey, shear modulus Gxy and Poisson's ratio have been computed in composite

2. DETERMINING MECHANICAL PROPERTIES USING THREE-POINT BEND TEST

From the fiber hemp plate, ten specimens have been cut according to SR EN ISO 14125:2000 and subjected to three-point bend test until break occur. The composite plate has been manufactured at Compozite Ltd., Brasov and tested in the Materials Testing Laboratory within Transilvania University of Brasov, Romania. The materials testing machine used in tests is a LR5KPlus type, produced by Lloyd Instruments, with following characteristics (Fig. 1):

- Force range: up to 5 kN;
- Test speed accuracy: < 0.2 %;
- Load resolution: < 0.01 % from the force cell;
- Extension resolution: < 0.1 microns.

Following main features have been determined:

- Stiffness (N/m);
- Young's modulus of bending(MPa);
- Flexural rigidity (Nm^2);
- Load/stress/strain at maximum load;
- Load/stress/strain at maximum extension;

- Load/stress at break;
- Work to maximum load/extension.

Test and specimens features are:
- Test speed: 5 mm/min;
- Median span: 64 mm;
- Median specimens width: 14,300 mm;
- Median specimens thickness: 5,5150 mm;
- Median cross-sectional area: 78,864 mm^2.

The materials testing machine allows determination of experimental results in electronic format by help of the NEXYGEN Plus software.

Figure 1. LR5KPlus materials testing machine

Figure 2. TA Plus materials testing machine in a three-point bend test

3. THREE- POINT BEND TEST RESULTS

Basic mechanical properties determined in three-point bend tests are presented in table 1.

Table 1. Basic mean mechanical properties of a hemp fiber composite

Feature	Value
Stiffness (N/m)	133440
Young's modulus of bending (MPa)	3559.4
Flexural rigidity (Nm^2)	0.72875
Load at maximum load (kN)	0.25351
Maximum bending stress at maximum load (MPa)	55.130
Extension at maximum load (mm)	2.7235
Maximum bending strain at maximum load (-)	0.022133
Work to maximum load (Nmm)	387
Maximum bending stress at maximum extension (MPa)	39.483
Extension at maximum extension (mm)	4,1705
Maximum bending strain at maximum extension (-)	0.033963
Work to maximum extension (Nmm)	428
Load at minimum load (kN)	0.00043888
Extension at minimum load (mm)	0.79157
Maximum bending stress at minimum load (MPa)	0.092288
Maximum bending strain at minimum load (-)	0.0065489
Work to minimum load (Nmm)	35
Load at minimum extension (kN)	0.000004152
Maximum banding stress at minimum extension (MPa)	0.00088132
Extension at minimum extension (mm)	0.000000176
Maximum banding strain at minimum extension (-)	0.0065489

The distribution shows a nonlinear tendency due to the nonlinearity of polyester resins and materials anisotropy. Load-extension distributions for all ten specimens subjected to three-point bend loads as well as other important characteristics are presented in Fig. 3-8.

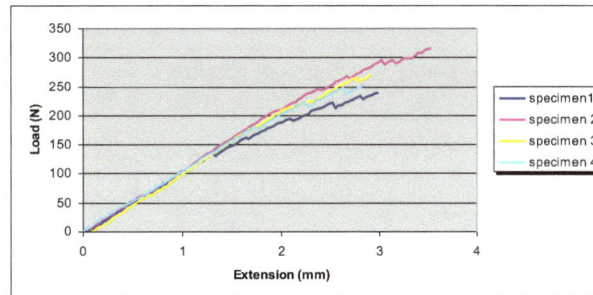

Figure 3. Load-extension distribution of specimens 1-4

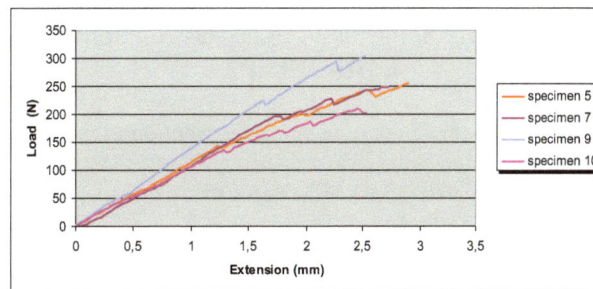

Figure 4. Load-extension distribution of specimens 5-10

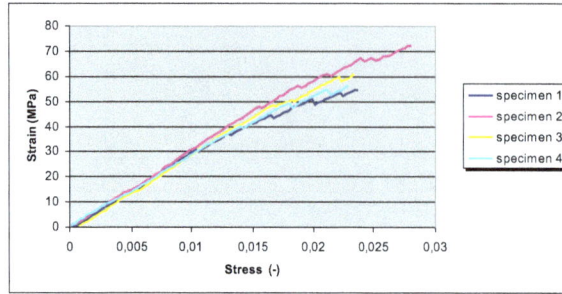

Figure 5. Strain-stress distribution of specimens 1-4

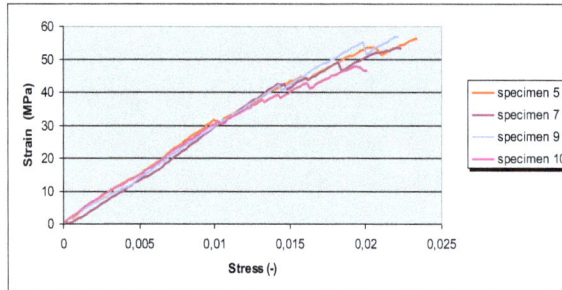

Figure 6. Strain-stress distribution of samples 5-10

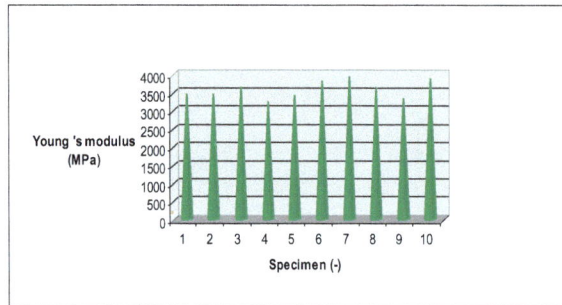

Figure 7. Young's modulus of bending distribution

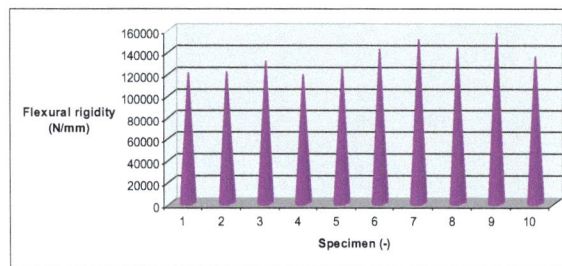

Figure 8. Flexural rigidity distribution

4. DISCUSSION

Regarding the maximum bending stress-maximum bending strain distribution of specimen's number 1-4 as shown in Fig. 5, in the strain range between 0 and 0.0130 the stress-strain distribution is almost linear. The stiffness decrease occurred at specimen 4 at 0.0142 strains and approximately 39.03 MPa stress stress due to core failure. Finally, the specimen 4 has been broken at 0.0022 strains and a stress of 55.96 MPa.

Regarding the maximum bending stress-maximum bending strain distribution of specimen's number 5-10 as shown in Fig. 6, in the strain range between 0 and 0.0100 the stress-strain distribution is almost linear. The stiffness decrease occurred at specimen 10 at 0.0101 strain and aproximately 31,016 MPa stress stress due to core failure. Finally, the specimen 10 has been broken at 0.0201 strains and a stress of 46.24 MPa.

Regarding the failure modes of the specimens, some of them presented delamination core at specific strain values and other specimens presented both delamination and core break at high strain values. Some failure modes of the hemp fiber structure are presented as side views in Figs. 9-10.

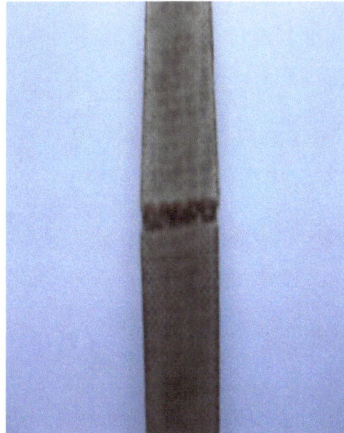

Figure 9. Typical failure mode in side view 1

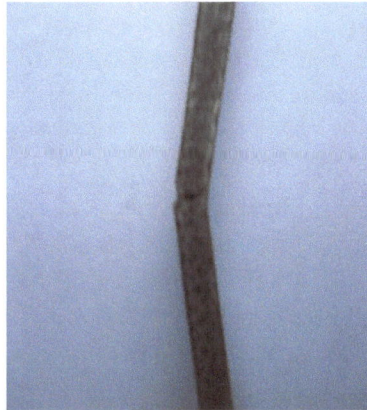

Figure 10. Typical failure mode in side view 2

ACKNOWLEDGEMENTS

This paper is supported by the Sectoral Operational Programme Human Resources Development (SOP HRD), financed from the European Social Fund and by the Romanian Government under the contract number POSDRU/89/1.5/S/59323.

REFERENCES

[1]. N.D. Cristescu, E.M. Craciun, E. Soos, Mechanics of elastic composites, Chapman & Hall/CRC, (2003).

[2]. H. Teodorescu-Draghicescu, S. Vlase, L. Scutaru, L. Serbina, M.R. Calin, Hysteresis Effect in a Three-Phase Polymer Matrix Composite Subjected to Static Cyclic Loadings, Optoelectronics and Advanced Materials – Rapid Communications (OAM-RC), 5, 3, March (2011).

[3]. S. Vlase, H. Teodorescu-Draghicescu, D.L. Motoc, M.L. Scutaru, L. Serbina, M.R. Calin, Behavior of Multiphase Fiber-Reinforced Polymers Under Short Time Cyclic Loading, Optoelectronics and Advanced Materials – Rapid Communications (OAM-RC), 5, 4, April (2011).

[4]. H. Teodorescu-Draghicescu, A. Stanciu, S. Vlase, L Scutaru, M.R. Calin, L. Serbina, Finite Element Method Analysis Of Some Fibre-Reinforced Composite Laminates, Optoelectronics and Advanced Materials – Rapid Communications (OAM-RC), 5, 7, July (2011).

[5]. H. Teodorescu-Draghicescu, S. Vlase, D.L. Motoc, A. Chiru, Thermomechanical Response of a Thin Sandwich Composite Structure, Engineering Letters, 18, 3, Sept. (2010).

NON-DESTRUCTIVE TESTING TO DETERMINE ACOUSTIC PROPERTIES OF LIGNOCELLULOSES COMPOSITES REINFORCED WITH WEAVE FABRICS OF FLAX FIBERS

M. D. Stanciu[1], I. Curtu[1], O. M. Terciu[1], C. Cerbu[1], S. Nastac[2]

[1] Transilvania University of Brasov, Brasov, Romania, mariana.stanciu@unitbv.ro
[2] "Dunarea de Jos" University of Galati, Galati, Romania, snastac@ugal.ro

Abstract: *The aim of this paper is to analyse the acoustic behaviour of lignocelluloses composites reinforced with weave fabrics of flax fibers in order to establish the proper application of them. The non-destructive used method is based on Kundt's tube in accordance with ISO 10534 standards. The experimental results in terms of absorption coefficient, reflection coefficient and impedance ratio are presented. According to variation curves of sound absorption coefficient and reflexive coefficient, the studied materials can be used in automotive components, noise reduction structures, buildings.*
Keywords: *acoustic, lignocelluloses composite, Kundt's tube, noise barriers, reflection, absorption*

1. INTRODUCTION

Designing composite materials obtained from waste or recyclable materials is a current trend in scientific research. Depending on the physical properties, mechanical and elastic that they need to hold, it can be design materials that meet the requirements of the application. This paper presents the study of acoustic properties of lignocelluloses composites reinforced with weave fabrics of flax fibers. Lignocelluloses fibres have a number of advantages compared with traditional glass fibres used to reinforce composite materials [1]. In previous work, the elastic, mechanical and dynamic properties were determined. Regarding the acoustical properties (absorption, reflection, impedance, sound transmission loss), literature review relieved numerous studies on different types of materials such polyester fibre, glass fibre and urethane foam, but about the new designed materials the references are very poor [2, 3]. To characterize the lignocelluloses composites reinforced with weave fabrics of flax fibers, the impedance tube was used in the experimental study.

2. EXPERIMENTAL SET-UP
2.1. Materials

The composite materials made of polyester resin and oak wood particles were studied. Different types of samples in term of thickness, number of layers and type of resin were performed. The plain weave fabric of flax fibres has a density per unit area of 225 g/cm2 and number of yarns per unit length is 14 yarns / cm for both directions of warp and weft yarns. The direction of the warp yarns being aligned with the length of the roll of fabric. The new lignocelluloses material was a laminate having 3 or 5 layers made of epoxy or polyester resin reinforced with plain weave fabric of flax fibres and wood sawdust of oak species. The composition of material can be seen in Figure 1. To manufacture the plate of composite material a lower forming pressure was used by hand lay-up process. The oak density was between 0.71 - 0.75 g/cm^3, but wood substance density (without cellular gaps) of the same wood species varies in the range 1.53 - 1.56 g/cm^3. For each type of composite, four samples were tested. For the experimental tests, the samples were cut into specimens with a diameter of 63.5 mm and the thickness in the range between 3 and 6 mm as can be seen in Figure 2.

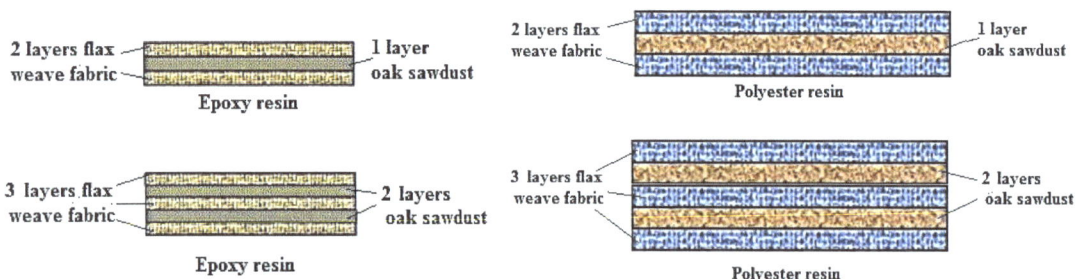

Figure 1: The structure of composite materials

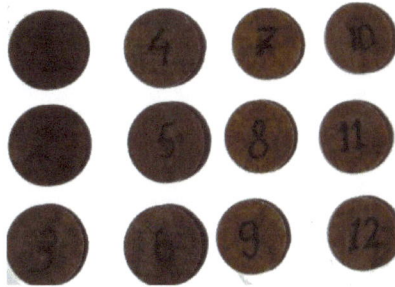

Figure 2: The tested samples

The principle of the impedance tube method is based on the measurement of transfer function between two signals of microphones mounted inside the tube. In accordance with measurement chain, a loudspeaker is placed at the end of the tube as can be seen in Figure 3 [4, 5]. In our experiment we used Kundt's tube [6, 8].

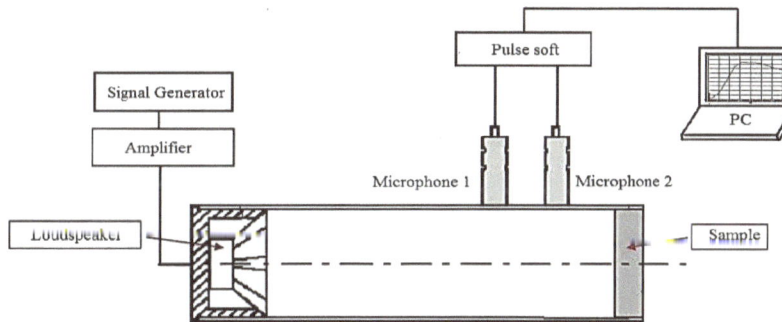

Figure 3: Schematic view of the experimental set-up

When the tube is fed by 1/3 octave frequency bands, a stationary plane wave is created and pressure measured with microphones can be decomposed into its incident and reflected components. First, the equipment without samples was prepared, in order to configure the microphones and to calibrate them using the calibration function from Pulse soft. [7] This operation is necessary because of phase and amplitude of the two microphones is not perfectly identical. In this sense the frequency response function is measured with the two microphones interchanged position. After calibration, each sample is properly inserted into the tube and the measurements started. The generated noise is connected to the amplifier and the tube filter emits the set signals. The emitted signal and reflected signal is captured by microphones and transmitted to Pulse hardware and displayed with the Pulse soft. The input data from the project set-up are established automatically by soft in the calibration stage.

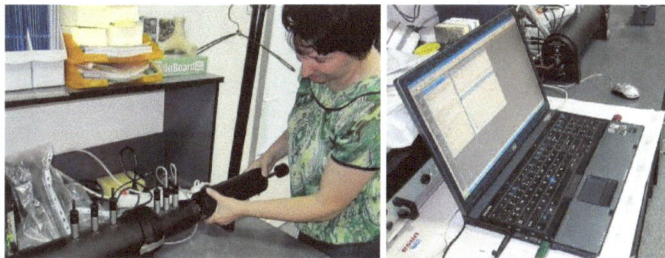

Figure 4: Capture image during the measurements

2.2. Results and discussion

The sound absorption coefficient indicates what amount of sound is absorbed in the actual material and depends of the frequency type. In Figure 5, the variation of average values of sound absorption coefficient against the frequency is presented, for different materials. Generally, all tested composite recorded low values of absorption coefficient. For high frequencies, it can be noticed that the absorption varied in accordance with type of resin and number of layers. So, the lignocelluloses composites with epoxy resin tend to absorb the high sound compare with the other one with polyester resin which tends to have a constant behavior.

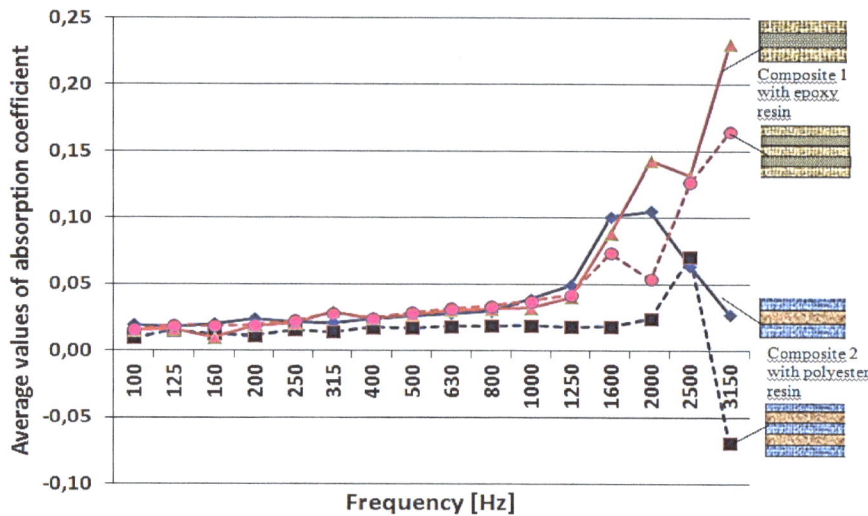

Figure 5: The average values of sound absorption coefficient of different types of materials

In Figures 6, 7, 8 and 9 the comparisons of reflection coefficients are presented. The acoustic reflection capacity characterized these lignocelluloses materials in terms of high values for frequency range between 100 and 3150 Hz. At high frequencies, it can be noticed the differences between composite with three and five layers (Figure 6).

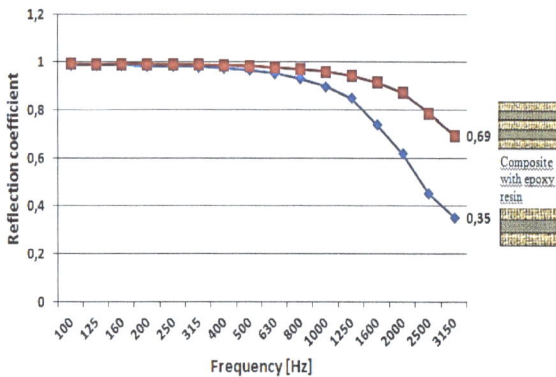

Figure 6: Comparison between the average values of sound reflection coefficient of composite with 3 and 5 layers with epoxy matrix

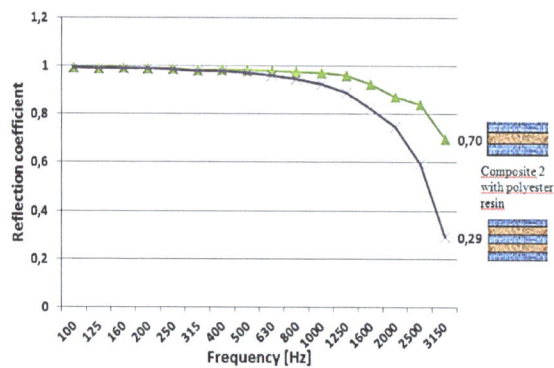

Figure 7: Comparison between the average values of sound reflection coefficient of composite with 3 and 5 layers with polyester matrix

The decreasing of reflection coefficient is more obviously for composite with five layers which express that at higher frequencies, the number of layers influences the acoustic behaviour. In Figure 8 is presented the comparison between different lignocelluloses composites in terms of type of used resin. If for the low and medium frequencies, the reflection behaviours is similarly for both types of materials, at high frequencies, the reflection of composite with epoxy resin is two time lower compared with the other one which contains polyester resin. Unlike the behavior of composites with three layers, the composites with five layers recorded the value of reflection coefficient almost twice higher for epoxy resin, as can be seen in Figure 9.

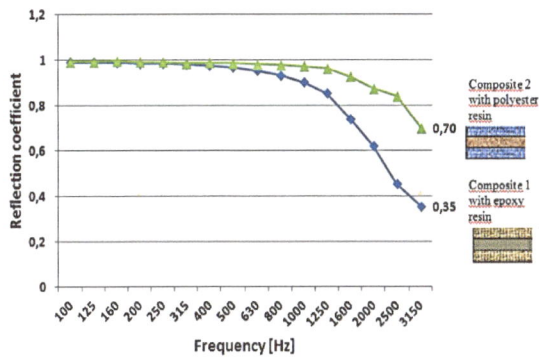

Figure 8: Comparison between the average values of sound reflection coefficient of composite with 3 layers with different matrix

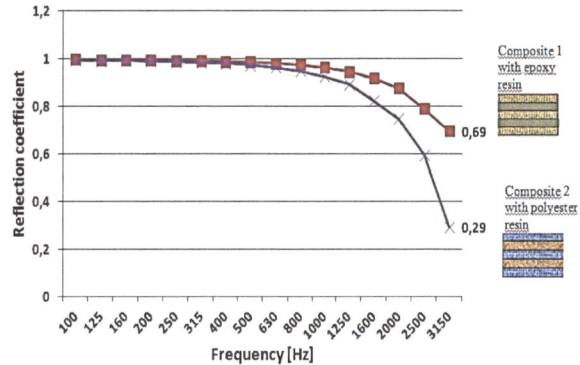

Figure 9: Comparison between the average values of sound reflection coefficient of composite with 5 layers with different matrix

3. CONCLUSION

The surface of separation between two different medium, reflection is always accompanied by a phenomenon of energy absorption. The tested composites recorded low the sound absorption at low and middle frequencies (100 - 1600 Hz). Being characterized by a good absorption coefficient for very high frequencies (ultrasonic), they are recommended for noise protection produced by industrial equipment and machine tools which emit high frequency. In this sense, it can be used in cabin structures, industrial wall and helmet or in combination with other soundproofing materials in construction of noise barriers or other structures with improved acoustic behavior.

4. ACKNOWLEDGMENTS

This paper is supported by the Sectorial Operational Programme Human Resources Development (SOP HRD), financed from the European Social Fund and by the Romanian Government under the contract number POSDRU POSTDOC-DD, ID59323- Transilvania University of Brasov, Romania.
We are also grateful to ICECON Romania, manager Prof. Polidor Bratu who facilitates the measurements.

REFERENCES

51. Terciu O. M, Curtu I., New Hybrid Lignocellulosic Composite made of Epoxi Resin Reinforced with Flax Fibers and Wood Sawdust, in Rev. Materiale Plastice, Vol 49 (2) (81-132), nr. 2, iunie 2012, ISSN 0025-5289, Bucuresti, Romania, pp. 114 –117.
52. Bratu M, Ropota I, Vasile O, Dumitrescu O, Muntean M., Research On The Absorbing Properties Of Somenew Types Of Composite Materials, In Romanian Journal Of Materials, Vol. 41, Nr. 2, 2011, P. 147- 154
53. Bratu P., Tracing Curves For The Sound Absorbing Characteristics In Case Of Composites Consisting Of Textile Materials, In Romanian Journal Of Acoustics And Vibrations, Vol 4, Nr. 1, 2007, P. 23-26
54. Jung S.S., Kim Y.T, Lee Y. B, Measurement Of Sound Transmission Loss By Using Impedance Tubes, In Journal Of The Korean Physical Society, Vol. 53, Nr. 2, 2008, P. 596-600.
55. Curtu I., Stanciu M. D., Cosereanu C., Vasile O., Assessment Of Acoustic Properties Of Biodegradable Composite Materials With Textile Inserts, In Rev. Materiale Plastice, Vol 49, Nr. 1, Ianuarie 2012, Issn 0025-5289, Bucuresti, Romania, Pp. 68 – 72.
56. Youneung Lee, Changwhan Joo, Sound Absorption Properties Of Recycled Polyester Fibrous Assembly Absorbers, Autex Research Journal, Vol. 3, No2, June 2003
57. Stanciu, M.D., Curtu I., Cosereanu C., Lica D., Nastac, S.: Research regarding acoustical properties of recycled composites, in Proceedings of the 8th International Conference of DAAAM Baltic Industrial Engineering, 19 – 21 April, 2012, Tallinn, Estonia, ISBN 978-9949-23-265-9, p. 741 - 746
58. ***SR EN ISO 10534-1 Determination Of Sound Absorption Coefficient And Acoustic Impedance With   The Interferometer, Part 1: Stationary Wave Ratio Method, 2002

SOLUTION OF THE THERMOELASTIC EQUILIBRIUM PROBLEM FOR CYLINDRICAL TUBES WITH BIG TORSION ANGLE IN THE CASE OF COMPOSITE MATERIAL STRUCTURES

M. Lupu[1], F. Isaia[2]

[1] Transylvania University of Brasov, Brasov, Romania, m.lupu@unitbv.ro, florin.isaia@unitbv.ro

Abstract: *In this paper, the coupled problem of the therm elastic equilibrium for cylindrical tubes with big torsion angle for composite materials is solved. The mathematical model of isotropic nanoelastic (auxetic) cylinders with big torsion angle is also presented. Using this new mathematical model, we actually solve the thermoelasticity problem for cylindrical tubes.*

Keywords: *thermoelasticity, elastic tube, torsion angle, state of stress.*

Consider an isotropic homogeneous cylindrical nanoelastic bar for composite materials. The lateral surface is free of external forces. The body forces are absent. We assume that the section is a bounded (simply or multiple connected) domain. We take the origin of coordinates at the centroid of the end section B_1 ($z = 0$), Oz axes parallel to the generators and Ox, Oy axes arbitrarily directed. The point B_2 is obtained for $z = L$, where L is the length of the bar (sufficiently long). The ends are acted on by distributed forces reduced to twisting moments M of opposite sense $\vec{M}_3^0(B_1) = -\vec{M}_3^0(B_2) = M\vec{k}$. This is the classical model of the torsion [1], [3], [5], [7], [9].

In this paper, we investigate the homogenized problem of the stationary thermoelasticity of a cylindrical tube with a big torsion angle, the deformation and the state of stress being caused by the torsion moment M (which is known) and by the heating of the boundary or by the change of temperature between the boundary and the environment.

1. THE MATHEMATICAL MODEL

Here, we present A. Y. Ishlinski's mathematical model (see [4]) for the torsion of a cylindrical bar in the case of nanoelastic composite materials with a big torsion angle.

Let's consider an element of length l located at distance r from the Oz-axis. Let α denote the specific torsion angle. After torsion, the generators of the cylindrical tube take the shape of a circular propeller of length

$$ds' = \sqrt{l^2 + r^2\alpha^2} \ . \tag{1}$$

If the expression $\dfrac{r\alpha}{l}$ is sufficiently small, the specific elongation is

$$\frac{1}{l}\left(\sqrt{l^2 + r^2\alpha^2} - l\right) \cong \frac{r^2\alpha^2}{2l^2} \ . \tag{2}$$

For a generator located at distance r from the cylindrical axis, we denote by χ the specific torsion coefficient:

$$\frac{r^2\alpha^2}{2l^2} = \chi r^2, \text{ where } \chi = \frac{\alpha^2}{2l^2} \ . \tag{3}$$

We take the expression χr^2 as a measure of the specific elongation to Oz-axis, therefore it will represent the component ε_{zz} of the deformation tensor. In cylindrical coordinates (r, φ, z), the nonzero deformations are (see [4]):

$$\varepsilon_{zz} = \chi r^2, \ \varepsilon_{rr} = \frac{\partial u}{\partial r}, \ \varepsilon_{\varphi\varphi} = \frac{u}{r} \ . \tag{4}$$

In (4), $u_r = u$ represents the radial displacement. Due to the condition of uniform torsion, the tangential displacement u_φ will be constant for any constant radius r and will not occur in the components of the deformation tensor.

The constitutive equations in the case of linear isotropic thermoelasticity, which generalize Cauchy and Hooke's equations, are

$$T_{ij} = \lambda \theta \delta_{ij} + 2\mu \varepsilon_{ij} - \beta T \delta_{ij},$$

$$\varepsilon_{ij} = \frac{1+\nu}{E} T_{ij} - \frac{\nu}{E} \Theta \delta_{ij} + \overline{\alpha} T \delta_{ij},$$

$$\theta - 3\overline{\alpha} = \frac{1-2\nu}{E} \Theta,$$

$$\beta = \frac{E}{1-2\nu} \overline{\alpha},$$

and they are due to Duhamel and Neumann (see [2], [8]). Here, λ and μ are Lamé's coefficients, E is Young's modulus, ν is Poisson's coefficient (see [1]), T is the temperature, $\overline{\alpha}$ is the coefficient of linear dilatation, $\theta = \varepsilon_{11} + \varepsilon_{22} + \varepsilon_{33}$, $\Theta = T_{11} + T_{22} + T_{33}$.

Due to the axial symmetry with respect to the Oz-axis and to the conditions which depends only on r, in cylindrical coordinates we have

$$\begin{aligned}
T_{zz} &= \lambda(\varepsilon_{zz} + \varepsilon_{rr} + \varepsilon_{\varphi\varphi}) + 2\mu\varepsilon_{zz} - \beta T(r), \\
T_{rr} &= \lambda(\varepsilon_{zz} + \varepsilon_{rr} + \varepsilon_{\varphi\varphi}) + 2\mu\varepsilon_{rr} - \beta T(r), \\
T_{\varphi\varphi} &= \lambda(\varepsilon_{zz} + \varepsilon_{rr} + \varepsilon_{\varphi\varphi}) + 2\mu\varepsilon_{\varphi\varphi} - \beta T(r), \\
T_{rz} &= T_{r\varphi} = T_{z\varphi} = 0.
\end{aligned} \tag{5}$$

Substituting (4) into (5) we get

$$\begin{aligned}
T_{zz} &= (\lambda + 2\mu)\chi r^2 + \lambda\left(\frac{\partial u}{\partial r} + \frac{u}{r}\right) - \beta T(r), \\
T_{rr} &= (\lambda + 2\mu)\frac{\partial u}{\partial r} + \lambda\left(\chi r^2 + \frac{u}{r}\right) - \beta T(r), \\
T_{\varphi\varphi} &= (\lambda + 2\mu)\frac{u}{r} + \lambda\left(\chi r^2 + \frac{\partial u}{\partial r}\right) - \beta T(r).
\end{aligned} \tag{6}$$

The equilibrium equation is

$$\frac{\partial T_{rr}}{\partial r} + \frac{T_{rr} - T_{\varphi\varphi}}{r} = 0, \tag{7}$$

By substituting (6) into (7), we get the linear differential equation of the radial displacement $u_r = u(r)$:

$$u'' + \frac{1}{r}u' - \frac{1}{r^2}u = -\frac{2\nu}{1-\nu}\chi r + \frac{(1+\nu)(1-2\nu)}{E(1-\nu)}\beta T'(r). \tag{8}$$

The torsion problems without heating for the cylinder and for the cylindrical tube, respectively, are developed in [4] and [6].

2. THE SOLUTION OF THE THERMOELASTIC EQUILIBRIUM PROBLEM

First, we solve the problem of the temperature distribution $T(r)$ for the annulus $a \leq r \leq b$. Due to the axial symmetry and to the boundary conditions (independent of φ), the heat equation will be

$$\frac{\partial^2 T}{\partial r^2} + \frac{1}{r}\frac{\partial T}{\partial r} = 0, \, a \leq r \leq b. \tag{9}$$

Theorem

If we associate to equation (9) the boundary conditions for $T(r)$ compatible in an exclusive manner, without heat sources, we get the situations (I-V) below, with the corresponding solutions:

(I) $\begin{cases} T(a)=0 \\ T(b)=T^* \end{cases} \rightarrow T(r)=\dfrac{T^*}{\ln\dfrac{b}{a}}\ln r - \dfrac{T^*\ln a}{\ln\dfrac{b}{a}}$,

(II) $\begin{cases} T(a)=T^* \\ T(b)=0 \end{cases} \rightarrow T(r)=\dfrac{T^*}{\ln\dfrac{a}{b}}\ln r - \dfrac{T^*\ln b}{\ln\dfrac{a}{b}}$,

(III) $\begin{cases} T(a)=0 \\ \dfrac{\partial T}{\partial r}(b)=T^* \end{cases} \rightarrow T(r)=T^* b \ln r - T^* b \ln a$,

(IV) $\begin{cases} T(a)=T^* \\ \dfrac{\partial T}{\partial r}(b)=0 \end{cases} \rightarrow T(r)=T^* a \ln r - T^* a \ln b$,

(V) $\begin{cases} T(a)=T_1 \\ T(b)=T_2 \end{cases} \rightarrow T(r)=k\ln r + k_1$,

where

$$k=\frac{T_2-T_1}{\ln(b/a)}, \quad k_1=\frac{T_1\ln b-T_2\ln a}{\ln(b/a)}.$$

Moreover, considering the change of temperature between the boundary and the environment, we have the general conditions

$$m_i T(r_i)+n_i\frac{\partial T}{\partial r}(r_i)=p_i, \quad i=1,2,$$

where $r_1=a$, $r_2=b$, with similar types of solutions as (I-V).

We denote by $T(r)=k\ln r+k_1$ the solution $T(r)$ corresponding to any case (I-V).

Therefore, $T'(r)=\dfrac{k}{r}$ in all these cases.

Consequently, equation (8) becomes

$$r^2 u'' + r u' - u = -\frac{2\nu}{1-\nu}\chi r^3 + \frac{(1+\nu)(1-2\nu)}{E(1-\nu)}\beta k r,$$

and its general solution will be

$$u(r)=c_1 r + c_2\frac{1}{r} - \frac{\nu}{4(1-\nu)}\chi r^3 + \frac{(1+\nu)(1-2\nu)}{E(1-\nu)}\beta k r. \tag{10}$$

Theorem 1

The solution of the thermoelastic equilibrium problem for cylindrical tubes with big torsion angle is

$$u(r)=-\frac{\nu\chi}{4(1-\nu)}\left[r^3+(1-2\nu)(a^2+b^2)r+\frac{a^2 b^2}{r}\right]+\frac{(1+\nu)(1-2\nu)}{2E}\beta r\left[\frac{1-2\nu}{1-\nu}k\frac{b^2\ln b-a^2\ln a}{b^2-a^2}-k+2k_1\right]$$

$$+\frac{(1+\nu)(1-2\nu)}{2E(1-\nu)}\beta k\frac{a^2 b^2(\ln b-\ln a)}{b^2-a^2}\frac{1}{r}+\frac{(1+\nu)(1-2\nu)}{2E(1-\nu)}\beta k r\ln r,$$

$$T_{rr}=\frac{E\nu\chi}{4(1-\nu^2)}\left[r^2-(a^2+b^2)+\frac{a^2 b^2}{r^2}\right]+\frac{1-2\nu}{2(1-\nu)}\beta k\left[\frac{b^2\ln b-a^2\ln a}{b^2-a^2}-\frac{a^2 b^2(\ln b-\ln a)}{b^2-a^2}\frac{1}{r}-\ln r\right],$$

$$T_{zz}=\frac{E\chi}{1-\nu^2}\left[r^2-\frac{a^2+b^2}{2}\nu^2\right]-\frac{1-2\nu}{1-\nu}\beta k\ln r+\frac{\nu(1-2\nu)}{1-\nu}\beta k\frac{b^2\ln b-a^2\ln a}{b^2-a^2}-\frac{\nu(1-2\nu)}{2(1-\nu)}\beta k-(1-2\nu)\beta k_1,$$

$$T_{\varphi\varphi} = \frac{Ev\chi}{4(1-v^2)}\left[3r^2 - (a^2+b^2) - \frac{a^2b^2}{r^2}\right] + \frac{1-2v}{2(1-v)}\beta k\left[\frac{b^2\ln b - a^2\ln a}{b^2-a^2} + \frac{a^2b^2(\ln b - \ln a)}{b^2-a^2}\frac{1}{r^2} - \ln r - 1\right].$$

Proof

Replacing $T(r) = k\ln r + k_1$ and $u(r)$ from (10) into T_{rr} from (6) and using the boundary conditions $T_{rr}(a) = 0$, $T_{rr}(b) = 0$ (the lateral surface is stress free), we can find the constants c_1, c_2 and then the above formulas for $u(r)$, T_{rr}, T_{zz}, $T_{\varphi\varphi}$.

Remark

For a given moment M, the constant χ can be found from the following identity

$$M = \iint_D T_{\varphi\varphi} r\,d\sigma, \quad D: a \le r \le b, 0 \le \varphi \le 2\pi,$$

we obtain

$$\chi = \frac{1-v^2}{Ev}\frac{15}{b^5 - a^5 - 5a^2b^2(b-a)}\left\{\frac{M}{2\pi} - \frac{1-2v}{2(1-v)}\beta k\left[\frac{4a^2b^2(\ln b - \ln a)}{3(a+b)} - \frac{2(b^3-a^3)}{9}\right]\right\}.$$

In particular, in the absence of the heat, the solutions for the cylindrical tube can be found by making $\beta = 0$ (see [6]):

$$u(r) = -\frac{v\chi}{4(1-v)}\left[r^3 + (1-2v)(a^2+b^2)r + \frac{a^2b^2}{r}\right],$$

$$T_{rr} = \frac{Ev\chi}{4(1-v^2)}\left[r^2 - (a^2+b^2) + \frac{a^2b^2}{r^2}\right],$$

$$T_{zz} = \frac{E\chi}{1-v^2}\left[r^2 - \frac{a^2+b^2}{2}v^2\right],$$

$$T_{\varphi\varphi} = \frac{Ev\chi}{4(1-v^2)}\left[3r^2 - (a^2+b^2) - \frac{a^2b^2}{r^2}\right].$$

Let us notice that, due to the heat, the deformations, the displacements and the stresses are different in comparison to the case of a simple torsion (by the occurrence of β). Various studies and diagrams can also be done for stresses and deformations in the thermoelastic case.

In the case of a simple torsion, the new deformed radii of the new tube become

$$a' = a + u(a) = a\left[1 - \frac{v\chi}{2}(a^2+b^2)\right],$$

$$b' = b + u(b) = b\left[1 - \frac{v\chi}{2}(a^2+b^2)\right].$$

By fixing the distance L between the ends of the tube, the inner radius a and the outer radius b, then the thickness $d = b - a$ becomes after torsion:

$$d' = b' - a' = d\left[1 - \frac{v\chi}{2}(a^2+b^2)\right],$$

with $d' < d$.

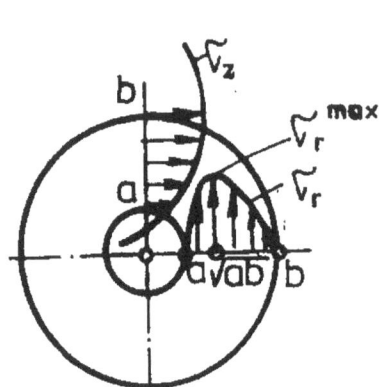

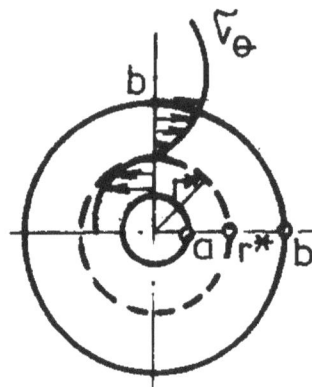

Figure 1   **Figure 2**

The diagram of the stresses: σ_r, σ_z, σ_θ

In the thermoelastic coupled case, we have

$$d' = d\left[1 - \frac{v\chi}{2}\left(a^2 + b^2\right) + \frac{(1+v)(1-2v)}{2E}\beta\left(2k\frac{b^2\ln b - a^2\ln a}{b^2 - a^2} - k + 2k_1\right)\right].$$

The variation of the thickness in the thermoelastic coupled case will depend on the terms containing β and on the inequality between T_1, T_2 (if $T(r)$ is taken from (V)).

REFERENCES

59.   Dragoş, L.: Principiile mecanicii mediilor continue, Ed. Tehnică, Bucureşti, 1973.

60. Ieşean, D.: Teoria termoelasticităţii, Ed. Academiei Române, Bucureşti, 1979.

61. Isaia, F.; Lupu, M.; Nicoara, D.: Analytical methods for the solution of the plane problem of the thermoelastic equilibrium, Proceedings of the 3rd International Conference on "Computational Mechanics and   Virtual Engineering" COMEC 2009, Vol. 1, p. 403-409, 29–30 October 2009, Brasov, Romania.

62. Ishlinski, A. Y.: Mechanics of elastic solid and rigid body, Nauka, Moscow, 1979.

63. Lupu, M.: Metode şi modele matematice in mecanica mediilor continue, Ed. Univ., Braşov, 1985.

64. Lupu, M.; Nicoara, D.: Study of the state of stress of elastic tube for big torsion angles, Bul. Univ. Tehn. "Gh. Asachi", Tom. XLVII, Fasc. 1-2, 95-101, 1997.

65. Lupu, M.; Isaia, F.: The solution of the thermoelastic equilibrium problem for cylindrical tubes with big torsion angle, Carpathian Journal of Mathematics, Vol. 21, No. 1-2, 77-82, 2005.

66. Teodorescu, P.P.: Asupra problemei plane a termoelasticitaţii, Bul. Şt. Acad. R.P.R., Secţ. Şt. Mat. Fiz., 2, 333-340, 1957.

67. Timoshenko, S.P., Goodier, J.N.: Theory of elasticity, Nauka, Moscow, 1979.

MICROSTRUCTURAL CHARACTERIZATION OF AlSi7Mg/AlN AND AlSi12Mg/SiC COMPOSITES OBTAINED BY REACTIVE GAS INJECTION METHOD

D. Mitrica [1], V. Soare [1], I. Constantin [1], F. Stoiciu [1], G. Popescu [2]

[1] National R&D Institute for Nonferrous and Rare Metals, Pantelimon, ROMANIA, dmitrica@imnr.ro, vivisoare@imnr.ro, iconstantin@imnr.ro, fstoiciu@imnr.ro

[2] University "POLITEHNICA" of Bucharest, Bucharest, ROMANIA, gabriela81us@yahoo.com

Abstract: *The paper studies the microstructural and chemical characterization of in-situ AlSi7Mg/AlN and AlSi12Mg/SiC composites obtained by the injection of reactive gases (CH_4 and N_2) in aluminum alloy melt. The samples were characterized using optical microscopy, X-ray diffraction and SEM. The chemical composition was determined by inductive coupled plasma spectrometry. Samples taken from different parts of the product revealed uneven compositional distribution of reinforcements and primary silicon in the alloy matrix. In the upper side of the products was found a significant enrichment of the reinforcing particles, for both composite systems, caused by the small dimensions of the particles and the upward movement of the injected gas, during the bubbling process. Optical micrographs revealed SiC and AlN particles of irregular shapes and small sizes ($<10\mu m$).*
Keywords: *composites, in-situ, SiC, AlN*

1. INTRODUCTION

Composite materials possess remarkable mechanical properties (yield strength, stiffness, hardness and resilience), high temperature stability and a reduced specific weight. The special properties of composite materials make them attractive for applications as structural and functional materials, if conventional materials do not reach the specific demands.

Particle reinforced aluminum metal matrix composites are known for their high strength to weight ratio. Examples of applications for these products range from expensive satellites and aerospace structures to sports products. Further development in manufacturing particle reinforced aluminum metal matrix composites created the possibility to obtain large concentrations exceeding 10% by volume of reinforcement (SiC, AlN, Al2O3, etc.). The improvement of the strength and wear resistance at elevated temperatures of the composites, made them appropriate for the use in automotive applications, such as the pistons, rotors, disc brakes, and propeller shafts [1]. But, in order to benefit from the advantages of the composite materials a reasonable production cost – performance relationship must be obtained [2]. Recent innovations in the manufacturing of aluminum-based metal matrix composites provided material quality improvements but no feasible solutions for lowering the present high price of the final product.

One solution that can produce new stable composites for advanced structural and wear applications is represented by the in-situ methods for manufacturing of metal-matrix composites (MMCs) [3]. In this type of process a part of the starting metal or alloy is used as a reactive element for the obtaining of secondary phase particles by the interaction with highly reactive gases or solids. Thermodynamically stable reinforcing particles are obtained by in situ processing and the interface incompatibilities are eliminated. A fundamental characteristic of the in situ methods is that the nucleation and growth of the reinforcing particles occur in the matrix. Depending on the nature of the base alloy a wide range of oxides, nitrides, carbides, borides and silicides can be obtained [4].

Various trials, with successful, have been performed for producing in-situ MMC by different methods (SHS, DIMOX, XD, PRIMEX, RD, Mixalloy, Osprey, Direct nitridation, Mixed Salt Reaction, [4,5,6]), reactive plasma synthesis in [7], casting [8], rapid solidification [9,10], DERP [11,12], magnetochemistry reaction [13], reactive gas injection (RGI) [14,15,16,17], etc..
The synthesis of aluminium metal matrix composites (AlMMC) by reactive gas injection (RGI) was first proposed by Kocsak and Kumar [14]. This is a relatively simple method were the reinforcing particles are obtained in-situ from the reaction of the injected gas with the molten matrix alloy. Various systems have been investigated in the past, with CH_4, N_2, NH_3, or a mixture of these as the reactive gas and the molten alloy was composed of Al, Si, Ti, Ta, etc. Wu and Reddy [15] studied the synthesis of Al-Si/SiC composites by RGI method. The research results indicated the presence of a large quantity of reinforcement (30 wt.%) in a solidified foam composite collected in a trey under the crucible,

composite obtained by overflowing of the melt during the bubbling process. The SiC reinforcing particles were agglomerated around the primary silicon formations, had irregular shapes and were of small sizes (5-10μm). Q. Zheng and R. G. Reddy [16] investigated the in situ processing of AlN particle reinforced aluminum composites by bubbling nitrogen gas in a pure molten aluminum matrix. The obtained in situ AlN particles were small in size (<5 μm) and were present in the top part of the product formed in the crucible. They concluded that nitrogen gas injection did not lead to the formation of significant AlN due to the deleterious effect of the trace oxygen impurities in the bubbling gas.

C. Borgonovo and D. Apelian [17] also obtained AlN reinforced composites, in aluminum melt with high magnesium additions. They observed that magnesium acted as an oxygen getter diminishing the content of oxygen in the melt by MgO formation and thus enhancing the nitridation reaction. The size of the particle embedded in the matrix varied from 1- 3 μm to sub-micron scale.

Present paper studies the chemical and microstructural characterization of AlSi7Mg/AlN and AlSi12Mg/SiC composites obtained in situ by injection of reactive gases (CH_4 and N_2) in the alloy melt. The research work is intended to produce in-situ composites without overflowing in a foam state (for SiC synthesis) and using purified nitrogen as precursor gas in conjunction with the addition of a small quantity of magnesium in the initial alloy (for AlN synthesis). The research work investigates the structural particularities of the samples for the two types of MMC systems.

2. EXPERIMENTAL

The schematic diagram of the experimental installation is given in figure 1. A vertical Carbolite resistance furnace with a working temperature of up to 1200°C was used in the experiments. The experiments took place in a sealed reaction chamber which was introduced in the furnace. The reaction chamber was provided with a water cooled lid, protective atmosphere, a thermocouple to record the process temperature, a pressure gauge and a gas flow meter. An eye hole was provided on the top of the reaction chamber for observing and monitoring the process.

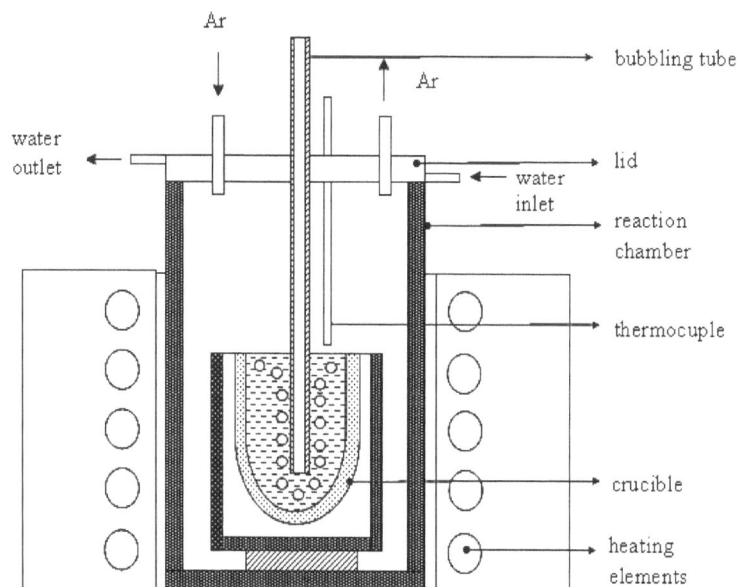

Figure 1. Experimental installation for the in-situ synthesis of AlSi7Mg/AlN and AlSi12Mg/SiC composites

The starting aluminum alloy Al-Si-Mg (300 grams, 7÷12% Si) was placed in a crucible, in the reaction chamber. Before the experiments, the system was evacuated and purged with argon for 30 minutes and was maintained at 0.1 bar during the whole process.

The alloy was melted and maintained at the process temperature (900 - 1200°C) for 30 minutes, before the beginning of the bubbling process. The bubbling tube (high density graphite, 30 cm long) was submersed into the melt close to the bottom of the crucible. The reactive gas was injected in the alloy melt at a constant flow rate, through four nozzles situated at the bottom of the bubbling tube. The entire process was monitored through the eye hole placed on the lid of the reaction chamber. The duration of the bubbling process was estimated from the stoichiometric relationship between initial materials and expected reinforcement percentage. At the end of the bubbling process, the resulting material was left in the furnace to cool at room temperature. The obtained material was then prepared for chemical and microstructural characterization.

The chemical analysis was performed using an optical emission plasma spectrometer – DCP, Spectraspan V-Beckman – Germany. The SiC and AlN quantity was determined by the investigation of the carbon and nitrogen content with a Leybold Heraeus CS 5003 gas analyzer and calculated from the stoichiometry of the SiC and AlN formation reactions.

The metallographic characterization of the samples was effectuated with a Zeiss Axio Scope A1m Imager microscope, with bright field, dark field, DIC and polarization capabilities, and high-contrast EC Epiplan 10X/50X/100X lenses. The micrographs were captured with a polarized camera provided with the equipment.

The SEM-EDAX analyses were performed using a FEI Quanta Inspect F scanning electron microscope, provided with field emission and a dispersive energy analysis system (EDS).

The X-ray diffraction characterization of the samples was performed using a BRUKER D8 DISCOVER X-ray diffractometer. The obtained data was processed using the FPM (Full Pattern Matching) module from the DIFFRACplus BASIC (Bruker AXS) program package and the ICDD PDF-2 Release 2006 database.

3. RESULTS AND DISCUSSIONS

Tables 1 and 2 represent the chemical composition of the two composite samples. The analyses performed on samples taken from the top part of the composites revealed significant concentrations of AlN and SiC. The variation of the magnesium contents with the height of the crucible is due to the bubbling process.

The obtained AlN and SiC quantities were calculated using the values obtained after the nitrogen and carbon determinations in the samples and the molecular ratios of the atomic masses. The analyses performed on samples taken from the top part of the composite products revealed higher concentrations of AlN, SiC, Mg and Si. This is explained by the low surface tension of these substances with the gas bubble, determining their entrainment in the gas flow in the liquid alloy

Table 1. Chemical composition (%) from different parts of the in-situ AlSi7Mg/AlN product

Level	Si	Mg	AlN	Fe	Cu	Mn	Ti	Al
top	6.87	0.40	4.47	0.10	<0.1	<0.01	<0.001	base
middle	6.27	0.12	2.73	0.12	<0.1	<0.01	<0.001	base
bottom	6.22	0.04	0.71	0.12	<0.1	<0.01	<0.001	base

Table 2. Chemical composition (%) from different parts of the in-situ AlSi12Mg/SiC product

Level	Si	Mg	SiC	Fe	Cu	Mn	Ti	Al
top	12.61	0.59	7.60	0.091	<0.1	<0.01	<0.001	base
middle	12.03	0.36	3. 57	0.093	<0.1	<0.01	<0.001	base
bottom	11.40	0.29	0.86	0.061	<0.1	<0.01	<0.001	base

The optical micrographs captured from the upper side of the AlSi7Mg/AlN composite samples (figure 2) indicated the presence of light color Al_{ss} dendrites, black pores generated by the gas, grey Al-Si interdendritic eutectic and AlN particle agglomerations. The reinforcement particles presented irregular shapes, small dimensions and were grouped as clusters and strings at the grain boundaries and in the areas with increased porosity.

In the samples taken from the upper side of the AlSi12Mg/SiC composite products (figure 3), the typical structure of the AlSi12 alloy can be observed (dark grey primary Si and Al-Si eutectic lamellae embedded in α-AlSi) and the presence of in-situ formed SiC particles, separated from the matrix by dark boundaries. The ceramic reinforcing particles were of small sizes (<10μm) and with angular contour.

The different distribution of the particles in the two composite systems is explained by the particle/matrix compatibility and the differences in the sizes of the particles.

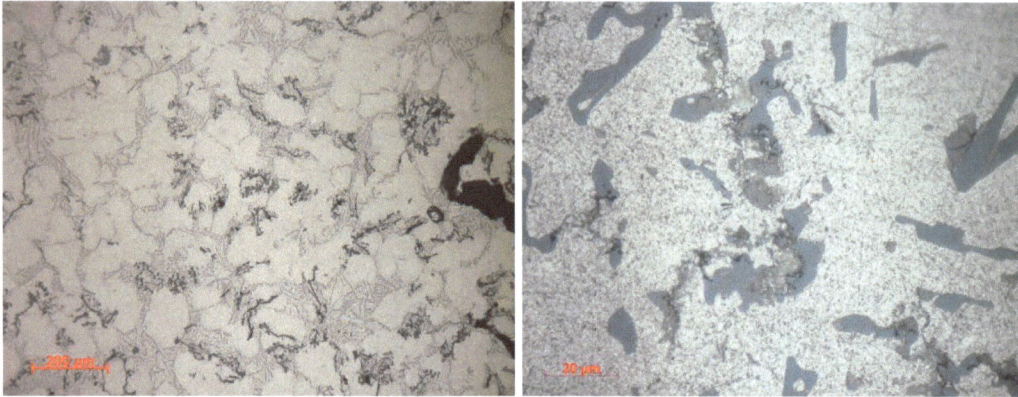

Figure 2 Micrographs of the top part of the AlSi7Mg/AlN composite sample

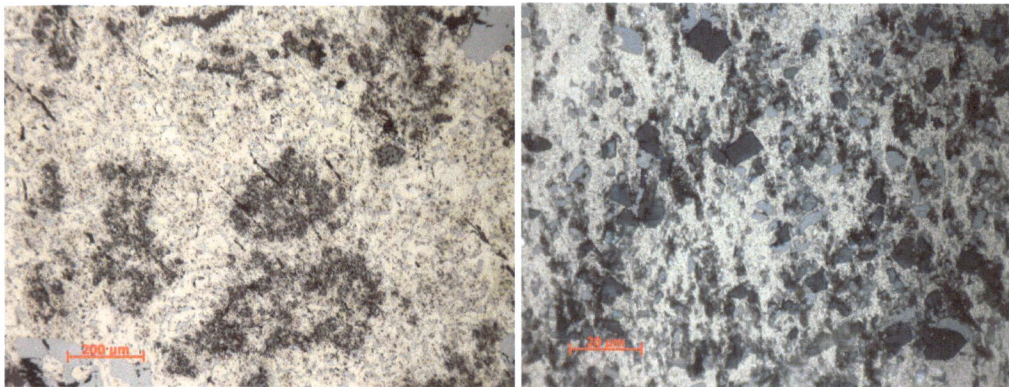

Figure 3 Micrographs of the top part of the AlSi12Mg/SiC composite sample

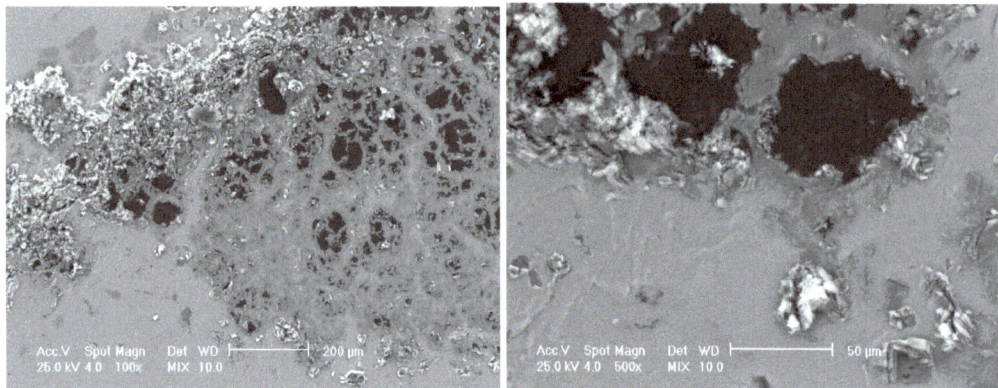

Figure 4. SEM images from top part of the AlSi7/AlN composite sample

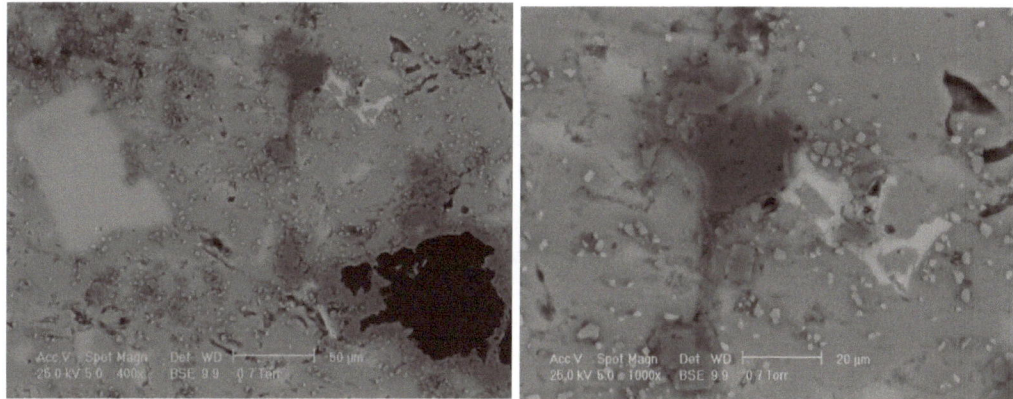

Figure 5. SEM images from the top part of the AlSi12Mg/SiC composite sample.

SEM images revealed dense agglomerations of AlN particles near the areas with high concentration of porosities, in the top part of the AlSi7Mg/AlN composite (Figure 4). The average particle size ranged from 1 to 5μm. SiC particles found in the samples taken from the top part of the AlSi12Mg/SiC composite (Figure 5) present a better distribution in the alloy mass, and they can be clearly distinguished even in lower magnification images.

The X ray diffraction patterns (figures 6 and 7) indicated a phase composition formed of Al-α solid solution, primary Si, Mg_2Si, MgO, spinels, SiC as moissanite and AlN. $MgAl_2O_4$ spinel is formed mainly as a result of the reaction between the trace oxygen as impurity from the injected gas and the magnesium contained in the melt. The phase composition for both composite systems is presented in table 3.

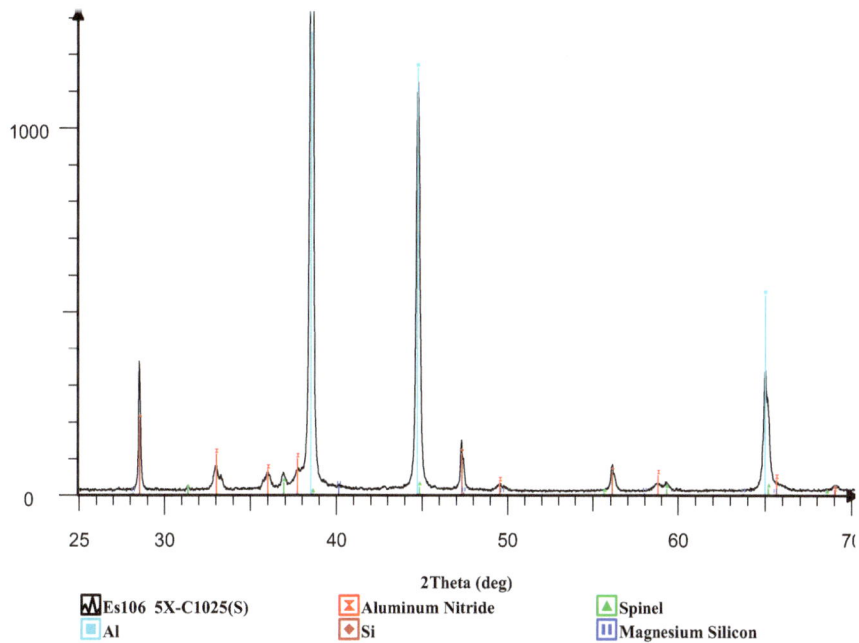

Figure 6 X-ray diffraction pattern of a sample from the top part of the AlSi7Mg/AlN composite sample

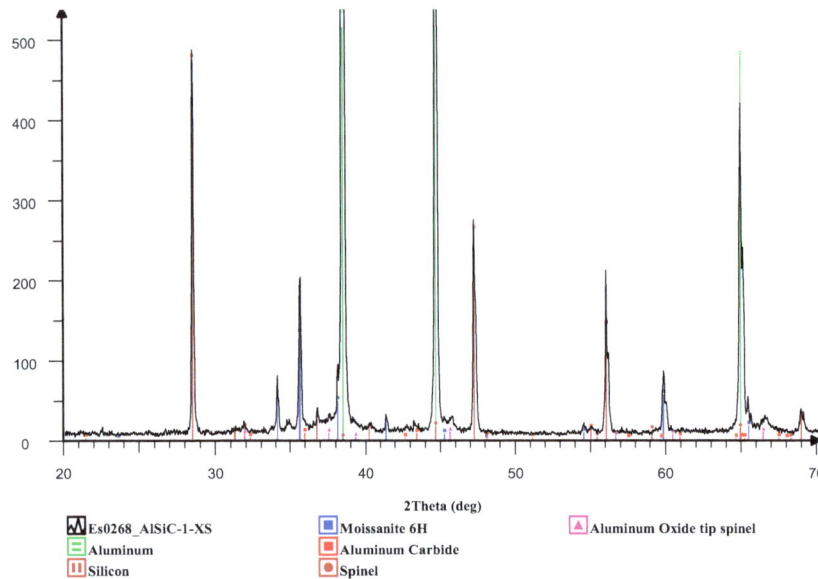

Figure 7. X-ray diffraction pattern of the superior and inferior part of the AlSi12Mg/SiC composite sample

Table 3. Phase composition of the top part of the samples, for both composite systems.

Composite system	AlSi/AlN					AlSi/SiC				
Phase	Al	Si	AlN	MgAl$_2$O$_4$	M$_2$Si	Al	Si	SiC	MgAl$_2$O$_4$	Al$_2$O$_3$
S-Q(wt.%)	82.8	5.2	9.3	2.2	0.5	71.3	12	11.8	1.7	1.7

4. CONCLUSIONS

Aluminum metal matrix composites reinforced with AlN and SiC, obtained in-situ by the reaction of a precursor gas (nitrogen and methane) with matrix alloy, were characterized by chemical analyses, optical microscopy, SEM and X-ray diffraction. The chemical and microstructural investigations indicated a higher concentration of reinforcing particles in the top part of the product. The obtained particles were distinguished as large agglomerations (for SiC) or cluster strings in the grain boundary area (for AlN) and presented small dimensions (<5μm for AlN and <10μm for SiC) and irregular shapes. SEM determinations show different distributions and morphologies of the particles for the two composite systems; and indicate that the agglomeration of AlN particles is more probable in zones with high concentration of micropores. XRD analyses confirmed the formation of AlN and SiC particles in the melt, toghether with spinels formed by the oxygen impurities from the injected gas.

REFERENCES

68. Kainer U. K., Metal Matrix Composites. Custom-made Materials for Automotive and Aerospace Engineering, WILEY-VCH Verlag GmbH & Co. KGaA, Weinheim, 2006.
69. Suresh S., Mortensen A., Needleman A., Fundamentals of metal matrix composites, published by Butterwor-Heinemann, 1993, pp. 3-22.
70. Koczak M.J., Premkumar M.K., Emerging technologies for the in-situ production of MMCs, JOM, vol.45, no.1 (1993), 44-48
71. Apelian D., Processing challenges of light weight, high specific strength metallic materials, Worchester Polytechnic Institute.
72. Cup C., Shen Y., Meng F., Review on Fabrication Methods of In Situ Metal Matrix Composites, Journal of Materials Science and Engineering, vol. 16, no.6 (2000), 619-626.
73. Surappa M. K., Aluminium matrix composites: Challenges and opportunities, Sadhana, vol. 28 (2003), parts 1 & 2 , 319–334.

74. Borisov V. G., Ljudmila P. Borisenko, Alexandr V. Ivanchenko, Alexandr P. Bogdanov, Metal for production of metal base composite material, US Patent 5305817, 1994.

75. Varin R.A., Intermetallic-Reinforced Light-Metal Matrix In-Situ Composites, Metallurgical and Materials Transactions A, vol. 33A (2002), 193-201.

76. Tong X.C., Fang H.S., Al-TiC Composites In Situ–Processed by Ingot Metallurgy and Rapid Solidification Technology: Part I. Microstructural Evolution, Metallurgical and Materials Transactions A, vol. 29A (1998), 875-891.

77. Tong X.C., Fang H.S., Al-TiC Composites In Situ–Processed by Ingot Metallurgy and Rapid Solidification Technology: Part II. Mechanical Behavior, Metallurgical and Materials Transactions A, vol. 29A (1998), 893-902.

78. Song I. H., Kim D. K., Hahn Y. D., Kim H. D., Synthesis of In-Situ TiC-Al Composite by Dipping Exothermic Reaction Process, Metals and Materials International, vol. 10, no. 3 (2004), 301-306.

79. Song I. H., Hahn Y. D., Kim H. D., Kyung K. D., Method for making high volume reinforced aluminum composite by use of dipping process, Patent US6406516, 2002.

80. Zhao Y. T., Zhang S. L., Chen G., Aluminum matrix composites reinforced by in situ Al2O3 and Al3Zr particles fabricated via magnetochemistry reaction, Transactions of Nonferrous Metals Society of China, vol. 20 (2010), 2129-2133.

81. Koczak M. J., Kumar K.S., In situ process for producing a composite containing refractory material, U.S. patent 4808372, 1989.

82. Wu B., Reddy R. G., In-situ Formation of SiC Alloy Composites Using Methane Gas Mixtures, Metallurgical and Materials Transactions B, vol. 33B (2002), 543-550.

83. Zheng Q., Reddy R. G., Mechanism of in situ formation of AlN in Al melt using nitrogen gas, Journal of Materials Science, vol. 39 (2004), 141– 149.

84. Borgonovo C., Apelian D., Aluminum Nanocomposites via Gas Assisted Processing, Materials Science Forum, vol. 690 (2011), 187-191.

CORRELATION OF THE HUMAN DENTINE HARDNESS AND ELASTIC PROPERTIES

V. Pomazan[1], C. L. Petcu[1]

[1] Ovidius University, Constanta, ROMANIA, vpomazan@univ-ovidius.ro, petculucian@univ-ovidius.ro

Abstract: *The paper presents a study performed on human dentine, in order to correlate the micro hardness values measured with elastic evolution of the indentation and mechanical behavior of the ceramic biocomposite. Statistical correlations were made between the indentations elastic recovery and the morphology and normalization formulae was used to establish a correct value for the Vickers micro hardness index .*

Keywords: *biocomposite, dentine, micro hardness, elastic recovery*

1. INTRODUCTION

Knowing the real macro and micro properties of the human tooth tissues, is very important for the choice of compatible building materials, with a behavior similar with the natural biocomposites. Of a great importance is the behavior at dynamic, cyclic loads or repetitive temperature variations. The actual elastic modulus and hardness values can be of a great help in understanding how the man-made materials can absorb and dissipate the stresses, both locally and in the tooth block.

This study aims to determine the human dentine hardness using the micro hardness Vickers index and to analyze the indentation force correlated with the indentation time, and to establish an effective method to take into account the elastic recovery phenomenon, when stating the Vickers index. This value is to be further linked with mechanical and chemical properties of the dentine.

2. THE ANALYSIS METHOD

Using an HV-1000 apparatus, and several forces with corresponding application durations (10gf, 25gf, 50gf for 10s,15s, 30s) [1], for each dentine sample incorporated in polymer support (figure 1) were made 10 indentations, at appropriate distance (larger than 2.5 max. diagonal) [1, 2]. The indentations were check for symmetry and filtered. Figure 1 presents the samples and indentations, figure 2 presents detailed view for sampled indentations for various load at the same measurement time. Figure 3 shows the distribution of the HV values on the measured length L, at 25gf/10s.

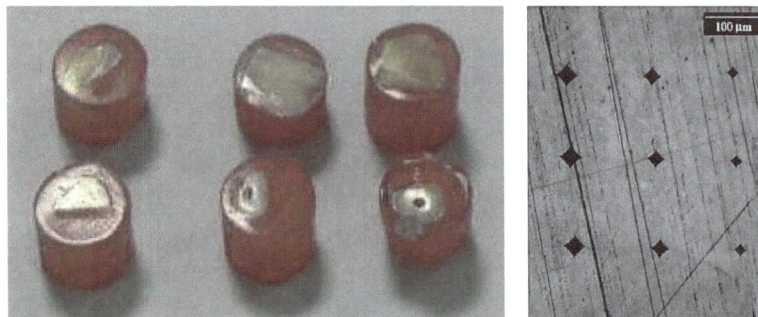

Figure 1: The dentine samples and a capture of the indentations

Figure 2: Indentations in human dentine for: 10gf/10s, 25gf/10s, 50gf/10s.

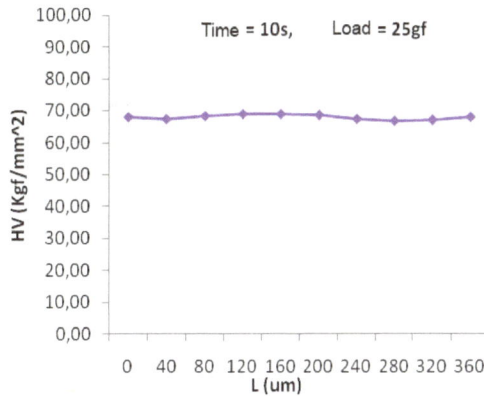

Figure 3: HV values on the measured length L, at 25gf/10s

For each sample, the arithmetic mean, standard deviation, asymmetry quotients were calculated (figure 4). The One Way ANOVA test [3] confirmed that there are not significant differences among hardness values for the ten indentations per sample, each set being obtained for different force and time values (F<Fcr, p>0,05).

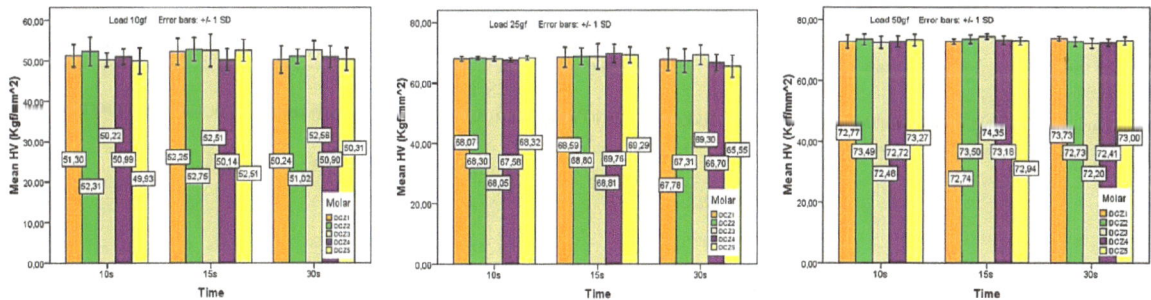

Figure 4: Comparison of the human dentine HV values, for 10, 25 and 50 gf at different load durations.

The result of this preliminary analysis stated that, for a given duration and load, the hardens values obtained for each sample can be associated in simple data sets of 50 values, fitted for further statistical analysis [4, 5]. The One Way ANOVA applied for these data sets showed that, for a given load, there are no significant differences among the hardness values (F<Fcr, p>0,05), depending of the indentation load of, respectively, 10s, 15s and 30s (figure 5).

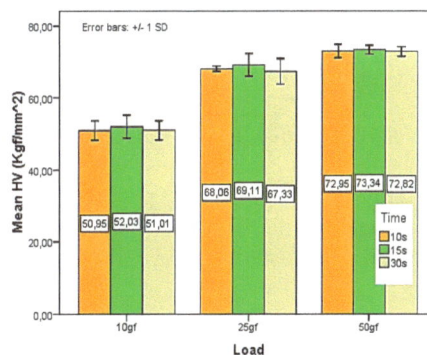

Figure 5: Comparison of the dentine HV values at a given load and different durations.

The unifactorial analysis One Way ANOVA applied to the 50 component data sets [5] revealed that, for a given indentation duration, there are significant differences amongst the HV values (F>Fcr, p<0,05), depending on the indentation loads of, respectively, 10gf, 25gf and 50gf. This conclusion is sustained also by Post-Hoc Multiple Comparisons with a probability of p< 0,001 in all cases, the largest value registered corresponding to the largest load of 50gf.

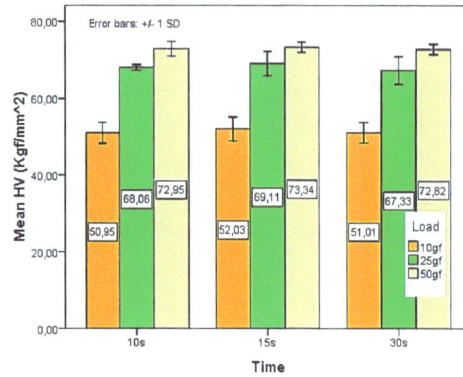

Figure 5: Comparison of the dentine HV values at a given duration and different loads.

The data were analyzed with Two Way ANOVA [6], with several hypotheses regarding: the global effect of the factors involved in hardness determination (load and duration), considered altogether, indistinct; principal effects, for each analysis factor and cumulative effect of the independent variables (force and load) on the dependent one (HV) (table 1).

Table 1: Tests of Between-Subjects Effects

Dependent Variable:HV (Kgf/mm^2)

Source	Type III Sum of Squares	df	Mean Square	F	Sig.	Partial Eta Squared
Corrected Model	39041,190[a]	8	4880,149	799,938	,000	,936
Intercept	1853378,493	1	1853378,493	303799,753	,000	,999
Load	38916,232	2	19458,116	3189,511	,000	,935
Time	100,866	2	50,433	8,267	,000	,036
Load * Time	24,093	4	6,023	,987	,414	,009
Error	2690,390	441	6,101			
Total	1895110,074	450				
Corrected Total	41731,581	449				

a. R Squared = ,936 (Adjusted R Squared = ,934)

Table 1 reveals that, although the global effect is significant, (F = 799,93 and p < 0.001), a detailed analysis shows that the dentine hardness values is separately intensely influenced by the load (F = 3189,5 cu p < 0,001), with an intensity of $\eta2 = 0,935$, while the indentation duration has a smaller influence (F = 8,26 cu p < 0,001) with an intensity of $\eta2 = 0,036$; the cumulative influence of the two factors is insignificant (F = 0,987 cu p = 0,414).

3. HUMAN DENTINE ELASTIC BEHAVIOUR

3.1. The elastic recovery index for the indentation depth correction

After the hardness test, important modification in the imprint geometry occur, if the material limits are not over passed. In order to take into account the elastic recovery of the biocomposite, when a micro hardness value is obtained via indentation, an supplementary index is needed – the elastic recovery index I_{er} (1), to reflect the imprint depth modification after the indentation [7].

$$I_{er} = \frac{H_t - H_{meas}}{H_t} \cdot 100\%$$

(1)

In formula 1, a theoretical depth is used H_t. Assuming that the diagonals remain unchanged and the recovery takes place mostly on the normal direction of the tested surface (up to 50%), the theoretical depth useful for a valid normalized quotient is:

$$H_t = \frac{d_1 + d_2}{2} \cdot ctg\varphi$$

(2)

where $d_{1,2}$ are the two diagonals of the imprint and φ is the half of the indenters pyramid angle of 74,05°.

The measured depth *Hmeas* is the largest of the measured imprint's depth, in absolute value. Table 1 reveals that the elastic recovery is more important at lower indentation loads.

Table 2: Measured and theoretical indentations depths

Load		N	Minimum	Maximum	Mean	Std. Deviation
10gf	Ht x 10-2 (mm)	150	,5079	,5775	,543818	,0153250
	Hmeas x 10-2 (mm)	150	,3100	,3800	,348617	,0194595
	Ier (%)	150	27,05	45,66	35,8287	4,24592
25gf	Ht x 10-2 (mm)	150	,7020	,7877	,745741	,0156773
	Hmeas x 10-2 (mm)	150	,4700	,5400	,508229	,0165789
	Ier (%)	150	25,02	38,90	31,8177	2,68540
50gf	Ht x 10-2 (mm)	150	,9942	1,0514	1,018389	,0108114
	Hmeas x 10-2 (mm)	150	,7769	,8400	,809393	,0123180
	Ier (%)	150	17,15	24,48	20,5147	1,40194

3.2. The hardness value normalization

As a consequence of the nonlinearity between the hardness and the elastic recovery, a 4th grade inverse dependence is fit to describe this relationship [8]:

$$HV_n = \frac{1,854 \cdot P}{d^2} \cdot \sqrt[4]{I_{er} \cdot C}$$

(3)

where C is a constant with a specified value of 2,5 [9], while P is the indentation load and d is the arithmetic mean of the diagonals if the imprinted pyramid. The hardness values, calculated before and after normalization, and the corresponding elastic recovery index are presented in table 3.

Table 3: HV values before and after normalization

Load		N	Minimum	Maximum	Mean	Std. Deviation
10gf	HV (Kgf/mm^2)	150	45,41	58,70	51,3288	2,89651
	HVn (Kgf/mm^2)	150	44,22	56,83	49,8226	2,33536
	Ier (%)	150	27,05	45,66	35,8287	4,24592
25gf	HV (Kgf/mm^2)	150	61,02	76,82	68,1664	2,85082
	HVn (Kgf/mm^2)	150	58,66	71,84	64,3014	2,24951
	Ier (%)	150	25,02	38,90	31,8177	2,68540
50gf	HV (Kgf/mm^2)	150	68,50	76,60	73,0342	1,54209
	HVn (Kgf/mm^2)	150	58,62	65,20	61,7673	1,19126
	Ier (%)	150	17,15	24,48	20,5147	1,40194

Table 4: Paired Samples Test

HV (Kgf/mm^2) - HVn (Kgf/mm^2)		Load		
		10gf	25gf	50gf
Paired Differences	Mean	1,50620	3,86501	11,26686
	Std. Deviation	1,54748	1,45777	1,19599
	Std. Error Mean	,12635	,11903	,09765
95% CI of the Diff.	Lower	1,25653	3,62981	11,07390
	Upper	1,75588	4,10021	11,45982
t		-6,486	7,335	11,921
df		149	149	149
p (Sig.) 2-tailed		,000	,000	,000

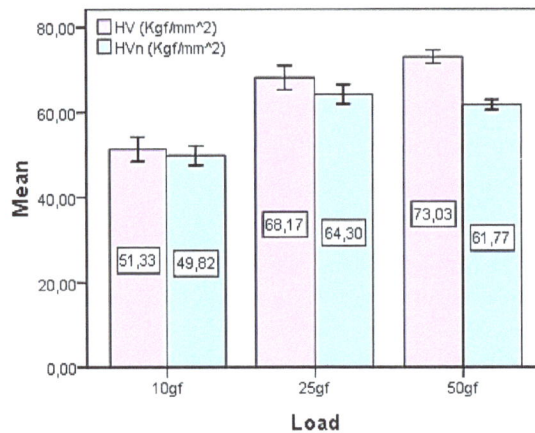

Figure 6: HV and HVn comparison

With Paired Samples Test (table 4), it becomes obvious that the hardness values measured after indentation HV (Kgf/mm^2) and the normalized ones HVn (Kgf/mm^2), are significant different with p (Sig.)< 0,001.

4 CONCLUSIONS

This study aimed to validate a simple method to take into account the complexity of the human dentine behavior and to include the elastic recovery in the hardness measurements. Because under the load the material is more deformed than after the indenter release, even the immediate measurements of the imprint can induce errors in hardness values. These errors can be avoided using a normalization quotient that takes into account the depth variation and the nonlinear 4th degree relationship between the hardness and the elastic recovery.

A thorough measurement sequence was established using a micro hardness tester, samples of dentine and corresponding indentation under specific standard loads and durations. After the statistical verification of the coherence of the measurements, cohorts of data were established for further analysis, to determine the most influent factor on the hardness (the load).

The elastic recovery takes place mostly on the normal of the indentation, while the transversal dimensions remain, practically, unchanged. The depth measurements immediately after the indentation and, respectively, after 12 hours showed depth recoveries up to 20-30%, as the indentation load grows, confirming that the dentine has better elastic properties at lower stresses and its capacity to reformulate it's structures decreases as local stresses increase.

The normalization decreases the HV values, less for the values obtained with 10gf loads and more for bigger loads, both being dragged near a mean value, which is to be taken into account for dentine mechanical characterization (table 5).

Table 5: Measured and normalized HV behavior for various indentation loads

Load		Mean	Std. Deviation
10gf	HV (Kgf/mm^2)	51,3288	2,89651
	HVn1 (Kgf/mm^2)	**49,8226**	2,33536
	Ier (%)	35,8287	4,24592
25gf	HV (Kgf/mm^2)	68,1664	2,85082
	HVn1 (Kgf/mm^2)	**64,3014**	2,24951
	Ier (%)	31,8177	2,68540
50gf	HV (Kgf/mm^2)	73,0342	1,54209
	HVn1 (Kgf/mm^2)	**61,7673**	1,19126
	Ier (%)	20,5147	1,40194

The statistical tests applied to the measured and normalized values confirmed the validity of the normalization procedure, with a high rate of confidence.

Further research can link the hardness value with other, non-mechanical properties of the dentine.

REFERENCES

[1] Meredith N, Sherriff M, Setchell DJ and Swanson SA. Measurement of the micro hardness and Young's modulus of human enamel and dentine using an indentation technique. Archives of Oral Biology. 1996; 41(6):539—545.

[2] John F. McCabe, Steven Bull , Sandra Rusby , Robert W. Wassell, Hardness measured with traditional Vickers and Martens hardness methods, Dental Materials, Volume 23, Issue 9 , Pages 1079-1085, September 2007, ISSN (electronic): 1879-0097.

[3] Chanya Chuenarrom, Pojjanut Benjakul, Paitoon Daosodsai, Effect of Indentation Load and Time on Knoop and Vickers Microhardness Tests for Enamel and Dentin, Materials Research, Vol. 12, No. 4, 473-476, 2009

[4] Chicot D, Mercier D, Roudet F, Silva K, Staia MH and Lesage J. Comparison of instrumented Knoop and Vickers hardness measurements on various soft materials and hard ceramics. Journal of the European Ceramic Society. 2007; 27(4):1905-1911.

[5] Fuentes V, Ceballos L, Osorio R, Toledano M, Carvalho RM and Pashley DH. Tensile strength and micro hardness of treated human dentin. Dental Materials. 2004; 20(6):522—529.

[6] Gutiérrez—Salazar M and Reyes-Gasga J. Microhardness and chemical composition of human tooth. Materials Research. 2003; 6(3):367—373.

[7] Kinney JH, Balooch M, Marshal SJ, Marshal GW, Weihs TR. Hardness and young's modulus of human peritubular and intertubular dentin. Arch Oral Biol 1996; 41: 9-13

[8] Kinney JH, Balooch M, Marshall GW et al. 1999. A micromechanics model of the elastic properties of humandentine. Archives of Oral Biology 44: 813822.

[9] Lowen, B., Elastic recovery at hardness intendations, Journal of Material Science, pp2745-2752, 1981.

THE BASALT-A MATERIAL USED IN MACHINE TOOLS PARTS MANUFACTURING

I.O. Popp

"Lucian Blaga" University, Sibiu, ROMANIA, ilie.popp@ulbsibiu.ro

Abstract: This paper presenst some results of the theoretical and experimental research made upon basalt parts which were submitted to specific static stress of machine-tools. This research is conducted according to specific parts loading. This work presents the basalt structures modeling used in (parallel) test, the final output data, specific comparative references between theoretical and experimental results and detached conclusions.

Keywords: materials (basalt), machine tools, FEM, experimental techniques

1. INTRODUCTION

Is ascertaining he course of new materials introducing for supporting parts in machine building. According to that, it was included the preoccupation to extended basalt practical applications in body or housing machine parts. It is imperative those big components must have the required static dynamic and thermal comportment as well as they must guarantee working precision of machine they are included in. Basalt is a material in big quantities available in actual natural reserves. It has a low price and can be used in various domains. As a raw material the basalt is a igneous rock it has a wild geographic location. It is relative mining accessible and can be not difficult processed. Technologically, basalt confers the following advantages [3]:

a) The present casting process permit to obtain relatively big parts having many shapes and good precision

b) Basalt casting does not require special conditions and can be performed on existing installations having low processing costs.

2. GENERAL REMARKS

During the machine is working, it is loaded by different stresses which awards the prevalent nature of load: orderly weight of stationary or moving (sub)assembly, of semi-fractured article or facilities actuating like stating loads. Static stiffness is a very important test indicator to evaluate stabile domain comportment of structural parts manufactured by different materials. Part or structure stiffness represents the capacity to impede the external forces which tends to deformation it. Mathematical, it is defined by the ratio between the force size which acts in assigned direction and the traveling size of respective structure. It comes from stiffness definition the general approaching mode used in test methods: taking measurement of displacement generated in closing paint by a force well defined in size, direction and sense.

While running the machine tool is stressed by various factors which confer the predominant form of stress. The most important factors are:

- the weight of supporting and mobile assembly, of workpiece rigs acting as a static working load;
- the cutting forces, the running mobile assemblies, acting as a dynamic working load;
- frictional loss, the heat recessed while chipping cutting determine thermal deformation.

Structural elements of the machine tool are stressed in different ways due to all previously presented stress factors. Related to body and housing machine parts building, basalt can be accepted as a construction possibility. Previously it must be studies the stress comportment and must elaborate an appropriate building technology. Technological is a 800 mm main dimension limit for actual basalt parts. A complex machine structure (like housing, body, trunk, and lug support, base) is composed by many basic parts (rungs (draw bar), plates) joint together [3]. Behavior basalt analysis was performed on a representative structural element. According to structural machines analysis and technological limits of basalt manufacturing, it was selected a natural embedded rectangle plate having dimensions: 250 x250x40 mm. The number and the coordinates of measuring points were selected as well as to confer an accurate interpretative routine of measurement results.

3. TESTS AND RESULT INTERPRETATION

To determine the static stiffness of tested parts the force loading system was generated using a hydraulic device. The continuous force measuring apparatus includes an inductive force transducer and amplifier bridge. While loading and unloading the real displacement in the selected points (of tested parts) is continuously measured using contact inductive transducers. The collected signal is amplified in a bridge amplifier. This signal is proportionate with loading force and deformation. A graphic X-Y printer records the measuring data. The loading force was applied consecutively on 3 points. Every time the structural deformation data was recorded.

For the performed test it was used the machine tools testing Laboratory apparatus [1]:
- 6 channel Hottinger amplifier bridge;
- Hottinger force transducer: output signal u- 60 mV, frequency = 7,5 KHz;
- displacement inductive transducer (Wit-ADT 10/8 type): output signal U - 50 mV, amperage = 0,1 mA, frequency = 7,5 KHz;
- Hewlett-Packard X-Y graphic plotter; signal convertor;
- Osciloscope.

Through performed test it was possible to determine static stiffness (static elasticity value) of structural parts used. In addition it was possible to obtain the static stiffness charts force [daN] - displacement [μm]

3.1. Static behaviour

The structure design is presented in figure 1. That includes particular load and contains the directions, senses and application points of the (loading) forces. In addition there is presented the directions and senses of measured deformation.

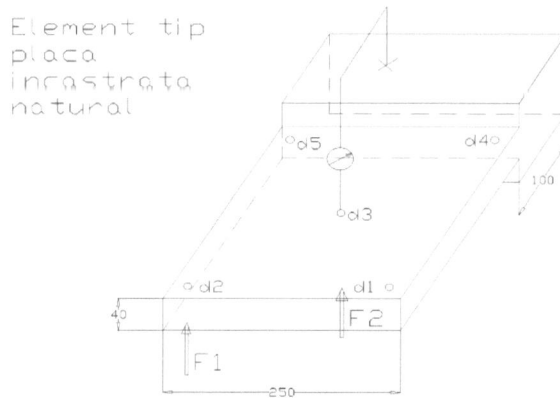

Figure 1: The basalt part

Loading forces were 250 daN and they were applied consecutively on the two points mentioned in the picture (F1 eccentric at 40 mm from the (piece's) edge and F2 centric at 125 mm from the table's edge). For both forces they were measured the produced deformation in everyone of those five symmetrical positioned points marked on the upside face of the structure. It was used the specific methodology to establish the indicators of directly static stiffness. The synthetic results of performed tests are presented in table 1 which encloses the values of static elasticity of flexure recorded in measuring points.

Table 1: The static elasticity values

Measuring points	Static elasticity (μm/daN)	
	F1	F2
d1	0,1917	0,1215
d2	0,3186	0,1161
d3	0,1242	0,0675
d4	0,0216	0,0108
d5	0,0108	0,0108

3.2. Modeling

Presented structure was design using finite element method (FEM) to compare and proof the performed test results. The complete set of equation which generally describes the structure comportment has subsequent relation [2]:

$$[M]\{ü\} + [C]\{u\} + [K]\{u\} = \{R\} \tag{1}$$

where: [K] represent the stiffness matrix;

[M] represent the mass matrix; it contains the terms which defines the net weight of component parts or other implicated mass in dynamic comportment of structure;

[C] represent the damping coefficient matrix;

{u} represent the branch point displacement vector;

{u} represent the branch point speed vector;

{ü} represent the branch point acceleration vector;

{R} represent the external loading vector.

Particular on static analysis the speed and acceleration vector are null and the terms of loading vector are constant. Static comportment analysis is applied to evaluate the values of structure deformations and stress. That can be performed by using MARC, a finite element computer program. The results obtained running this program contains complete information about dimensional variation and about normal and tangential stress distribution. The initial conditions:

- plate dimensions 250x250x40 mm (supported: 250x100x100 mm);
- material: basalt with elasticity modules $E = 85230$ MPa, contraction coefficient $\varepsilon = 0,33$;
- load $F = 250$ daN applied in the two points indicated in figure 1.

The static elasticity modules value was experimentally measured using the electrical strain ganged method [1].

Due to the geometrical symmetry and of the loading and supporting mode the analyzed structure was transformed in a finite element mesh containing hehaedric three-dimensional solid elements resulting 1131 mash point having 3393 degree of freedom. The FEM performed analysis results for both tests are listed in figure 2, 3.

They have subsequent meaning:

- displacement u_y [μm]
- comp. σ_{11} of Stress [MPa].

The analytical solution for the main stress component on OX direction in F2 loading case is:

$$\sigma_{11} = \frac{M}{W} = \frac{Fl}{\frac{bh^2}{6}} = \frac{125 \times 2500}{250 \times 40^2} 6 = 4,687\,MPa \tag{2}$$

The analithycal solution of d3 point displacement equation is:

$$f_{(3)} = u_{y(3)} = \frac{1}{3}\frac{Fl^3}{EI_z} = \frac{1}{3}\frac{2500 \times 125^3}{85230 \times \frac{250 \cdot 40^3}{12}} = 0,014322\,mm$$

FEM determinate solution: $u_y = 17,977$ μm.

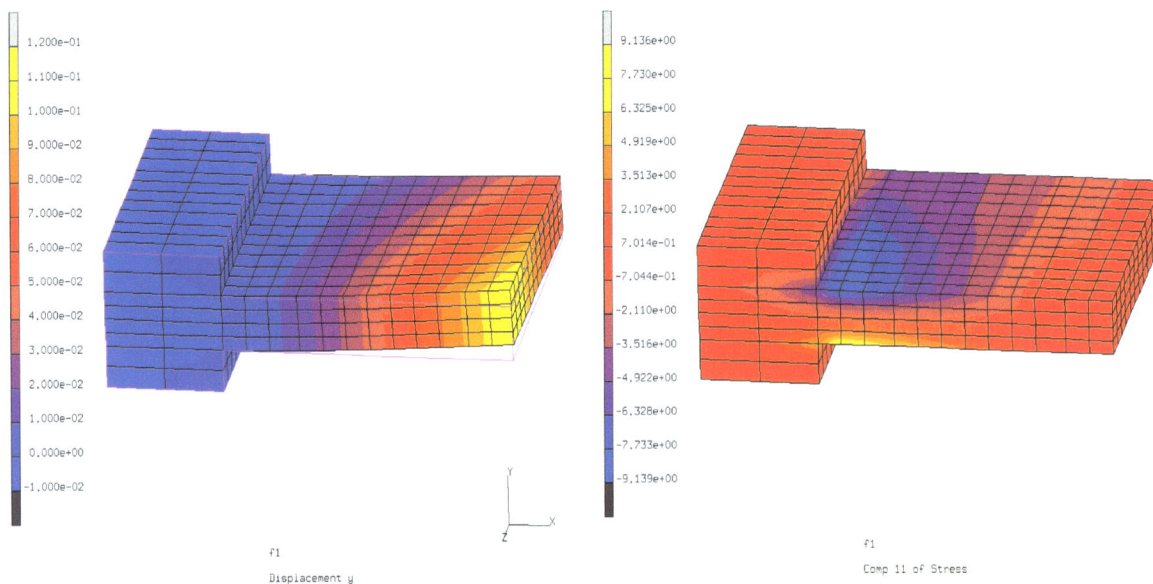

Figure 2: The FEM analysis results for F1

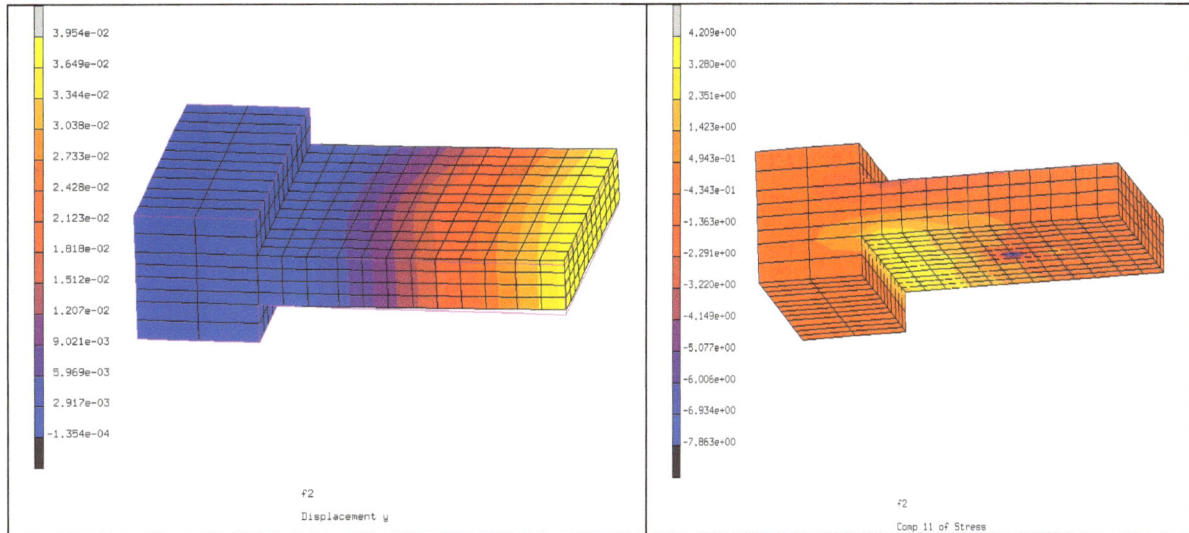

Figure 3: The FEM analysis results for F2

The maximum determinate value of σ_{11} performing FEM is 4,209 MPa. The difference may be caused by the stress located phenomena in restraining zone where the geometrical structure is different by ideal structure. The displacement values are influenced by elastic modules E and structural geometry. Table 2 contain the displacement values in every one of the 5 measuring points obtained performing FEM and experimental tests.

Table 2 The displacement values

Measuring points	Displacement u_y [μm]			
	FEM		Experimental	
	F1	F2	F1	F2
d1	50,780	31,143	47,925	30,375
d2	85,516	31,143	79,651	29,025
d3	31,142	17,977	31,051	16,875
d4	6,95	2,925	5,413	2,715
d5	3,015	2,925	2,712	2,715

4. CONCLUSION

Melted and recrystallized basalt comportment was studied as a material used in machine construction. According to analyzed output parameters it was drafted three experimental test programs. Each program has the same experimental development including; testing methodology, specific measuring apparatus used, experimental test data records, data processing and resulting conclusions. During performed test they were considered all disturbing factors who could affect the measuring results.

The experimental researches lead to following conclusions:

The deformation of tested elements can be approximate as linear. In the same way are acting end restrained basalt plate type elements. Simple shape basalt plates (parallelepiped type) have a nonlinear deformation. There are realistic the experimental test results and they are confirmed by the FEM pattern results.

Maximum values of deformation amplitude were obtained in those areas positioned on the direction which contains application points of loading force, on their proximity.

Static stiffness has a linear variation on restraint plate structures.

When eccentric forces (F1) applied, the deformation increase approximate 2.5 times.

When centric forces (F2) applied the deformations are symmetrical.

REFERENCES

[1]. Popp, I.O. - *Cercetări privind comportarea complexă a elementelor de maşini unelte din bazalt,*, Referatul 3 al tezei de doctorat, Universitatea Lucian Blaga din Sibiu, 1999.

[2]. Pascariu, I. - Elemente finite. Concepte. Aplicaţii, Editura Militară, Bucureşti, 1995.

[3]. * * * - *Bazaltul în industrie*, Intreprinderea pentru Lianţi, Braşov, 1990.

COMPOSITE MATERIALS AND MODERN TECHNOLOGIES TO WATERPROOFING REHABILITATION

F.L. Tămaş[1], I. Tuns[2]
[1] Transilvania University, Brasov, ROMANIA, florin.tamas@gmail.com
[2] Transilvania University, Brasov, ROMANIA, ioan.tuns@unitbv.ro

Abstract: The emergence and development of composite materials enabled their successful use in civil engineering domain. The aim of this article is to present some of the materials and technologies used to waterproofing rehabilitation for building infrastructure. Therefore, there are indicated the technical characteristics of the finished product as long with technologies with certified results over time.
Keywords: HDPE, brick walls, insulation sheet, waterproofing, bicomponent solution

1. INTRODUCTION

The drainage practice of buildings affected by capillary moisture represents a complex issue and allows a technical approach from multiple points of view. In old buildings (such as castles, cathedrals, churches) was found in many cases an increase of capillary moisture at basement and foundations level, with consequences in degradation of finishes, plaster and even walls. Several causes that have led to these deficiencies are: heating some areas that were design to work unheated, different construction works that created barriers to the moisture movement, the raise of water table or inadequate work for rainwater capture and removal process [1]. To remedy these deficiencies different materials and technologies are used. In terms of materials, they have developed with the evolution of composites field.

2. COMPOSITE MATERIALS

Waterproofing rehabilitation for building infrastructure involve different composite materials which appear as liquid solutions, gels, sheet or plastic materials. Some of them will be mentioned as follows.

2.1. Liquid solutions

One of the methods most applied in order to remove capillary moisture from old buildings brick walls is DryKit. Its efficiency has been proven over time, as a result of numerous objectives rehabilitated using mentioned technology. This system acts on masonry through the formation of a chemical barrier which is guaranteed and unalterable in time and can be obtained by use of different formulas (solutions), as described bellow:
- TRE 128 - specifically formulated siloxane microemulsion-based solvents in heteropolar hydrolysates for walls of any type of material or thickness to be applied by insertion of diffuser tube shaped made from pressed cellulose at a series of holes passing near, prepared at 15 cm from the floor.
- TRS 114 - formulated specifically based on polysiloxanes in aliphatic solvent, for walls of any material or thickness to be applied by insertion of diffuser tube shaped made from pressed cellulose at a series of holes passing near, prepared at 15 cm from the floor.
- TRX 118 - monomeric silane component formulation with high penetration of any masonry material or thickness to be applied as mentioned above.
- TRA 115 - silicone formulated in deionized water suitable for masonry or stone, for compact brick masonry with a thickness greater than 40-50 cm, to be applied as mentioned above.
- TRF 135 - formulated specifically based on modified polysiloxane solvents super rectified, for the treatment of frescoed walls, to be applied by insertion of speakers tubopress of cells at a series of holes almost loops, drawn 15 cm from the floor share [2].

2.2. Sheets and plastic materials

The other method which is subject for this article is Comer. This technology has two systems that can be applied depending on the particularities of each work: Isolcomer and Igrostop. The first one, Isolcomer, uses the following composite materials:
- Insulation sheets (fig. 1a), made of compound resin-polyester plastics reinforced with fiberglass or pure polyethylene type plastics and specifically designed for use as insulation against rising damp [3]. They can be sandblasted on one side or both sides (the latter being mainly used in areas with significant seismic applications, which have provided better grip in the cut).
- Anchoring wedges (fig. 1b), produced with differentiated thickness by using thermal hardening resins and have a high resistance to traction, shocks, pressure and compression.

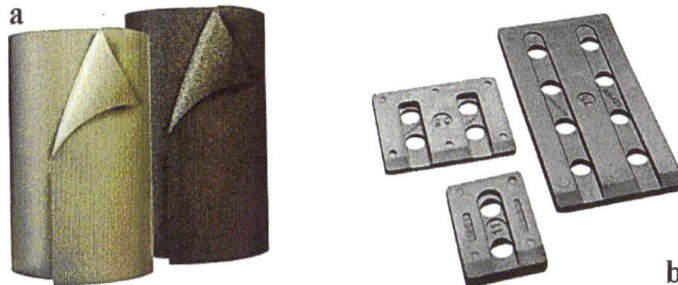

Figure 1: Insulation sheets (a) and anchoring wedges (b)

In order to solve the rising damp problem for good, different measures are adopted. These are conducted in three directions: the implementation of a barrier at floor level at the entire section of the wall with effect in breaking the rising moisture; ventilation works on the exterior foundation, using high-end products such as high-density polyethylene (HDPE) and ventilation works on the interior side of floor and foundation. The material used in some case studies mentioned in this article is the Tefond (fig. 2) high-density polyethylene waterproofing sheet, well known for its strength to impact and corrosion [4].

Figure 2: Tefond high-density polyethylene waterproofing sheet

3. CASE STUDIES

Some of the main objectives where modern methods and materials were used for waterproofing building infrastructure are presented as follows. It is about constructions with a great value for our national historical heritage, such as churches, medieval castles or memorial houses.

3.1. Waterproofing brick walls of „Buna Vestire" church from Jina village, Sibiu County

Jina village is located in the southwestern county of Sibiu, on the boundary with the Alba County and part of the famous folkloric area The Edge of Sibiu, being representative of ports and related traditional grazing habits. The church was built in 1782 and painted by Vasile Munteanu of Laz in 1802. According to the technical expertise [5], measures of intervention on interior and exterior walls, in order to eliminate capillary moisture, were established. The rising damp barrier was obtained by using DryKit technology throughout liquid components mentioned above in the article. Were also executed works in order to eliminate the floor effect and to ventilate foundations both interior and exterior side of the church. Few images after implementing specific measures are presented below (fig. 3; fig. 4).

Figure 3: Using Tefond high-density polyethylene sheet to ventilate foundations

Figure 4: Positive effect to the wall, after applying DryKit technology

3.2. „Ilie Birt" memorial house, Brasov

At the „Ilie Birt" memorial house from Brasov [6], DryKit system was also applied for creating the horizontal chemical barrier into the walls. Some images that illustrate this technology and its way to applied at exterior and interior walls are presented in figure 5 and figure 6.

Figure 5: Applying DryKit technology to an exterior wall

Figure 6: Introducing DryKit bicomponent formula into the wall

3.3. Rising damp treatment at Karolyi castle from Carei, Satu Mare department

At Karolyi castle, interventions aims to correct the problems of strength and stability, removing moisture from the capillary walls, cleaning basements, exterior and interior plaster, roof restoration, decorative items, etc. Related to drying up measures for building walls, Comer method was used based on its high rate of success. Work began in late 2009 and the implementation of insulating sheets along with anchoring wedges is according to figure 7.

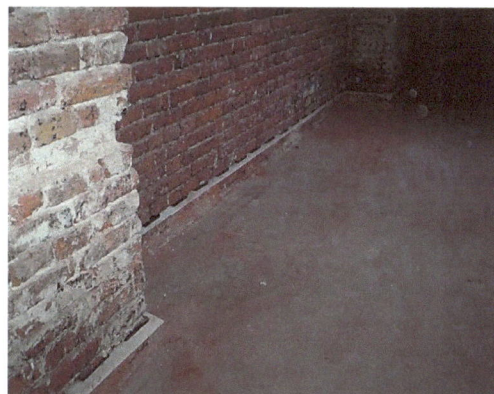

Figure 6: Introducing DryKit bicomponent formula into the wall

4. CONCLUSION

Modern methods and materials used to waterproofing rehabilitation for building infrastructure shows that applied according to the technological specifications and technical expertise specific to each case, it can lead to the desired and certified results.

REFERENCES

[1]   Tămaş F.-L., Software optimization of a technological solution for waterproofing rehabilitation of rising damp affected walls, Proceedings of the International Conference Deducon – Sustainable Development   in Civil Engineering, Iaşi 2011.
[2]   *** www.tecnored.eu.
[3]   ***www.comerspa.com.
[4]   *** www.tegola.pl.
[5]   Streza T., Technical expertise for removing capillary moisture from walls of Ortodox Church in Jina,   Sibiu county 2007.
[6]   Tămaş, F.-L., Streza T., Technical expertise for removing capillary moisture from the basement walls of   "Ilie Birt" memorial house, Braşov, 2006.
[7]   Tămaş F.-L., Tuns I., Streza T., Modern methods for waterproofing rehabilitation of existing building International Scientific Conference CIBv 2010, 12-13 November, Braşov.

COMPARATIVE NUMERICAL STUDY OF FEM AND SPH METHOD FOR BULLET-MULTILAYERED PLATE IMPACT SIMULATION

V. Năstăsescu[1], S. Roateşi[2]

[1] Military Technical Academy, Bucharest, ROMANIA, nastasescuv@gmail.com
[2] Military Technical Academy, Bucharest, ROMANIA, sroatesi@yahoo.fr

Abstract: *This paper presents a numerical simulation of the impact problem of a bullet into a two layered composite plate representing an armor used for the human protection. The approach of this problem is performed using both classical Finite Element Method (FEM) and Smoothed Particle Hydrodynamics (SPH). The aim of the work is to point out that the SPH method is a valuable numerical approach of the impact problems of the composites. The penetration and perforation of a composite that could occur as a result of an impact process are a complex problem which implies many technical and design aspects. This numerical study deals mainly with the impact velocity effect on the delaminating area of the impact zone. The numerical solutions obtained by both methods are very close and the good investigating potential of the SPH method is pointed out.*

Keywords: *SPH, FEM, Impact, Composites, Numerical Simulation*

1. INTRODUCTION

The penetration and perforation of the composite could occur as a result of the impact process. The impact response of materials can range from low (large mass) velocity to high/ballistic (small mass) velocity regimes. A significant development on impact mechanics is presented in [1] and particularly on composite structures in [2].

Large mass impact, known as low velocity impact, results from conditions arising for instance, from tool drops on a product which typically occur at velocities below 10 m/s, while intermediate velocity impact regime occurs for instance for secondary blast debris, hurricane and tornado debris, in the 10m/s to 50 m/s range.

High velocity (ballistic) impact is usually a result of small arms fire, explosive warhead fragments or space debris on a spacecraft and it range from 50 m/s to 1500 m/s. In hyper velocity impact > 2-5 km/s, the projectile is moving at very high velocities and the target material behaves like a fluid.

High velocity impact response is dominated by stress wave propagation through the thickness of the material, in which the structure does not have time to respond. Boundary condition effect can be ignored because the impact event passes before the stress waves reach the boundary.

This paper deals with an application referring to the bullet-armor impact, usually called ballistic impact.

In studying ballistic impacts, it is important to determine the residual velocity of the projectile accurately. From the experimental testing point of view, this is a difficult task because many small particles, fibers, and shear plugs are pushed out by the projectile during penetration. This material can trigger the speed-sensing device being used and yield erroneous values. Moreover, even from a numerical point of view, the particles spread represents an obstacle in obtain the adequate solutions.

Considering the balance of energy reveals important features of ballistic impact, including the effects of laminate thickness, projectile size, shape, and initial velocity. The application of bullet-armor impact presented in this study is performed alternatively by Smoothed Particle Hydrodynamics (SPH) method and Finite Element Method (FEM) in order to estimate and validate SPH method to solve impact problems of bullet targeting a composite plate simulating the armor.

SPH is a numerical simulation meshless method proposed by Lucy in 1977, see [3]. The first applications of this method were connected to astrophysical problems. The method was extended to fluid simulation, especially with free-surface by Monaghan in 1992 and to other fields, see [4] and [6]. The field of applied mechanics is the last one, but it is extensively studied and significant advances have been made, see [5]-[8].

The last preoccupations are focused on coupling SPH with standard numerical procedures, such as the FE method or other meshless techniques because they offer new possibilities to solve complex problems in engineering. The SPH method is validated as a numerical method in fluid mechanics, but the applied mechanics is concerned it is still in progress.

The SPH advantages seem to be greater then disadvantages, from a lot points of view, especially in some fields, like fluid mechanics and even applied mechanics for those problems that involve large displacements or for materials with a fluid-like behavior or for brittle materials with special properties like ceramics, glass etc.

Some SPH programs exist, but next to these, the SPH method is implemented in the most powerful programs, like ANSYS, beginning with 10th version and the last one is used in this study.

2. SPH FORMULATION

The SPH method is a meshless method, in which the investigated domain is represented by a number of nodes, representing the particles of this domain, having their material and mechanical (mass, position, velocity etc.) characteristics. Each particle represents an interpolation point on which the material properties are known.

The boundary conditions have to be imposed to some of particles, according to the problem analyzed, like in the case of finite element method.

The problem solution is given by the computed results, on all the particles, using an interpolation function. We can say that the fundamentals of SPH theory consist in interpolation theory; all the behavior laws are transformed into integral equations. The kernel function, or smoothing function, often called smoothing kernel function, or simply kernel, gives a weighted approximation of the field variable (function) in a point (particle).

Integral representation of a function $f(x)$, used in the SPH method starts from the following identity:

$$f(x) = \int_\Omega f(x')\delta(x-x')dx' \tag{1}$$

where f is a function of a position vector x, which can be an one-, two- or three-dimensional one; $\delta(x-x')$ is a Dirac function, having the properties:

$$\delta(x-x') = \begin{cases} 1 \to x = x' \\ 0 \to x \neq x' \end{cases} \tag{2}$$

In equation (1), Ω is the function domain, which can be a volume, that contains the x, and where $f(x)$ is defined and continuous. By replacing the Dirac function with a smoothing function $W(x-x',h)$ the integral representation of $f(x)$ becomes:

$$f(x) = \int_\Omega f(x')W(x-x',h)dx' \tag{3}$$

where W is the smoothing kernel function, or smoothing function, or kernel function.

The parameter h, of the smoothing function W, is the smoothing length, by which the influence area of the smoothing function W is defined (Figure 1 and Figure 2).

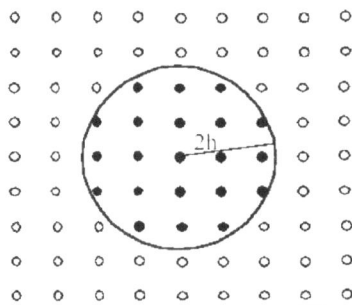

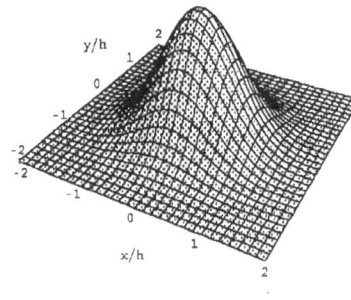

Figure 1: Support domain of W **Figure 2:** Graphical representation of 2D-Kernel function

As long as Dirac delta function is used, the integral representation, described by equation (1), is an exact (rigorous) one, but using the smoothing function W instead of Dirac function, the integral representation can only be an approximation. This is the reason for the name of kernel approximation. Using the angle bracket $\langle \ \rangle$ this aspect is underlined and the equation (3) can be rewritten as:

$$\langle f(x) \rangle = \int_\Omega f(x')W(x-x',h)dx' \tag{4}$$

The smoothing function W is usually chosen to be an even one, which has to satisfy some conditions. The first condition, named normalization condition or unity condition is:

$$\int_\Omega W(x-x',h)dx' = 1 \tag{5}$$

The second condition is the Delta function property and it occurs when the smoothing length approaches zero:

$$\lim_{h\to 0} W(x-x',h) = \delta(x-x') \tag{6}$$

The third condition is the compact condition, expressed by:

$$W(x-x',h) = 0 \text{ when } |x-x'| > kh \tag{7}$$

where k is a constant related to the smoothing function for point at x, defining the effective non-zero area.

As the particle approximation is concerned, the continuous integral approximation (4) can be converted to a summation of discrete forms, over all particles belonging to the support domain. Changing the infinitesimal volume dx' with the finite volume of the particle ΔV_j, the mass of the particles m_j can be written,

$$m_j = \Delta V_j \rho_j \tag{8}$$

and finally, relation (3) becomes:

$$\langle f(x) \rangle = \sum_{j=1}^{N} \frac{m_j}{\rho_j} f(x_j) W(x-x_j,h) \tag{9}$$

The particle approximation of a parameter described by a function, for particle i can be expressed by,

$$\langle f(x_i) \rangle = \sum_{j=1}^{N} \frac{m_j}{\rho_j} f(x_j) W_{ij} \tag{10}$$

where,

$$W_{ij} = W(x_i - x_j, h), \tag{11}$$

being the kernel function.

3. NUMERICAL MODELS

The numerical study is accomplished comparatively by FE and SPH models (Figures 3 and 4), for solving the main aspects regarding to the impact problem. Such a problem is a very difficult one, by many reasons.

Among these reasons, some characteristics have to be emphasized: the very short time for developing of some physics phenomena, material and large displacement nonlinearities, behavior of the material beyond the elastic limit, the strain rate and many others. All these aspects are not the aim of this work, but for a part of them, some synthetic presentations will be made in connection of the model construction.

Figure 3: FE model   **Figure 4:** SPH model

The geometric model consists of a 7.62 caliber bullet impacting a 10cm×10cm×0.6cm composite (multilayered) plate, simulating the armor, made up of two isotropic layers: an exterior one of steel and interior one of aluminum. In the case of SPH model, the exterior steel layer consists of particles. Such a problem (normal impact of a bullet with a plane plate) can be considered a structure with two symmetric planes or an axisymmetric structure. A simple model can be adopted, represented only by a half or by a quarter of a 3D structure (using 3D finite elements), or represented by a plane structure (using 2D axisymmetric elements).

Because the considered example is not a large one, we preferred a 3D completed models, presented in the Figures 3 and 4. The 3D finite element model consists of 1131 nodes and 1008 elements (SOLID3D) for bullet, and of 30603 nodes and

20000 elements (SOLID3D) for each layer. The 3D SPH model is a model, which consists in finite elements for bullet and aluminum plate and SPH particles for steel plate. A model using only SPH particles is possible, but not every time is the best solution; from different reasons, not discussed here, the most used models consists in finite elements and SPH elements, like the models adopted by us.

The impact velocity of 420 m/s is considered, being the case of a medium distance of firing. As the material models are concerned, three material models were used: plastic-kinematic (Cowper-Symonds) material model for the bullet, elastic-plastic-hydro material model for the aluminum plate and Johnson-Cook material model for the steel plate. For each material models specific constants were used according to the experimental results and the values presented in the technical literature.

4. MATERIAL MODELS

One of the most used material model, adopted for the target (often for the projectile or hammer too), is the Elastic Plastic with Kinematic Hardening Model, being strain rate dependent plasticity for isotropic materials. The strain rate is taken into account by Cowper-Symonds model using the coefficients C and P, having the same name. The yield function σ_y is given by [9]:

$$\sigma_y = \left[1 + \left(\frac{\dot{\varepsilon}}{C}\right)^{\frac{1}{P}}\right]\left(\sigma_0 + \beta E_p \varepsilon_p^{ef}\right) \tag{12}$$

where σ_0 is the initial yield stress, ε_p^{ef} is the effective plastic strain, E_p is the plastic hardening modulus which is given by:

$$E_p = \frac{E_T E}{E - E_T}, \tag{13}$$

β being the hardening parameter that can vary between 0 and 1 depending on plasticity type (0 for kinematic and 1 for isotropic respectively) and E_T is the tangent modulus. For this model, the user has to specify the failure strain for which elements will be eliminated.

According to the flow rule, the direction of plastic straining can be evaluated. The hardening rule describes the changing of the yield surface with progressive yielding, so that the conditions for subsequent yielding can be established.

Another material model equally used for specimen (target) and hummer (projectile) modeling is Johnson and Cook Plasticity Model, [9], which express the flow stresses:

$$\sigma_y = \left(A + B\bar{\varepsilon}^{p^n}\right)\left(1 + c\ln\dot{\varepsilon}^*\right)\left(1 - T^{*m}\right) \tag{14}$$

where A, B, C, n, and m are user defined input constants, $\bar{\varepsilon}^p$ is the effective plastic strain, and:

$$\dot{\varepsilon}^* = \frac{\dot{\bar{\varepsilon}}^p}{\dot{\varepsilon}_0} \tag{15}$$

is the effective plastic strain rate for $\dot{\varepsilon}_0 = 1$ s^{-1}, and:

$$T^* = \frac{T - T_{room}}{T_{melt} - T_{room}} \tag{16}$$

T being the temperature (by empirical assumption T represents 90% of the plastic work, T_{room} is the room environment temperature and T_{melt} is the melting temperature of the material). The strain at fracture is given by:

$$\varepsilon^f = \left[D_1 + D_2 \exp D_3\sigma^*\right]\left[1 + D_4 \ln\dot{\varepsilon}^*\right]\left[1 + D_5 T^*\right] \tag{17}$$

where $D_1 \ldots D_5$ are input constants and σ^* is the ratio of pressure divided by effective stress:

$$\sigma^* = \frac{p}{\sigma_{eff}} \tag{18}$$

The fracture occurs when the damage parameter D, relation (19), reaches the value 1.

$$D = \sum \frac{\Delta \bar{\varepsilon}^p}{\varepsilon^f} \tag{19}$$

The elastic-plastic-hydro material model often is used when the material has a hydrodynamic behavior, specially under strong shock like an impact. This behavior is up to material type and velocity level. The yield strength calculus depends on the values of effective plastic strain (EPS) and effective stress (ES) are defined or not. If ES and EPS are undefined, the yield strength is calculated as [9]:

$$\sigma_y = \sigma_0 + E_h \bar{\varepsilon}^p + (a_1 + a_2 p) \max[p, 0] \tag{20}$$

The quantity E_h is the plastic hardening modulus defined in terms of Young's modulus, E, and the tangent modulus, E_t [9]:

$$E_h = \frac{E_t E}{E - E_t} \tag{21}$$

If ES and EPS are specified, the yield stress is given by a relation (22), which is obtained by mathematical and graphic manipulation.

$$\sigma_y = f(\bar{\varepsilon}^p) \tag{22}$$

Johnson-Cook Plasticity Model, like many others, is accompanied by an equation of state (EOS). In our examples a Gruneisen EOS was used. For compressed materials, the Gruneisen equation of state, with cubic shock velocity-particle velocity defines a pressure [9]:

$$p = \frac{\rho_0 C^2 \mu \left[1 + \left(1 - \frac{\gamma_0}{2} \right) \mu - \frac{a}{2} \mu^2 \right]}{\left[1 - (S_1 - 1)\mu - S_2 \frac{\mu^2}{1 + \mu} - S_3 \frac{\mu^3}{(1 + \mu)^2} \right]^2} + (a\mu + \gamma_0)E \tag{23}$$

Many others EOS (linear-polynomial, JWL, ratio of polynomials, tabulated etc.) are implemented and the using of each one of them depends on some aspects, which are not the subject of this paper.

For projectile modeling, the Rigid Material Model [9] is often used. Such approximation of a deformable body is a preferred modeling technique in many real work applications, because the calculus time can be significant smaller. In many cases we are interested in what happens with the target and fewer with the projectile.

The elements which are considered rigid are bypassed in the element processing and no storage is allocated for storing history variables, so the rigid material model is a very cost efficient one.

Some material properties have to be given by the user. Young's modulus E and Poisson's ratio υ being used for determining sliding interface parameters if the rigid body interacts in a contact definition. Density ρ is necessary for calculus of the inertial properties.

In all cases, unrealistic values of the material constants may cause some solving difficulties.

5. NUMERICAL RESULTS

A comparison between the FE and SPH model for the case of 420 m/s bullet velocity can be performed studying the figures 5 and 6 representing the field of von Mises stresses. It is observed a similar distribution of effective stress in both cases and a certain zone of detachment of the two layers with a bigger displacement of the aluminium one as it was expected.

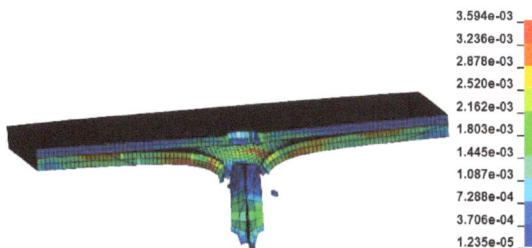

Figure 5: FE model - von Mises stress field

Figure 6: SPH model - von Mises stress field

As we can see in Figure 5 and 6, the maximum stress values (3.594e-3 and 3.569e-3 gcm/μs^2, or 359.40 and 356.90 MPa) are very closed, the error for SPH model being only -0.69%. The fundamental measure units were [g] for mass, [cm] for length and [s]e-6 for time.

Concerning the bullet velocity, a similar profile is obtained by those two methods as it is noticed in figures 7 and 8, respectively. At the same time (100e-6 seconds), the residual bullet velocities are also very closed (355.25 and 353.70 m/s^2). The error regarding residual bullet velocities calculated using FE and SPH models is -1.69%.

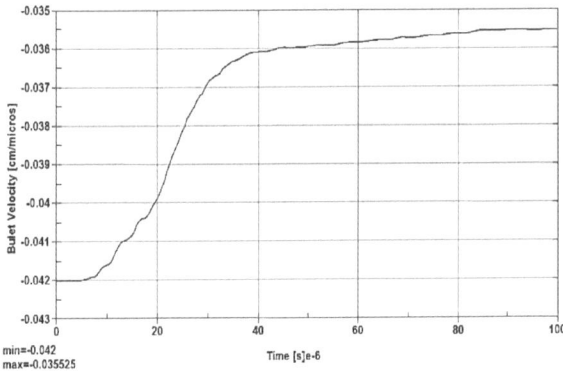

Figure 7: FE model – time bullet velocity   **Figure 8:** SPH model – time bullet velocity

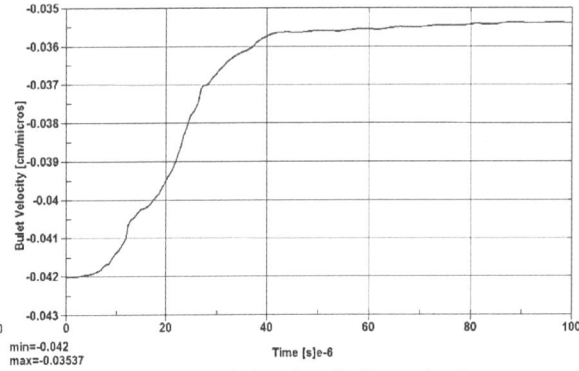

Many other parameters like displacements, accelerations, different forms of energies (internal, kinematic or total) in connection with each plate or with the bullet can also be calculated, represented as a function of time and compared. All these parameters obtained by FEM or SPH method are in a good concordance.

6. CONCLUSION

This paper dealing with the SPH method represents a validation of this method for using in the applied mechanics, in special dynamic problems, like an impact problem of a bullet into a multilayered composite plate.

Many other aspects can be also studied, numerical simulated but for each aspect a more space requires.

The SPH method is also fitted for a right numerical simulation of some materials, like glass, ceramics etc., adopting also appropriated material models.

The case of plate layers which can be deformed only together, (layers without friction or moving between them) can be also treated, in a similar way, but the problem of the contact and the friction between layers does not exist and the computer time is shorter.

This paper did not present all the advantages of SPH method, but we consider it to be an incentive to be used in a larger area, specially in applied mechanics. In our country, this method is still little used by different reasons.

EFERENCES

[1]   Stronge W.G., Impact Mechanics, Cambridge Univ. Press, 2000.

[2]   Abrate S., Impact on Composite Structures, Cambridge Univ. Press, 1998.

[3]   Lucy, L., B., A numerical approach to the testing of the ssion hypothesis, Astrophysical Journal, (82), pp.1013-1024, 1977.

[4]   Monaghan, J., J., Smoothed Particle Hydrodynamics, In Ann. Rev. Astron. Astrophys., 1992, pp. 543-574

[5]   Liu, G. R., Liu, M. B., Smoothed Particle Hydrodynamics, World Scientific Publishing Co.Pte.Ltd., 2009

[5]   Buruchenko, S., K., Smoothed Particle Hydrodynamics: Some Results, Вычислительные, технологии, Том 7, No. 1, 2002.

[6]   Nastasescu, V., Barsan, Gh., Metoda SPH, Editura Academiei Forţelor Terestre, Sibiu, 2012

[7]   Ma, W., L., Pang, B., J., Zhang, W., Gai, B., Z., Numerical Simulation of Major Debris-Cloud Features Produced by Projectile Hypervelocity Impact on Bumper, Hypervelocity Impact Research Center, Harbin Institute of Technology, P.O. Box 3020, No.2, Yikuang Street, Harbin 150080, P. R. China,   2008.

[8]   Tanaka, K., Numerical Study on the High Velocity Impact Phenomena by Smoothed Particle, Hydrodynamics (SPH), 1-1-1 Umezono, Centtral 2 Tsukuba, Ibaraki 305-8568 JAPAN, 2005.

[9]   LS-DYNA Keyword User's Manual, Version 971, May 2007.

ECOLOGICAL THERMAL INSULATION COMPOSITE SYSTEMS USED IN CONSTRUCTION

D. Fiat[1], M. Lazăr[1], M. Prună[1]

[1] Research Institute for Construction Equipment and technology - ICECON SA

Abstract: The paper presents a thermal insulation composite system based on wood fiber boards, putty, reinforcement mash, decorative plaster and a study on the durability of such products under laboratory biodegradation conditions caused by Xylophages biological agents (fungi, insects).

Keywords: wood fiber boards, durability, biodegradation, biological agents

1. INTRODUCTION

Wood structure housing can use thermal insulation ecological systems based on wood fiber boards, cellulose flakes, putty, reinforcement mash and a finishing layer (decorative plaster).

Wood fiber boards used in sandwich type modules or in external thermal insulation systems
In Figures 1a and b are shown examples of buildings using thermal insulation wood fiber boards.

Such boards are used on the external face of the buildings, in sandwich type modules where the middle is made of cellulose flakes that are blown in a closed space formed by internal wood fiber boards, or plaster boards. The boards are fixed on the wood structure by mechanical screws and washers.

After being fixed on the structure and the sandwich type modules are formed, the external wood fiber boards are finished similar to the ETICS (External Thermal Insulation Composite Systems).

Figure. 1a   **Figure. 1b**
Figure. 1a, b - Building made with exterior wood fiber boards on wood structure

Alternative constructions (examples)
Alternative Type a, wood fiber boards used in sandwich type module, having the following structure, from the inside out (Figure 2)

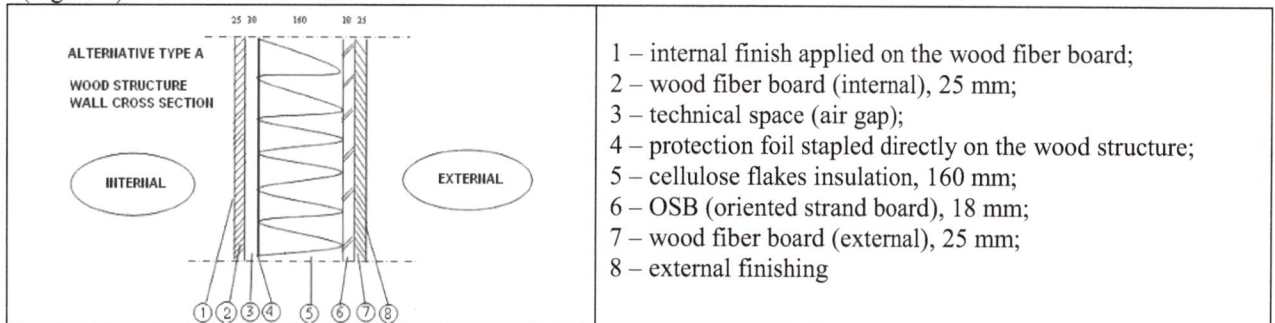

1 – internal finish applied on the wood fiber board;
2 – wood fiber board (internal), 25 mm;
3 – technical space (air gap);
4 – protection foil stapled directly on the wood structure;
5 – cellulose flakes insulation, 160 mm;
6 – OSB (oriented strand board), 18 mm;
7 – wood fiber board (external), 25 mm;
8 – external finishing

Figure 2 - Alternative Type A

Alternative Type B, wood fiber boards used in thermal insulation of existing houses, having the following structure, from the inside out (Figure 3)

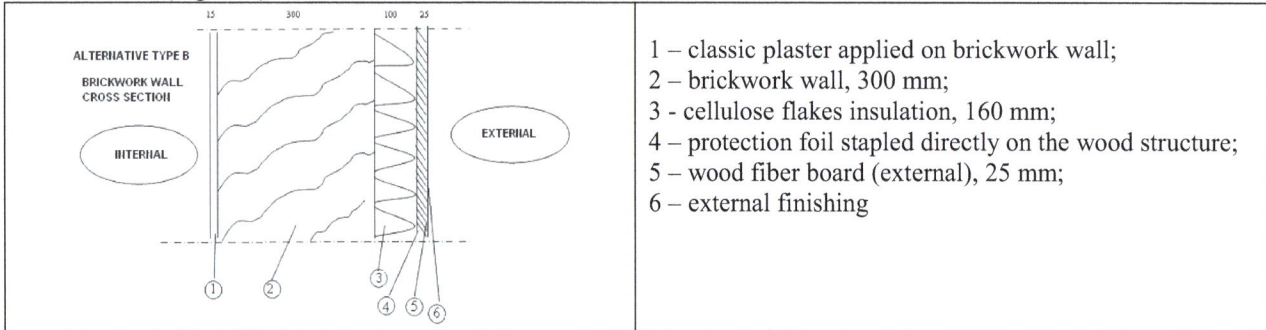

ALTERNATIVE TYPE B BRICKWORK WALL CROSS SECTION INTERNAL	1 – classic plaster applied on brickwork wall; 2 – brickwork wall, 300 mm; 3 - cellulose flakes insulation, 160 mm; 4 – protection foil stapled directly on the wood structure; 5 – wood fiber board (external), 25 mm; 6 – external finishing

Figure 3 - Alternative Type B

The cellulose flakes insulation is performed „in situ". Cellulose is an ecological material obtained by paper recycling and modified with additives in order to become fireproof and antifungal.

The external finishing of wood fiber boards is similar to the ETICS (External Thermal Insulation Composite Systems), according to Figure 4.

Figure 4 - External thermo insulation

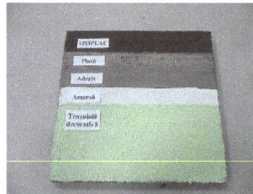

Figure 5 - Composition of the thermo-system with decorative plaster final layer

The composition of the thermal insulation composite system is the following: wood fiber board, putty and reinforcing mash, specific plastering primer, decorative plaster (final finishing layer).

Composite system for internal and external buildings thermal insulation

The system described in this paper is based on wood fiber board with low density (211 kg/m^3) and thermo insulating proprieties (thermal conductivity λ_{10} = 0,044 W/m^{2}K), finishing materials such as putty, reinforcement mash and decorative plaster. Such materials are designed to be used in constructions exploitation class 1... 3, in order to decrease the volumic mass of the construction materials, to reduce the materials consumption and aiming to decrease the energy losses caused by transfer from inside to outside. The materials used in the thermal insulation composite systems contain natural materials, based on wood fiber and inorganic compositions, without synthesis polymeric compounds, dangerous chemical compounds and biocides. The system resistance, realized in this manner, is based on the porosity high degree that allows a high air volume to be incorporated, do not cause condensation and do not retain Xylophages microorganisms. The material sterilization effect, in its final use, it is not caused by a long lasting natural durability, but as a result of an effect induced by the resistance due to constant drying condition, thanks to the continuously aerated structure. The wood fiber board's ecological character is revealed by biological tests, in extreme exposure conditions, of high humidity and optimum temperatures for mould, rot fungi growth and for insects that attack wood. In Table no. 1 are presented the pictures of wood fiber boards and the thermal insulating composite system that are using them, together with the ecological finishing.

Table 1 - Wood fiber boards and the thermal insulating composite system

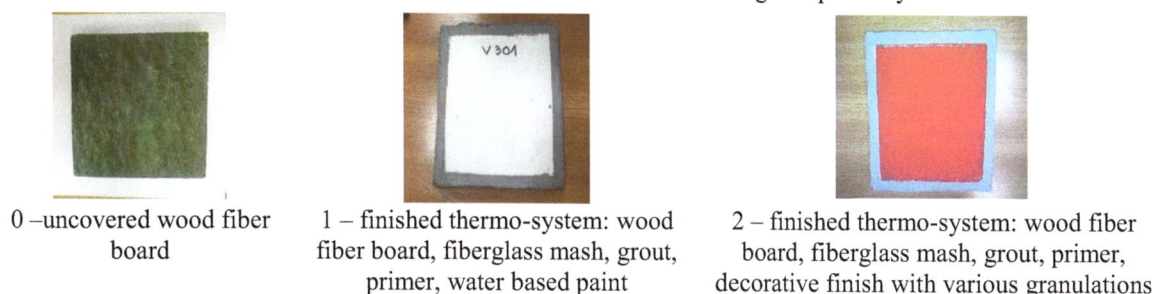

| 0 –uncovered wood fiber board | 1 – finished thermo-system: wood fiber board, fiberglass mash, grout, primer, water based paint | 2 – finished thermo-system: wood fiber board, fiberglass mash, grout, primer, decorative finish with various granulations |

While for veneers large sheets are detached from the row wood and for chipboards the wood is chipped under splitters shape, in wood fiber boards processing the wood structure is decomposed, in such a manner that the individual fibers and the small fibers bundles form the starting point for the new material to be produced. Different types of boards distinguish between them according to their degree of compression. While the insulation boards are subject to a heating and soft pressing technology, the same material hard boards are subject to heating and pressing under a much higher compression regime. Depending on the specific applied pressure, the boards with different densities are obtained with different properties induced by the pressing force.

As a row material for the manufacturing of wood fiber boards are used wood scraps from cutting-off classified as technological scraps, such as: branches, round wood heads, billets/cleft logs, laths and small size assortments, or inferior sorts, and so on. Wood fiber boards are manufactured using wet or dry methods. The manufacturing process is similar to the one used for paper manufacturing. In wood grinding the wood substance is in water slurry. Wood compomposition, in this case do not reaches the lignin removal, the wood being just spread its fibers, respectively in fibers bundles. The production of such boards stated, in the beginning, as a recycling process of the wood scraps that aimed to manufacture construction boards.

2. PROPERTIES

Similar to the massive wood, in the wood fiber boards there is a hygroscopic balance between the relative humidity of the free air and board's humidity. Initially it was thought that this hygroscopic balance would have the same values as in solid wood. Based on a series of researches it could be proven that in the same ambient conditions the wood fiber boards humidity is slightly smaller than that of solid wood. Insulation boards when taken from the drying unit have a humidity of about 5%. Due to the loose structure of these boards, they can be stored and processed in this state without any treatment. On the other hand, hard boards require a conditioning. Hard boards when taken from the hot press unit have a humidity of about 1%. For these boards, hygroscopic balance in a normal environment (20°C and 60% relative air humidity) is 7% to 9%. In order to reach such humidity - after being removal of the hot press unit, the boards are passed through a conditioning chamber. In this chamber the temperature range is 40 - 50 ° C and the relative humidity rage is 95 - 100%. Table 2 presents the wood fiber boards classification on types, according to density

Table 2 - Wood fibre boards classification type

Item no.	Types of wood fiber boards	Apparent specific density[kg/m^3]
Insulation wood fiber boards, bounding agents content up to 12% of the dry weight		
1	Very porous insulating wood fiber boards	Up to 230
2	Porous insulating wood fiber boards	Up to 230 … 400
Hard wood fiber boards, bounding agents content up to 12% of the dry weight		
1	Medium hard wood fiber boards	650 … 850
2	Hard wood fiber boards	Over 850
3	Very hard wood fiber boards	Over 900

Beside the type classification, the importance of specific volume weight is increasing taking into account that for different types of boards the specific volume weight allows to draw conclusions about the other proprieties.
Table 3 presents the wood fiber boards characteristics determined by measurements.

Table 3 - Wood fiber boards characteristics

Characteristic	Very hard boards	Hard boards 1	Hard boards 2	Medium hard wood fiber boards	Insulation wood fiber boards
Specific volume weight, [kg/m^3]	1000-1050	950-1050	800-900	500-750	250-400
Bending resistance, [kg/cm^2]	500-650	350-450	200-250	150-200	20-40
Stretching resistance, [kg/cm^2]	300-450	150-300	90-150	70-100	5-10
Ball pressing on spheres hardness, [kg/cm^2]	5.0-6.0	4.2-5.0	3.0-4.0	1.5-3.0	-
Wear after 10.000 travels, [mm]	1.0-1.5	1.5-2.0	2.0-2.5	2.5-3.0	-
Resilience, [kgm/cm^2]	2.0-3.0	2.0-2.5	1.8-2.3	1.5-2.0	0.5-0.8
Water absorption, [%]	10-15	15-25	15-30	20-30	30-100
In 24 hours in thickness swelling, [%]	7.5-12	12-18	10-20	15-20	10-20

While for the hard boards the most important property is resistance, for the insulation wood fiber boards the decisive role rests in thermal and acoustic insulation. An insulation wood fiber board having the thickness of 10 mm and the specific volume weight of 300 kg/m^3 is equivalent from the insulating capacity point of view with a brick wall of about 15 cm thickness.

Thermal insulating capacity of different wood fiber boards is determined by establishing of the thermal conductivity coefficient. Due to their loose structure, these boards have a thermal conductivity coefficient even lower [0.8 kcal/mh] than solid wood and thus a higher thermal insulating capacity.

Wood fiber boards characteristics

As a main component of a thermal insulation composite system, they can have a variety of internal and external uses, according to 1 to 3 use classes (presented in Table 4), as they are definite by EN 335-1 standard, for exposure to humidity conditions and the frequency of biological attack risk, that wood and wood fiber boards used in construction are exposed to.

Table 4 - Use class and general use conditions

Use class	General use condition	In use humidity exposure	Biological aggents appearance[a]			
			Fungi	Coleoptera[b]	Termites	Marine borers
1	covered internal	dry, maximum	–	U	L	–
2	internal, or covered	occasionally, > 20%	U [c]	U	L	–
3	external, or above ground, protected	occasionally, > 20%	U [c]	U	L	–
	external, or above ground, unprotected	frequently, > 20%	U [c]	U	L	–
4	external in contact with soil and/or freshwater	predominantly or permanently, > 20%	U [d]	U	L	–
	outdoor in contact with soil (severe) and/or freshwater	permanently,> 20%	U [d]	U	L	–
5	in salt water	permanently, > 20%	U [d]	U [e]	L [e]	U

U = present everywhere in Europe;   L = locally present in Europe

[a] Due to local exposure risk and the necessity of a target requirement, it is locally possible for specific location, a second level of classification of biological agents

[b] The risk of attack can be insignificant in some specific situations and specific geographical locations

[c] Fading mushrooms + decay fungi

[d] Fading mushrooms + decay fungi + soft rot fungi.

[e] Part of construction elements placeed above the water, that can be exposed to Xylophages insects, including termites

As shown in Table 4, humidity plays an important role in the development of biological factors, and the constructive designed solution do not favor a threshold of 20% humidity inside or on the surface of wood fiber board to be reached, as a result of their own structure, as well as of the subsequent applied finishing system.

Briefly, wood fiber boards are classified according to use classes, in four types, having the physical and mechanical characteristics according to Table 5:

Table 5 - Wood fibre boards type and characteristics[3]

Characteristic	UM	Wood fibre boards for internal use		Wood fibre boards for external use		Wood fibre boards for flooring use	Wood fibre boards for roof framing use
Thickness	mm	12 ± 1.2	25 ± 1.8	12 ± 1.2	25 ± 1.8	7.4 ± 0.3	25 ± 1.8
Boards per pallet	peaces	90	45	90	45	280	42
Square meter per pallet	m²	291.60	145.80	291.60	145.80	201.60	78.75
Width	mm	1200 ± 5	1200 ± 5	1200 ± 5	1200 ± 5	$600...1200 \pm 3$	750 ± 5
Length	mm	2700 ± 5	2700 ± 5	2700 ± 5	2700 ± 5	2500 ± 3	2500 ± 5
Density	kg/m³	≥ 230	≥ 220	≥ 240	≥ 230	≥ 240	≥ 230
Thermal conductivity	W/mK	≤ 0.045	≤ 0.045	≤ 0.05	≤ 0.05	≤ 0.05	≤ 0.05
Bending strength	N/mm²	≥ 1.7	≥ 1.3	≥ 1.2	≥ 1.8	≥ 2.5	≥ 1.3
Bending modulus	N/mm²	-	-	≥ 140	≥ 120	-	-
In thickness swelling	%	-	-	≤ 6	≤ 6	-	≤ 6
Water permeability, Δp 100 Pa	m³/m²sPa	-	-	$\leq 25 \times 10^{-6}$	$\leq 25 \times 10^{-6}$	-	$\leq 25 \times 10^{-6}$
Vapours permeability	kg / m²sPa	-	-	1.58×10^{-9}	1.0×10^{-9}	-	1.0×10^{-9}

3. OBTAINED RESULTS

3.1. In tables 6 and 7 are presented the fire tests results obtained in the laboratory.

Table 6 - Fire test results - cigarette and match

Test method	Conditions / test criteria	Obtained results		Observations
		source 1*	source 2*	
EN 1021-1 Furniture - Assessment of the ignitability of upholstered furniture - Part 1: Ignition source smouldering cigarette EN 1021-2 Furniture - Assessment of the ignitability of upholstered furniture - Part 2: Ignition source match flame equivalent	Combustion criteria dangerous trend (3.1. a)	Yes	Yes	Findings: During the ignition source contact with the test specimen are detected: melting deformation: Yes, carbonization: Yes, smoke emanation: Yes. Conclusions: Wood fiber boards do not have combustion resistance proprieties (exothermic oxidation not accompanied by flame)
	entire assembly consumed (3.1. b)	Yes	Yes	
	up to the ends (3.1. c)	Yes	Yes	
	crossing the entire thickness (3.1. c)	Yes	Yes	
	more than 1 h (3.1.d)	Yes	Yes	
	present to the end (3.1.e)	Yes	Yes	
	Ignitability criteria circumstances (3.2)			
	dangerous trend (3.2. a)	Yes	Yes	
	entire assembly consumed (3.2.b)	Yes	Yes	
	up to the ends (3.2. c)	Yes	Yes	
	crossing the entire thickness (3.2. c)	Yes	Yes	
	more than 120 seconds (3.2.d)	Yes	Yes	

Table 7 - Fire test results - mass loss

Test method	Conditions / test criteria	Obtained results for mass loss		Observations
		Board 1*	Board 2*	
SR 652	Test specimen dimensions	86.96	84.62	Findings:

(Romanian standard) Determination of efficiency of fireproofing	(400 x 150 x gr.) mm Test specimen conditioning: 22,6...23,4°C ; 45...50% relative humidity Application and the duration of ignition source contact: 1.75 l/min, 20 minutes Wood fiber board – 25 mm			For wood fiber board (25 mm), after 8 minutes are present: intense flame, flame penetration and test specimen bending Conclusions The high value of the mass loss is determined by the lack of a fireproofing product. This is specific to untreated wood combustible materials

The wood fiber boards used in internal conditions do not require fireproofing when using concealed under a fire-resistant building material cover (mortar, plaster, wood fireproof, fireproof plasterboard).

In the case of thermal insulation composite systems, the wood fiber boards are under layers of plaster and silicates based primer and other inorganic salts that are resistant to ignition.

3.2. Biological tests results

3.2.1. Efficiency of antiseptic treatment against mould attack, according to Romanian standard STAS 8022-91

The fungi species used in the test were: Chaetomium globosum Kunze; Alternaria tenuis Ness; Stachybotrys atra Corda; Paecilomyces variotti Bainier; Trichoderma viride Persom ex Fries

Table 8 - Efficiency of antiseptic treatment against mould attack

Product *	Test specimen no.	Development of spores and mycelium growth	Development degrees	Antiseptic treatment efficiency *
Wood fiber boards 25 mm	1/2/3/4/5/6	Strong development of the fungus mycelium on the test sample and in the growth medium.	3	Poor

*According to Romanian standard STAS 8022, there are 4 growth degrees (0, 1, 2, and 3) that determine the efficiency of antiseptic treatment (very good, good, mild and low).

3.2.2. Wood protection products. The efficacy against Reticulitermes species (European termites) threshold determination (laboratory method), according to EN 117.

Table 9 - The efficacy against Reticulitermes species

Test specimen type	Test specimen no.	Surviving workers,%	Surviving soldiers and/or pupas, %	Visual examination
Wood fiber boards - 25 mm	1,2,3	0	0	0**
Witness test specimen - untreated pine	1,2,3	70	60	4***

The test specimens dimensions: (40 x 20 x 10 mm); **no attack; *** severe attack with in depth erosion, greater than 1 mm and less than 3 mm over an area greater than one tenth of the test specimen surface.

The effectiveness threshold is between the average product retention value and the lowest retention value achieved in practice. The validation takes place when the virulence witness test specimens show a level 4 attack and a minimum of 50% survivors.

3.2.3. The preventive efficacy against lignicola basidiomycetes fungi determination, according to EN 113 standard. Application by in depth treatment

Mass loss determination (p.m). Corrective factor C values obtained in the rotting test:

Table 10 - Corrective factor C value

Item no.	Test sample exposed on un-seeded medium	Corrective factor C value, %
1	Wood fiber boards – 25 mm	2.45

Medium mass losses obtained in the rotting test:

Table 11 - Corrective mass (final results)

Item no.	Fungus species	uncorrected p.m, %	corrected p.m, %
1	*Coriolus versicolor, susa CTB 863*	16.20	13.75
2	*Poria placenta, susa 125c BAM*	19.33	16.88
3	*Lenzites trabeum, susa 109 BAM*	16.24	13.79
4	*Lentinus lepideus, susa 20 BAM*	18.15	15.70

The results validation was done by complying with the conditions imposed by the method standard

4. CONCLUSIONS

The values obtained in biological tests highlight that the wood fiber boards are attacked by vegetal regnum microorganisms (imperfect fungi and basidiomycetes) but do not present interest to isoptera insects. In terms of biodegradation, the uncovered wood fiber boards in use class 1 may be used without risk of biological attack caused by fungi and Xylophagous insects. In use class 2 (internal, covered with occasional insects and fungi risk) and use class 3 (external, covered, with frequent moisture frequency and biological risk of attack) wood fiber boards are sensitive to vegetal regnum microorganisms. In order to increase the resistance to biological organisms, the wood fiber boards used in thermal insulation composite systems, are protected by applied adherent layers of mortar and primer. They are creating a physical barrier to moisture, any bio-degradation sources and ignition sources.

In the urban-industrial environment characterized by concerted action of *physical* stress factors (temperature, fire, light, humidity, noise), *chemical* (ionic composition, salinity, oxygen, pH, industrial chemical pollutants) and *biological* (bacteria, fungi, insects, birds, animals) it can be considered that thermal insulation wood fiber boards used in different thermal insulation composite systems present a number of advantages compared with other building materials used in similar applications, as follows: they are made of wood, a renewable and biodegradabile material, from the chemical and structural point of view, contain wood fibers that are rich in carbon, helping to reduce the amount of greenhouse effect substances in the atmosphere (CO_2 including), responsable for global warming, the low resistance to fire and Xylophages biological organisms highlights the lack of hazardous chemicals and biocides. From this point of view, wood fiber boards fulfills the necessary requirement to be classified as a ecological, clean, environment friendly material. The wood fiber boards use in thermal insulation composite systems together with mortars, plasters and waterproofing finishing introduce on the construction products market, a material_that_increase the buildings energy economy and heat retention. Wood fiber boards performances indicate the use for internal and external applications in walls, ceilings and floors covering.

REFERENCES

[1]   Ecological products for construction. The use and behavior of natural polymer based products and environmentally friendly finishing products on thermal insulated surfaces "- Contract no. 414/2009-MDRT;

[2]   www.isocell.at

[3]   Technical Data Sheet Wood fiber boards - IZOPLAC.

THERMAL CHARACTERISTICS OF WOOD-BASED MIXTURES BIOMASS

D. Şova 1, L. Costiuc 1, D. Cioranuranu 2, C. Enăşoae 2

[1] Transilvania University, Faculty of Mechanical Engineering, Braşov, ROMANIA, sova.d@unitbv.ro, lcostiuc@unitbv.ro

[2] Transilvania University, Faculty of Wood Engineering, Braşov, ROMANIA, danna_cioranu@yahoo.com, cornel_bigest@yahoo.com

Abstract: *In this paper, five different mixtures of wood (sawdust of soft- and hardwood and bark), agricultural (straw) and food (husks from sunflower seeds) residues are considered. They were processed into fine waste and then transformed into pellets. The higher and lower heating values were determined, firstly by experimental means using the XRY-1C Oxygen Bomb Calorimeter. By means of two equations and the initial mass-based composition, the heating values were then calculated. The experimental results are close to the calculated ones, validating the equations used and the elemental composition. For the thermal characterization of the combustible mixtures, there were determined, by using the combustion calculations, the theoretical air volume, the theoretical combustion products volume, the combustion products enthalpy and the combustion temperature. From the experimental data, the combustion time has been also evaluated.*

Keywords: *biomass, wood, heating value, combustion products enthalpy, combustion temperature*

1. INTRODUCTION

Biomass ranks as the fourth source of energy in the world, representing approximately 14% of world final energy consumption, a higher share than that of coal (12%) and comparable to those of gas (15%) and electricity (14%) [1]. Alternative biomass fuel is obtained from lignocellulosic biomass that includes forest and agricultural residues (trunks, branches, straw, vines, trees), food residues (fruit seeds), industrial residues (wood chips or sawdust, bark, pulp and paper processing) and municipal wastes (waste from gardens and parks) [2, 3].   Due to the various sources, biomass varies in composition, quality and quantity. The biomass is transformed into briquettes and pellets, improving thus some properties, like energy density increase, moisture content decrease and handling properties enhancing. When biomass is converted into compact briquettes, the moisture content must be 10% to 18% (20%) and the granulation up to 5 mm [2]. Biomass pellets are produced in pellet mills by pressing the biomass through cylindrical shaped press channels in which the biomass is exposed to high pressure and heat that arises from the high friction between the biomass and the press channel walls.

Fuel pellets are frequently made from beech, spruce and straw, which represent the three most common classes of biomass used for fuel pellet production, i.e. hardwoods, softwoods and grasses, respectively [4]. Wood pellets are used for space heating in households, public and other large buildings and in all sizes of combustion plants: pellet stoves and small boilers in single family houses, small block central heating, medium and large district heating plants and large power plants [5].

The combustion properties of biomass pellets are evaluated by use of calorimeters [6]. The use of a cone calorimeter can trace combustion properties and the whole combustion behavior of biomass pellets [3]. The authors [3] have reported relationships between physical properties of wood pellets and combustion behavior. In what regards the heating value, bark has a lower heating value than wood [7].

 Some physical properties (bulk density, true density and durability) of different biomass pellets and their relationship with moisture content, particle size of biomass and die thickness were determined by Theerarattananoon et al. [8]. Biomass differs from coal in what regards energy content and physical properties. Comparative to coal, biomass has generally less carbon, more oxygen and hydrogen, larger volatile matters and lower heating value. It behaves similarly to low-rank coals [1].

The objectives of the paper were to study the thermal characteristics of five lignocellulosic biomass mixtures made from beech and fir sawdust, pear bark, husks of sunflower seeds and wheat straw, which are the heating value, the air and combustion products volumes, the combustion products enthalpy, combustion temperature and time.

2. MATERIALS AND METHODS

Five biomass types, which are beech and fir sawdust, pear bark, husks from sunflower seeds and wheat straw were used as raw materials for the study. The beech and fir sawdust was obtained on a 4 kW circular saw; the pear bark, the husks and straw were grinded by using a 1.7 kW hammer mill with 2.5 mm sieve openings. The grindings were stored in sealed plastic bags at room temperature.

The moisture content of the biomass was determined in different ways. For the beech and fir sawdust, a Feutron moisture meter (Model F10) was used. The air temperature was 20 °C. For the other biomass types, two samples of about 2 g of each one were dried in an oven (Feutron) at 103 °C for 24 h. The samples were weighed several times a day until a constant weight was achieved. The moisture content was calculated on dry basis as follows:

$$MC = \frac{m_1 - m_2}{m_2} \times 100 \, [\%]$$ (1)

where m_1 is the mass of the moist biomass (g) and m_2 is the mass of dry biomass (g).

Equal parts of each biomass were mixed into five combinations of 1.5 g each one. For each combination there were made three samples. The components and mixtures masses were measured with two balances of different precision, a technical balance (Kern, 0.01g) and an analytical balance (Vietzke, 0.1 mg).

The biomass mixtures were transformed into pellets by using a hand-driven press with the die size of 12 × 40 mm (hole diameter × length). The pellets were afterwards weighted.

The biomass pellets were submitted to the experimental determination of the heating value by using the XRY-1C Oxygen Bomb Calorimeter (Shanghai Changji Geological Instruments).

The dry-basis (db) and ash-free (a-f) composition of hard- and softwood, husks of sunflower seeds and wheat straw that was used in following calculation is indicated by [9] and that of bark is indicated by [10], respectively. Bark composition varies slightly according to different authors [1, 10 and 11]. The initial mass-based composition (i-b) was calculated.

The composition of the biomass mixtures was calculated considering the number of components participating in the mixture.

Based on the pellets composition, two equations for the determination of the higher and lower heating values (HHV and LHV) were used [12, 13], which are

$$Q_{i1} = 33900 \cdot c + 119850 \cdot \left(h - \frac{o}{8} \right) + 10470 \cdot s - 2500 \cdot u$$ (2)

$$Q_{s1} = 33900 \cdot c + 142350 \cdot \left(h - \frac{o}{8} \right) + 10470 \cdot s$$ (3)

$$Q_{i2} = 33900 \cdot c + 120120 \cdot \left(h - \frac{o}{8} \right) + 9250 \cdot s - 2510 \cdot u$$ (4)

$$Q_{s2} = 33900 \cdot c + 142710 \cdot \left(h - \frac{o}{8} \right) + 9250 \cdot s$$ (5)

The theoretical air volume and the theoretical combustion products volume were calculated from the stoichiometric combustion equations [12], by using the biomass mixtures composition. It was made the assumption that combustion runs in theoretical conditions, thus the excess-air coefficient is $\lambda = 1$.

The combustion time was calculated for each biomass mixture based on the time intervals indicated by the experimental records obtained during the heating values measurements. The time interval between two temperature measurements during the main period (combustion) is 30 s.

Combustion products enthalpy was determined according to the energy conservation law applied to 1 kilogram of fuel burned in a furnace [12] and accordingly, the combustion temperature was obtained by using H-t diagram that was represented for each biomass mixture.

3. RESULTS AND DISCUSSION

The average value of the moisture content for each biomass type is presented in Table 1.
The masses of the biomass pellets are shown in Table 2.
The experimentally determined higher and lower heating values are indicated in Table 3.

Table 1: Moisture content

Biomass	Moisture content (%) (mean ± SD)
Beech sawdust	7.6
Fir sawdust	8.8
Pear bark	17.355 ± 0.355
Husks of sunflower seeds	2.98 ± 0.48
Wheat straw	6.024 ± 4.575

Table 2: Biomass masses

Biomass mixture	Biomass components	Mass (g) Sample		
		S_1	S_2	S_3
M_1	Beech sawdust (50%) Pear bark (50%)	1.4233	1.4279	1.4172
M_2	Beech sawdust (33%) Fir sawdust (33%) Pear bark (33%)	1.4289	1.4948	1.4599
M_3	Beech sawdust (50%) Husks of sunflower seeds (50%)	1.4993	1.4911	1.4913
M_4	Beech sawdust (50%) Wheat straw (50%)	1.47	1.4627	1.4633
M_5	Fir sawdust (33%) Husks of sunflower seeds (33%) Wheat straw (33%)	1.4919	1.4997	1.4739

Table 3: Experimental heating values

Biomass mixture	Higher heating value (MJ/kg) (mean ± SD)	Lower heating value (MJ/kg) (mean ± SD)
M_1	17.143 ± 0.14	16.703 ± 0.13
M_2	17.014 ± 1.2	16.576 ± 1.18
M_3	17.542 ± 0.07	17.102 ± 0.07
M_4	17.056 ± 1.86	16.617 ± 1.33
M_5	17.65 ± 0.22	17.209 ± 0.22

The dry-basis (db) and ash-free (a-f) composition of hard- and softwood, husks of sunflower seeds and wheat straw, as indicated by [9] and of bark, as indicated by [10], respectively is shown in Table 4, comparative to the initial mass-based composition (i-b) that was calculated.

Table 4: Elemental composition

	Hardwood		Softwood		Bark		Husks of sunflower seeds		Wheat straw	
	db/ a-f	i-b	db/ a-f	i-b	db	i-b	db/ a-f	i-b	db/ a-f	i-b

Carbon (%)	50.5	46.157	51	46	52	42.975	51.5	49.451	50	44.08
Hydrogen (%)	6.1	5.576	6.15	5.548	5.51	4.554	5.9	5.665	6.2	5.467
Oxygen (%)	42.8	39.119	42.25	38.11	36.065	29.806	41.9	40.232	43.1	38.0039
Nitrogen (%)	0.6	0.548	0.6	0.542	0.56	0.463	0.5	0.48	0.6	0.529
Sulphur (%)	0	0	0	0	0.045	0.037	0.2	0.192	0.1	0.0881
Chlorine (%)	-	-	-	-	0.14	0.116	-	-	-	-
Water content (%)	-	7.6	-	8.8	-	17.355	-	2.98	-	6.024
Ash (%)	-	1	-	1	5.68	4.694	-	1	-	5

The elementary composition of the biomass mixtures was calculated considering the number of components participating in the mixture. It is shown in Table 5.

Table 5: Biomass mixtures composition

	M_1	M_2	M_3	M_4	M_5
Carbon (%)	44.566	45.044	47.804	45.1225	46.511
Hydrogen (%)	5.065	5.226	5.6205	5.5215	5.56
Oxygen (%)	34.4625	35.678	39.6755	38.5615	38.782
Nitrogen (%)	0.5055	0.51767	0.514	0.5385	0.517
Sulphur (%)	0.0185	0.01233	0.096	0.0441	0.0934
Chlorine (%)	0.058	0.058	-	-	-
Water content (%)	12.4775	11.2517	5.29	6.812	5.935
Ash (%)	2.847	2.23133	1	3	2.3334

The heating values were calculated with relations (2)-(5) and the mean values are shown in Table 6.

Table 6: Calculated heating values

Biomass mixture	Higher heating value (MJ/kg) (mean ± SD)	Lower heating value (MJ/kg) (mean ± SD)
M_1	17.023 ± 0.0064	16.407 ± 0.0011
M_2	16.42 ± 0.0014	15.957 ± 0.00081
M_3	17.157 ± 0.0006	16.876 ± 0.000041
M_4	16.3 ± 0.00091	15.972 ± 0.00034
M_5	16.792 ± 0.00071	16.482 ± 0.000095

The experimental and calculated higher and lower heating values are represented in Figs. 1 and 2.

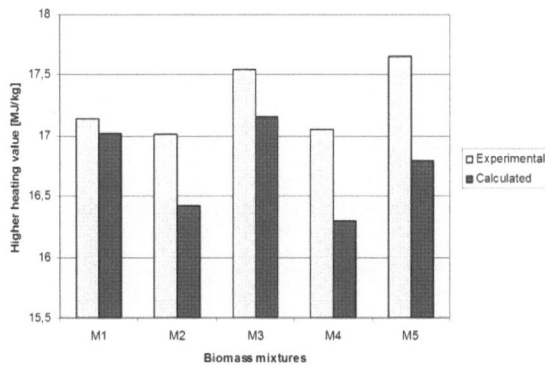

Figure 1: Experimental and calculated

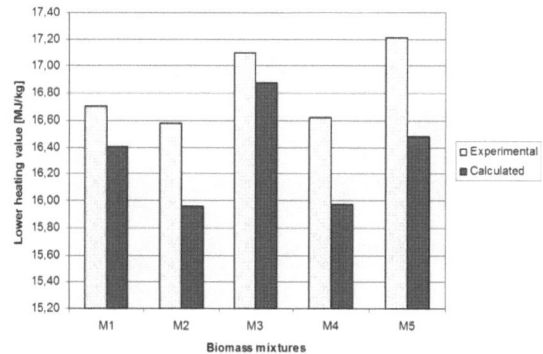

Figure 2: Experimental and calculated

higher heating value   lower heating value

Biomass mixtures M_5 and M_3 have high experimental heating values, which can be explained by low moisture and ash content of sunflower seeds husks and high carbon, hydrogen and oxygen mass fractions of the same biomass type. Lower heating values have mixtures M_2 and M_4, due to the high moisture and ash content of bark, high ash content of straw and low carbon mass fraction of straw and bark. The calculated heating values are lower than the experimental ones. Mixture M_3 has the highest calculated heating value and mixture M_4 the lowest. The moisture content values of mixtures M_3 and M_5 are very close and their compositions are comparable. Larger differences between experimental and calculated higher and lower heating values are met at those mixtures that contain wheat straw (M_4: HHV-4.31%, LHV-3.88%; M_5: HHV-4.86%, LHV-4.22%). They can be explained by some moisture content variations between the moisture content measurement and the heating value measurement. According to Kamikawa [3] the lower heating value of softwood xylem and bark mixtures ranges from 15.47 and 16.78 MJ/kg. Straw has the gross calorific value of 15.354 MJ/kg [1].

In Fig. 3, the volume of combustion air is indicated and in Fig. 4, the combustion products volume. The results are similar. High volumes of air and combustion products are obtained for mixture M_3 that contains large amounts of carbon and hydrogen and low amounts of ash and moisture. Mixture M_1 has low volumes because the same components have opposed mass fractions to those of mixture M_3.

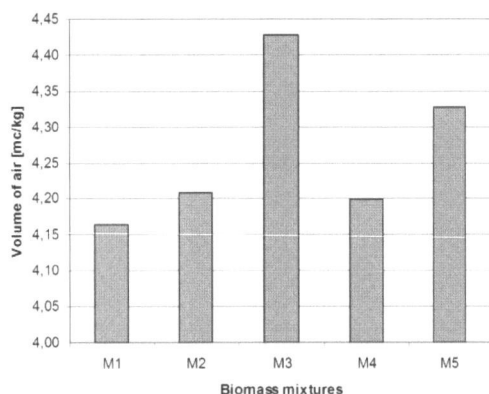

Figure 3: Volume of combustion air

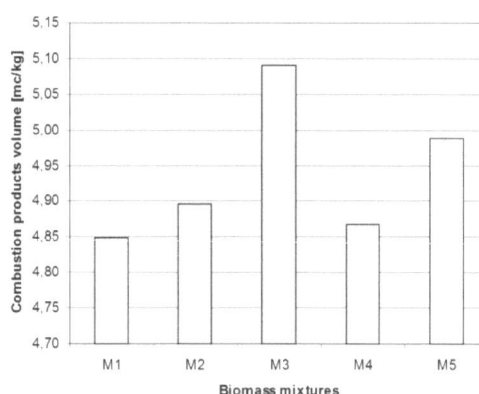

Figure 4: Combustion products volume

The combustion time is indicated in Fig. 5. There is a lower combustion time for mixture M_3 and a higher time for M_4. The combustion dynamics is inverse proportional to the heat of combustion. Kamikawa [3] made the observation that bark pellets showed slower combustion than wood pellets. If comparing the results of the burn-out time with the total amount of heat release, the reversed proportionality can be remarked too in [3]. Combustion products enthalpy is shown in Fig. 6. There is the same distribution of the biomass mixtures like in Figs. 1 and 2, showing a good correlation between the heating value and heat of the combustion products.

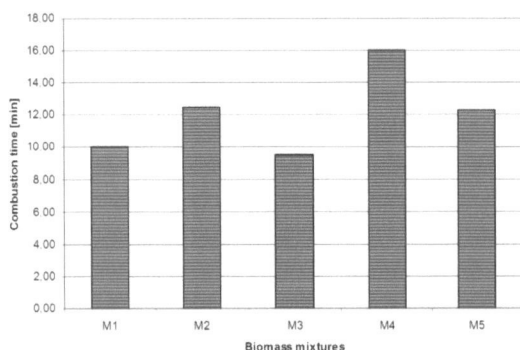

Figure 5: Combustion time

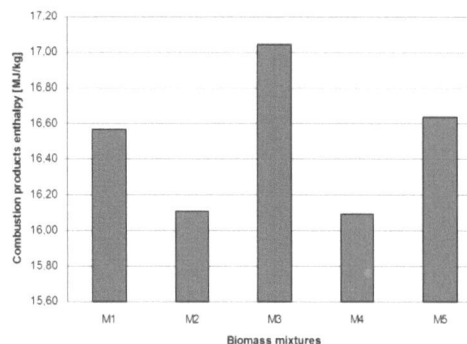

Figure 6: Combustion products enthalpy

In Fig. 7 H-t diagrams are represented. The curves are very close. Combustion temperature is presented in Fig. 8. It ranges from 1738 °C (M_2) to 1799 °C (M_1).

4. CONCLUSIONS

In this study we obtained thermal characteristics (heating value, air and combustion products volumes combustion products enthalpy, combustion temperature and time) of five lignocellulosic biomass mixtures, composed of beech and fir sawdust, pear bark, husks of sunflower seeds and wheat straw. The higher and lower heating values were obtained experimentally and were also calculated by use of the initial mass-based composition (ultimate analysis) of the mixtures. The results are very close, except the mixtures that contain wheat straw. Mixtures M_3 (beech sawdust and husks of sunflower seeds) and M_5 (fir sawdust, husks of sunflower

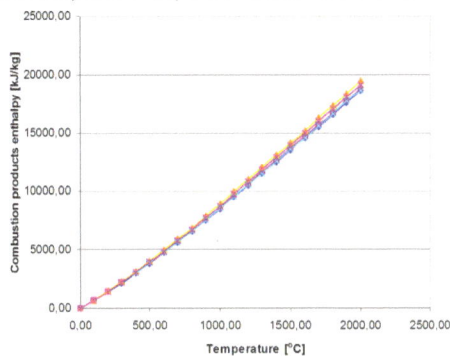

Figure 7: H-t diagrams	**Figure 8:** Combustion temperature

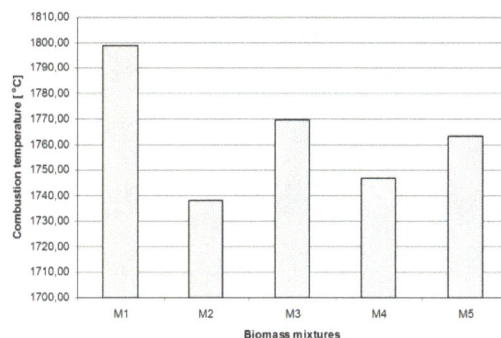

seeds and wheat straw) have high heating values because they have high carbon and hydrogen mass fractions and low moisture and ash content. Mixtures M_2 (beech sawdust, fir sawdust and pear bark) and M_4 (beech sawdust and wheat straw) instead, have low heating values. The volumes of combustion air and combustion products are similarly distributed on the five biomass mixtures, being influenced by their composition in the same way like the heating value. High volumes of air and combustion products are obtained for mixture M_3 and low volumes for mixture M_1 (beech sawdust and pear bark). The combustion time is inverse proportional to the heat of combustion. Thus, mixture M_3 has the lowest time and M_4 the highest time of combustion. The combustion products enthalpy is directly correlated with the heating value. The values of the combustion temperature are close. They range from 1738 ^{o}C (M_2) to 1799 ^{o}C (M_1).

Biomass derives from different renewable sources and varies therefore considerably in composition. The thermal characteristics of biomass enhance if the moisture and ash content is low and the carbon and hydrogen mass fractions are high.

REFERENCES

[1]   Saidur, R. et al., A review on biomass as a fuel for boilers, Renewable and Sustainable Energy   Reviews, 15: 2262–2289, 2011.

[2]   Đerčan, B. et al., Possibility of efficient utilization of wood waste as a renewable energy resource in Serbia, Renewable and Sustainable Energy Reviews 16:1516-1527, 2012.

[3]   Kamikawa, D. et al., Evaluation of combustion properties of wood pellets using a cone calorimeter, J Wood Sci., 55: 453-457, 2009.

[4]   Stelte, W. et al., A study of bonding and failure mechanisms in fuel pellets from different biomass resources, Biomass and Bioenergy 35:910-918, 2011.

[5]   Scarlat, N. et al., An overview of the biomass resource potential of Norway for bioenergy use,   Renewable and Sustainable Energy Reviews, 15:3388– 3398, 2011.

[6]   Ungureanu, V. B. et al., Termodinamică. Aplicaţii practice, Editura Universităţii Transilvania din Braşov, 2010, p. 112-131.

[7]   Klašnja, B., Kopitovič, Š., Quality of wood of some willow and robinia clones as fuelwood, Drevársky Výskum, 44(2):9-18, 1999.

[8]   Theerarattananoon, K. et al., Physical properties of pellets made from sorghum stalk, corn stover, wheat   straw and big bluestem, Industrial Crops and Products, 33:325-332, 2011.

[9]   Popa, B. et al., Manualul inginerului termotehnician, Editura Tehnică, Bucureşti, 1986, p. 538, 540.

[10]   Wilk, V. et al., Gasification of waste wood and bark in a dual fluidized bed steam gasifier, Biomass

Conv. Bioref. 1:91-97, 2011.

[11]   Johansson, L. et al., Emission characteristics of modern and old–type residential boilers fired with wood   logs and wood pellets, Atmospheric Environment, 38:4138-4195, 2004.

[12]   Şova, D., Termotehnică, vol.1, Editura Universităţii Transilvania din Braşov, 2000, p. 65.

[13]   Marinescu, M., Ştefănescu, D., Chisacof, A., Adler, D., Instalaţii de ardere. Culegere de probleme pentru ingineri, Editura Tehnică, Bucureşti, 1985, p. 8.

ON THE NUMERICAL ANALYSIS OF COMPOSITE MATERIAL

D.D. Nicoara

Transilvania University, City Brasov, ROMANIA, tnicoara@unitbv.ro

Abstract: *In this paper we analyze laminate composite materials by numerical methods. We write a Matlab program that assists the user to find out the ABBD stiffness matrix of a laminate composite. The main objective of the paper is to show the advantages and ease use of Matlab software in composite materials analysis. To demonstrate the capability of the program a lot of numerical examples was presented.*
Keywords: *composite laminate, stiffness matrix, constitutive equations, MATLAB.*

1. INTRODUCTION

Numerous papers have been published on the analysis and design composite material. Because composite materials are produced in many combinations and forms the designer engineer must consider many design cases.

In study of structural response composite material is often analyzed by analytical methods and by numerical methods [2], [3], [8], [9]. This requires a large amount of calculations that depend on many parameters, Calculations of macro-mechanical proprieties and calculations of the constitutive equations involve many matrix manipulations. The manual calculations would take long time. Solutions was to write computers programs. Numerical computing package MATLAB was used as a basis for the programs because the facilities offered in matrix computations [1], [4], [6], and [7]. The main objective of this paper is to show the advantages of using MATLAB to analysis of composite material.

We illustrate these features by analyzing orthotropic laminate composite material. Fiber-reinforced composites are analyzed by two-dimensional theories. The Kirchoff –Love hypothesis is used. The objective is to write a program to determine the laminate constitutive equations for multi-layered composites. This approach follows the conventional methods for designing composite structures. When an orthotropic materials is in a plane stress state the relationship between the stresses and strain involve the four elastic constants E_1, E_2, ν_{12},and G_{12}. In section 2 we present the characteristics of unidirectional composite material as functions of the characteristic of the fibres and the matrix. In section 3 we will review basis assumptions for study the constitutive relations for fiber-reinforced composite and calculate stiffness matrices for laminate, ABBD matrix. In sections 4 details for computational approach have been outlined. How to coding in MATLAB software has been outlined step-by-step procedure. Numerical applications are presented in section 5. In section 6 we present the conclusions.

2. EVALUATION OF ELASTIC CONSTANTS

Stiffness and strength is the basic concept for underlying the mechanics of fiber-reinforced advanced composite materials. This aspect of composite materials technology is sometimes terms micromechanics, because it deals with the relations between macroscopic engineering properties and the microscopic distribution of the material's constituents, namely the volume fraction of fiber. The purpose of this section is to predict the material constants (also called elastic constants) of a composite material by studying the micromechanics of the problem, i.e. by studying how the matrix and fibers interact.

There are three different approaches that are used to determine the elastic constants for the composite material based on micromechanics. These three approaches are [2], [3]:

1. Using numerical models such as the finite element method.
2. Using models based on the theory of elasticity.
3. Using rule-of-mixtures models based on a strength-of-materials approach.

Terminology used in micromechanics:

- E_v , E_m -Young's modulus of fiber and matrix
- G_f , G_m -Shear modulus of fiber and matrix
- ν_f , ν_m -Young's modulus of fiber and matrix

- V_f, V_m -Volum fraction of fiber and matrix;

There are four elastic constant of a unidirectional lamina:

- Longitudinal Young's modulus: E_1
- Transverse Young's modulus : E_2
- Major Poisson's ratio v_{12}
- In-plane shear modulus: G_{12}

Using the strength-of-materials approach and the simple rule of mixtures, we have the following relations for the elastic constants of the composite material in local coordinates systems, [1], [5].

Longitudinal Young's modulus in the 1-direction (the longitudinal stiffness):

$$E_1 = E_f V_f + E_m V_m \tag{1}$$

Transvers Young's modulus in the 2-direction (also called the transverse stiffness):

$$\frac{1}{E_2} = \frac{V_f}{E_f} + \frac{V_m}{E_m} \tag{2}$$

Poisson's ratio v_{12} in the 1-2 plane:

$$v_{12} = v_f V_f + v_m V_m \tag{3}$$

where v_f and v_m are Poisson's ratios for the fiber and matrix, respectively.

Shear modulus in the 1-2 plane G_{12}:

$$\frac{1}{G_{12}} = \frac{V_f}{G_f} + \frac{V_m}{G_m} \tag{4}$$

While the simple rule-of-mixtures models used above give accurate results for E_1 and v_{12}, the results obtained for E_2 and G_{12} do not agree well with finite element analysis and elasticity theory results. Therefore, we need to modify the simple rule-of-mixtures models shown above.

In Matlab file we using the elasticity solution give the following formula for G_{12}:

$$G_{12} = \frac{G_m(G_f + G_m + (G_f - G_m)V_f)}{G_f + G_m - (G_f - G_m)V_f} \tag{5}$$

3. THE STRESS-STRAIN RELATION

This approach follows the conventional methods for designing composite structures. Equations relating the stresses and strains have been developed and are available from various tests [2], [3], [5], and [8].

Fiber-reinforced composite are analyze by two-dimensional theories. The Kirchoff-Love hypothesis is used.

Using the assumption of plane stress, the relationship between the stress-strain involve the four elastic constants: E_1, E_2, v_{12}, G_{12}, determined in section 2. It is seen that the stress-strain relations for a single 2D orthotropic lamina are:

$$\{\varepsilon\} = [S]\{\sigma\} \quad \begin{Bmatrix} \varepsilon_1 \\ \varepsilon_2 \\ \gamma_{12} \end{Bmatrix} = \begin{bmatrix} S_{11} & S_{12} & 0 \\ S_{12} & S_{22} & 0 \\ 0 & 0 & S_{66} \end{bmatrix} \begin{Bmatrix} \sigma_1 \\ \sigma_2 \\ \sigma_3 \end{Bmatrix} = \begin{bmatrix} \dfrac{1}{E_1} & \dfrac{v_{12}}{E_1} & 0 \\ \dfrac{v_{12}}{E_1} & \dfrac{1}{E_2} & 0 \\ 0 & 0 & \dfrac{1}{G_{12}} \end{bmatrix} \begin{Bmatrix} \sigma_1 \\ \sigma_2 \\ \tau_{12} \end{Bmatrix} \tag{6}$$

$$\{\sigma\} = [Q]\{\varepsilon\} \quad \begin{Bmatrix} \sigma_1 \\ \sigma_2 \\ \tau_{12} \end{Bmatrix} = \begin{bmatrix} Q_{11} & Q_{12} & 0 \\ Q_{12} & Q_{22} & 0 \\ 0 & 0 & Q_{66} \end{bmatrix} \begin{Bmatrix} \varepsilon_1 \\ \varepsilon_2 \\ \varepsilon_3 \end{Bmatrix} = \begin{bmatrix} \dfrac{E_1}{1 - v_{12}v_{21}} & \dfrac{v_{12}E_2}{1 - v_{12}v_{21}} & 0 \\ \dfrac{v_{12}E_2}{1 - v_{12}v_{21}} & \dfrac{E_2}{1 - v_{12}v_{21}} & 0 \\ 0 & 0 & G_{12} \end{bmatrix} \begin{Bmatrix} \varepsilon_1 \\ \varepsilon_2 \\ \gamma_{12} \end{Bmatrix} \tag{7}$$

where Q_{ij} are the reduced stiffness constants.

The 1-2 co-ordinate system can be considered to be local co-ordinates based on the fibre direction. However this system is inadequate as fibres can be placed at various angles with respect to each other and the structure. Therefore a new co-ordinate system needs to be defined that takes into account the angle the fibre makes with its surroundings. This new system is referred to as global co-ordinates (*x-y* system) and is related to the local co-ordinates (1-2 system) by the angle θ, Fig.1.

To find the stress and strain in the (x,y,z) global coordinate system a simple rotational transformation is need.

The transformation relation from local stresses to the global stresses is:

$$\begin{Bmatrix} \sigma_1 \\ \sigma_2 \\ \sigma_3 \end{Bmatrix} = \begin{bmatrix} m^2 & n^2 & 2mn \\ n^2 & m^2 & -2mn \\ -mn & mn & -n^2 \end{bmatrix} \begin{Bmatrix} \sigma_x \\ \sigma_y \\ \sigma_z \end{Bmatrix} \tag{9}$$

where m= $\cos\theta$; n = $\sin\theta$.

Similar transformation relations hold for the strains.

The transformed reduced stiffness matrix and the stress-strain relations to a global (*x, y, z*) coordinates are:

$$[\overline{Q}] = [T]^{-1}[Q][T] \; ; \quad \begin{Bmatrix} \sigma_x \\ \sigma_y \\ \sigma_{xy} \end{Bmatrix} = \begin{bmatrix} \overline{Q}_{11} & \overline{Q}_{12} & \overline{Q}_{16} \\ \overline{Q}_{12} & \overline{Q}_{22} & \overline{Q}_{26} \\ \overline{Q}_{16} & \overline{Q}_{26} & \overline{Q}_{66} \end{bmatrix} \begin{Bmatrix} \varepsilon_x \\ \varepsilon_y \\ \gamma_{xy} \end{Bmatrix} \tag{10}$$

Note that the following relations hold $[\overline{Q}] = [\overline{S}]^{-1}$; $[\overline{S}] = [\overline{Q}]^{-1}$.

Equations (1) - (10) are used to calculate the stresses and strain for a single layer.

Fiber-reinforced materials consist usually of multiple layers of material to form a laminate.

Consider a plate of total thickness h composed of N orthotropic layers with the principal material coordinates oriented at angles $\theta_1, \theta_2, ..., \theta_N$.

We omit the details, and summarize the laminate constitutive equations for multi-layered composites. The laminate constitutive equations relate the force and moment resultants (N_i , M_i) to the vector $\{\varepsilon_x^0 \quad \varepsilon_y^0 \quad \gamma_{xy}^0\}^T$ of the mid-plane strains and to the vector of the mid-plane curvatures $\{\kappa_x^0 \quad \kappa_y^0 \quad \kappa_{xy}^0\}$.

The constitutive equations are:

$$\begin{Bmatrix} N \\ M \end{Bmatrix} = \begin{bmatrix} A & B \\ B & D \end{bmatrix} \begin{Bmatrix} \varepsilon^0 \\ \kappa^0 \end{Bmatrix} \tag{11}$$

where the 6x6 laminate matrix consisting of the components A_{ij}, B_{ij}, D_{ij} (i, j = 1, 2, 6) is laminate stiffness matrix, also called ABBD matrix.

The sub-matrix [A] is called the extensional stiffness matrix, sub-matrix [B] is called the coupling stiffness matrix and sub-matrix [D] is called the bending stiffness matrix.

The stiffness coefficients are determined by the following equations:

$$A_{ij} = \sum_{k=1}^{N} (\overline{Q}_{ij})_k (h_k - h_{k-1}) = \sum_{k=1}^{N} (\overline{Q}_{ij})_k t_k \tag{12}$$

$$B_{ij} = \frac{1}{2} \sum_{k=1}^{N} (\overline{Q}_{ij})_k (h_k^2 - h_{k-1}^2) = \sum_{k=1}^{N} (\overline{Q}_{ij})_k t_k \overline{h}_k \tag{13}$$

$$D_{ij} = \frac{1}{3} \sum_{k=1}^{N} (\overline{Q}_{ij})_k (h_k^3 - h_{k-1}^3) = \sum_{k=1}^{N} (\overline{Q}_{ij})_k (t_k \overline{h}_k^2 + \frac{t_k^3}{12}) \tag{14}$$

where N is the number of layers, $(\overline{Q}_{ij})_k$ are the elements of $[\overline{Q}]$ matrix for he k-th layer, h_k is the distance from the mid-plane of laminates, Fig. 2.

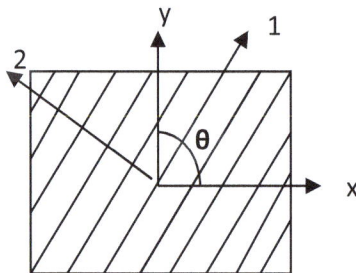

Figure 1: Local / global coordinate systems

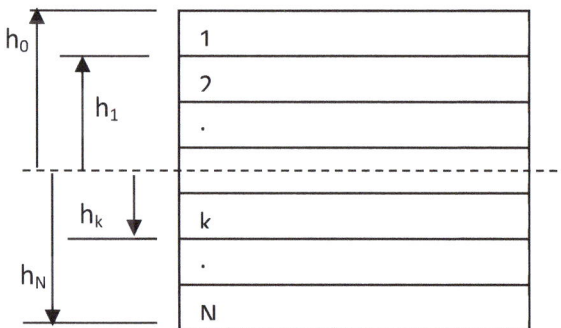

Figure 2: Cross-section input data

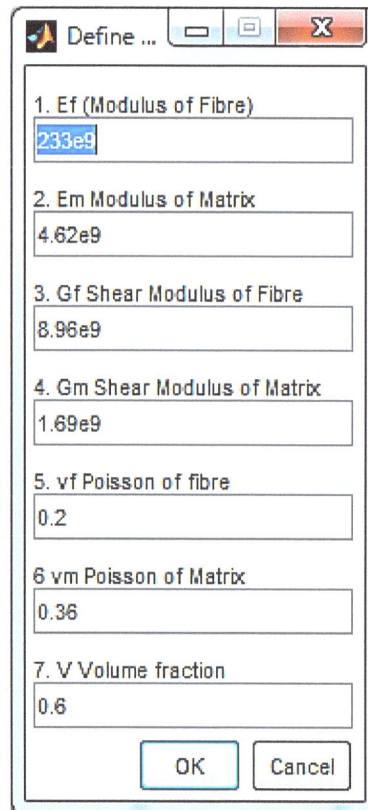

Figure 3: Interactive interface input data

4. MATLAB PROGRAM

Manual calculations would take a long period therefore the laminate analysis outlined above has been implemented in a MATLAB code. This program finds the overall laminate proprieties.
The outputs of the program are $[A]$, $[B]$ and $[D]$ matrices.

If we are going to make a laminated structure we must know the material proprieties, number of fiber layers, the thickness of each ply, the angle of each ply (in degrees) from the top of layer down, these are the inputs. A user interactive interface input data was made, Fig. 3.

In the Matlab script the calculations are done in the following step:

1. Micromechanics:

- enter the fiber and matrix proprieties (from the user interface) $E_f, E_m, G_f, G_m, V_f, V_m$;

- calculate ply elastic proprieties E_1, E_2, v_{12}, G_{12} using Equations (1) - (5);

2. Enter the laminate characteristic:

- angle of fibers and thickness of each layer;
- number of layers;
- material proprieties of each layer.
- total height of the laminate

3. Calculate the stiffness matrix $[Q]$ for each layer using Eq. (7).

4. Calculate the transformation matrix $[T]$ for each layer.

5. Calculate the transformed reduced stiffness matrix $[\overline{Q}]$ for each layer using Eq. (10).

6. Calculate $[A]$, $[B]$ and $[D]$ matrices via Eq. (12)-(14).

This approach follows the commonly found methods laid out in the various texts [2], [3], [5], and [9].

Program consist in a main script file and various functions. These Matlab functions correspond to the steps of the program described above.

4. NUMERICAL EXAMPLE

The following example is taken to demonstrate the capability of the program. This example demonstrates the use of Matlab code in determining the stiffness of a four-ply layup. The material chosen for the composite plies was Graphite/Epoxy (Gr/E). This material consists of carbon (graphite) fibers in a standard epoxy matrix. Table 1 summarizes the composite properties. Values displayed in the table are standard values for Gr/E composite with a 0.55 volume fraction.

The ply angle orientation $[0 \setminus 90 \setminus 90 \setminus 0]$ was considered.

Table 1: Material Proprieties

Material	E_1	E_2	G_{12}	v_{12}	t
Gr/E Composite	155	12.10	4.40	0.248	0.5

The outputs of the program for 4-ply systems are $[A]$, $[B]$ and $[D]$ stiffness matrices, listed below:

$$A = \begin{bmatrix} 1.6791e+011 & 6.0306e+009 & -2.1009e-008 \\ 6.0306e+009 & 1.6791e+011 & 8.8133e-006 \\ -2.1009e-008 & 8.8133e-006 & 8.8e+009 \end{bmatrix}$$

$$D = \begin{bmatrix} 9.1866e+010 & 2.0102e+009 & -1.7507e-009 \\ 2.0102e+009 & 2.0071e+010 & 7.3444e-007 \\ -1.7507e-009 & 7.3444e-007 & 2.9333e+009 \end{bmatrix}$$

$$B = \begin{bmatrix} 0 & 0 & 0 \\ 0 & 0 & 0 \\ 0 & 0 & 0 \end{bmatrix}$$

$$[A\ B;\ B\ D] = \begin{bmatrix} A & B \\ B & D \end{bmatrix} =$$

$$\begin{bmatrix} 1.6791e+011 & 6.0306e+009 & -2.1009e-008 & 0 & 0 & 0 \\ 6.0306e+009 & 1.6791e+011 & 8.8133e-006 & 0 & 0 & 0 \\ -2.1009e-008 & 8.8133e-006 & 8.8e+009 & 0 & 0 & 0 \\ 0 & 0 & 0 & 9.1866e+010 & 2.0102e+009 & -1.7507e-009 \\ 0 & 0 & 0 & 2.0102e+009 & 2.0071e+010 & 7.3444e-007 \\ 0 & 0 & 0 & -1.7507e-009 & 7.3444e-007 & 2.9333e+009 \end{bmatrix}$$

In order to ensure the validity of the program, the result was compared with those reported by Fiber Reinforced Composites (FRC) calculator in the website efunda.com [9], listed below:

The extensional stiffness matrix [A]:

$$[A] = \sum_{k=1}^{N} \left(\overline{C}_{ij} \right)_k \left(z_k - z_{k-1} \right) = \sum_{k=1}^{N} \left(\overline{C}_{ij} \right)_k t_k$$

$$= \begin{vmatrix} 167.9 & 6.031 & 0 \\ 6.031 & 167.9 & 0 \\ 0 & 0 & 8.800 \end{vmatrix} \quad \text{GPa-mm}$$

The bending stiffness matrix [D]:

$$[D] = \frac{1}{3} \sum_{k=1}^{N} \left(\overline{C}_{ij} \right)_k \left(z_k^3 - z_{k-1}^3 \right) = \sum_{k=1}^{N} \left(\overline{C}_{ij} \right)_k \left(t_k \overline{z}_k^2 + \frac{t_k^3}{12} \right)$$

$$= \begin{vmatrix} 91.87 & 2.010 & 0 \\ 2.010 & 20.07 & 0 \\ 0 & 0 & 2.933 \end{vmatrix} \quad \text{GPa-mm\^{}3}$$

5. CONCLUSIONS

The objective of this paper was to show the advantages of using MATLAB software to study various laminates composite. In this work, go through the first stage of analysis of composite materials, which consist of stiffness matrix calculations. The program written in MATLAB allows the analysis of different laminates, by introducing different material proprieties and inner ply orientations from interface input. The input are the material properties, number of fiber layers, and thickness and fibre orientation of each layer. The output of the program are $[A]$, $[B]$ and $[D]$ stiffness matrices. The use of the program will greatly reduce the analysis and design time of laminate fiber – reinforced composite.

6. REFERENCES

[1]   Chen, K., Giblin, P., Irving, A., Mathematical Explorations with MATLAB, Cambridge University Press, 1999.

[2]   Daniel, I., Ishai, O., Engineering Mechanics of Composite Materials, Oxford University Press, 1994.

[3]   Decolon, C., Analysis of Composite Structure, Hermes Penton Ltd., London, 2002.

[4]   Dukkipati, R.V., MATLAB for Mechanical Engineers, New Age International (P) Limited, New Delhi, 2008.

[5]   Gibson, R., Principles of Composite Material Mechanics, McGraw-Hill, 1994.

[6]   Higham, D. , Higham, N., MATLAB Guide, SIAM, 2000.

[7]   MediaWiki MATLAB, 2008. http://en.wikipedia.org/wiki/MATLAB.

[8]   Voyiadjis, G.Z., Kattan, P.I., Mechanics of Composite Materials with MATLAB, ISBN-10 3-540-24353-4 Springer Berlin Heidelberg New York 2005.

[9]   http://www.efunda.com/formulae/solid_mechanics/composites/calc_ufrp_abd_match.cfm.

[10]   http:/mathworks.com/products/matlab.

DESIGN OPTIMIZATION PROCESS WITH APPLICATION TO THE COMPOSITE STRUCTURAL ELEMENT

E. Kormaníková[1]

[1] Technical University of Košice, Civil Engineering Faculty, Košice, SLOVAKIA, eva.kormanikova@tuke.sk

Abstract: *The paper deals with a numerical approach of modelling of laminate plates and with their optimal design. It provides bases for the modelling of mechanical behaviour of laminates by reviewing general assumptions of classical laminate theory (CLT)). Elements of optimization of laminate plates are also discussed. The thicknesses of layers with the known orientation, referred as the thickness variables, will be used as design variables. The optimization problem with strength constraints will be formulated to minimize the laminate weight. Analytical and numerical approaches outlined in this paper are accompanied by computer generated example. There are depicted distributions of numerical results during the optimization process.*

Keywords: *laminate plate, sizing optimization problem, the Modified Feasible Direction method, the Sequential Linear Programming method*

1. INTRODUCTION

The rapid growth in the use of composite materials in structures has required the development of the theory of mechanics of composite laminates and the analysis and optimization of structural elements made of composite laminates. In this paper there are included an explanation of the concepts involved in the analysis and optimization of laminates, the mechanics needed to translate those concepts into a mathematical representation of the physical reality, and a explanation of the solution of the resulting boundary value problems by using Finite Element Analysis software.

2. MODELLING AND ANALYSIS OF LAMINATE PLATES, CLASSICAL LAMINATE THEORY

In the classical laminate theory the Kirchhoff hypotheses of the classical plate theory remains valid [1-3]. These assumptions imply that the transverse displacement w is independent of the thickness coordinate z, the strains γ_{xz}, γ_{yz} and ε_z are zero and the curvatures κ_i are given by

$$\kappa = -\left(\frac{\partial \psi}{\partial x}, \frac{\partial \varphi}{\partial y}, \frac{\partial \psi}{\partial y} + \frac{\partial \varphi}{\partial x}\right) \tag{1}$$

The equilibrium equations will be formulated for a plate element $dxdy$ and yield three force and two moment equation

$$\frac{\partial N_x}{\partial x} + \frac{\partial N_{yx}}{\partial y} = -p_1 \qquad \frac{\partial N_{xy}}{\partial x} + \frac{\partial N_y}{\partial y} = -p_2 \qquad \frac{\partial V_{xz}}{\partial x} + \frac{\partial V_{yz}}{\partial y} = -p_3$$

$$\frac{\partial M_x}{\partial x} + \frac{\partial M_{xy}}{\partial y} = V_{xz} \qquad \frac{\partial M_{yx}}{\partial x} + \frac{\partial M_y}{\partial y} = V_{yz} \tag{2}$$

The transverse shear force resultants V_{xz}, V_{yz} can be eliminated and the five equations (2) reduce to three equations. The in-plane force resultants N_x, N_y and N_{xy} are uncoupled with the moment resultants M_x, M_y and M_{xy}. The three equilibrium equations are

$$\frac{\partial N_x}{\partial x} + \frac{\partial N_{yx}}{\partial y} = -p_1 \qquad \frac{\partial N_{xy}}{\partial x} + \frac{\partial N_y}{\partial y} = -p_2 \qquad \frac{\partial^2 M_x}{\partial x^2} + 2\frac{\partial^2 M_{xy}}{\partial x \partial y} + \frac{\partial^2 M_y}{\partial y^2} = -p_3 \tag{3}$$

The equations are independent of material laws and present the static equations for the undeformed plate element. In-plane reactions can be caused by coupling effects of unsymmetric laminates or sandwich plates. Putting the constitutive equations

$$\begin{pmatrix} N \\ M \end{pmatrix} = \begin{pmatrix} A & B \\ B & D \end{pmatrix} \begin{pmatrix} \bar{\varepsilon} \\ \kappa \end{pmatrix} \tag{4}$$

into the equilibrium (3) and replacing using the in-plane strains ε and the curvatures κ by

$$\begin{pmatrix} \varepsilon_x \\ \varepsilon_y \\ \gamma_{xy} \end{pmatrix} = \begin{pmatrix} \dfrac{\partial u}{\partial x} \\[2mm] \dfrac{\partial v}{\partial y} \\[2mm] \dfrac{\partial u}{\partial y}+\dfrac{\partial v}{\partial x} \end{pmatrix} \qquad \begin{pmatrix} \kappa_x \\ \kappa_y \\ \kappa_{xy} \end{pmatrix} = - \begin{pmatrix} \dfrac{\partial^2 w}{\partial x^2} \\[2mm] \dfrac{\partial^2 w}{\partial y^2} \\[2mm] 2\dfrac{\partial^2 w}{\partial x\partial y} \end{pmatrix} \tag{5}$$

we get the differential equations for general laminate plates [2].

3. DESIGN OPTIMIZATION AND SENSITIVITY ANALYSIS

Design optimization refers to the automated redesign process that attempts to minimize or maximize objective function subject to limits or constraints on the response by using a rational mathematical approach to yield improved designs. A feasible design is a design that satisfies all of the constraints. A feasible design may not be optimal. An optimum design is defined as a point in the design space for which the objective function is minimized or maximized and the design is feasible. If relative minima exist in the design space, other optimal designs can exist.

3.1. Numerical aspects of optimization process

The basic problem is the minimization of a function subject to inequality constraints.

$$Z = F(X) \rightarrow \min$$
$$\overline{X}_i^L \leq X_i \leq \overline{X}_i^U \qquad i = 1, 2, ..., N_d$$

$$g_j(X) \leq 0 \qquad j = 1, 2, ..., N_c \tag{6}$$

where X is design variable.

Linear, quadratic, cubic, or quadratic cross-terms may be selected for the polynomial approximation depending on the approximation type. They are as follows

$$F = a_0 + \sum_{i=1}^{N_d} a_i X_i + \sum_{i=1}^{N_d} b_i X_i^2 + \sum_{i=1}^{N_d-1} \sum_{j=i-1}^{N_d} c_{ij} X_i X_j + \sum_{i=1}^{N_d} d_i X_i^3 \tag{7}$$

where:

N_d is number of design variables,

X_i is i^{th} design variable,

a_i, b_i, c_{ij}, d_i are coefficients to be determined.

Singular value decomposition (SVD) is used for regression analysis.

After the objective function and constraints are approximated and their gradients with respect to the design variables are calculated based on the approximation, we are able to solve the approximate optimization problem. One of the algorithms used in the optimization module is called the Modified Feasible Direction method (MFD). Using the Modified Feasible Direction method (MFD) [4] the solving process is iterated until convergence is achieved:

1- $q = 0$, $X^q = X^m$.
2- $q = q+1$.
3- Evaluate objective function and constraints.
4- Identify critical and potentially critical constraints $\overline{N}_c$.
5- Calculate gradient of objective function $\nabla F(X_i)$ and constraints $\nabla g_k(X_i)$, where $k = 1, 2, ..., \overline{N}_c$.
6- Find a usable-feasible search direction S^q.
7- Perform a one-dimensional search $X^q = X^{q-1} + \alpha S^q$.
8- Check convergence. If satisfied, make $X^{m+1} = X^q$. Otherwise, go to 2.
9- $X^{m+1} = X^q$.

Within the Kuhn-Tucker conditions the Lagrangian multiplier method was used.

By using the Lagrangian multiplier method, we define the Lagrangian function as the following

219

$$L = F(X_1,...,X_n) + \sum_{j=1}^{k} \lambda_j h_j(X_1,...,X_n) + \sum_{j=1}^{m} \mu_j \left[g_j(X_1,...,X_n) + s_j^2 \right] \qquad (8)$$

where $\lambda_j, j = 1, ...,k$ and $\mu_j, j=1, ..., m$ are Lagrangian multiplicators and s_j is a slack variable which measures how far the jth constraint is from being critical.

Differentiating the Lagrangian function with respect to all variables we obtain the Kuhn-Tucker conditions which are summarized as follows

$$\frac{\partial F}{\partial X_i} + \sum_{j=1}^{k} \lambda_j \frac{\partial h_j}{\partial X_i} + \sum_{j=1}^{m} \mu_j \frac{\partial g_j}{\partial X_i} = 0, \qquad i = 1, ..., n \qquad (9)$$

Stationarity with respect to $\lambda_j, j = 1, ... ,k$ gives following restrictions

$$h_j(X_1, ..., X_n) = 0, \qquad j = 1, ..., k. \qquad (10)$$

Stationarity L with respect to s_j, gives $\mu_j s_j = 0$ and $\partial^2 L / \partial s_j^2$ for minimum of F implicates $\mu_j \geq 0, j = 1,..., m$.

Finally we get the following equations:

$$\mu_j = 0, \qquad \text{if} \quad g_j(X_1, ..., X_n) < 0, \qquad j=1, ..., m$$
$$\mu_j \geq 0, \qquad \text{if} \quad g_j(X_1, ..., X_n) = 0 \qquad (11)$$

The physical interpretation of these conditions is that the sum of the gradient of the objective function and the scalars λ_j times the associated gradients of the active constraints must vectorally add to zero as shown in Figure 1.

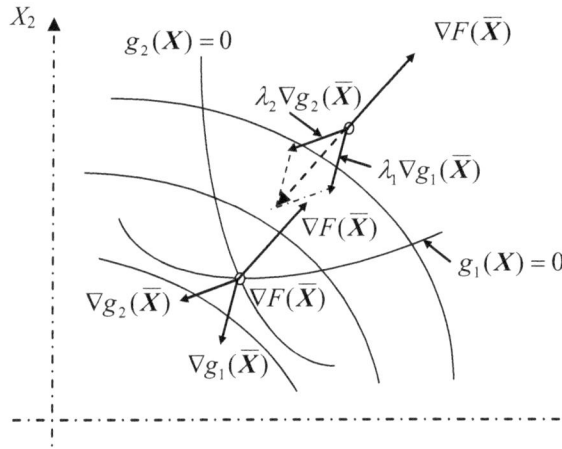

Figure 1: Kuhn-Tucker conditions at a constrained optimum

The Kuhn-Tucker conditions are also sufficient for optimality when the number of active constraints is equal to the number of design variables. Otherwise, sufficient conditions require the second derivatives of the objective function and constraints (Hessian matrix) similar to the unconstrained one. If the objective function and all of the constraints are convex, the Kuhn-Tucker conditions are also sufficient for global optimality.

4. MINIMUM WEIGHT OF ALAMINATE PLATE SUBJECT TO STRENGTH CONSTRAINT

Design of a laminate plate (Fig. 2) with orientation of angle $[0/45/-45/90]_S$ under loading $N_x = 725.6$ kN/m, $N_y = 181.4$ kN/m. Properties of the layers correspond to that of AS4/3501-6 Carbon/Epoxy material. The maximum strain failure limits for the material are $\varepsilon_1^t = \varepsilon_1^c = 0.0115$, $\varepsilon_2^t = \varepsilon_2^c = 0.00535$ and $\gamma_{12}^s = 0.02$.

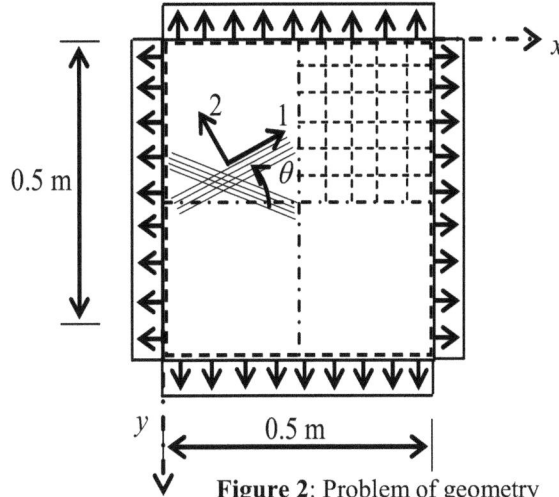

Figure 2: Problem of geometry

Table 1: Design set number: 1

Design Variable	Value
1	0.2000000E-03
2	0.2000000E-03
3	0.2000000E-03
4	0.2000000E-03

Objective	F. E. Value
Function	0.8400004E+01

Table 2: Design set number : 44

Design Variable	Value
1	0.3521971E-03
2	0.1680379E-03
3	0.1696104E-03
4	0.4895812E-04

Objective	F. E. Value
Function	0.7757434E+01

We consider the optimal design of a symmetric laminate with fixed orientation of angles. Because of the laminate symmetry, only the thicknesses $t_k, k = 1,...,I,$ of one-half of the total number of layers, $I = N/2$, are used as design variables. The laminate is considered to be under the action of combined uniform in-plane stress resultants N_x and N_y. The optimization problem is formulated in the following form

minimize

$$W = \sum_{k=1}^{I} 2\rho_k t_k \qquad (12)$$

subject to

$$g_{kj} = \left(P_j^{(k)} \varepsilon_{1k} + Q_j^{(k)} \varepsilon_{2k} + R_j^{(k)} \gamma_{12k} \right) - 1 \leq 0 \qquad (13)$$

for $\qquad k = 1,...,I, \qquad j = 1,...,J,$

where ρ_k and t_k are the density and the thickness, respectively, of the kth layer, $P_j^{(k)}$, $Q_j^{(k)}$, $R_j^{(k)}$, are coefficients that define the jth boundary of a failure envelope for each layer in the strain space, and the $\varepsilon_{1k}, \varepsilon_{2k}, \gamma_{12k}$ are the strains in the principal material direction in the kth layer. For a maximum strain criterion, which puts bounds on the values of the strains in the principal material directions, the failure envelope has four facets with P and Q defined as a inverse of the normal failure strains in the longitudinal and transverse directions to the fibers, once in tension and once compression. The coefficient R is the inverse of the shear failure strain for positive shear and for negative shear. The nonlinear programming problem is transformed to a linear by the help of sequential linear programming. The strength constraint of Eq. (13) is a nonlinear function of the thickness variables and, therefore, is linearized as

$$g_{kjL}(t_k) = g_{kj}(t_{k0}) + \sum_{i=1}^{I} (t_i - t_{0i}) \left(P_j^{(k)} \frac{\partial \varepsilon_{1k}}{\partial t_i} + Q_j^{(k)} \frac{\partial \varepsilon_{2k}}{\partial t_i} + R_j^{(k)} \frac{\partial \gamma_{12k}}{\partial t_i} \right) \qquad (14)$$

where $\partial\varepsilon_{1k}/\partial t_i$, $\partial\varepsilon_{2k}/\partial t_i$, $\partial\gamma_{12}/\partial t_i$ are the derivatives of the principal material direction strains in the kth layer with respect to the thickness of the ith layer. For a specified in-plane loading, the derivative of the laminate strains with respect to the thickness variables can be determined by differentiating in-plane part of stress-strain relation

$$\frac{\partial N}{\partial t_i} = \frac{\partial A}{\partial t_i}\varepsilon^0 + A\frac{\partial\varepsilon^0}{\partial t_i} = 0 \tag{15}$$

The derivatives of the mid-plane strains are

$$\frac{\partial\varepsilon^0}{\partial t_i} = -A^{-1}\overline{Q}_{(i)}\varepsilon^0 \tag{16}$$

where

$$\frac{\partial A}{\partial t_i} = \overline{Q}_{(i)} \tag{17}$$

The derivatives of the strains in the fiber and transverse to the fiber are calculated from

$$\frac{\partial\varepsilon_{(k)}}{\partial t_i} = T_{(k)}\frac{\partial\varepsilon^0}{\partial t_i} \tag{18}$$

the linear approximations to the strain constraints can be constructed using Eq. (13) at any step of the sequential linearizations.

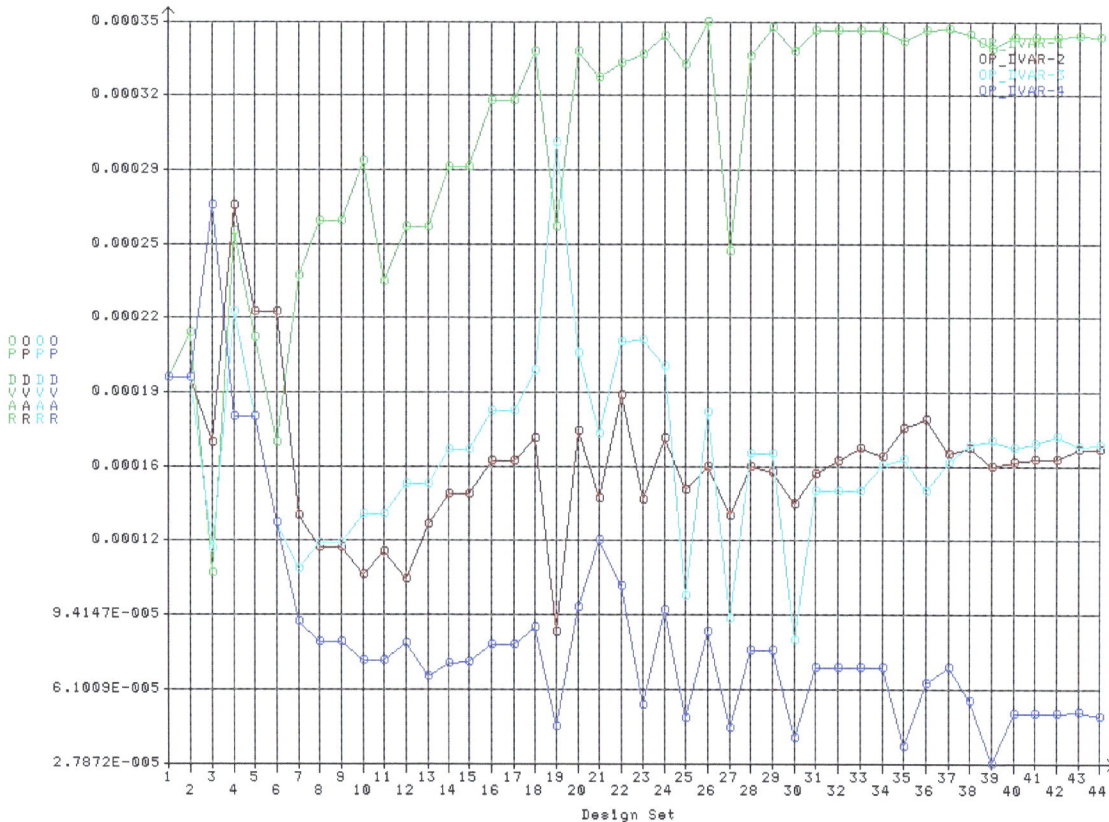

Figure 3: Design variables during the optimization process

222

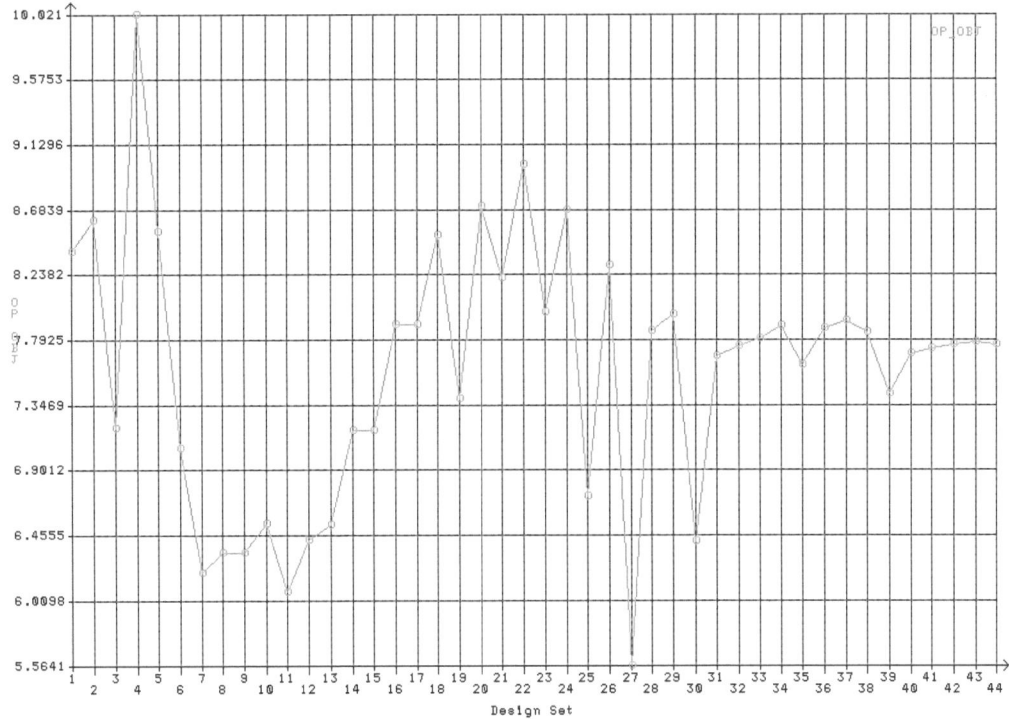

Figure 4: Objective function during the optimization process

5. CONCLUSION

The example combines describing of laminate plate modelling with optimization techniques that enable to find the best design.

For the modelling and analysis we used the classical laminate theory. We assumed the assumptions according to the Kirchhoff´s classical plate theory.

The general optimization contains [5, 6]:

1- Initial analysis with input dates.
2- Mathematical optimization problem as follows (Eqs. 12-18).
3- Linear approximation of objective function and constraints.
4- Own algorithm of MFD method with convergence criteria.
5- Convergence or termination checks of general optimization.

The maximum number of MFD iterations was 100. The general optimization process was stopped after 44 design sets (Figs. 3, 4), because the difference between the current value and the one or two previous designs was less than tolerance. Results of the optimization process following (Figs. 3,4) are listed in the Table 2.

This research has been supported by the Ministry of Education of the Slovak Republic under the Research Project No.: VEGA 1/0201/11.

REFERENCES

[1] Gürdal, Z., Haftka, R.T., Hajela, P. Design and Optimization of Laminated Composite Materials. J. Wiley   & Sons, 1999.

[2] Altenbach, H., Altenbach, J., Kissing. W. Structural analysis of laminate and sandwich beams and plates. Lublin: 2001.

[3] Dický, J., Tvrdá, K. Optimal design of non-homogeneous plates. Proc. of International Conference VSU´   2008 – Vol. 1, 2008, pp. I-100 – I-105, Sofia.

[4] Structure Research and Analysis Corp. – COSMOS Design Star Product, User´s Guide and Tutorial, Los Angeles, 2011.

[5] Šejnoha, M., Kormaníková, E., Száva, I., et al. Selected Chapters of Mechanics of Composite Materials II, Czech Technical University in Prague, 2012, 336p., ISBN 978-80-01-05068-2.

[6] Šejnoha, M., Kormaníková, E., Száva, I., et al.: Mechanics of Composite Materials II, Lectures, CD-rom, Czech Technical University in Prague, 2012, ISBN 978-80-01-05069-9.

COATING PROPERTIES OBTAINED BY THERMAL DEPOSITION USED IN CAMS/CAMS FOLLOWERS APPLICATIONS

G.N. Basescu[1], G.L. Pintilei[1], M. Benchea[1], A.C. Barbanta[1], I.V. Crismaru[1], C. Munteanu[1*]

[1] "Gheorghe Asachi" Technical University of Iasi, Faculty of Mechanical Engineering,
61-63 Prof. dr. doc. D. Mangeron Blvd, 700050, Iasi, Romania basescunarcis@yahoo.com

Abstract: *High inertial forces which occur in the distribution mechanism and certain operating regimes leading to deficitary lubrication can cause wear of contact in mechanism cam / cam followers. In this paper we present a new concept of thermal deposition of a 40Cr130 wear resistant material on a 18MnCr11R50 steel through a electric spark process. Friction and wear tests were performed using the CETR UMT-2 tribometre. To highlight the obtained results scanning electron microscopy analyzes were performed by using the QUANTA 200 3D DUAL BEAM electron microscope. To determine the profile and roughness of the wear trace the Form Talysurf I50 was used.*
Keywords: *CETR UMT-2 tribometre, friction, were, roughness, 40Cr130*

1. INTRODUCTION

Cams followers are components which insure the mechanical transmission of motion to the valve or to the drive rod as well as the takeover of the lateral forces produced by the cam. The cam followers used in current engines have a flat contact surface, for this reason they are called flat cam followers. In the gas distribution system the pourpose of the camshaft is to drive the valves through the cams motion in accord with the engine thermal cycle for every cylinder.

The forces acting on the camhaft lead to bending, torsion and squeeze tensions at the cam level.

The pourpose of applying a ceramic coating on cams is to reducing the coefficient of friction between the cam and cam followers significantly reducing the contact wear. The thermal deposition through a electric spark process leads to the possibility of obtaining layers with different thickness composition and surface quality.

2. EXPERIMENTAL PROCEDURE

This paper presents a new concept of thermal deposition which consists of a 40Cr130 deposited layer by electric arc using the Smart Arct 350 from Sulzer Metco. This layer has the pourpose of increasing the wear resistance especially the pitting, surface distress, delamination and spalling.

The QUANTA 200 3D DUAL BEAM electron microscop was used to highlight the obtained resultes. After each test cycle, the samples were removed, cleaned with a special solution in ultrasonic bath.

Electric arc deposition parameters are presented in Table 1.

Table 1: Technical parameters

Smart Arc 350	40Cr130
U	28V
I	252A
Air pressure	60PSI

3. EXPERIMENTAL RESULTS

The two samples were mounted on the device shown in Figure 1. The tests were performed in dry and in lubricated conditions with SAE 5W-30 Audi Original oil, with a viscosity of 51.7 mm^2/s at 40°C in according to the ASTM D-445 standard.

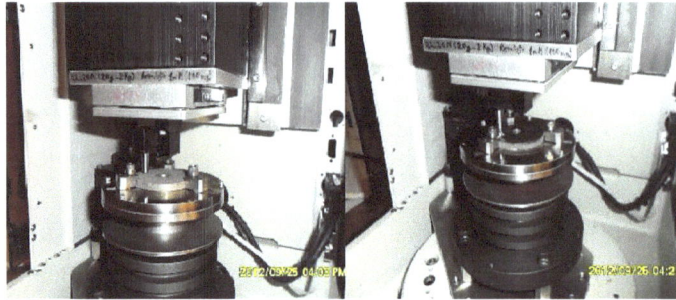

Figure 1: The way how the sampled is mounted on the CETR UMT-2 tribometre:
a) the samples subjected to friction wear without lubrication;
b) the samples subjected to friction wear lubrication.

The experimental tests were performed with CETR UMT-2 tribometer with a pin-disc test system [1]. We used a pin with a diameter of 6,3 mm and a disc sprayed with 40Cr130. At a radius of 10 mm from the center of the disc a normal pressures forces with a value $F_z = 10$ N was applied through the pin, for a period of 10 minutes. Tests were performed both under dry friction wear and lubrication conditions.

Since the deposited layer surface roughness is very important, it was measured before and after the abrasion test to estimate all the wear resulting from friction [2]. In Figure 2 is presented the mounting of the sample in order to determine the roughness of the layer.

Figure 2: The mounting of the sample on the stand to determine the roughness of the profile

In Figure 3 is presented the mounting of the sample in the working room of the microscope [3].

Figure 3: The mounting of the sample in the working room of the microscope

Images were acquired with the microscope on used samples in different areas, including on the wear trace. The microscope worked in the High Vacuum module, with a pressure of $5.33 \cdot 10^{-3}$ Pa, using the ETD (Everhardt -Thornley Detector) and an accelerating tension of the electron beam of 20kV. The magnifying power of the microscope was in the range of 80X - 5000X, with a working distance (the distance between the sample and the electron cannon) of 15 mm. Apart of the SEM images there also EDS elemental chemical analyses were made, for the pourpose of evaluating the state of the coating wear due to the friction test.

3.1. The analyses performed on the samples subjected to gry friction wear

In the figure 4-a it can be noticed that the width of the running track is between 685 μm - 731 μm, and in the Figure 4-b is observed that in the surface layer changes occur in the meaning of pitting traces. In Figure 4-c its shown the pitting wear trace, at the magnifing power of 5000X [6].

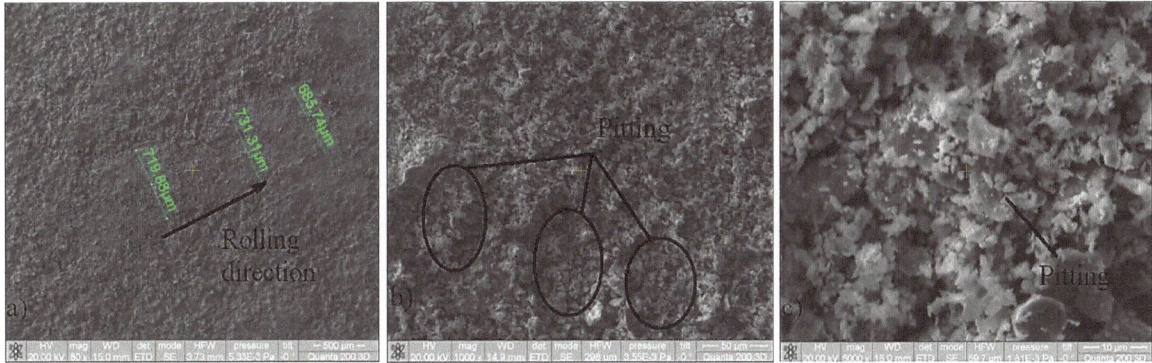

Figure 4: The appearance and dimension of the wear trace resulted after dry friction test:
a) 80X, b) 1000X si c) 5000X

In the distribution map from Figure 5 it can be seen that besides the wear produced on the sample, on its surface there are present traces of iron, this happens because of the particles resulting from the wear of the pin which adhere on the samples surface.

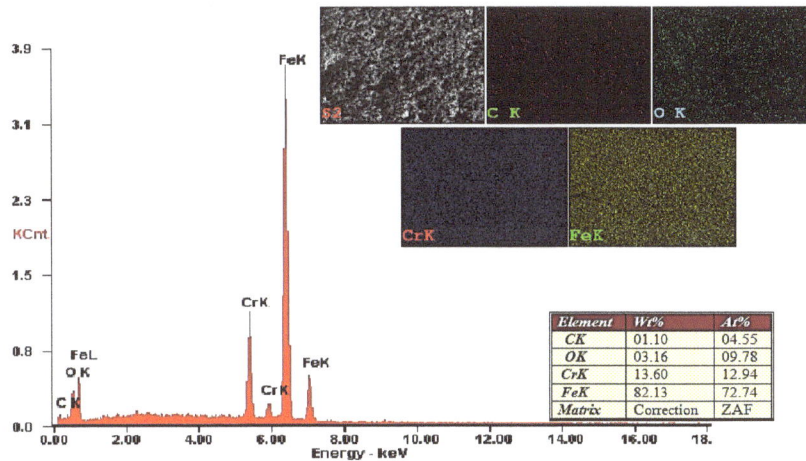

Element	Wt%	At%
CK	01.10	04.55
OK	03.16	09.78
CrK	13.60	12.94
FeK	82.13	72.74
Matrix	Correction	ZAF

Figure 5: Map distribution of the elements and EDS analyses of the wear trace

In the Figure 6 it's presented the variation of the friction coefficient under dry frictrions conditions with a apllied normale force of 10N. In the begining of the test the friction coefficient has a accelerated downward tendency down to a value of 0,17, then has an upward tendency until it reaches the value of 0,3, at which it stabilizes, and in the final part of the test it rises to the value of 0,42 [3].

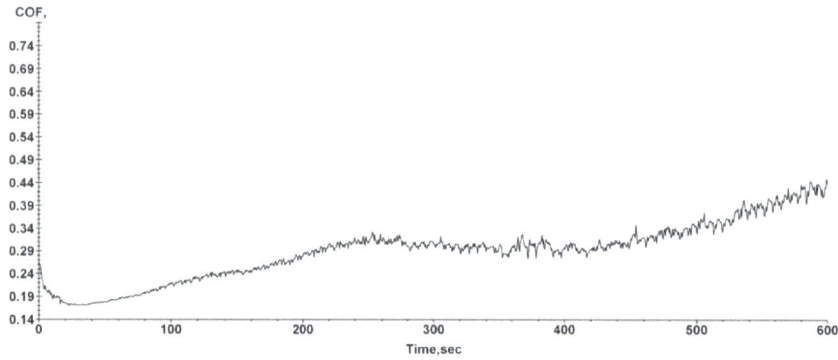

Figure 6: The variation of the friction coefficient from the pin-disk test with dry friction conditions

Figure 7 presents the scanned tested sample after the dry friction test. A wear depth of 80 µm can be observed in the picture.

The shape of the channel due to wear is achieved with the Login Form Talysurf 50 profilometer, produced by Taylor Hobson, England.

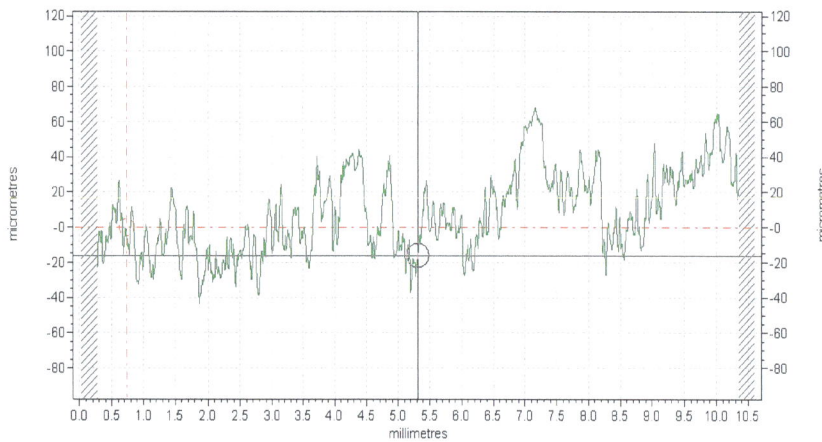

Figure 7: The shape of the wear trace under dry friction condition

Figure 8 shows the roughness analysis on a LS-Line profile. The arithmetic mean deviation of the roughness profile Ra is evaluated to 12,4745 µm. The standard deviation of the assessed roughness profile Rq is 15,2887 µm and the average height of the roughness profile is measured to Rz = 70,1730 µm [6].

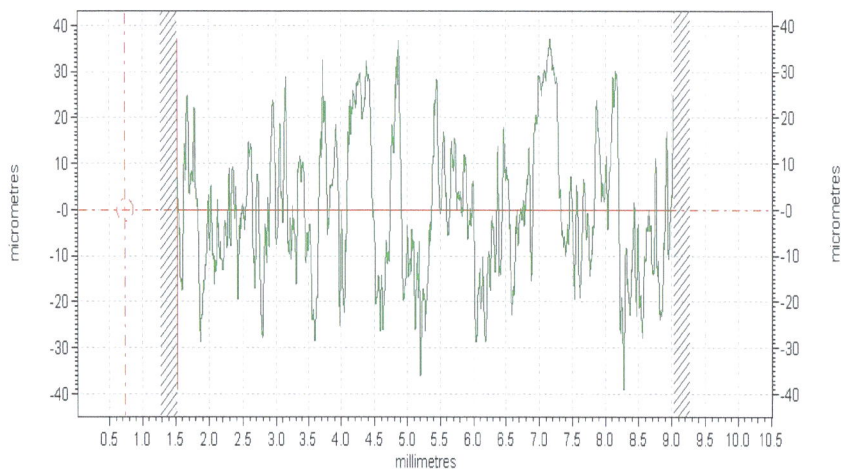

Figure 8: The roughness values of the sample in dry friction condition

228

3.2. The analyses performed on the sample subjected to lubricate with ester-containing oil sae 5w-30 audi original

In Figure 9-a is observed that the width of the wear track is between 336 μm - 374 μm, but on the surface of the layer, shown in Figure 9-b, a smoothing appears. In Figure 9-c it can be seen that the layer has ompacted as a result of the wear test. There are no traces of pitting on the layer [6].

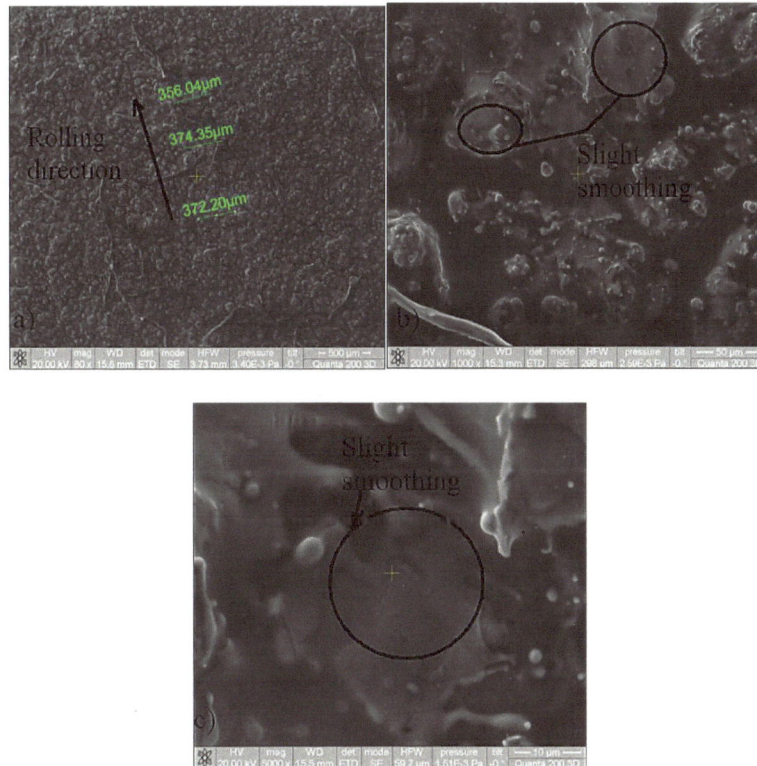

Figure 9: The appearance and dimension of the wear trace resulted after lubrificated friction test:
a)   80X, b) 1000X si c) 5000X

The fact that wear didn't occur in the deposited layer is also shown by the EDS chemical analysis presented in Figure 10, in which the presence of Cr, O_2, C and Fe in the analyzed areas can by observed. The distribution map presents a uniform distribution of elements. Also this time in the surface of the deposited layer traces of iron are present, this is due to the fact that particles resulting from wear of the pin adhered to the sample surface.

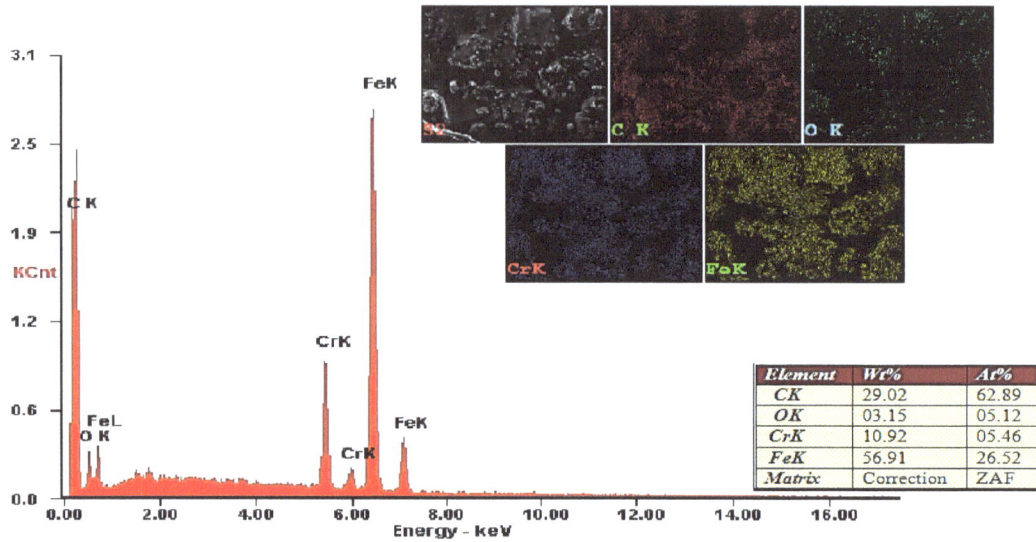

Element	Wt%	At%
CK	29.02	62.89
OK	03.15	05.12
CrK	10.92	05.46
FeK	56.91	26.52
Matrix	Correction	ZAF

Figure 10: The distribution map of the chemical elements and the EDS analyses

In Figure 11 can be seen the evolution of the coefficient of friction during the friction test in the presence of lubricant. The friction coefficient presence a downward tendency in the beginning of the test from the value of 0,16 to 0,13, after this value it is approximately constant until the end [3].

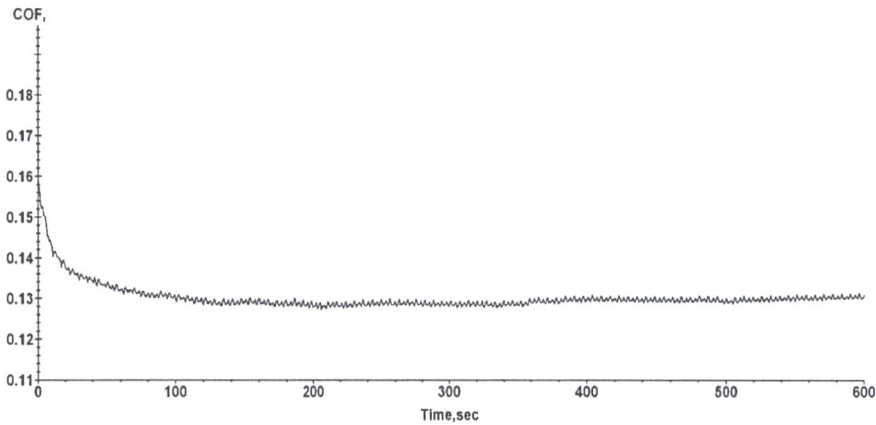

Figure 11: The variation of the friction coefficient from the pin-disk test in lubricant friction conditions

In Figure 12 is presented the scann of the tested sample. In the case of the sample subjected to friction with lubricant in the contact area the depth of the wear trace is very low, fact that is show in the profilometry. Wear depth is only 60 μm. Compared with the shown in Figure 7 a very good adherence and stability in presents for the layer tested with lubricant. The depth of wear is reduced by half.

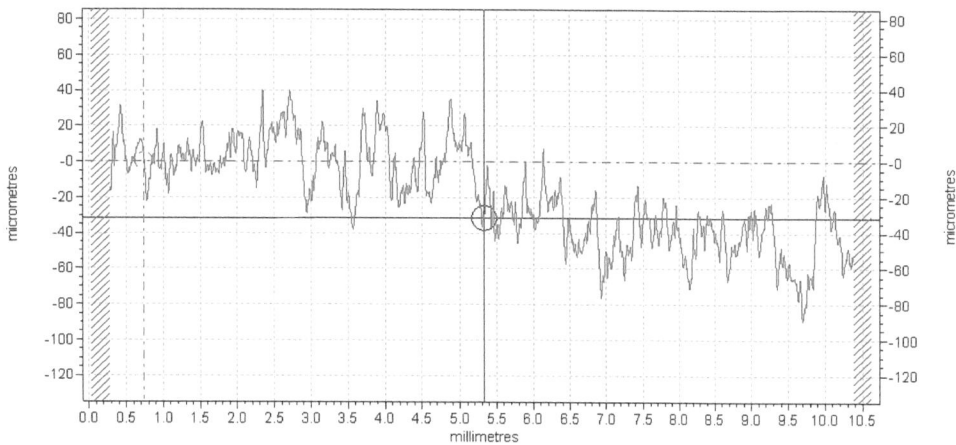

Figure 12: The shape of the wear trace under lubricant friction condition

Figure 13 shows the roughness analysis of a LS-Line type profile. The arithmetic mean deviation of the roughness profile R_a is evaluated to 11,3891 μm. The standard deviation of the assessed profile roughness R_q = 14,0036 μm and the average height of roughness profile is R_z = 67,1976 im.

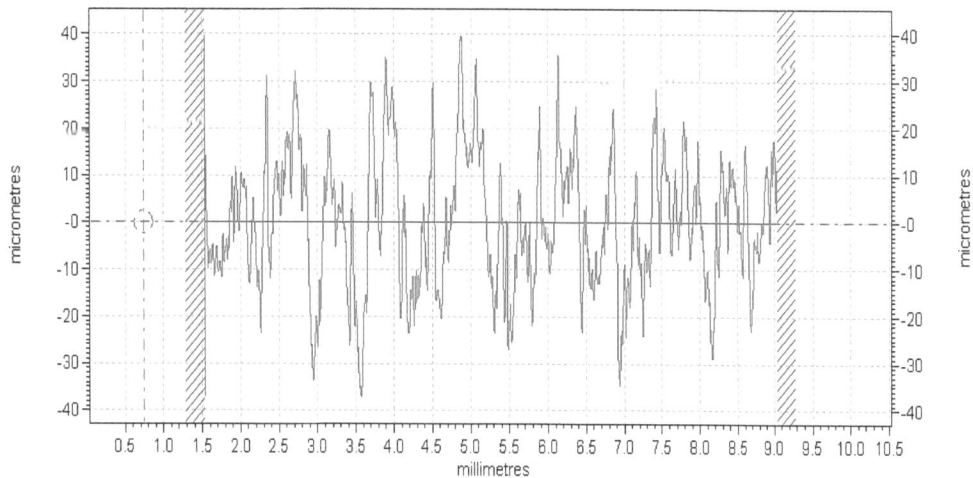

Figure 13: The roughness values of the sample in lubrificant friction condition

4. CONCLUSIONS

In the case of dry friction pitting wear had occurred with the thickness of the wear trace between 685 μm - 731 μm. For the sample subjected to friction in presents of the SAE 5W-30 Audi Original lubricant with a viscosity of 51,7 mm^2 /s at 40°C, the wear on the surface is present but has a very small value.

From the elemental chemical analyses and the distribution map performed on the sample in cam by concluded that in both cases wear did occurred and material from the pin adhered to the layer.

The variation of the friction coefficient is higher for the sample subjected to dry friction at the beginning of the test with a downward trend accelerated up to a maximum value of 0,17, and has a tendency to rise to around 0.3 at which is stable. In the last part of the test the coefficient of friction increases to around 0,42. In the case of friction with lubricant the friction coefficient tends to decrease rapidly from 0,16 to 0,13, and then remains constant around this value.

Following the performed tests it can be concluded that the samples subjected to friction in the presence of a lubricant behaved very well to wear, hence we conclude that a coating based on 40Cr130 deposited on the camshaft would lead to an improvement in its functioning in terms of contact wear due to the friction to the cam follower.

ACKNOWLEDGEMENTS

I give thanks to the Faculty of Mechanics for the support and for making available the science of materials and tribology laboratories for performing the necessary tests for the paper.

We acknowledge that this paper was realized with the support of EURODOC "Doctoral Scholarship for Research Performance at European level" project financed by the European Social Found and Rumanian Government.

We thank to SC Plasma Jet SA for the support in the plasma jet deposition of the analyzed layer from this paper.

This paper was realized with the support of POSDRU CUANTUMDOC "Doctoral Studies for European Performances in Research and Innovation" ID79407 project funded by the European Social Found and Romanian Government.

REFERENCES

[1]   Manual de utilizare, Microdurimetru UMTR 2M-CTR.
[2]   Manualul utilizatorului Profilometrul Form Talysurf I50
[3]   Manualul utilizatorului pentru sistemul QUANTA 200 3D, Fei Company Olanda, 2006
[4]   Dumitru Olaru, *Fundamente de lubrificaţie,* Ed. "Gh. Asachi" Iaşi, 2002.
[5]   Sp. Creţu, *Mecanica contactului,* Ed. "Gh.Asachi" Iaşi, 2002.
[6]   Stefan Ghimisi, *Elemente de tribologie,* ISBN:9736859037, Editura:Matrixrom.
[7]   Rodica Cozma, *Elemente fundamentale de tribologie,* ISBN13 – 2000000304489, Editura   Matrixrom.

CONTRIBUTIONS TO NONCONTACT ULTRASONIC EXAMINATION OF COMPOSITE MATERIALS USED IN CIVIL AND INDUSTRIAL CONSTRUCTION

G. Amza[1], C. Florică[1], G.D. Taşcă[1], I. Ciobanu [1]

Polytechnic University of Bucharest, Romania, amza@camis.pub.ro

Abstract: During operation of a commercial or industrial structure several defects may occur: micro cracking and cracking of resistance elements due to fatigue, corrosion of columns and beams and others; earthquakes may contribute to degradation of electrical installations and embedded piping.
This paper presents a method of non-contact ultrasound for the determination of such defects in beams, fittings and embedded pipes.
Keywords: ultrasound, non-contact, examination, pipes, structures

1. INTRODUCTION

The main defects that may arise or affect any civil or industrial construction are defects in the installation, material defects and operational faults | 1 |, | 2 |, | 3 |.

If earthquakes occur under the action of important physical wear demands on materials that are part of the resistance structure (internal cracks and fissures, breaks and damage resistance pillars and beams, cracking the plates and beams, joints in masonry or between masonry and plates embedded piping degradation, degradation of electrical installations etc..).

Most of the defects listed above can be highlighted using ultrasound method in several ways: pulse reflected and transmitted pulse echoes repeated.

During operation of a commercial or industrial structure may show up a number of defects: micro cracking and tear resistance elements due to fatigue applications, corrosion of reinforcement in columns or beams of resistance and other defects that are very difficult to determine the classical technologies.   Ultrasound methods have a number of disadvantages, related to very high sensitivity of that method, the need for an acoustic coupler, essential dependence on operator interpretation of results and failure control by two or more opposite sides of a wall, beam or column.|3|

To remove some of these disadvantages, the research contained in this doctoral thesis try to combine with another method ultrasonic nondestructive investigation - thermography infrared - and to propose a new technology for investigation and control without contact between the object and instrument control .

2. TECHNICAL REQUIREMENTS

Since the acoustic coupling problem is difficult to solve the investigation of civil and industrial construction elements, the issue of using air to achieve acoustic coupling between the ultrasonic transducer and controlled environment.

Because attenuation of ultrasound in air, in MHz is very large compared to the medium investigated, i.e. when ultrasound moves from an environment with very low acoustic impedance, acoustic impedance in an environment with very high only a fraction of ultrasonic energy is transmitted in rural investigated.

Transmission coefficient T, the ultrasonic transducer investigated environment is determined by the relationship:

$$T = 4 Z_1 Z_2 / (Z_1 + Z_2)^2 \tag{1}$$

where: Z1 is the acoustic impedance of the ultrasonic carrier environment (eg air non-contact control) and Z2 - acoustic impedance test environment.
Transmission coefficient is defined as the ratio of transmitted acoustic energy V (measured in volts) and V0 input energy, it is refracted longitudinal wave at the angle of incidence of 0 ° on the interface between the two areas, namely|2|:

$$T = V^2/V_0^2 \tag{2}$$

This relationship can be described as logarithmic scale:

$$T = 20\log V/V_0 \; [dB] \tag{3}$$

So, the energy transferred to propagation medium Et is calculated by the formula:

$$E_t = 20\log T \tag{4}$$

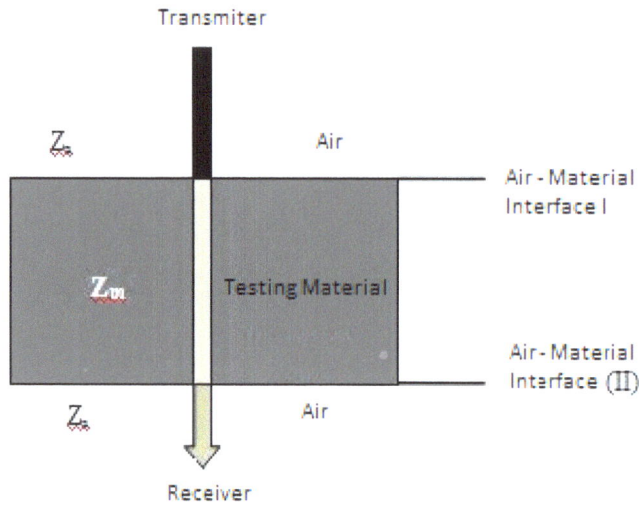

Figure 1: Areas to be traversed by ultrasound (shown by arrow) in non-contact mode of transmission ultrasound to propagate through a material testing.
Area "I" corresponds to air-material transmission (acoustic impedance Za of the Zm) and the surface "II" corresponds sending material to air (acoustic impedance Zm of the Za).

Ultrasound for non-contact examination must be spread in the air in the test material and then again in the air so that the transmitted wave can be detected by a receiver (Figure.1). Therefore, high energy loss at the air-material loss is accompanied by additional material-air interface.|3|,|4|,|5|

In relative mode, the attenuation of ultrasound in air is intrinsically higher in comparison with attenuation of solids or liquids. Since attenuation increases in an environment based on the fourth power of the frequency, transmission megahertz frequency ultrasound with air becomes almost impossible.

Effectiveness of an ultrasonic transducer coupling coefficients depend on other electromechanical properties of piezoelectric material. Also depends on the mechanism by which ultrasound piezoelectric material is transferred from the environment should be propagated ultrasound. Non-contact mode, this medium is air. Since the acoustic impedance of the piezoelectric material is a few degrees higher magnitude than air, it is usually necessary to insert the transition layer (suitable acoustic impedance) of different materials before piezoelectric material.

Finally, the last layer characteristics determine the transduction efficiency of a transducer device. Significance final acoustic impedance matching layer of non-contact transducers cannot be overstated. Since the piezoelectric properties of a given material can be considered constant for a given device, the final transfer of ultrasonic energy in the air is wholly controlled acoustic characteristics of the final layer piezoelectric material adapted.

Because ultrasound transmission in air to be as close to the truth, the final layer should be composed of piezoelectric polymers fine. Polymer layer that builds these transducers can be porous or non-porous and can be embedded inside hollow spheres (the polymer layer). To simplify, we will identify all polymers suitable acoustic impedance transducers using an air gap. These transducers emit ultrasound in air at -58 ... -54 dB, which corresponds to the propagation of ultrasound in some parts ~ 2 MHz contactless method. Using this non-contact transducers with final layer of a polymer built a stand that experimental measurements were made on several material (Fig. 3), SECU-01FC model.

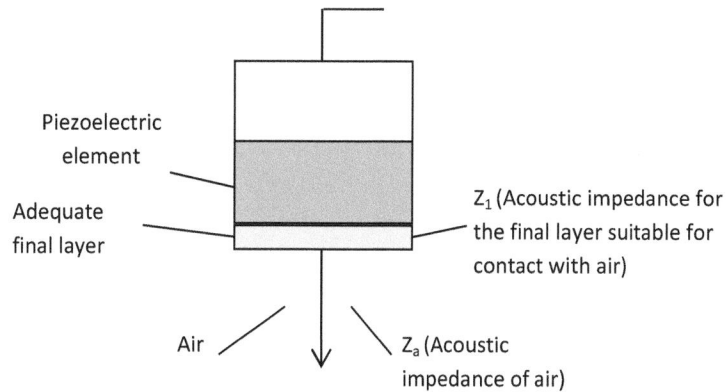

Figure 2. An ultrasonic transducer diagram indicating adequate final layer of acoustic impedance relative to critical and piezoelectric coupling medium, air.

3. EXPERIMENTAL CONTRIBUTIONS OF THE ULTRASOUND EXAMINATION OF BURIED PIPELINES THROUGH WHICH THE FLUID PRESSURE

Examination or monitoring of corrosion / erosion, to establish lifetime of pipes through which the fluid at different temperatures and / or pressure is a very important activity for both the designer and the owner, because knowing 'health "their time is extremely important. Existence of defects possible occurrence of cracks can lead to accidentally run out, sometimes with serious consequences and most significant damage.Therefore periodic examination of buried piping systems or by circulating different fluids (water, air, gas, steam) at different temperatures and pressures is required to know the life and establish the necessary measures to prevent various accidents.

For testing were chosen most important parts of a pipeline where possible occurrence of defects and where corrosion may be greater. Thus, we have established methods of examination for parts such as elbows to 900, sections of pipe welded pipe portions reduction and seamless pipe. For elbow 900, made of OLT60 with nominal thickness of 10 mm, and the nominal diameter of 145 mm (Fig. 4) found the results shown in Table 1.

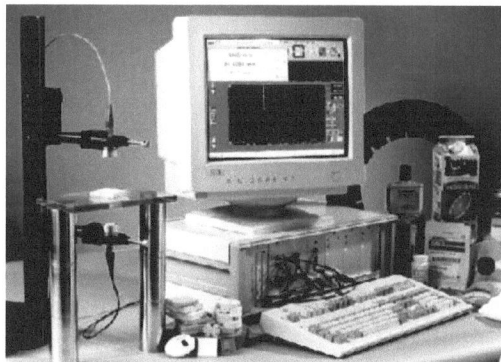

Figure 3: Stand SECU 01-FC experimental model of non-contact ultrasonic testing, which uses contactless sensors and a screen that displays the material thickness and speed testing

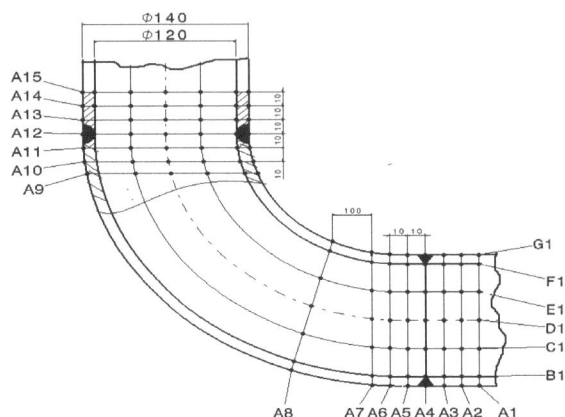

Figure 4: Ultrasonic control scheme of an elbow 900:
A1 ... A15, B1 ... B15, C1 ... C15, D1 ... D15, E1 ... E15, F1 ... F15, G1 ... G1

Table 1: Experimental results obtained from ultrasound examination of a side 900 (excerpt)

No. Crt.	Measuring point	Date of measurement	Thickness measured	The next	Thickness measured	Observations
0	1	2	3	4	5	6
1	A1	14.08.2010	9,54	20.08.2011	9,05	0,49
2	A2	14.08.2010	9,53	20.08.2011	9.01	0,52
3	A3	14.08.2010	9,95	20.08.2011	9,41	0,54
4	A4	14.08.2010	9,93	20.08.2011	9,37	0,56
5	A5	14.08.2010	9,47	20.08.2011	8,82	0,65
6	A6	14.08.2010	9,46	20.08.2011	8,93	0,53
7	A7	14.08.2010	9,52	20.08.2011	9,03	0,49
8	A8	14.08.2010	9,51	20.08.2011	9,12	0,39
9	A9	14.08.2010	9,52	20.08.2011	9,21	0,31
10	A10	14.08.2010	9,47	20.08.2011	8,97	0,50
11	A11	14.08.2010	9,92	20.08.2011	8,99	0,93
12	A12	14.08.2010	9,89	20.08.2011	9,28	0,61
13	A13	14.08.2010	9,74	20.08.2011	9,60	0,14
14	A14	14.08.2010	9,73	20.08.2011	9,28	0,45
15	A15	14.08.2010	9,48	20.08.2011	9,02	0,46
16	B1	14.08.2010	8,82	20.08.2011	8,40	0,42
17	B2	14.08.2010	8,83	20.08.2011	8,37	0,46
18	B3	14.08.2010	8,93	20.08.2011	8,17	0,76
19	B4	14.08.2010	8,90	20.08.2011	8,22	0,68
20	B5	14.08.2010	9,28	20.08.2011	8,75	0,53
21	B6	14.08.2010	9,26	20.08.2011	9,08	0,18
22	B7	14.08.2010	9,32	20.08.2011	8,88	0,44
23	B8	14.08.2010	9,66	20.08.2011	8,98	0,68
24	B9	14.08.2010	9,73	20.08.2011	9,13	0,60
25	B10	14.08.2010	9,17	20.08.2011	8,62	0,56
26	B11	14.08.2010	9,10	20.08.2011	8,56	0,54
27	B12	14.08.2010	8,95	20.08.2011	8,47	0,52
,,,	,,,	,,,	,,,	,,,	,,,	,,,
82	F7	14.08.2010	9,36	20.08.2011	8,86	0,50
83	F8	14.08.2010	9,42	20.08.2011	8,72	0,70
84	F9	14.08.2010	9,38	20.08.2011	8,68	0,70
85	F10	14.08.2010	9,28	20.08.2011	8,78	0,50

| 86 | F11 | 14.08.2010 | 9,17 | 20.08.2011 | 8,93 | 0,24 |
| 87 | F12 | 14.08.2010 | 9,42 | 20.08.2011 | 8,82 | 0,50 |

Table 1 (Continued)

88	F13	14.08.2010	9,31	20.08.2011	8,92	0,29
89	F14	14.08.2010	9,29	20.08.2011	8,62	0,67
90	F15	14.08.2010	9,12	20.08.2011	8,61	0,51
91	G1	14.08.2010	9,41	20.08.2011	9,06	0,35
92	G2	14.08.2010	9,37	20.08.2011	9,21	0,16
93	G3	14.08.2010	9,56	20.08.2011	9,32	0,24
94	G4	14.08.2010	9,22	20.08.2011	9,08	0,14
95	G5	14.08.2010	9,12	20.08.2011	8,93	0,19
96	G6	14.08.2010	9,17	20.08.2011	8,98	0,19
97	G7	14.08.2010	9,48	20.08.2011	9,17	0,31
98	G8	14.08.2010	9,53	20.08.2011	9,72	0,11
99	G9	14.08.2010	9,47	20.08.2011	9,32	0,15
100	G10	14.08.2010	9,46	20.08.2011	9,28	0,28
101	G11	14.08.2010	9,28	20.08.2011	9,01	0,27
102	G12	14.08.2010	9,33	20.08.2011	9,08	0,25
103	G13	14.08.2010	9,48	20.08.2011	9,17	0,31
104	G14	14.08.2010	9,53	20.08.2011	9,21	0,32
105	G15	14.08.2010	9,61	20.08.2011	0,32	0,29

4. CONCLUSIONS

The analysis results can be drawn the following conclusions:
- Elbow minimum thickness was measured at the point B3, which was 8.17 mm;
- In sections A1 ... G15, thickness of 9.12 was measured at point G5;
- Maximum corrosion occurred at the point E12, which was 0.93 mm result is explained because the heat affected zone, where there is already internal thermal stresses introduced during the welding process;
- Minimum corrosion G8 is the point where was 0.11 mm, resulting understandable because the point is in an area with no possibility of turbulence and possible impact without solid impurities entrained fluid.

REFERENCES

[1]   Amza Gheorghe, Ultrasound of high energy - Ed. Academiei, Bucureşti, 1984.
[2]   Amza Gheorghe, Systems ultraacustice – Ed. Tehnică, Bucureşti, 1989.
[3]   Amza, Gh. Borda C. , Marinescu M., Arsene D., Design of Ring – Type Ultrasonic Motor – Scientific Session of the University "Petru Maior" - Targu Mures, 27-28 oct. 2001, vol.6, pa.7-14, Petru Maior University Press - Targu Mures , 2001, ISBN 973-8084-10-5, vol.6ISBN 973-8084-19-0.
[4]   Berlin, A.A., et al. – Principles of Polymer Composite, Ed. Springer – Verlag, New York 1986   (Polymer – Properties and Aplications, vol. 10).
[5]   Dry M. Carolyn, Sottos, R. Nancy, Passive smart self-repair inpolamer matrix composite materials-Smart Structures abd Materials 1993, Albuquerque, N.M. USA- Proceedings of SPIE – The International Society for Optical Engineering v 1916 1993, Publ by Society of Photo Optical Instrumentation Engineers, Belligham, WA USA, p- 438-444.
[6]   Gandhi M.V. and Thompson B.S. , A New Genertaion of Revolutionary Ultra Advanced Compostes Materials Faturing Electro-Reological Fluide, U.S Army Research Office Workshop on Smart Materials Structures and Mathematical Issues, 1988.

STRENGTH ANALYSIS OF THE SHIP DECK DELAMINATED PLATES AT EXPLOSIONS

I. Chirica[1], E.F. Beznea[2]

[1] University Dunarea de Jos, Galati, ROMANIA, ionel.chirica@ugal.ro
[2] University Dunarea de Jos, Galati, ROMANIA, elena.beznea@ugal.ro

Abstract: In the paper a nonlinear FEM analysis on the protective capacity of ship hull structures made of laminated composite materials with imperfections (a circular central delamination) subjected to blast loading is treating. The methodology for the blast pressure charging and the mechanism of the blast wave in free air are given. The space pressure variation is determined by using Friedlander exponential decay equation. Various scenarios (parametric calculus) to evaluate the behavior of the ship structure composite plate to blast loading are presented: explosive magnitude, distance from source of explosion, position of the delamination.

Keywords: blast loading, composite laminated plates, delaminations

1. INTRODUCTION

Nowadays, in the marine and shipbuilding industry composites have an ever growing importance. The main reason for the fabrication of composite structures is to provide a composite with high bending stiffness with overall low density. These composites are a very good material for ship hull structures such as shells, decks and bulkheads, where structures must be lightweight and strong.

Due to different accidental or intentional events, blast loads induced by explosion within or immediately nearby ship hull can cause catastrophic damage on the ship structure and shutting down of critical life safety systems. Loss of life and injuries to crew and passengers can result from many causes, including direct blast-effects, structural collapse, debris impact, fire and smoke. To provide adequate protection against explosions, the design and construction of ship hull are receiving renewed attention of structural engineers. Difficulties that arise with the complexity of the problem, which involves time dependent finite deformations, high strain rates, and non-linear inelastic material behavior, have motivated various assumptions and approximations to simplify the models.

In United States the government distributed the simple blast program ConWep. Users can input a charge size and standoff distance and receive pressure for that point in relation to time as output. It also allows users to receive pressure data after interaction with simple structures such as plates and shells ([1]).

An explosion within or around a ship hull can have catastrophic effects, damaging and destroying internal or external portions of the ship hull structure. Explosion damage to the ship structure depends on the type and layout of the structure, material used, range of the located explosive device, and the charge weight (equivalent TNT mass). If a structure is designed for a blast loading, its behavior will be better, having more mass, more damping and energy-absorbing capacity. In this case, certain bomb scenarios have to be done during design stage.

The response of elastic structures to time-dependent external excitations, such as sonic boom and blast loadings, is a subject of much interest in the design of marine vehicles [[2], [3]]. For the case of blast loadings, various analytical expressions have been proposed and discussed in [4].

The particular equation used for the pressure-time history of the load is often chosen to best match the particular phenomenon; considered. The time history of overpressures due to explosions is often represented by the modified Friedlander exponential decay equation [4].

The methods presented in [4] use experimental data from explosive tests to develop expressions for the blast overpressure as a function of time and distance from the blast, as well as charge weight and other important blast parameters. Very few papers make use of such a realistic blast load. Most of the literature available concerning impulsively loaded plates considers a linear solution for isotropic plates. There are also many linear solutions available for impulsively loaded composite plates, and some of the references listed thus far are of this type ([5]).

Recent work on polymer matrix fiber-reinforced composite has focused on the advantages of this type of construction under blast loads when the deformations remain dominantly elastic. The relative advantage of composite plates over solid plates has not been firmly established for metal construction when strong blast loads require both high strength and energy absorption ([6]).

Predicting the structural response to such an explosion requires accurate prediction of the applied pressures and a solution procedure that is adequate for such transient phenomena.

The work presented here focuses on the composite laminated plates with imperfections response to such close proximity explosions. In particular, the structures considered include orthotropic composite and contacting plates subject to mine blasts.

2. BLAST WAVE PROPAGATION IN AIR

In a blast analysis, one can find the size and location of the explosion to protect against. By using the relationship that the intensity of a blast decays in relation to the cube of the distance from the explosion one can adopt an idealized blast wave and at the target. The positive phase duration of the blast wave is compared with the natural period of response of the structure or analyzed element.

The response of the structural element can be defined by two possible extreme loadings: a) Impulsive loading, occurring in the case of load pulse is short compared to the natural period of structure vibration; b) quasi-static or pressure occurring in the case of long duration of the load, compared to the natural period of structure vibration. The case with a regime where the load duration and structural response times are similar, the loading is of dynamic or pressure time type ([7]).

The explosion is characterized by transient air pressure wave moving further from the source of explosion. The peak overpressure (the pressure above the normal atmospheric pressure) and duration of the overpressure vary with the distance from the centre of explosion. As the wave moves further from the source of explosion the peak overpressure drops. Stand-off distance is a fundamental parameter when determining the blast pressure experienced by a structure. Values of components such as high-mass, long span beams and floors can absorb a great part of the energy delivered by a blast load.

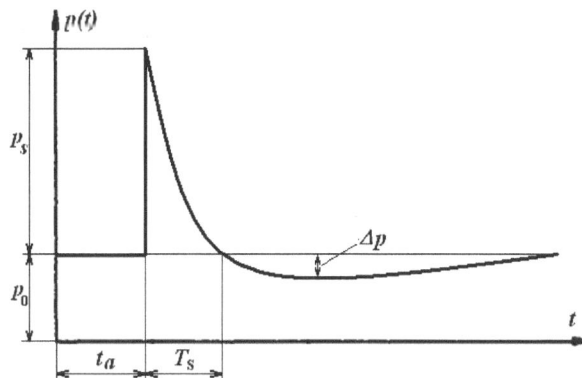

Figure 1: Typical blast pressure - time profile

The pressure at a specific point in air in the path of an explosion over time will follow the same general pattern, so long as does not exist any reflection from nearby objects.

This pattern, called an overpressure curve (Fig.1) has main components: the detonation, arrival time (t_a), peak pressure (p_s), and time duration (T_s). The detonation can be considered as time 0, while the arrival time is the time that it takes for the pressure wave to reach the point of interest. Once the peak pressure is reached, it immediately starts to decay to the normal pressure during the time duration. As the material in the blast wave expands outward it can leave a void, creating a region with pressure lower than normal atmospheric pressure. The size, shape and material of the charge, as well as the stand off distance determine the magnitude and shape of this curve. In addition to the above factors, the blast wave and the pressure involved can reflect off of surfaces in various directions, and cause further fluctuations in pressure at a single point.

For describing the pressure-time history of a blast wave, certain equations have been developed and used in the calculus. The most frequently used is the modified Friedlander equation, describing the pressure after its arrival

$$p(t) = p_0 + p_s\left(1 - \frac{t}{T_s}\right)e^{-\frac{bt}{T_s}} \tag{1}$$

In Eq. (1), p_s denotes the peak reflected pressure in excess of the ambient one; T_s denotes the positive phase duration of the pulse measured from the time of impact of the structure and b denotes a decay parameter which has to be adjusted to approximate the overpressure signature from the blast tests.

Blast wave parameters for conventional high explosive materials have been the focus of a number of studies during the 1950's and 1960's. Estimations of peak overpressure ps (in kPa) due to spherical blast based on scaled distance Z, were introduced in [8] as:

- for $p_s > 1000$ kPa

$$p_s = 670/Z^3 + 100,$$

- for $10 < p_s < 1000$ kPa

$$p_s = 97.5/Z + 145.5/Z^2 + 585/Z^3 - 1.9$$

where Z is the dimensional distance parameter (scaled distance)

$Z = R/W^{1/3}$

R is the actual effective distance from the explosion and W is generally expressed in kilograms. Scaling laws provide parametric correlations between a particular explosion and a standard charge of the same substance.

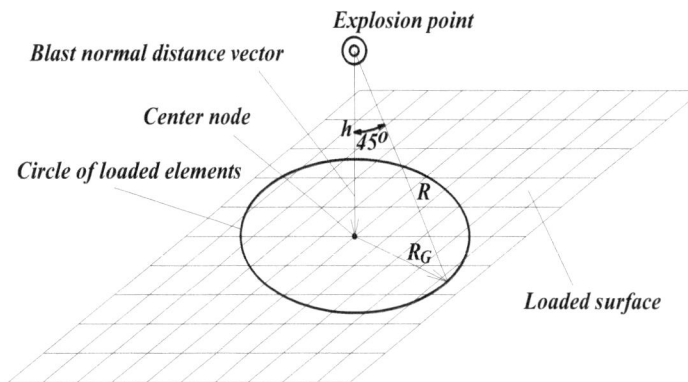

Figure 2: Blast pressure evaluation model

As it is shown in the figure 2, the methodology and model to determine the pressure from blast loading is presented. According to the actual effective distance from the explosion R (Fig. 2), elements within 45 degrees of the blast normal vector are divided into groups based upon their average distance to the center node. Only elements within the 45 degree cone are loaded. The area within the cone is divided into a number of 10 rings to determine the pressure acting on the elements of the mesh. The distribution of the blast load on the plate using 10 load rings is shown in the figure 3.

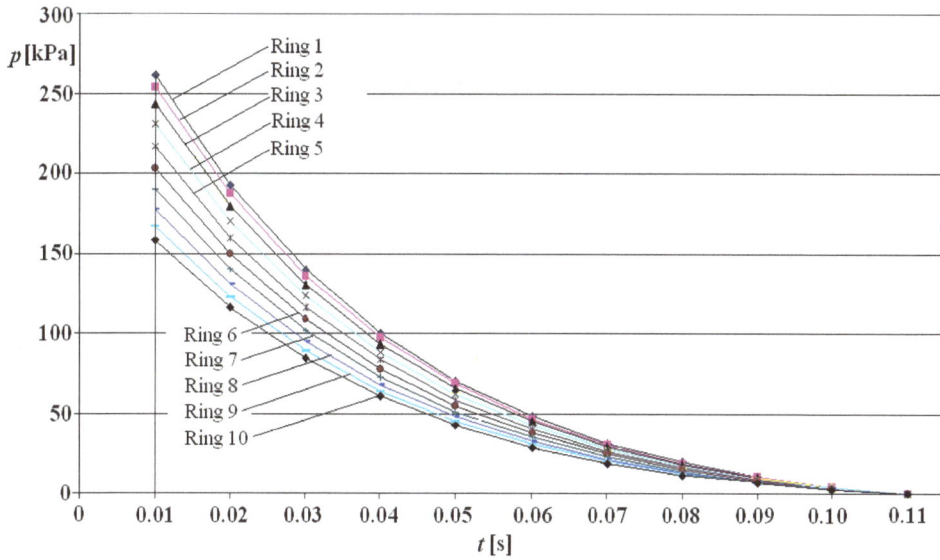

igure 3: Pressure - time histories for all rings

According to the methods used in this paper an individual pressure-time history to each element based on its distance from the blast is assigned. Each ring has its own pressure-time history as it is shown in figure 3.

The time duration Ts is determined from a natural vibration calculus, being equal to natural period of response of the structure ([9]).

3. DELAMINATION MODEL

The finite element delamination analysis was carried out using COSMOS/M finite element software. There are several ways in which the panel can be modeled for the delamination analysis. For the present study, a 3-D model with 4-node SHELL4L composite element of COSMOS/M is used. The panel is divided into two sub-laminates by a hypothetical plane containing the delamination. For this reason, the present finite element model would be referred to as two sub-laminate model. The two sub-laminates are modeled separately using 4-node SHELL4L composite element, and then joined face to face with appropriate interfacial constraint conditions for the corresponding nodes on the sub-laminates, depending on whether the nodes lie in the delaminated or undelaminated region.

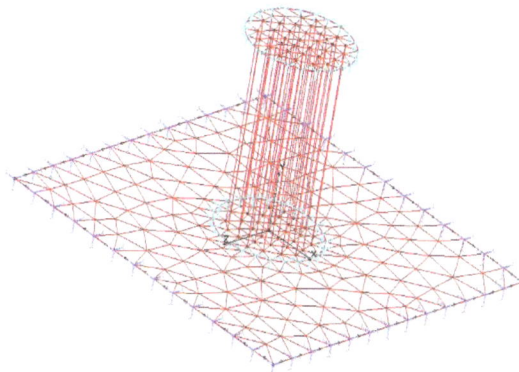

Figure 4: Delamination model

The delamination model has been developed by using the surface-to-surface contact option (Fig. 4). In case of surface-to-surface contact, the FE meshes of adjacent plies do no need to be identically. The contact algorithm of COSMOS/M has possibility to determine which node of the so-called master surface is in contact with a given node on the slave surface. Hence, the user can define the interaction between the two surfaces.

In the analysis, the certain layers are intentionally not connected to each other in ellipse regions. The condition is that the delaminated region does not grow. In COSMOS/M these regions were modeled by two layers of elements with coincident but separate nodes and section definitions to model offsets from the common reference plane. Thus their deformations are independent. At the boundary of the delamination zones the nodes of one row are connected to the corresponding nodes of the regular region by master slave node system.

In the calculus done in this paper, the delamination has a circular shape, with diameter of 50 mm, placed in the middle of the plate, between layers 1 and 2.

4. FEM ANALYSIS OF THE PLATE

The applied blast impulse is calculated using the plate mesh. The studies were carried out on a square plate from the ship side shell placed between two pairs of web stiffeners. So, the plate can be considered as being clamped on the all sides. The area of each of the ring zones is easily calculated from the formula for the cross sectional area of a cylinder. These individual areas are then multiplied by the corresponding impulse per unit area to obtain the total applied impulse. Using the ideal and applied impulse values, a percent error for the applied impulse can be determined.

The material is E-glass/polyester having the symmetric stack. The stack of the shell is according to the topologic code [A/B]3s.

The layers made of material A, have the thickness of 0.25 mm and characteristics:

E_x=80 GPa, E_y=80 GPa, G_{xy}=10 GPa, μ_{xy}=0.2

The layers made of material B, have the thickness of 0.1 mm and characteristics:

E_x=3.4GPa, E_y=3.4GPa, G_{xy}=1.3GPa, μ_{xy}=0.3.

Due to the double symmetry, one quarter of plate was studied.

In the non-linear calculus, Tsai-Wu criterion was considered for the limit state stresses evaluation.

Time variation of the maximum von Mises stress obtained in the point placed on the middle of the side plate are presented in figure 5 (material with damping characteristics) and figure 6 (material without damping characteristics). As it is seen, the maximum stress in the case of damping is 2.5 times bigger than the stress in the case of the plate without damping.

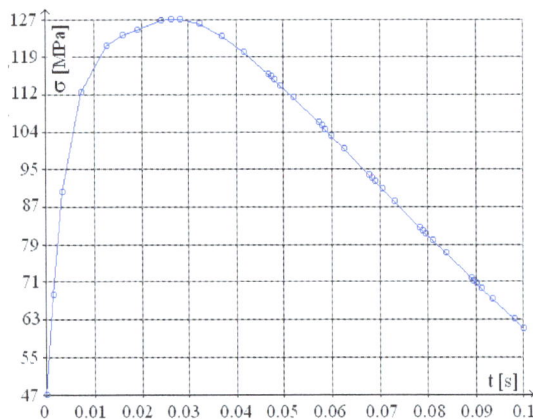

Figure 5: Maximum von Mises stress on the sides of the plate with damping

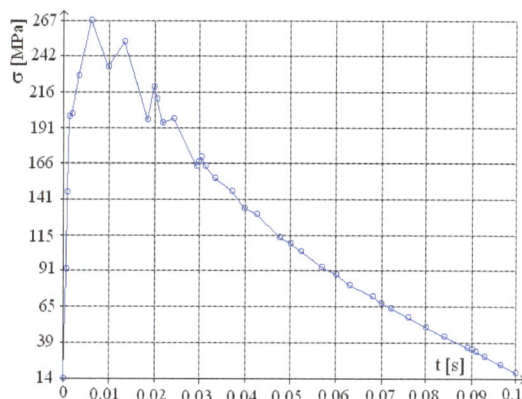

Figure 6: Maximum von Mises stress on the sides of the plate without damping

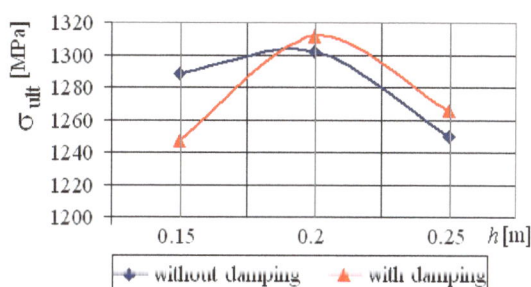

Figure7: The variation of ultimate stress (Fail 1 - tension) for t=2.1mm, in the case W=0.7kg

5. CONCLUSION

For high-risks facilities such as navy and commercial ships, design considerations against extreme events (bomb blast, high velocity impact) are very important. It is recommended that guidelines on abnormal load cases and provisions on progressive collapse prevention should be included in the current ship hull structure design norms.

Requirements on ductility levels also help improve the structure performance under severe load conditions.

Dynamic calculus was done with COSMOS/M soft package using specific elements SHELL4L.

According to the parametric calculus, the material damping model used in the analysis leads to the decreasing of the maximum stress occuring in the plate up to 0.5 from the value obtained in the model without damping. For example, in the figure 7 the variation of maximum stress versus equivalent TNT mass W for plate thickness of 2.1mm and blast normal distance h of 0.25mm for the both cases of damping is presented.

In all analyzed cases, for equivalent TNT mass W lesser than 0.2 kg, the stress obtained in the plate is almost constant, for various distances h and various plate thicknesses t.

The blast wave is instantaneously increases to a value of pressure above the ambient atmospheric pressure. This is referred to as the side-on overpressure that decays as the shock wave expands outward from the explosion source. After a short time, the pressure behind the front may drop below the ambient pressure (Fig. 1). During such a negative phase, a partial vacuum is created and air is sucked in. This is also accompanied by high suction winds that carry the debris for long distances away from the explosion source. In the paper this phase is considered as equal to zero.

The values of the pressure acting on the plate have small differences for W=0.1kg and W=0.2 kg.

For the values of the equivalent TNT mass W lesser than 0.7 kg, the fails do not occur in the material and so the integrity of the plate is not affected. In the case of W=0.7kg in all cases of distance h and thickness t the variations of failure criterion for tension (Fail 1) are presented in figure 7. As it is seen, generally speaking, the damping "is helping" the plate integrity: the stress when the Fail 1 (case of tension) is occurring in the case of damping is greater than the plate without damping.

According to the analysis, the developed blast simulation model and optimal design system can enable the prediction, design and prototyping of blast-protective composite structures for a wide range of damage scenarios in various blast events, ranging from plate damage, localized structural failure. From the studies, the proposal of a composite structure with special damping system can help the structure to sustain blast load. The inclusion of a damping material in the composite structure can absorb energy under blast load and help to reduce the force transmitted to the main structure. Also, the damping material helps to reduce stress concentration in the plate material.

ACKNOWLEDGEMENT

The authors acknowledge the financial support received for this work from the Romanian UEFISCDI Project 168 EU (2012-2014).

REFERENCES

[1]   Black G., Computer Modeling of Blast Loading Effects on Bridges, Report, University of Delaware, Newark, Delaware, 2006.

[2]   Houlston, R., Slater, J. E., Pegg, N., Des Rochers, C. G., On the analysis of structural response of ship panels subjected to air blast loading, Computers and Structures, 21, 273-289, 1985.

[3   ] Gupta, A.D., Gregory, F.H. and Bitting, R.L., Dynamic Response of a Simply Supported Rectangular   Plate to an Explosive Blast, Proc. XIII Southestern Conf. on Theoretical and Appl. Mech. 1, 385-390,   1985

[4]   *** Risk Management Series, Reference Manual to Mitigate Potential Terrorist Attacks Against   Buildings. Providing Protection to People and Buildings, FEMA, 426, 2003.

[5]   Cooper P., Explosive engineering, Wiley-VCH, New York, 1996.

[6]   Xue Z., Hutchinson J.W., Preliminary assessment of sandwich plates subject to blast loads,   International Journal of Mechanical Sciences, vol. 45, pp. 687–705, 2003.

[7]   Mays G., Smith P.D., Blast Effects on Buildings, Thomas Telford Publication, London, 2001.

[8]   Brode, H.L., Numerical solution of spherical blast waves, Journal of Applied Physics, American   Institute   of Physics, New York, 1955.

[9]   Son J., Performance of cable supported bridge decks subjected to blast loads, doctoral thesis, University   of California, Berkeley, 2008.

A COMPARATIVE STUDY OF THE PROPERTIES OF ZINC-SiO$_2$ AND ZINC-Al$_2$O$_3$ COMPOSITE LAYERS

I. Constantinescu[1], F. Oprea[1] ,O. Mitoseriu[1]

[1]Faculty Metallurgy Material Science and Environment, "Dunărea de Jos" University of Galați,111 Domnească Street, 800201, email:ionel53gl@yahoo.com

Abstract: *The aim of this work was the electrodeposition of SiO$_2$ and Al$_2$O$_3$ particles into zinc matrix in order to improve the surface properties.The effect of loading particles in bath on composition,morphology, structure of deposits and their influence on the microhardness,roughness and corrosion resistance of zinc-SiO$_2$ and zinc-Al$_2$O$_3$ composite layers was investigated. Another objective of the present study is to develop the plating baths with suitable compositions and establish the optimum electrodeposition conditions for obtaining good quality zinc-SiO$_2$ and zinc-Al$_2$O$_3$ composites.*
Keywords*: SiO$_2$, Al$_2$O$_3$, electrochemical deposition, composites.*

1.INTRODUCTION

Metallic matrix composites with a wide range of matrix materials such as copper, aluminum, nickel, chromium, cobalt, zinc, etc and second-phase ceramic particles have been produced. The metallic matrix has ductility and toughness properties combined with ceramic characteristics (high strength, hardness) leading to attractive physical and mechanical properties for the new composite material. The reinforced phase includes particulates, whiskers or short fibers [1,2].

The properties of metallic matrix composites can be controlled by the size and volume fraction of the dispersed phase (DP) as well as by the nature and properties of the matrix material. An optimum set of mechanical properties can be obtained when fine, thermally stable ceramic particulates are dispersed uniformly in the metal matrix.[3,4,5]

Metallic matrix composites with ceramic oxides as DP can be manufactured by electrodeposition method obtaining thin-film layers with improved wear resistance, good high-temperature stability and improved friction properties which are important characteristics for industrial applications.

2. EXPERIMENTAL CONDITIONS

The composite layers made using the electrochemical method can be obtained using different methods like: depositing in a centrifugal field, depositing on a centrifugal cathode, direct current or alternative current depositing, depositing with or without the recirculation of the electrolyte, etc[6].

In our experiments we used the direct current electrodeposition technology in order to obtain the composite layers. For obtaining composites with adherence, smoothness, good mechanical and corrosion properties must be respected some experimental parameters such as pH, current density, temperature, stirring, deposition time and electrolyte composition; these parameters are presented in Table 1:

Table 1: The working parameters for electrodeposition of Zn-SiO$_2$ and Al$_2$O$_3$ and composite coatings

Electrolyte composition	$ZnSO_4 \cdot 7H_2O=315gl^{-1}$; $Na_2SO_4 \cdot 10H_2O=75gl^{-1}$ $Al_2(SO_4)_3 \cdot 18H_2O=40gl^{-1}$
pH	3-4
Temperature (°C)	20
Current density (A·dm^{-2})	2,3,4
Magnetic stirring (rpm)	300
Electrodeposition time (min.)	60

The surfaces morphologies of the electrodeposits were characterized with ZEISS DSM-960A scanning electron microscopy(SEM) and EDX analysis using Philips XL-30FEG.The microhardness of the samples was analyzed by a CV-400DAT2 NAMICON durimeter with a down force fixed at 50 grams and layers roughness with NAMICON TR 100.

3. RESULTS AND DISCUSSIONS

Detailed investigation on the morphology of the surface of composite coatings deposited at different current densities and different particle loadings in electrolyte has been done. The deposit quality was first identified by visual examination watching to be uniform and adherent. Then we've made the next step-EDX and SEM characterization of the composite layers.

The dispersed phases characterization

The dispersed phase used in the codeposition were SiO_2 and Al_2O_3 microparticles with the average size range between 1-5 μm. The aspect of the used microparticles is shown in Figure 1:

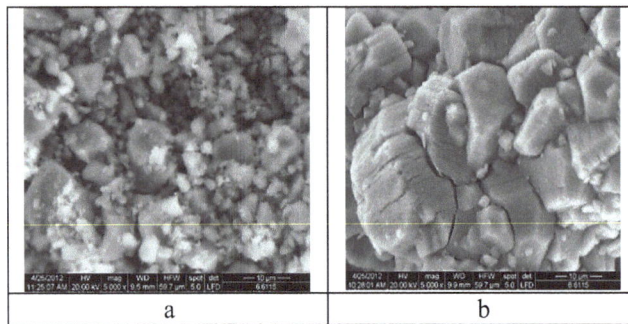

Figure1.SEM analysis of the SiO_2 microparticles(a) and Al_2O_3 microparticles(b)

The EDX analysis and DP embedding in zinc layers

The presence of the dispersed phase in the composite layer has been highlighted by EDX spectra. Both ceramic phases are present in zinc matrix(Figure 2).Oxigen is also present in EDX spectra proving that the second phase is embedded in zinc matrix in the form of oxides.

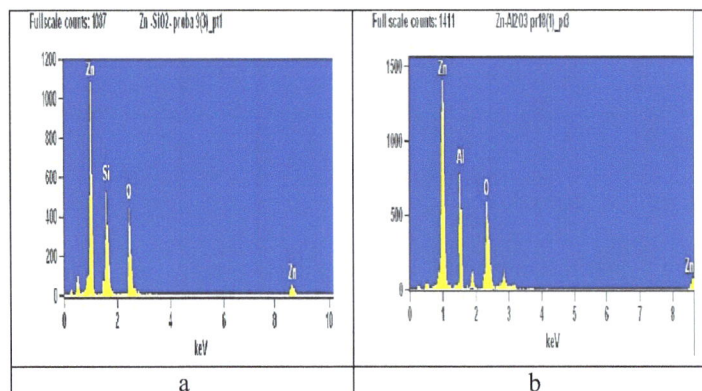

Figure2: The EDX analisys for a)zinc-SiO_2 and
b) zinc-Al_2O_3 composites(30gl^{-1}) obtained with the following parameters: 3Adm^{-2},60min,300rpm

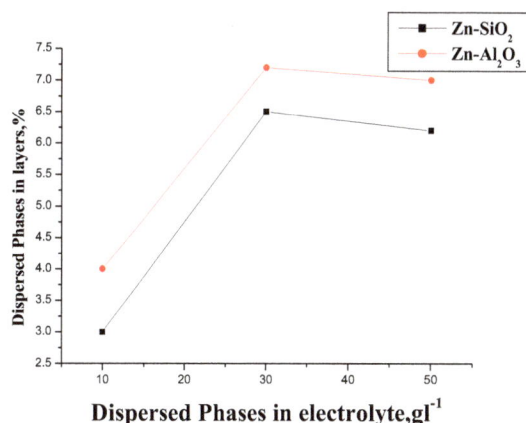

Figure 3.Variation of dispersed phases in electrolyte in accordance with dispersed phases embedded in layers using the following parameters:$3A/dm^2$, 60min, 300rpm,$30gl^{-1}$

It has been observed that the amount of embedded DP increases with increasing concentration of suspended particles in the electrolyte. The highest value for the embedded Al_2O_3 DP in zinc matrix is 6.5% mass percentage obtained for $30gl^{-1}$ particles loading in electrolyte, 3 Adm^{-2} current density and electrolyte agitation of 300rpm; as for $Zn-SiO_2$ composites maximum of embedded phase is 7.2% mass percentage obtained for $30gl^{-1}$ particles loading in electrolyte, 3 Adm^{-2} current density and electrolyte agitation of 300rpm. SiO_2 particles are embedded easier in zinc matrix than Al_2O_3 particles.

The microstructure of zinc-SiO_2 and zinc-Al_2O_3 composite layers

The microstructure of zinc-Al2O3 composite layers is different compared to pure zinc layer. Oxide particles dispersed in matrix and crystallization of zinc grains in co-deposition process makes relevant changes in depositing structure.

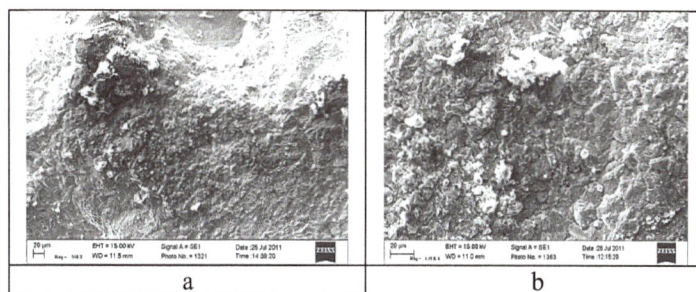

Figure 4: SEM Micrographs of zinc matrix composites obtained at: 3 Adm^{-2}, $30gl^{-1}$ DP ,60 min ,300rpm.
Figure a: $Zn-SiO_2$ composites and figure b: $Zn-Al_2O_3$ composites

The SEM analysis indicates an uniform deposit accomplished for 3 Adm^{-2},$30gl^{-1}$DP in zinc matrix, 60 min. deposition time and 300rpm stirring.

Microhardness of the electrodeposited layers

The presence of particles in the deposited layers changed both their structure and properties. As for microhardness a comparative study has been done. The study shows that the microhardness values of composite layers are higher than microhardness values of pure zinc layers (Figure 5):

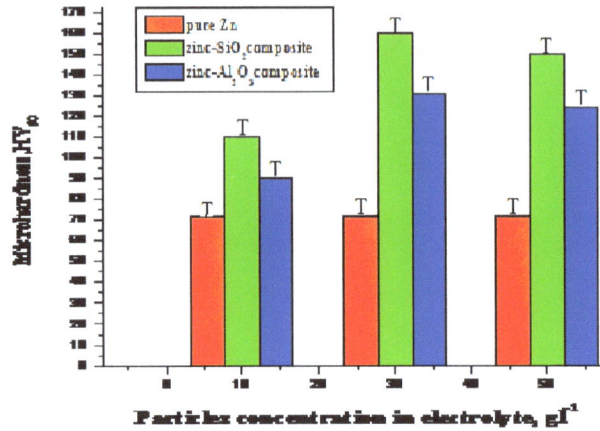

Figure 5:Average microhardness values for samples obtained by electrodeposition according to the particles concentration in electrolyte

The highest average value (158 HV_{50}) is obtained for zinc-SiO_2 at 3Adm^{-2}, 60min, 300rpm, and 30gl^{-1}electrocodeposition parameters. At the same parameters, for Zn-Al_2O_3 the highest average value is 131 HV_{50}. Zinc-SiO_2 composite layer is harder than zinc-Al_2O_3 layer.

Roughness of the electrodeposited layers

Another change in properties made by the ceramic oxides dispersed phases is the roughness of the layers.

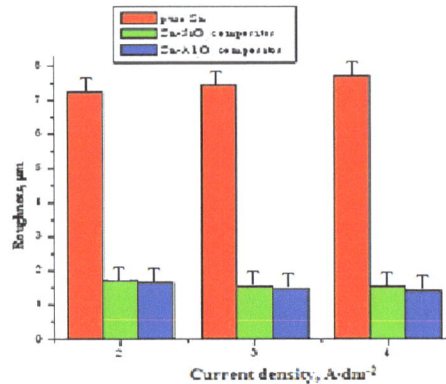

Figure 6:Roughness variation of Zn layers and Zn composite layers for different current densities

In contrast to the microhardness data, the smoothing effect is produced by the amount of DP embedded in layers along with zinc. The lowest roughness average values for composite coatings were obtained at the following parameters:4A/dm^{-2}, 60min, 300rpm and 30gl^{-1}particles in electrolyte for zinc-roughness value of pure zinc layers is higher than roughness values of zinc composites. The smoothest composite surface is obtained for zinc-Al_2O_3 composites which is 1.3μm Ra.

Corrosion comparative study of Zn coatings

The corrosion behaviour of Zn coatings(composite layers and pure zinc layers) was investigated by electrochemical methods in a special facility prepared in order to determine the potentiodynamic corrosions. From the experimental data obtained from measurements we chose representations in the form of Tafel polarization curves. The analysis of the graphical representations made it possible to characterize the corrosion of pure zinc and composite samples made at 3Adm^{-2}, 60min, 300rpm deposition parameters. The results are given in Figure 7:

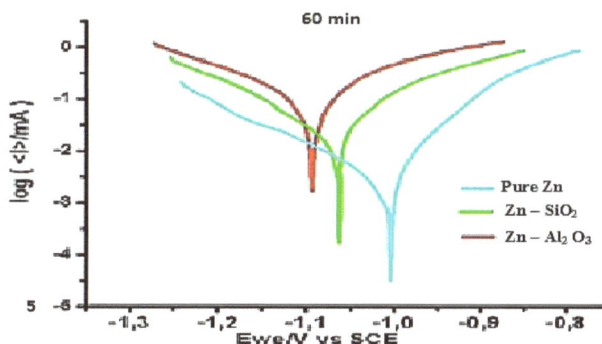

Figure 7: Tafel polarization curves for pure Zn; Zn- SiO_2 and Zn-Al_2O_3, composite layers

Corrosion behovior is appreciated by electrochemical tests showing lower resistance of pure zinc layer compared with Zn composites (corrosion potential of pure Zn is higher than corrosion potential of the composite coatings).The best corrosion resistance is obtained for Zn-Al_2O_3 composite layers.

4.CONCLUSIONS

We have characterized zinc composite coatings where SiO_2 and Al_2O_3 particles are embedded in the metal matrix as dispersed phases. Zinc composites presents good adhesion on substrate and uniform appearance. The composite coatings quality analysis was performed using EDX and SEM analysis which shows that the embedded oxides second phase in zinc matrix depends on their concentration in electrolyte and current density.
Microhardness of zinc composites was improved compared with Zn layers. The highest average microhardness value was obtained for Zn-SiO_2 composites.

Zn layers present a large variation of roughness values depending on the amount of embedded particles in layers; the lowest roughness average values were obtained for the following parameters:4Adm^{-2}, 60min, 300rpm, and 30gl^{-1} particles in electrolyte for zinc-Al_2O_3 composites. Pure zinc layers had the highest roughness average values.

The corrosion behavior of composite layers in Zn matrix was studied by potentiodynamic method; the results show an improvement of corrosion behavior of Zn composites compared with pure zinc layers.

REFERENCES

[1]   Schwarzacher W.: Electrodeposition; A Technology for the Future, Electrochemical Society Interface, volume 15,no. 1,pag.32–35,2006.

[2]   Binder L.,Kordesch K.,:Electrodeposition of Zinc Using a Multi-Component Pulse Current, Electrochimica Acta,volume 31,issue 2, pag.255–262,1986.

[3]   Socha R.P., Fransaer J.,:Mechanism of Formation of Silica-Silicate Thin Films on Zinc,Thin Solid Films, 488,pag.45,2005.

[4]   Azizi M.,Schneider W.,Plieth W.,Electrolytic co-deposition of silicate and mica particles with zinc, Journal of Solid State Electrochemistry,volume 9, issue 6, pag. 429-437,2005.

[5]   Satoshi O., Hiroaki n.,Shigeo K., Electrodeposition of Zn-Al_2O_3 Composite from Non-   Suspended Solution Containing Quaternary Ammonium Salt, Journal of the Surface Finishing   Society of Japan,ISSN:0915-1869, Volume 53,no.12,page.920-925,2002.

[6]   Khan T.R., Erbe A., Auinger M.,:Electrodeposition of Zinc–Silica Composite Coatings:Challenges in Incorporating Functionalized Silica Particles into a Zinc Matrix,Science and Technology of Advanced Materials,doi:10.1088/1468-6996/12/5/055005, published 14 September 2011.

[7]   Lin Z.F., Zhang D.,Wang Y.,Wang P.,:A Zinc/Silicon Dioxide Composite Film: Fabrication and Anti-corrosion Characterization,Materials and Corrosion, volume 63,Issue 5, pag. 416–420,2012.

[8]   Khorsand S.,Raeissi K., Golozar M.A.: An Investigation on the Role of Texture and Surface   Morphology in the Corrosion Resistance of Zinc Electrodeposits, Corrosion Science,no.53,pag. 2676- 2678,2011.

[9]   Fontenay F.,Andersen L.B.,Moller P.,:Electroplating and Characterisation of Zinc Composite Coatings, Galvanotechnik, 92(4),pag.928,2001.

[10]   Hashimoto S.,Masaki A.,:The Characterization of Electrodeposited Zn-SiO$_2$ Composites before and   after Corrosion Test, Corrosion Science, volume 36, issue 12, pag.2125–2137,1994.

[11]   Kondo K., Tanaka A.O.Z.,:Electrodeposition of Zinc-SiO$_2$ Composite, Journal of the Electrochemical Society, 147(7): pag.2611,2000.

[12]   T.H. Muster, A. Bradbury, A. Trinchi, I.S. Cole, T. Markley, D. Lau, S. Dligatch, A. Bendavid, P. Martin;   The atmospheric corrosion of zinc: The effects of salt concentration, droplet size and droplet   shape, Electrochimica Acta, 56, pag.1866-1873, 2011.

ANALYSIS OF COMPOSITE RECTANGULAR PLATES BASED ON THE CLASSICAL LAMINATED PLATE THEORY

I. Sprinţu

Military Technical Academy, Bucharest, Romania, sprintui@yahoo.com

ABSTRACT: In this paper are developed some analytical solutions for static analysis of thin orthotropic plates, based on the Classical Laminated Plate Theory (CLPT).

The considered reasons for the solutions were to exactly satisfy the boundary conditions and to verify as close as possible the differential equation of the plate. The weighted residue method was considered to optimise the chosen analytical solutions. Interesting evaluations were performed for different types of functions, especially with respect to the orthotropic answer of the plate.

Article contains a comparative analysis of analytical results with the numerical (FEM) and those obtained experimentally.

KEY WORDS: composite plates, orthotropic, static analysis, simply supported edge, clamped edge.

1.ASSUMPTIONS

To establish the constitutive equation of flat plates, is considering Kirchhoff Love hypothesis, "weak" as follows: We accept as working hypotheses to determine the constitutive equation of composite flat plate, assumptions of Kirchhoff-Love, extended as follows [5]:

a. During deformations, the normal to the median of the plate remain straight and normal to the deformed middle surface ($\gamma_{xz} = \gamma_{yz} = 0$ wich implies $\tau_{xz} = \tau_{yz} = 0$). Although efforts unit τ_{xz}, τ_{yz} are very small compared with the normal, we

extend the assumption considering their gradient high enough $\dfrac{\partial \tau_{xz}}{\partial z} \neq 0$, $\dfrac{\partial \tau_{yz}}{\partial z} \neq 0$.

b. During deformation of the plate its thickness is constant, equivalent with $\varepsilon_z = 0$.

c. We accept the hypothesis of small strains, wich approximates:

$\varepsilon_x = -z \cdot \dfrac{\partial^2 w}{\partial x^2}$, $\varepsilon_y = -z \cdot \dfrac{\partial^2 w}{\partial y^2}$, $\gamma_{xy} = -2 \cdot z \cdot \dfrac{\partial^2 w}{\partial x \partial y}$, where z is the distance from the point where displacements are been computed and the midplane.

2. ANALYSIS FOR COMPOSITE PLATE EQUATIONS

2.1. Case of clamped edge

For a rectangular plate, on which pressure p is evenly distributed, using the assumptions from the first chapter, we obtain a generalization of Sophie-Germain equation in particular case, when ox-axis is oriented along the fiber direction, [5],[8]:

$$Q_{11} \cdot \frac{\partial^4 w}{\partial x^4} + Q_{22} \cdot \frac{\partial^4 w}{\partial y^4} + 2 \cdot (Q_{12} + 2 \cdot G_{12}) \cdot \frac{\partial^4 w}{\partial x^2 \partial y^2} = \frac{12 \cdot p}{h^3}, \tag{1}$$

where E_1, E_2 are elasticity moduli in the longitudinal and transversal directions, respectively, G_{12} is the shear modulus in the plane of the ply, and v_{12} is the Poisson coefficient,

$$Q_{11} = \frac{E_1}{1 - v_{12} \cdot v_{21}}, \quad Q_{22} = \frac{E_2}{1 - v_{12} \cdot v_{21}}, \quad Q_{12} = \frac{v_{21} \cdot E_1}{1 - v_{12} \cdot v_{21}} = \frac{v_{12} \cdot E_2}{1 - v_{12} \cdot v_{21}}.$$

For example, we consider a flat plate with size $a = 300mm, b = 200mm$, thickness $h = 1.5\,mm$, plate made by incorporating the 8 layers of glass fiber fabric, type EWR $300\,g/mp$, Polylite 440-M888 resin, with technology

VARTM (Vacuum Assisted Resin Transfer Molding), at SC STRAERO SA. For each layer of fabric fibers have been removed on one direction (2 of 4 fibers). (Fig. 1(a))

Sails were arranged to have the same warp and weft direction, laminate being treated as orthotropic material. The main directions of elasticity are the warp and weft. To determine the elasticity modules were made standard flat specimens, the size 25×300.

Following tensile testing (SC STRAERO SA), were obtained values $E_1 = 22263 MPa$, $v_{12} = 0.18$, and $E_2 = 19035 MPa$.

In the same time, the orthotropic plate with clamped edge is required to stretching and bending.

Using weights of 1.2 kg and flat specimens size (24×100), were obtained values $E_1 = 22065 MPa$, $E_2 = 17079 MPa$.

For this, we used the relations:

$$v = \frac{F \cdot \ell^3}{48 \cdot E \cdot I}, \quad I = \frac{c \cdot h^3}{12}, \text{ which implies } E = \frac{F \cdot \ell^3}{4 \cdot v \cdot c \cdot h^3} = \frac{M \cdot g \cdot \ell^3}{4 \cdot v \cdot c \cdot h^3}.$$

The orthotropic plate was driven by a pressure $p = 0.00465 \, MPa$, uniformly distributed over the entire surface.

To ensure better clamped conditions, contour plate was reinforced (fig. 1(b)).

In table. 1 are presents the comparative results between the experimental measurements and the numerical results (using ANSYS- shell Elastic 4 node 63, mesh 10), taking into account the elasticity modules in stretching.

In table 2 are presents the comparative results between the experimental measurements and the numerical results, taking into account the elasticity modules in bending.

For a better approximation, we consider an average for modules of elasticity, respectively $E_1 = 22164 MPa$, $E_2 = 18057 MPa$. (table.3)

Mathematic, clamped edge for $x = 0$, $x = a$ and $y = 0$, $y = b$ follows:

$$\begin{cases} w(0,y) = w(a,y) = 0, & \dfrac{\partial w}{\partial x}(0,y) = \dfrac{\partial w}{\partial x}(a,y) \\ w(x,0) = w(x,b) = 0, & \dfrac{\partial w}{\partial y}(x,0) = \dfrac{\partial w}{\partial y}(x,b) = 0 \end{cases} \tag{2}$$

In [8], Reddy proposed solution for equation (1), as:

$$w(x,y) = \sum_{i=1}^{m} \sum_{j=1}^{n} c_{ij} \cdot \left(\frac{x}{a}\right)^{i+1} \cdot \left(\frac{y}{b}\right)^{j+1} \cdot \left(1-\frac{x}{a}\right)^2 \cdot \left(1-\frac{y}{b}\right)^2, \text{ and solved it for } m = n = 1.$$

In this paper, using Ritz method, we look for a solution as

$$w(x,y) = \sum_{i=1}^{n} \sum_{j=1}^{m} c_{ij} \cdot x^{i+1} \cdot (a-x)^{i+1} \cdot y^{j+1} \cdot (b-y)^{j+1}. \tag{3}$$

The coefficients $\left(c_{ij}\right)_{i,j}$ will be determined using the weighted residue method.

For $p = 0.00465 \, MPa$, $E_1 = 22164 MPa$, $E_2 = 18057 MPa$, we get coefficients (Maple):

$c_{11} = 1.29378 \cdot 10^{-16}$, $c_{12} = -3.2159 \cdot 10^{-21}$, $c_{13} = 2.09204 \cdot 10^{-26}$, $c_{21} = -3.3471 \cdot 10^{-21}$,

$c_{31} = 1.5746 \cdot 10^{-26}$, $c_{22} = 1.2466 \cdot 10^{-25}$, $c_{33} = -4.9444 \cdot 10^{-35}$,

for $m = n = 3$.

In this case, the maximum value of the arrow is $w\left(\dfrac{a}{2}, \dfrac{b}{2}\right) \approx 3.053 \, mm$.

Using finite element method (shell Elastic 4 node 63, mesh 10, ANSYS) we obtain a maximum displacement $w_{max} = w\left(\dfrac{a}{2}, \dfrac{b}{2}\right) \approx 3.055 \, mm$.

In table 4 are presented results comparison between the analytical (Maple) and numerical solution.

To validate the solutions we performed measurements on the same plate, driven uniformly with different pressures ($p = 0.00305 MPa$, table 5, and $p = 0.004 MPa$, table 6).

In fig. 2 (a,b,c) are experimental graphics solutions for requests made.

Given the need for several sets of measurements of orthotropic plate arrows, depending on the conditions imposed on the shape and position in plan, I chose the option to apply a pressure plate perpendicular, evenly distributed over the entire surface with a pressurized air chamber.

Solution readily allows repeated measurements, and application of pressure change.

Measurement of air pressure in the chamber and so the pressure plate application is made with a liquid manometer. To measure arrows of deformed plate, we use inductive displacement transducer.

In these conditions, the entire system covers the following modules:
- Application module orthotropic plate
- Displacement transducer module in the horizontal plane xoy
- Arrow measuring module
- Acquisition and storage module measurements (steper electric motor EM-257, 17PM-K212-P1T for moving the transducer 12 on the ox and EM-464, 2Y28AD2 for moving the transducer 12 on the oy). .

Air chamber pressure was made from a thick $1\,mm$ rubber sheet $340 \times 240\,mm$ and, on where were bonded two valves, one for charging and one for connection to the manometer.

Deformable wall of the chamber is a plastic foil. (Fig. 1(c))

2.2. Tables

Table 1: ($p = 0.00465MPa$, $E_1 = 22263MPa$, $E_2 = 19035MPa$)

ANSYS (mm)	EXPERIMENTALLY (mm)	Difference (%)
0.0208	0.303	-1.0
0.401	0.439	-3.8
1.004	1.056	-5.1
1.619	1.7	-8.1
2.145	2.248	-10.3
2.539	2.651	-11.2
2.788	2.861	-7.3
2.894	2.893	0.1
2.859	2.878	-1.9
2.682	2.807	-12.5
2.36	2.48	-12.0
1.898	2.001	-10.3
1.319	1.383	-6.4
0.691	0.701	-1.0
0.164	0.166	-0.2

Table 2: ($p = 0.00465MPa$, $E_1 = 22065MPa$, $E_2 = 17079MPa$)

ANSYS (mm)	EXPERIMENTALLY (mm)	Difference (%)
0.0222	0.0303	-0.8
0.431	0.439	-0.8
1.066	1.055	3.1
1.764	1.7	6.4
2.348	2.248	10.0
2.788	2.651	13.7
3.07	2.861	20.9
3.19	2.893	29.7
3.15	2.878	27.2
2.949	2.807	14.2
2.587	2.48	10.7
2.071	2.001	7.0
1.431	1.383	4.8
0.746	0.701	4.4

0.176	0.166	1.0

Table 3: ($p = 0.00465 MPa$, $E_1 = 22164 MPa$, $E_2 = 18057 MPa$)

ANSYS (mm)	EXPERIMENTALLY (mm)	Difference (%)
0.0215	0.0303	-0.9
0.416	0.439	-2.4
1.044	1.055	-1.1
1.689	1.7	-1.1
2.242	2.248	-0.6
2.658	2.651	0.7
2.871	2.861	1.0
2.923	2.893	3.0
2.871	2.878	-0.7
2.809	2.807	0.2
2.469	2.48	-1.1
1.981	2.001	-2
1.373	1.383	-1
0.717	0.701	1.6
0.17	0.166	0.4

Table 4: ($p = 0.00465 MPa$, $E_1 = 22164 MPa$, $E_2 = 18057 MPa$)

ANSYS	0.022	0.415	1.689	2.242	2.658	2.923	2.998	2.809	2.469	1.98	0.717	0.17
Maple	0.022	0.416	1.688	2.241	2.657	2.921	2.996	2.808	2.467	1.98	0.717	0.17

Table 5: ($p = 0.00305 MPa$, $E_1 = 22164 MPa$, $E_2 = 18057 MPa$)

ANSYS (mm)	EXPERIMENTALLY (mm)	Difference (%)
0.014	0.015	0.11
0.272	0.31	3.8
1.109	1.105	-0.3
1.471	1.508	3.7
1.743	1.723	-2
1.917	1.916	-0.1
1.966	1.982	1.6
1.843	1.845	0.2
1.619	1.647	2.8
1.299	1.267	-1.2
0.47	0.508	3.8
0.111	0.137	2.6

Table 6: ($p = 0.004 MPa$, $E_1 = 22164 MPa$, $E_2 = 18057 MPa$)

ANSYS (mm)	EXPERIMENTALLY (mm)	Difference (%)
0.019	0.000	1.8
0.617	0.619	-0.2
1.453	1.423	3.0
2.123	2.119	0.4
2.514	2.501	1.3
2.615	2.607	0.8
2.417	2.44	-2.3
1.929	1.921	0.8

1.181	1.198	-1.7
0.357	0.328	2.9

2.3. Figures

Figure.1(a)

Figure.1(b)

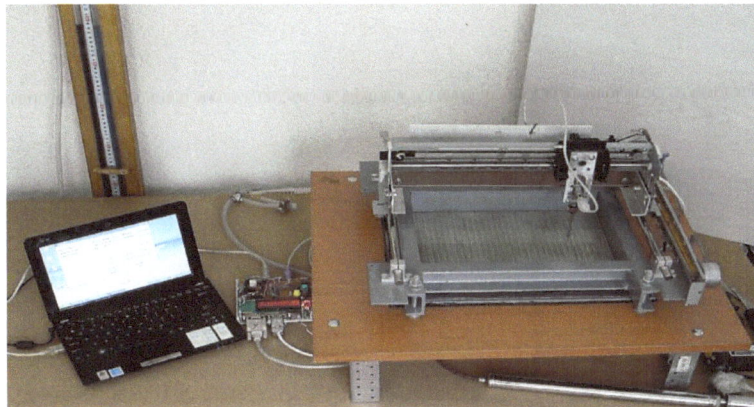

Figure.1(c)

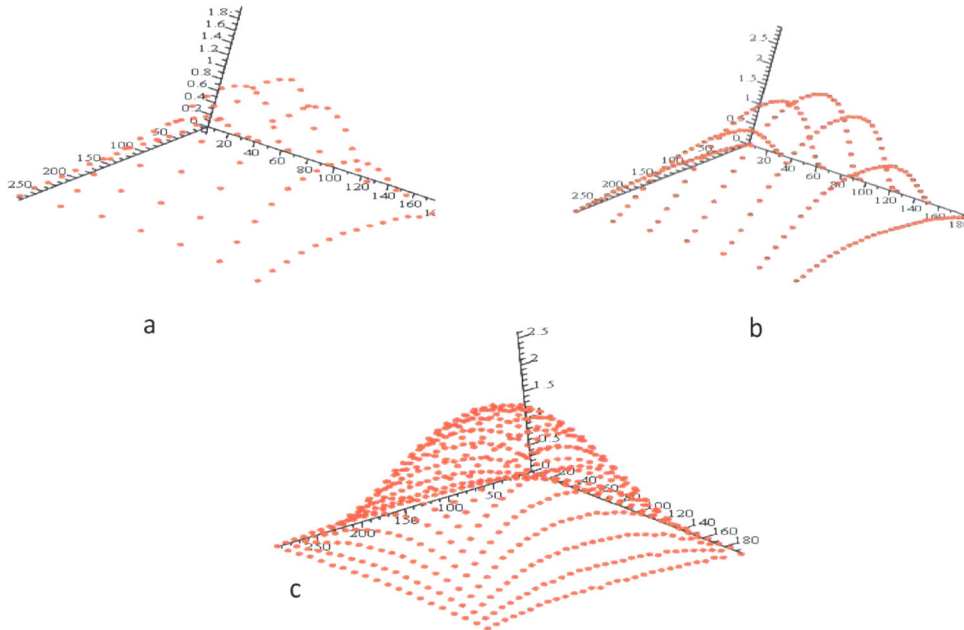

Figure. 2

3. CONCLUSION

This paper is an analysis of orthotropic plates, based on CLPT, characterizing thin plates, assuming small deformations. I proposed an analytical solution for case of clamped edge. After comparing the results with those treated in the literature, as well as the results from finite element, and experimentally analysis, we conclude that the solution presented in this paper approximates well plate subjected to compressive deformation.

Mechanical device allows measurements for other conditions imposed on the outline of the rectangular plate, simply supported, articulated and free.

REFERENCES

[1]   Alămoreanu, E., Negruţ, C., Jiga, G., Calculul structurilor din materiale compozite, U.P.B., 1993.
[2]   Ambartsumyam S.A., Theory of Anisotropic Plates, Tehnomic Publishing, 1969.
[3]   Araujo, A.L., Mota Soares, C.M. , Moreira de Freitas, M.J. ,
       Pedersen, P., Herskovits, J., Combined numerical-experimental model for the identification of   mechanical properties oflaminated structures Composite Structures, 2000.
[4]   Buzdugan, Gh., Rezistenţa materialelor, Ed. Academiei, Bucureşti, 1987.
[5]   Cristescu, N., Crăciun, M., Soos, E., Mechanics of Elastic Composites, Chapman&Hall/CRC, 2004.
[6]   Jones, R.M., Mechanics of Composite Materials, Scripta Book, Washington D.C., 1975.
[7]   Lekhnitskii, S. G., Anisotropic Plates. Gordon and Breach Science Publishers, New York, 1968.
[8]   Reddy, J.N., Mechanics of Laminated Composite Plates- Theory and Analysis, CRC Press, 1997.
[9]   Springer, G.S., Kollar, L.P., Mechanics of Composite Structures, Cambridge University Press, 2003.
[10   ] Vasiliev, V.V, Morozov, E.V., Mechanics and Analysis of Composite Materials, Elsevier, 2001.
[11]   Voyiadjis, G.Z., Kattan, P.I., Mechanics of Composite Materials with MATLAB, Springer, 2005.

DYNAMICAL ANALYSIS OF COMPOSITE RESERVOIR

K. Kotrasová[1]

[1] Department of Structural Mechanics, Institute of Structural Engineering Faculty of Civil Engineering, Technical University of Košice, Košice, SLOVAK REPUBLIC, e-mail: kamila.kotrasova@tuke.sk

Abstract: *This paper provides the theoretical background of simplified seismic design of liquid storage cylindrical ground -supported tanks. It takes into account impulsive and convective (sloshing) actions of the liquid in concrete tanks fixed to rigid foundations. This paper follows the influence of filling level and category of sub-soil of concrete cylindrical tank on total base shear V and overturning moment M.*
Keywords: *tank, liquid, earthquake*

1. INTRODUCTION

Large-capacity ground-supported tanks are used to store a variety of liquids, e.g. water for drinking and for fighting, petroleum, chemicals, and liquefied natural gas. Satisfactory performance of tanks during strong ground shaking is crucial for modern facilities. Tanks that were inadequately designed or detailed have suffered extensive damage during past earthquakes [2-4, 6].

2. PROPERTIES OF IRREGULARLY REINFORCED COMPOSITE WITH SHORT FIBERS

We assume the composite with unidirectionally oriented short fibers. We can write the longitudinal and transverse modulus of these composites with help of so-called Halphin-Tsai equations

$$E_1 = E^{(m)} \frac{1 + \frac{l}{d} \zeta_E \eta_L \xi}{1 - \eta_L \xi} \qquad E_2 = E^{(m)} \frac{1 + \zeta_E \eta_T \xi}{1 - \eta_T \xi} \qquad (1)$$

where

η_L and η_T are described in [5],

l is length of the fibre, d is diameter of the cross section, ξ is fibre volume fraction, ζ_E depends on the shape of cross section of the fibre.

For the modulus of elasticity for irregularly reinforced composite with short fibres we can write the empirical equation [1,7]

$$E = \frac{3}{8} E_1 + \frac{5}{8} E_2 .$$

3. SEISMIC DESIGN OF LIQUID STORAGE TANKS

Seismic design of liquid storage tanks has been adopted in [5, 8, 9]. When a tank containing liquid vibrates, the liquid exerts impulsive and convective hydrodynamic pressure on the tank wall and the tank base, in addition to the hydrostatic pressure. The dynamic analysis of a liquid – filled tank may be carried out using the concept of generalized single – degree – of freedom (SDOF) systems representing the impulsive and convective modes of vibration of the tank – liquid system. For practical applications, only the first convective mode of vibration need to by considered in the analysis (Figure 1). The impulsive mass of liquid m_i is rigidly attached to tank wall at height h_i (or h_i').

Similarly convective mass m_c is attached to the tank wall at height h_c (or h_c') by a spring of stiffness k_c. The mass, height and natural period each SDOF system are obtained by the methods described in [5, 8, 9]. For a horizontal earthquake ground motion, the response of various SDOF systems may be calculated independently and then combined to give the net base shear and overturning moment. For most tanks have got slimness of tank γ, whereby $0,3 < \gamma < 3$. Tank's slimness is given by relation $\gamma = H/R$, where H is the height of filling of fluid in the tank and R the tank radius [5, 9].

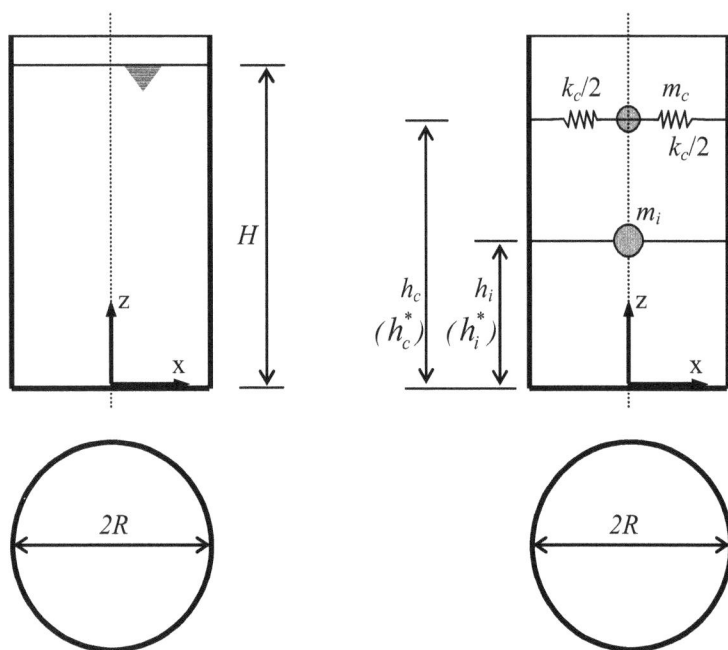

Figure 1: Two single – degree – of freedom systems for ground supported cylindrical tank

For a ground supported cylindrical tank, in which the wall is rigidly connected with the base slab, the naturals period of the impulsive mode of vibration T_i, in seconds, is given by

$$T_i = C_i \frac{H\sqrt{\rho}}{\sqrt{s/R}\sqrt{E}} \qquad (2)$$

where
H – height to the free surface of the liquid;
R – tank's radius;
s – equivalent uniform thickness of the tank wall;
ρ – mass density of liquid of tank material;
E – Modulus of elasticity of tank material;
Ci – the dimensionless coefficient, which is obtained from Figure 2 [5, 8, 9].
For a ground supported cylindrical tank, in which the wall is rigidly connected with the base slab, the natural period of the convective mode of vibration T_c, in seconds, is given by

$$T_c = C_c \sqrt{R} \qquad (3)$$

where
R – tank's radius;
C_c – the coefficient is expressed in $s/\sqrt{m}$, which is obtained from Figure 2 [5, 8, 9].
Total base shear of ground supported tank at the bottom of the wall can by also obtained by base shear in impulsive mode and base shear in convective mode:

$$V = (m_i + m_w + m_r)S_e(T_i) + (m_c)S_e(T_c) \qquad (4)$$

Total base shear of ground supported tank at the bottom of base slab is given also by base shear in impulsive mode and base shear in convective mode:

$$V' = (m_i + m_w + m_b + m_r)S_e(T_i) + (m_c)S_e(T_c) \qquad (5)$$

The overturning moment of ground supported tank immediately above of the base plate is given also by

$$M = (m_i h_i + m_w h_w + m_r h_r)S_e(T_i) + (m_c h_c)S_e(T_c) \qquad (6)$$

and the overturning moment of ground supported tank immediately below of the base plate is given also by

$$M = \left(m_i h_i' + m_w h_w + m_b h_b + m_r h_r\right) S_e(T_i) + \left(m_c h_c'\right) S_e(T_c) \tag{7}$$

where

m_i – the impulsive mass of fluid, given in Figure 3 [5, 8, 9];

m_c – the convective mass of fluid, given in Figure 3 [5, 8, 9];

h_i – height of wall pressure resultant for the impulsive component, given in Figure 4 [5, 8, 9];

h_c – height of wall pressure resultant for the convective component, given in Figure 5 [5, 8, 9];

h_i' – height resultant of pressures on the wall and on the base plate for the impulsive component, given in Figure 4 [5, 8, 9];

h_c' – height resultant of pressures on the wall and on the base plate for the convective component, given in Figure 5 [5, 8, 9];

$S_e(T_i)$ – impulsive spectral acceleration, is obtained from a 2% damped elastic response spectrum for steel and prestressed concrete tanks, or a 5% damped elastic response spectrum for concrete and masonry tanks;

$S_e(T_c)$ – convective spectral acceleration, is obtained from a 0,5% damped elastic response spectrum;

m_w – mass of the tank wall;

m_b – mass of the tank base plate;

m_r – mass of the tank roof;

h_w – the height of center of gravity of wall mass;

h_b – the height of center of gravity of base plate mass;

h_r – the height of center of gravity of roof mass.

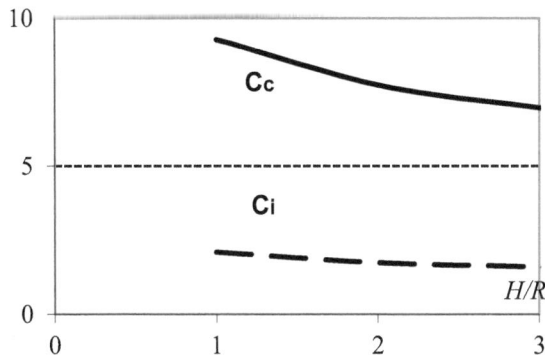

Figure 2: Impulsive and convective coefficients C_i and C_c

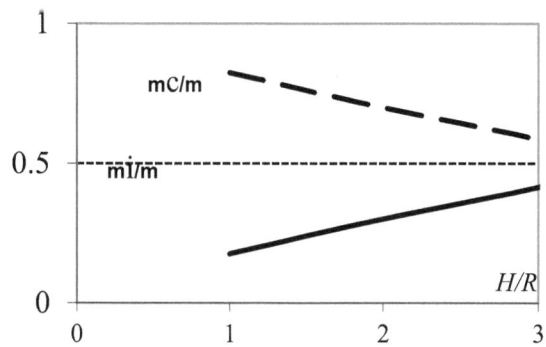

Figure 3: Impulsive and convective masses as fractions of the total liquid mass in the cylindrical tank

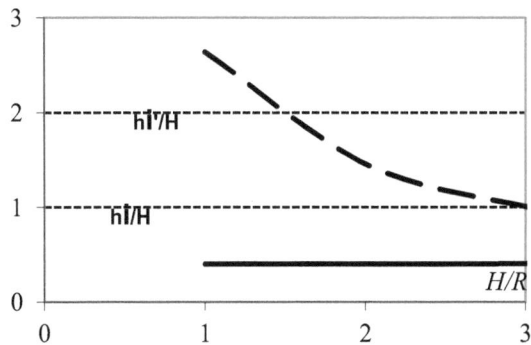

Figure 4: Impulsive heights as fraction of the height of the liquid in the cylindrical tank

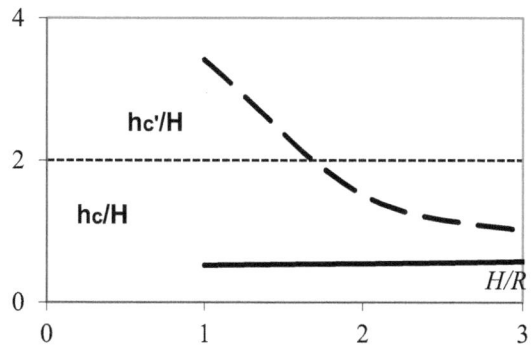

Figure 5: Convective heights as fraction of the height of the liquid in the cylindrical tank

4. NUMERICAL EXPERIMENT

A ground supported cylindrical tank of 1000 m³ capacity has plan dimension of $2R$ = 14 m and height H_w = 7,0 m. Wall has uniform thickness s = 0,25 m. The base slab is d = 0,4 m thick. There is no roof slab on the tank. from irregularly reinforced concrete with short steel fibers, l = 5cm, d = 4mm, ξ = 0,25. Tank is located on hard soil. The reservoir is filled with water (H_2O) to level from 0,5 m into 6,5 m. Seismic excitation is along x -direction. We consider only horizontal seismic load. Elastic response spectrum [5, 8] used for determination of impulsive spectral acceleration $S_e(T_i)$ and convective spectral acceleration $S_e(T_c)$. Elastic response spectrum is determinate for region of seismic risk is 2, category of sub-soil A, B, C and D [8]. Calculation had been realized by using [5, 8, 9] and the values m_i, m_c, h_i, h_i', h_c a h_c' were computed by using Figures 3 - 5.

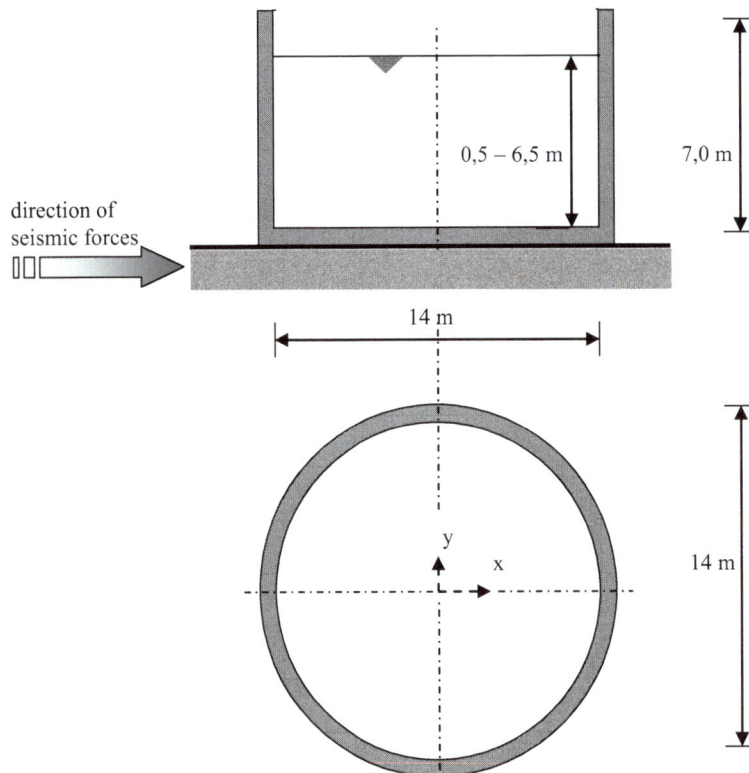

Figure 6: Details of tank geometry

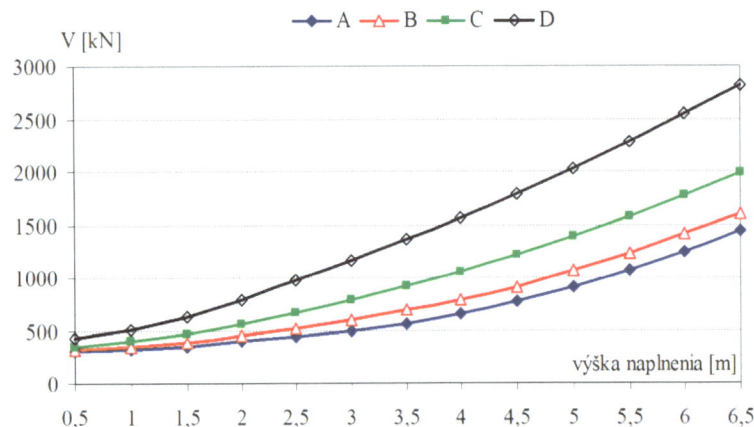

Figure 7: Total base shear at the bottom of the wall V in dependency on filling level and category of sub-soil

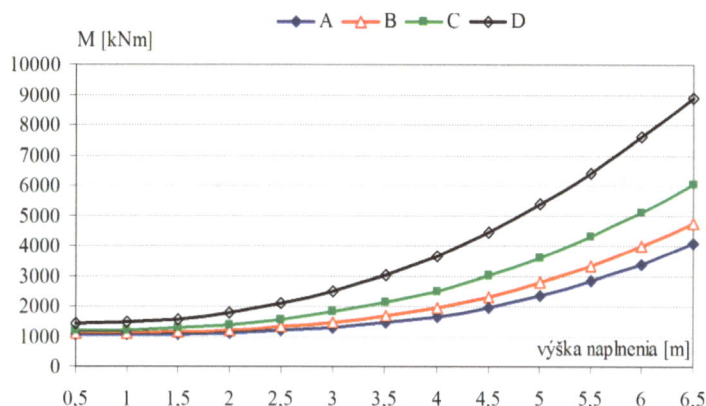

Figure 8: Total bending moment immediately above of the base plate M in dependency on filling level and category of sub-soil

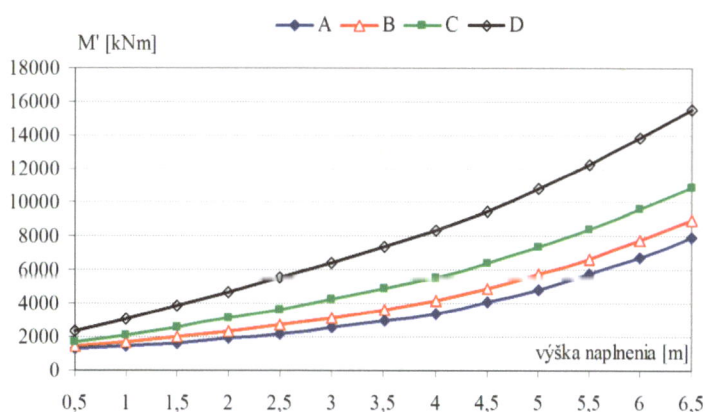

Figure 9: Total overturning moment immediately bellow of the base plate M' in dependency on filling level and category of sub-soil

5. CONCLUSION

From the Figures 7 - 9 is evident, that the maximum values of total base shears an overturning moments are getting, when the tank is completely filled with fluid. Value of total base shears and overturning moments of the reservoir fixed to the rigid foundations to stored on the more flexible sub – soil are getting to greater value.

ACKNOWLEDGEMENTS

This research has been supported by the Scientific Grant Agency of the Ministry of Education of Slovak Republic and the Slovak Academy of Sciences under Project 1/0201/11.

REFERENCES

[1]   Agarwal,B.D., Broutman, L.J.: Vláknové kompozity, Praha, 1987.

[2]   Juhásová, E., 1985: Pôsobenie seizmických pohybov na stavebné konštrukcie, VEDA, vydavateľstvo Slovenskej akadémie vied, Bratislava, 1985

[3]   Králik, J.: Probability seismic hazard loads for Mochovce site. Zborník z konferencie „Modelování v mechanice", Ostrava, 2005

[4]   Juhásová, E., Benčat, J., Krištofovič, V., Kolcún, Š., 2002: Expected seismic response of steel water   tank, In: 12th European Conference on Earthquake Engineering, Paper reference 595, London 2002.

Malhotra, P. K., Wenk, T., Wieland, M., 2000: Simple procedure for seismic analysis of liquid-Storage tanks, Structural Engineering International, No. 3, 2000, s. 197-201,
http://www.dec.fct.unl.pt/seccoes/S_Estruturas/Dinamica/Util/SeismicAnalysisTanks.pdf

[5]   Melcer, J., Lajčáková, G. Dynamický výpočtový model asfaltovej vozovky In: Stavebné a environmentálne inžinierstvo Roč. 7, č. 1 (2011), s. 2-12 ISSN: 1336-5835.

[6]   Kormaníková, K.:   Krátkovláknové kompozity s náhodne orientovanými vláknami. In: Nové trendy v prevádzke výrobnej techniky 2003. Prešov. TU-FVT, 2003 S. 69-74. - ISBN 8080730598

[7]   STN EN 1998-1, 2005: Design of structures for earthquake resistance - Part 1: General rules, seismic actions and rules for buildings, SÚTN, Bratislava, 2005.

[8]   EN 1998-4: 2006 Eurocode 8. Design of structures for earthquake resistance. Part 4: Silos, tanks and pipelines. CEN, Brussels, 2006.

BEHAVIOR TO HIGH SPEEDS HEATING-COOLING OF CERAMIC MULTILAYER MULTIFUNCTIONAL STRUCTURES BASED ON THE ZIRCONIE PARTIAL STABILIZED WITH YTTRIA

L.G. Pintilei[1], V. Manoliu[2], G. Ionescu[2], G.N. Basescu[1], S.C. Iacob Strugariu[1], C. Munteanu[1*]

[1] "Gheorghe Asachi" Technical University of Iasi, Faculty of Mechanical Engineering,
61-63 Prof. dr. doc. D. Mangeron Blvd, 700050, Iasi, Romania, laura_rares082008@yahoo.com
[2] National Institute for Aerospace Research "Elie Carafoli", Bucuresti, Romania, vmanoliu@yahoo.com

Abstract: Advanced materials intended for application in aerospace, power, metallurgical industries impose composite functional materials and new methods and installations to test them. Thermal barrier coatings (TBCs) are widely used to protect components and remain the most effective thermal insulation approach and their development was focused on partial stabilized zirconia. In this paper the used material is zirconia partial stabilized with 20% yttria in place of the classic TBC stabilized with 7-8%. In order to evaluate the behavioral assessment of TB protection, attesting installation was designed and realized by the INCAS, for cooling-heating speeds up to $100^0 C/s$, installation included in the experimental program. In the view of the evidence regarding interfacial structural changes induced by thermal shock due to high thermal gradients the results of SEM-EDS electronic microscopy investigations were presented.
Keywords: SEM, $ZrO_2/20\% Y_2O_3$, thermal shock

1. INTRODUCTION

Gas turbine aircraft engines are operating under severe thermal, mechanical and chemical conditions. The turbine technology depends on the development of suitable materials that can operate in extreme working environments. Of all the factors acting simultaneously and causing wear, the most important is thermal shock. In the case of the hot parts which compose the turbines, the temperature varies depending on operating conditions: takeoff, landing, cruising speed, landing failure, engine shutdown during flight, etc. The use of thermal barrier coatings allows the increase of the operating temperature without increasing the temperature of the base material and the reducing of the amount of air necessary for cooling.

Taking into account the extreme operational conditions of the "hot parts" of the turbo engine it is necessary to study the material behavior at high rate speeds of heating-cooling cycles and thermal shock.

2. MATERIALS, METHODS AND INSTRUMENTATION

For experiments the following multilayer samples were used:
- Nimonic 90 support;
- NiCrAlY bonding layer (AMDRY 962) having as chemical composition Ni,Cr,Al,Y;
- $ZrO_2/20\% Y_2O_3$ TBC layer;
- The thickness of the TBC deposition is between 0,15 and 0,47 mm;
- The samples have rectangular shape with the following dimensions 2.15x30x50 mm; 2.28x30x50 mm; 2.47x30x50 mm.

The protecting layers were obtained by successive deposition. Both the bonding layer and the ceramic layer were deposited by atmospheric plasma spraying method on a 7MB type METCO installation. The spray parameters used for the deposition installation are presented in Table 1.

Table 1: Parameters of deposition

Technological parameters	NiCrAlY	YSZ (202NS)
Spray distance, (mm)	120	120
Injector	1,8	1,8
Plasma gas intensity, (A)	600	600
Arc voltage (U)	62	65
Speed of rotation (rot/min)	55	55
Argon flow (m^3/h)	50	40

In the picture below is presented the QTS2 installation, designed and built by INCAS (National Institute for Aerospace Research "Elie Carafoli") for testing materials in extreme heating-cooling cycle conditions, Figure 1.

Figure 1: QTS2-Installation for material testing in extreme thermal conditions

The Quanta 200 3D electron microscope was used to perform secondary electron images and EDS analysis, working in the Low Vacuum module at pressures ranging from 50 to 60 Pa and using the LFD (Large Field Detector) detector. The voltage used to accelerate the electron beam had the value of 30kV and a working distance varied from 12 to 15 mm.

3. THERMAL SHOCK RESISTANCE TEST

The aim of the thermal shock resistance test is to reveal micro structural changes of the tested samples. The thermal shock test is completed when macroscopic exfoliation appears and the damage is more than 25% of the TBC surface [1]. The thermal cycling has been performed at 1200°C temperature. The tested samples coating thickness was of 100 µm, 200 µm and 400 µm. The samples were tested to a sufficient number of cycles so the coating is exfoliated.

The oven is heated at the test cycling temperature. The sample is moved from the environment temperature into the oven. The heating speed of the specimen is variable depending on the specimen size, type of material, single layer or multilayer. The specimen is moved from inside the oven to the cooling area where is cooled till about 40°C.

In Figure 2 are presented images of the samples with the three thicknesses, before and after the thermal shock test.

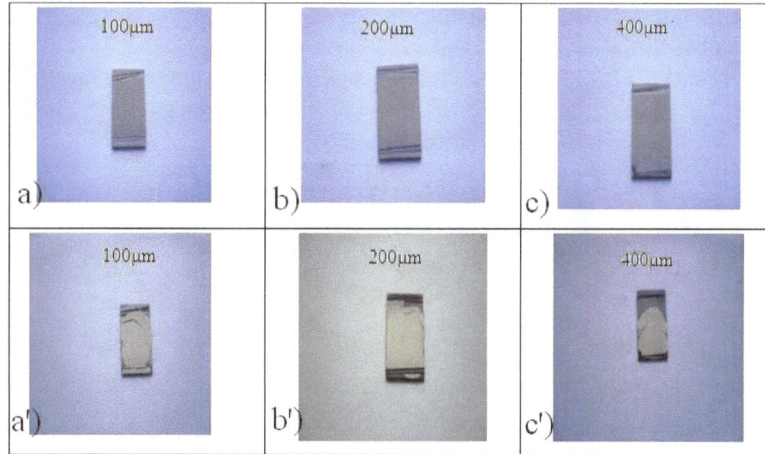

Figure 2: Samples with the coating thickness of 100 μm, 200 μm and 400 μm before and after the thermal shock test at 1200 °C: a;b;c- before thermal shock test; a';b';c'- after thermal shock test at 1200°C

In Figure 3 is shown the resulting chart after the thermal shock test at a temperature of 1200 °C, and in Figure 4 the images captured during the heat shock test using the Lab View software for the sample with the coating thickness of 100 μm.

The parameters used for the rapid thermal shock test are: the average speed of heating of the sample in the first 10 seconds is which 75,57 °C/s, the average speed of cooling of the sample in the first 10 seconds is which 69,34 °C/s for 60 s of cooling, while maintaining the sample in the oven 5 minutes and the duration of the test in of 6 minutes. The maximum pressure of the compressed air during cooling is of 8,7 bar and the minimum pressure has the value of 7,13 bar.

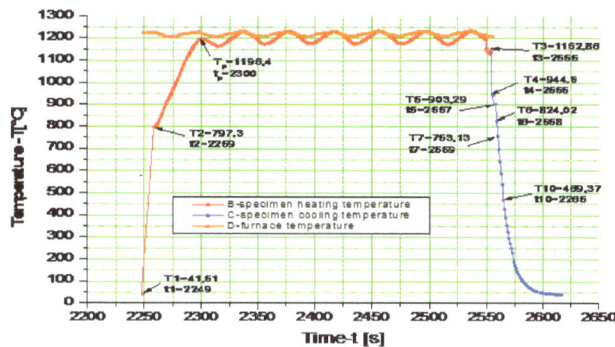

Figure 3: The chart resulted from the thermal shock test at 1200 °C for the sample with the coating with the thickness of 100 μm

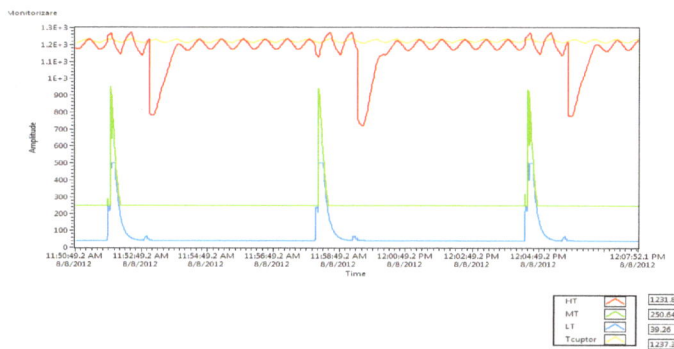

Figure 4: Images captured with the Lab View software during the thermal shock test

Following the thermal shock test electron microscopy investigations were performed and compared to the untested samples in order to observe the structural changes that occurred.

In Figure 5-a is presented a SEM cross-section image of a blank sample where the good adhesion between coating and substrate can be observed. Following the heat shock test at the temperature of 1200 °C on the sample a crack was identified at the interface between coating and substrate (Figure 5-b) [2].

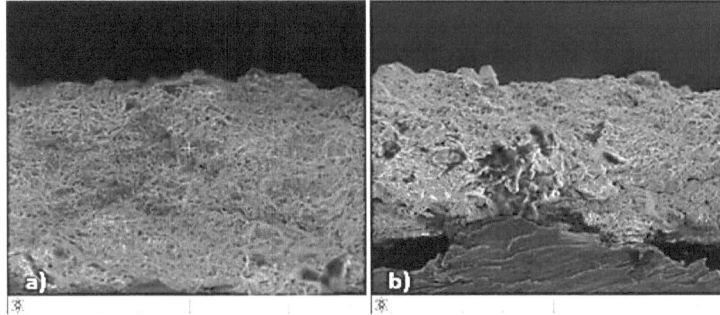

Figure 5: SEM images in cross-section of the YSZ coating:
a) before the thermal shock test; b) after the thermal shock test

Making comparison between the microstructure of the deposited layer in cross-section on the witness sample and the one of the sample subjected to thermal shock at the temperature of 1200 °C, it can be concluded that a sintering of the splats had occurred due to temperature (Figure 6-a/b).

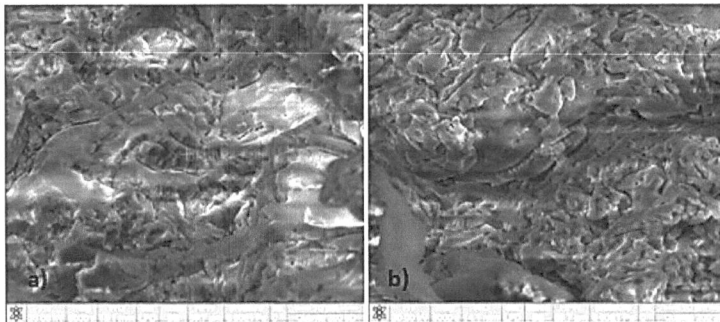

Figure 6: SEM images in cross-section of the YSZ coating:
a)   before the thermal shock test; b) after the thermal shock test.

In Figure 7 is shown the chart after the thermal shock test at a temperature of 1200 °C and in Figure 8 the images captured during the heat shock test using the Lab View software on the second sample with the thickness of 200 µm.

The parameters which were used on the installation for the thermal shock test are: the average speed of heating in the first 10 seconds of the sample is which 70,25 °C/s, the average speed of cooling of the sample in the first 10 seconds is which 69,52 °C/s and the time of cooling is which 60 s. The maintaining time in the oven of the sample is 5 minutes. The test duration is 6 minutes, the highest pressure of the compressed air during the cooling process is 8,7 bar and the minimum pressure is 7,13 bar.

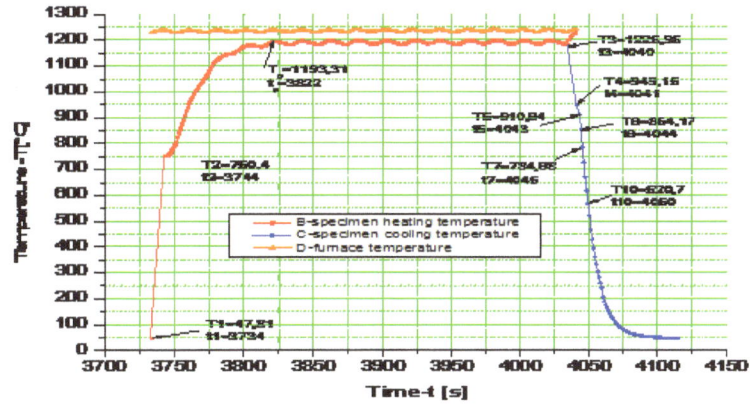

Figure 6: The resulted heat shock graph conducted at a temperature of 1200 ° C for the sample with the coating thickness of 200 μm

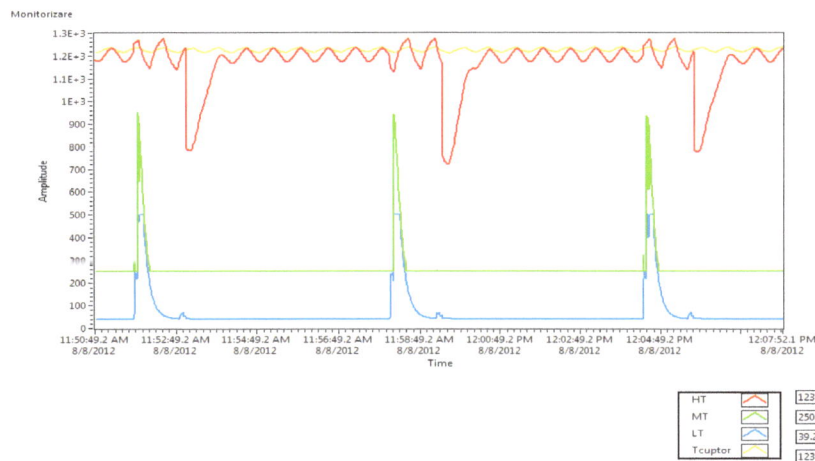

Figure 7: Lab View captured image during the thermal shock test

In Figure 8-a are presented the performed investigations using electron microscopy on the sample with the ceramic layer with a thickness of 200 μm. It can be seen from the SEM cross-section image of a blank sample the good adhesion of the layer with the substrate [3]. Following the thermal shock test a crack appeared at the interface between coating and substrate of about 50 mm, Figure 8-b. The columnar structure observed in Figure 9-b indicates the preferred growth direction of the splats after the thermal gradient direction formed during the thermal shock test cycles.

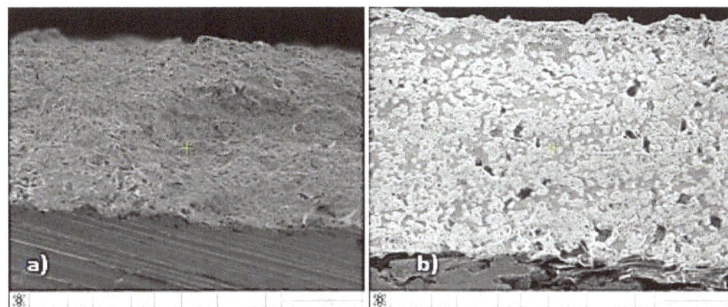

Figure 8: SEM images in cross-section of the YSZ coating:
a) before the thermal shock test; b) after the thermal shock test.

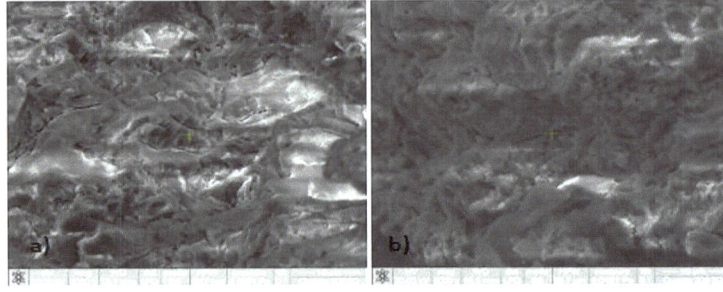

Figure 9: SEM images in cross-section of the YSZ coating:
a)   before the thermal shock test; b) after the thermal shock test.

Figure 10 shows the chart after the thermal shock test at a temperature of 1200 ° C and in Figure 11 the image captured during the heat shock test using the Lab View software on the third sample.

The parameters used for the thermal shock test are: the average speed of heating of the sample in the first 10 seconds which is 69,76 °C/s, the average speed of cooling of the sample in the first 10 seconds which is 60,52 °C/s and the time for cooling which is 60 s. The sample was maintained in the oven for 5 minutes and the test lasted 6 minutes. The highest pressure for the compressed air used for cooling is 8,7 bar and the minimum pressure is 7,13 bar.

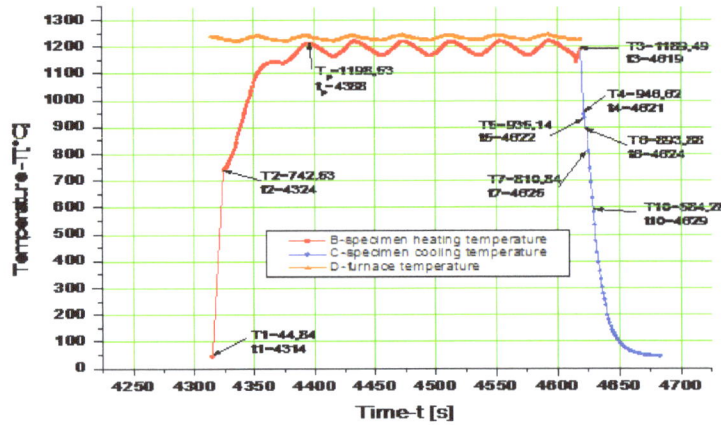

Figure 10: The graph resulted from the heat shock test conducted at a temperature of 1200 ° C for the sample with the coating thickness of 400 μm

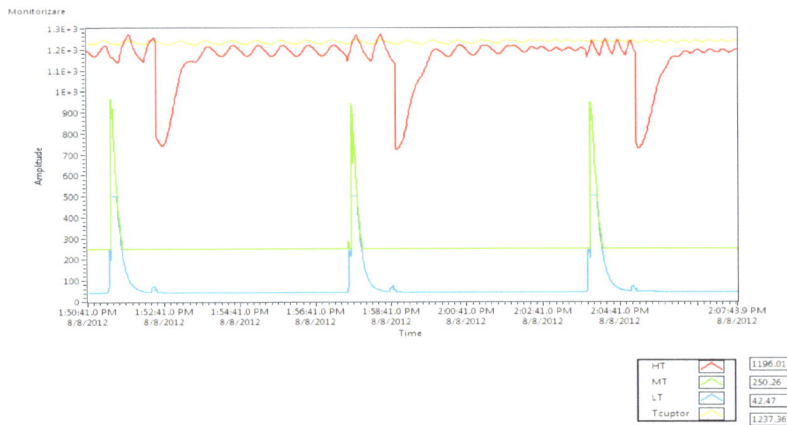

Figure 11: Lab View image captured during the thermal shock test

In Figure 12 are images of electron microscopy investigations of the sample with the coating thickness of 400 μm subjected to the thermal shock test compared with the control sample.

In Figure 12 shows the SEM cross-section image of the blank sample with the coating thickness of 400 μm. The good adhesion of the layer with the substrate can be observed from the picture. Following the heat shock performed at a temperature of 1200 °C, in the sample a crack parallel to the surface of the substrate can be identified which can also be caused by cutting of the sample (Figure 12-b). The crack opening is about 100 μm. Also a diffusion between the ceramic layer and the substrate may also be observed.

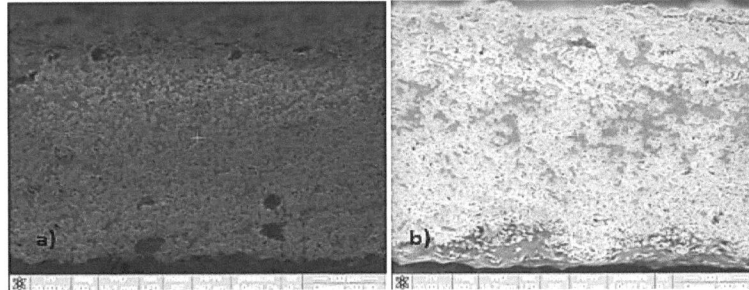

Figure 12: A SEM cross-sectional image which shows the microstructures of the YSZ surface layer:
a) before the heat shock test, b) after the heat shock test at a temperature of 1200 °C.

Because of the surface diffusion between the two deposited layers, isolated areas can be observed with splats which are specific to the $ZrO_2/20\%$ Y_2O_3 ceramic layer (Figure13).

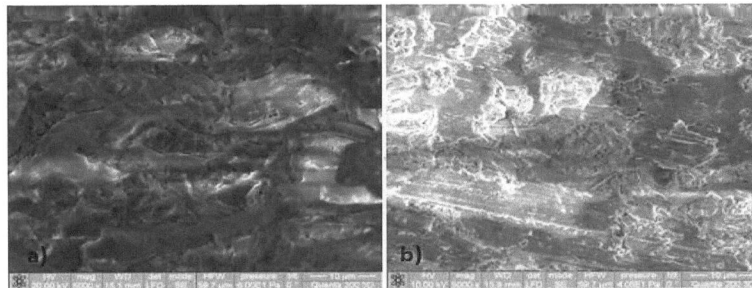

Figure 13: A SEM cross-sectional image which shows the microstructures of the YSZ surface layer:
a) before the heat shock test, b) after the heat shock test.

4. CONCLUSIONS

It could be concluded that after the thermal shock test for the three samples the one with the coating thickness of 400 μm behaved better for both heating and cooling than the other two layers with different thickness, due to a better insulation. Also the samples with the coating thickness of 100 μm and 400 μm lasted for 14 test cycle in comparison to the one with the thickness of 200 μm which lasted for only 11 cycles.Taking into account the operating conditions of the gas turbo engines in terms of thermal shock a coating with the thickness of 400 μm is more effective.

ACKNOWLEDGEMENTS

We acknowledge that this paper was realized with the support of **EURODOC** "Doctoral Scholarship for Research Performance at European level" project financed by the European Social Found and Rumanian Government.

We thank to the **INCAS** (National Institute for Aerospace Research "Elie Carafoli") research center for their support in performing the thermal shock tests.

We thank to **SC Plasma Jet SA** for the support in the plasma jet deposition of the analyzed layer from this paper.

REFERENCES

[1]. ISO 14188 N985

[2]. Bangdao Chena, Hongzhong Liua,*, Hongtao Wangb, Fan Fana, Li Wanga, Yucheng Dinga, Bingheng Lua, Thermal shock induced nanocrack as high efficiency surface conduction electron emitter, Elsevier B.V. All rights reserved, 2011.

[3]. Zhang Lingfeng1, a, Wang Yanzhi1, b, Xiong Yi1, c and Lu Jinzhong2, d, Phases Transformation Analysis   of Zro2 Ceramics after Laser Shock Processing, Trans Tech Publications, (2011).

[4]. Jinsheng Xiao1,2, Kun Liu1,2, Wenhua Zhao3 and Weibiao Fu3, Thermal Shock Experiment and   Simulation of Ceramic/Metal Gradient Thermal Barrier Coating, Trans Tech Publications, (2007).

STUDY ON DETERMINING THE MOMENTS REGRESSION RELATIONS AT DRILLING IN MINERAL COMPOSITE MATERIALS WITH 2% CONCENTRATION OF GLASS FIBER

A. Patrascu[1], C. Opran[2]

[1] University POLITEHNICA of Bucharest, Bucharest, Romania,
patrascu_vi_alexandru@yahoo.com, constantin.opran@ltpc.pub.ro

Abstract: Mineral composite with fiber reinforcements consist of two or more physically combined different natural solid minerals with fiber in order to create a new material whose properties are superior to those of the original material. Mineral composite materials raise several problems when they are machined. The key issues appear when the tool enters and exits from the drilled material. This article presents the obtained results and their evaluation for drilling in mineral composite materials with 2% concentration of glass fiber. The objective of this research is to obtain the dependence mathematical relation for cutting moments as a function of the cutting process parameters.

Keywords: drilling, mineral composite materials, moments, regression relations.

1. INTRODUCTION

When the mineral composite materials are machined the temperature increases and the material properties changes (changes are small or large) [1]. Therefore, the target is to obtain a balance [2], that don't affect the properties of the mineral composite material, between the drilling tool and the drilled composite material. The control of the process parameters is mandatory [3, 4] when mineral composite materials products are machined.

A lot of applications and products were created in the last years using mineral composite materials.

Bridges represent the products in which mineral composite materials have the most applicability. The JecComposites [5], a worldwide composite material magazine, had presented many articles with the benefits of using mineral composite materials, like: the June 2012 number "China first all composite public bridge built" (the future development of bridge technology), the May 2012 number "Life Insu-Shell wins best of the best Life environments award" (for the contributions to - textile reinforced concrete - a new life cycle).

The study's on the drilling process of composite materials are made to prevent: delamination, fiber pullout, thermal damage and hole diameter error. A study of the dry high speed drilling process was performed by Sanjay Rawat [6] how presented a machinability maps approach, characterizing the drilling process of woven composites. Using genetic algorithms Ramon Quiza Sardinas studied the optimization of cutting parameters for drilling laminate composite materials [7].

Although, experimental investigation has been done for many composite materials [8, 9], because the large variety on the market, a pattern can't be given for the drilling process.

2. EXPERIMENTAL WORK

Experiment programming involves a careful study and well documented of the process studied. This paper uses two specialized software that allow us to choose an experimentation program and the determination of the regression model.

The two software are: REGS (use for the determination of experimental programs, polynomial regression relations, optimization of the relations and static errors) and DOE PRO XL (use for the determination of experimental programs, polynomial regression relations, optimization of the relations, plotting 2D and 3D, Pareto diagram etc.).

During the experiments we used the following equipment:

- the machine tool: drilling CNC machine FIRST MCV 300 with power P = 11 kW, continuous range of revolutions, continuous range of tool travels;

- the cutting tool: DIN 8039 ISO 5468, high-speed SDS, covered with tungsten type YG8 (equivalent ISO K20) and the diameters: ø5, ø8 and ø12, peak angle 160°, producer ALPEN/MAZKESTAG;

- the processed material: mineral composite material reinforced with 2% concentration glass fiber, concrete matrix B 250, glass fiber type E (OCV^{TM} – P207);
- dynamometric structure: the drilling moments of the mineral composite material were captured with a Kistler dynamometric structure type-9257B.

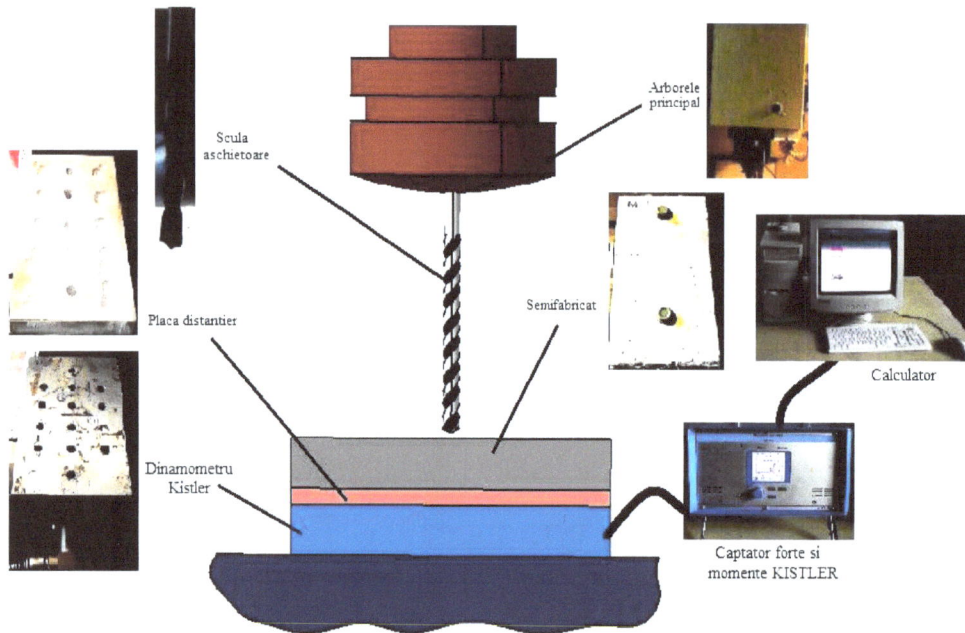

Figure 1: The schematic representation of the determination stand

The experiments were made in the composite laboratories from University Polytechnic of Bucharest faculty of Engineering and Management of Technological System Romania. The schematic representation of the determination stand can be observed in Figure 1.

3. EXPERIMENTAL RESULTS AND DATA PROCESSING

In order to have a good experimental program that we can rely on we have used three independent natural variables with three levels of variation: tool diameter D [mm], feed rate f [mm/rot] and drilling speed v_c [m/min], Table 1.

Table 1: Independent natural variable with three levels of variation

Natural variable	Cod	Codification level		
		-1	0	1
tool diameter D [mm]	X_1	5	8	12
feed rate f [mm/rot]	X_2	0,08	0,12	0,16
drilling speed vc, [m/min]	X_3	9,424	16,86	30,159

For every software used we have created one experimental program. In Table 2 we have the experimental program (with three independent natural variable and three codification levels) for REGS and in Table 3 we have the experimental program (with three independent natural variable and two codification levels) for DOE PRO XL.

Table 2: REGS experimental program

nr. exp. \ X_i	X_1	X_2	X_3
1	1	-1	-1
2	-1	1	-1
3	-1	-1	1

4	1	1	1
5	0	0	0
6	0	0	0

Tabel 3: DOE PRO XL experimental program

X_i nr. exp.	X_1	X_2	X_3
1	-1	-1	-1
2	-1	-1	1
3	-1	1	-1
4	-1	1	1
5	1	-1	-1
6	1	-1	1
7	1	1	-1
8	1	1	1

3.1. Experimental results obtained with REGS

The starting point equation used for REGS software in the analysis for the cutting moments at drilling in mineral composites materials with 2% concentration glass fiber, [10, 11]:

$$M_z = C_M \cdot D^{x_M} \cdot f^{y_M} \cdot v_c{}^{z_M}, \text{ in Nm} \tag{1}$$

where:

M_z – axial moment, in Nm;

D – tool diameter, in mm;

f - feed rate, in mm/rot;

v_c - drilling speed, in m/min;

In order to estimate the C_M constant and the x_M, y_M, z_M polytropic exponents the equation (1) has been linear zed by using the logarithm:

$$lg\, C_M + x_M\, lg\, D + y_M\, lg\, f + z_M\, lg\, v_c = lg M_z \tag{2}$$

In the Table 4 we have the results obtained according the experimental program used for REGS:

Table 4: Experimental results REGS

Experimental program			Tool diameter:	Feed rate:	Drilling speed: v_c,	M_z [Nm]
D	f	v_c	D, [mm]	f, [mm/rot]	[m/min]	
1	-1	-1	12	0.08	9.42	1.07
-1	1	-1	5	0.16	9.42	0.32
-1	-1	1	5	0.08	30.16	0.23
1	1	1	12	0.16	30.16	0.99
0	0	0	8	0.12	16.86	0.55
0	0	0	8	0.12	16.86	0.55

After the regression analysis of REGS software we have obtained the next regression relations at drilling in mineral composite materials with 2% concentration of glass fiber:

$$M_z = 0,05818 \cdot D^{1,5289} \cdot f^{0,1984} \cdot v_c{}^{-0,18}, \text{ in Nm} \tag{3}$$

3.2. Experimental results obtained with DOE PRO XL

The starting point equation used for DOE PRO XL software in the analysis for the cutting moments at drilling in mineral composites materials with 2% concentration glass fiber, [12]:

$$M_z = a_0 + a_1 D + a_2 f + a_3 v_c + a_{12} Df + a_{13} Dv_c + a_{23} fv_c + a_{123} Dfv_c, \text{ in Nm} \tag{4}$$

where:

M_z – axial moment, in Nm;
D – tool diameter, in mm;
f - feed rate, in mm/rot;
v_c - drilling speed, in m/min;
a_0, a_1, a_2, a_3, a_{12}, a_{13}, a_{23}, a_{123} – politropic exponets
In the Table 5 we have the results obtained according the experimental program used for DOE PRO XL:

Table 5: Experimental results DOE PRO XL

Experimental program			Tool diameter: D, [mm]	Feed rate: f, [mm/rot]	Drilling speed: v_c, [m/min]	M_z [Nm]
D	f	v_c				
-1	-1	-1	5	0.08	9.42	0.28
-1	-1	1	5	0.08	30.16	0.23
-1	1	-1	5	0.16	9.42	0.32
-1	1	1	5	0.16	30.16	0.26
1	-1	-1	12	0.08	9.42	1.07
1	-1	1	12	0.08	30.16	0.88
1	1	-1	12	0.16	9.42	1.23
1	1	1	12	0.16	30.16	1.01

After the regression analysis of DOE PRO XL software we have obtained the next regression relations at drilling in mineral composite materials with 2% concentration of glass fiber:

$$M_z = -0.259 + 0.103D - 0.595f + 0.002v_c + 0.230Df - 0.0008Dv_c + 0.0025fv_c - 0.0017Dfv_c, \text{ in Nm} \quad \textbf{(5)}$$

4. ANALYSIS OF THE RESULTS

The diagrams of the variation of the moment are shown in Figures 2 to 7, according the regression relations determined (3) and (5). These only apply to mineral composite materials with 2% concentration glass fiber.

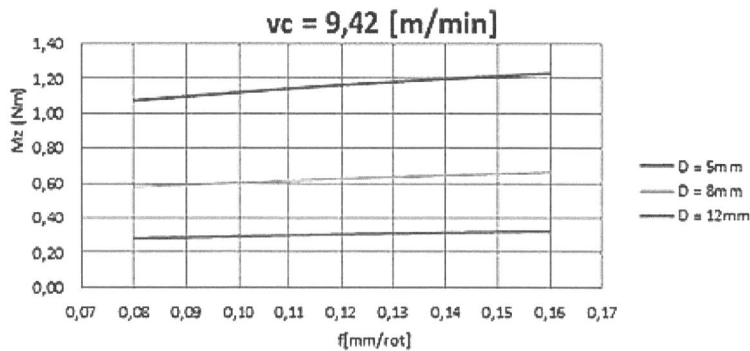

Figure 2: The variation of moment as a function of feed rate, for different tool diameter, (v_c=constant)

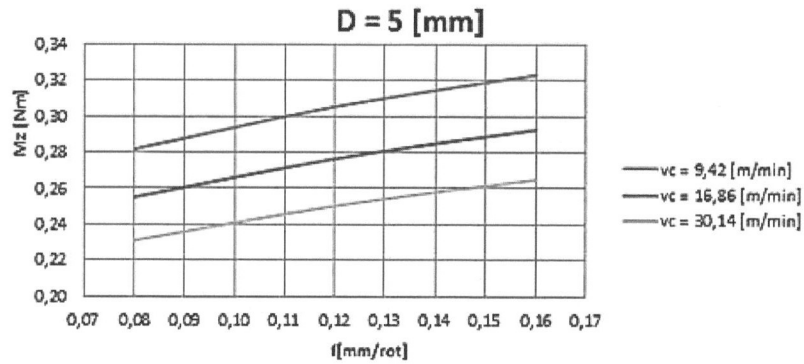

Figure 3: The variation of moment as a function of feed rate, for different drilling speed, (D=constant)

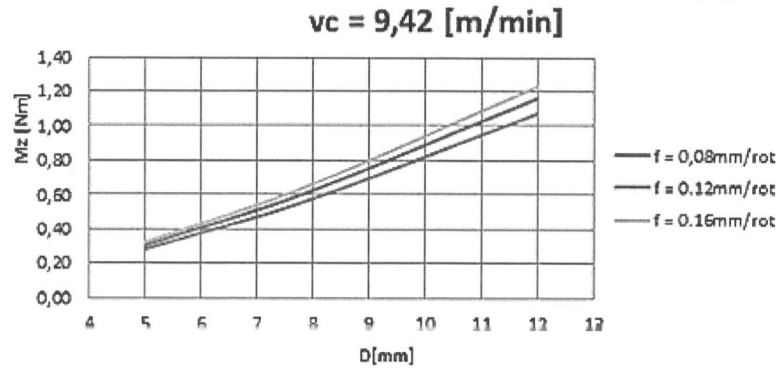

Figure 4: The variation of moment as a function of tool diameter, for different feed rate, (v_c =constant)

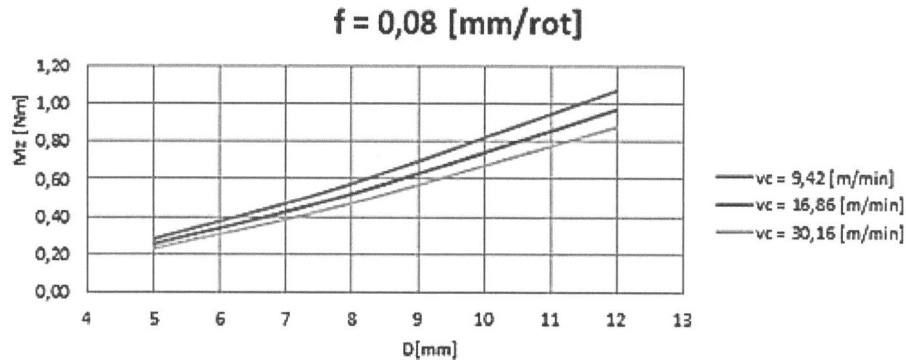

Figure 5: The variation of moment as a function of tool diameter, for different drilling speed, (f =constant)

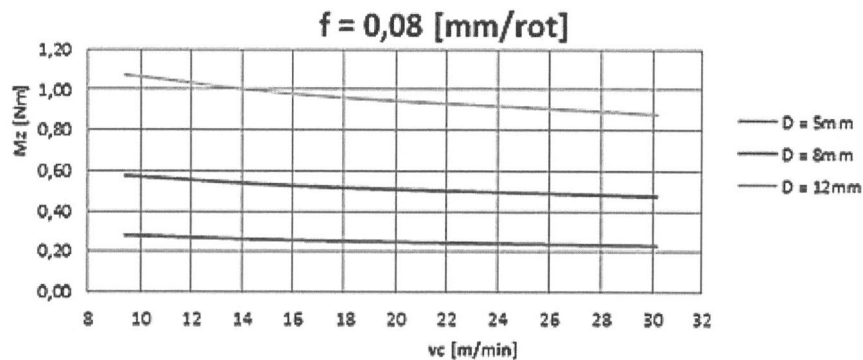

Figure 6: The variation of moment as a function of drilling speed, for different tool diameter, (f =constant)

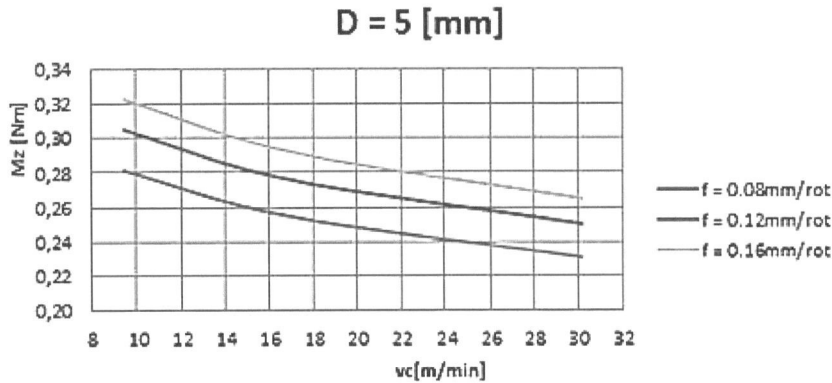

Figure 7: The variation of moment as a function of drilling speed, for different feed rate, (D =constant)

Analiasing the 2D figure of dependence, from Figure 2 to 7, betwen the drilling moment $M_z[Nm]$ and a parameter of the drilling process the others remeining constant, for products made from mineral composite materials with 2% concentration glass fiber we can see:
- the drilling moment $M_z[Nm]$ is increesing with the tool diameter and the feed rate;
- when the drilling speed is increasing the values recorded of the drilling moment $M_z[Nm]$ are decresing.

5. CONCLUSION

This paper presents a study on determining the moments regression relations at drilling in mineral composite materials with 2% concentration of glass fiber. Analyzing the results obtained we can see that:
- the drilling moment $M_z[Nm]$ is increesing with the tool diameter and the feed rate;
- when the drilling speed is increasing the values recorded of the drilling moment $M_z[Nm]$ are decresing;
- all the parameter taken in consideration have semnificativ influence on the values registreted;
- the process parameters with bigest influence are de tool diameter, the feed rate and the drilling speed;
- the drilling moments values recorded are similar to the ones of other types of composite materials (biocomposite, with polimeric matrix etc.).
The results presented in this paper can be taken into consideration in the educational studies and in the theoretical technical research. Also, they can be implemented in the manufacturing activity of these materials.

ACKNOWLEDGEMENTS

The work has been funded by the Sectoral Operational Programme Human Resources Development 2007-2013 of the Romanian Ministry of Labour, Family and Social Protection through the Financial Agreement POSDRU/88/1.5/S/61178.

REFERENCES

[1]   Patrascu Alexandru, Opran Constantin, Gheorghe Tudor Vlad: Research regarding the influence of r reinforcement elements to the vibration characteristics of the mineral composite products, 3rd   International Conference – Advanced composite materials engineering, COMAT 27-29 October 2010, Brașov, Romania.
[2]   Bivolaru Catalina,Opran Constantin, Murar Diana: Milling of polymeric sandwich composite products,   22nd DAAM International, World Symposium, Uno City-Austria, Vienna, 2011.11.23\26, pp. 1295-   1296.
[3]   Murar Diana, Opran Constantin, Bivolaru Catalina: Drilling of biocomposite polymeric products, Annals   of DAAM for 2011 & Proceedings of 22nd International DAAM Symposium, Volume 22, No1,   ISSN   1726-9679, DAAM International, Vienna, Austria.
[4]   Bivolaru Catalina, Opran Constantin, Murar Diana: Milling of polymeric sandwich composites   products; 22nd DAAM INTERNATIONAL; World Symposium, Uno City-Austria, Vienna, 2011-11-   23/26; pp.1295-1296.
[5]   JecComposite, http://www.jeccomposites.com/news, accessed September 2012.

[6]   Sanjay Rawat, Helmi Attia: Characterization of the dry high speed drilling process of woven   composites using Machinability Maps approach, Elsevier, CIRP Annals – Manufacturing Technology   58, 2009, pp. 105–108.

[7]   Ramon Quiza Sardinas, Pedro Reis, J. Paulo Davim: Multi-objective optimization of cutting   parameters for drilling laminate composite materials by using genetic algorithms, Elsevier, Composites   Science   and Technology 66, 2006, pp. 3083–3088.

[8]   Constantin OPRAN: Research concerning polymeric laminate composite materials products under impact conditions; PROCEEDINGS CNC TEHNOLOGIES 2011; Fifth Edition; ISSN: 2068-2093; Politehnica Press; Bucharest; May 12-13; 2011; 5pag; pp.115-120.

[9]   TSAI S.W.: Strength & life of composites, Editor Aeronautics & Astronautics Stanford University, SUA, 2008.

[10]   Adriana DAMIAN, Ovidiu BLĂJINĂ, Aurelian VLASE: Determination of the dependence relation for   the forces at drilling in composite materials with polymeric matrix and 24% concentration of glass   fibers,   U.P.B.   Sci. Bull., Series D, Vol. 73, Iss. 2, 2011, ISSN 1454-2358;

[11]   Adriana DAMIAN, Ovidiu BLĂJINĂ, Aurelian VLASE: Determination of the dependence relation for   the moments at widening a 36% glass fiber reinforced polymeric composite material, U.P.B. Sci. Bull.,   Series D, Vol. 73, Iss. 3, 2011, ISSN 1454-2358;

[12]   Murar Diana Anca: Cercetări privind prelucrabilitatea prin găurire a unor produse din materiale   biocompozite polimerice, teză de doctorat conducător ştiinţific Prof. dr. ing. Aurelian VLASE;   POLITEHNICA   Bucuresti, 2010, România;

PARTICULAR BEHAVIOUR OF CERTAIN COMPOSITES RELATED TO BASIC DAMAGE DETECTION TECHNIQUES

S. Nastac[1]

[1] "Dunarea de Jos" University of Galati, Engineering Faculty in Braila, Research Center for Mechanics of Machines and Technological Equipments, Braila, ROMANIA, e-mail: snastac@ugal.ro

Abstract: *This paper deals with dynamic damage detection involved on composites based structures analysis. It contains a briefly presentation of some particular and inadequate aspects of a certain composites regarding their ordinary behaviour under dynamic loads during the exploitation time and the implication on the damage monitoring process. The particularities of demotion processes and the selectiveness of damages initiation/developing into the basic array materials can lead to inappropriate evolution during exploitation for some structural systems subjected to external dynamic loads. In these cases the damage process monitoring can be significant affected by missing of essential information related to qualitative and quantitative changing of the dynamic parameters. The previous hypothesis must be added by taking into account the serviceable and the disposable means used for system status evaluation.*

Keywords: *damage, detection, composites, structure, dynamics.*

1. INTRODUCTION

Monitoring of bridges health level and evaluation of their structural integrity necessitate an appropriate method which has to be able to assure a set of essential requirements as follows: evaluation of main parameters regarding the bridge dynamics and significant changes of these during the exploitation time, prominence of direct influence of ageing about the damage imminence both into the deck bridge structure, and into the bearings, prominence of external perturbation factors influences about the bridge structure and the dynamic isolation systems, evaluation of global damage level in respect with its reference limit and mean time to fail values. Final purpose of these evaluations consists to an efficient maintenance procedure especially for dynamic isolation devices.

The analysis proposed in this paper start from the basic method of the impact test of deck bridges with truck tip movement over a regular bump. The instrumental tests of dynamic behaviour dignify the spectral characteristic of the structure and reveal the possible resonance area. Taking into account the rubber-based bearings it offers the premises to evaluate the isolation performances against vibration and shocks due to road traffic and also seismic actions [1...7].

Utilization of truck tips for dynamic excitation is able to induce different forces into the bridge structure as a function of truck masses geometry and loads. This type of external excitation can be assimilates such a random or such a deterministic perturbation, depending the way to generate the signal. Also, the spectral composition of excitation signal can lead to a weighted response - when the excitation can be assimilates with an impulsive load, or to a harmonic response - when the excitation is suppose to be a simplified (much theoretical) harmonic signal [5...7].

The theoretical approaches in this paper presents uncommon situations which can appears in case of impact test serviceable procedure. It has to be mentioned that these phenomenon can be either efficacious or inefficacious depending the type of excitation and of the response wanted to use into the experimental tests.

2. THE BASIC MODEL

This paragraph presents the basic model of the proposed method for dynamic evaluation of the bridge structural response. The scheme in Fig. 1 depicts the physical model of a bridge section with a truck tip moving over and with a regular bump mounted at a certain distance from one of the section end.

Resulting from the model in Fig. 1 the schematization supposes an elastic beam insulated at both ends with a regular bump mounted at L_{bump} from one of the end. The truck model supposes a rigid mass with two degrees of freedom linked with k_i rigidities by the beam. The position of this mass L_{ex} relative to the beam end is variable in respect with the constant velocity v of the truck. In Fig. 1 schematic diagram $u(x,t)$ denotes the beam deflection, z_i denote the vertical displacements of the wheels on the road surface, y_i are the displacements of the two ends of the truck model, y and φ

denote the independent coordinates of truck mass respectively in its center of gravity situated at a and b distances from each end.

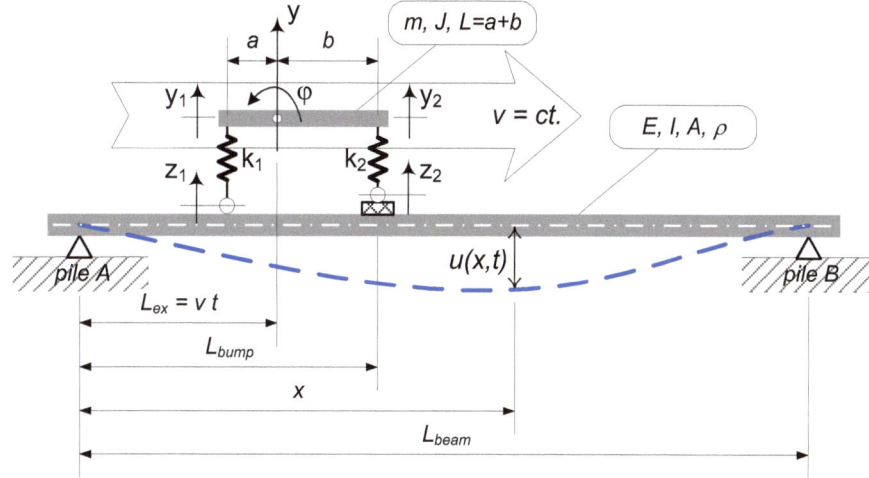

Figure 1: Schematic diagram of deck bridge dynamic testing

The differential equations systems of the model can be assembled based on the equation of transversal wave in beams and using Laplace 2nd order equation for the truck mass. Hereby the movement of the bridge section can be evaluate with the partial differential equation

$$[E(x)\,I(x)\,u''(x,t)]'' + m_v(x)\,A(x)\,\ddot{u}(x,t) = q(x,t) \tag{1}$$

where E, I, m_v, A denote Young's modulus, moment of inertia, volumetric specific mass, respectively sectional area of the deck bridge as variable parameters in respect with longitudinal position x, and $q(x,t)$ denotes the external excitation term.

The movement of the truck tip over the bridge section can be simulated with the next differential equations system

$$\begin{cases} m\,\ddot{\delta}_i\,\dfrac{L-\eta_i}{L} + k_i\delta_i = -m\left(\ddot{z}_i\,\dfrac{L-\eta_i}{L} + \ddot{u}(x_i,t)\right) \\[2mm] \dfrac{J}{L}(-1)^i\ddot{\delta}_i + (-1)^i k_i\,\eta_i\,\delta_i = -\dfrac{J}{L}(-1)^i(\ddot{z}_i + \ddot{u}(x_i,t)) \end{cases} \tag{2}$$

where m, J, L denote the mass, moment of inertia, respectively total length of the truck tip, η_i is the distance between the center of gravity and each end of truck, and δ_i denote the deformations of each wheel. The equations of system (2) suppose summation for $i=1,2$ according the two axles of the truck. Simplified notations in (2) have the significance as follows

$$\begin{cases} a = \eta_1 \\ b = \eta_2 = L - \eta_1 \\ \delta_i = y_i - z_i - u(x_i,t)\big|_{i=1,2} \end{cases} \tag{3}$$

The external perturbation function $q(x,t)$ contains both static and dynamic terms, and assure the linkage between the beam evolution and the truck dynamics. The expression of $q(x,t)$ can be written as follows

$$q(x,t) = q_B(x,t) + q_{ex\_st}(x,t) + q_{ex\_dyn}(x,t) =$$

$$= m_v(x)\,A(x)\,g + \begin{cases} \dfrac{mg}{L}(L-\eta_i) + \dfrac{k_i\delta_i}{L}, & x = L_{ex} \mp \eta_i \\ 0, & otherwise \end{cases} \tag{4}$$

The terms in (4) correspond by beam weight, static and dynamic external loads, and also respect the summation for $i=1,2$ according the two axles of the truck.

The hypothesis of constant characteristics of material and masses geometry for bridge section leads to a simplified form of dynamic equations as follows

$$\begin{cases} EI\,(u'')'' + m_v\,A\,\ddot{u} = m_v\,A\,g + \begin{cases} \dfrac{mg}{L}(L-\eta_i) + \dfrac{k_i\delta_i}{L}, & x = L_{ex} \mp \eta_i \\ 0, & otherwise \end{cases} \\[4mm] m\,\ddot{\delta}_i\,\dfrac{L-\eta_i}{L} + k_i\delta_i = -m\left(\ddot{z}_i\,\dfrac{L-\eta_i}{L} + \ddot{u}_i\right) \\[4mm] \dfrac{J}{L}(-1)^i\,\ddot{\delta}_i + (-1)^i k_i\,\eta_i\,\delta_i = -\dfrac{J}{L}(-1)^i\left(\ddot{z}_i + \ddot{u}_i\right) \end{cases}$$

(5)

with simplified notations

$$\begin{cases} a = \eta_1 \\ b = \eta_2 = L - \eta_1 \\ \delta_i = y_i - z_i - u(x_i,t)\big|_{i=1,2} \\ \ddot{u}_i = \ddot{u}(x_i,t) \end{cases}$$

(6)

3. THE ANALYSIS OF PERTURBATION

Taking into account the sub-model of the excitation system (truck tip equipment) results that analysis of perturbation signal structure can be developed with the help of two degrees of freedom dynamic model (see Fig.2).

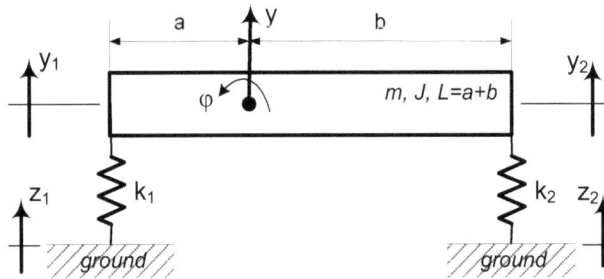

Figure 2: Schematic diagram of perturbation generating process

The natural pulsations of the simplified model like that depicted in Fig. 2 can be evaluated with the homogeneous differential equations system, as follows

$$m\left[\ddot{\delta}_1\,(1-\xi) + \ddot{\delta}_2\,\xi\right] + k_1\delta_1 + k_2\delta_2 = 0$$

$$\frac{J}{L}\left(-\ddot{\delta}_1 + \ddot{\delta}_2\right) - k_1\,L\,\xi\,\delta_1 + k_2\,L\,(1-\xi)\,\delta_2 = 0$$

(7)

where the center of gravity coordinates was written in respect with variable changes as follows

$$\begin{cases} a = L\,\xi \\ b = L\,(1-\xi) \end{cases}$$

(8)

Supposing the solutions of (7) as follows

$$\begin{cases} \delta_1 = a_1 \sin(\omega_n t + \theta) \\ \delta_2 = a_2 \sin(\omega_n t + \theta) \end{cases}$$

(9)

making derivatives of (9), replacing into (7) and grouping the terms results an algebraic equations system with two unknown variables as a_1, a_2

$$\begin{cases} a_1\left[k_1 - m\omega_n^2(1-\xi)\right] + a_2\left(k_2 - m\omega_n^2\xi\right) = 0 \\ -a_1\left(k_1 L\xi - \frac{J}{L}\omega_n^2\right) + a_2\left[k_2 L(1-\xi) - \frac{J}{L}\omega_n^2\right] = 0 \end{cases} \tag{10}$$

which have the non-zero solutions in case of null determinant ($\Delta_* = 0$)

$$\Delta_* = \begin{vmatrix} k_1 - m\omega_n^2(1-\xi) & k_2 - m\omega_n^2\xi \\ -k_1 L\xi + \frac{J}{L}\omega_n^2 & k_2 L(1-\xi) - \frac{J}{L}\omega_n^2 \end{vmatrix} \tag{11}$$

With the ratio between rigidities noted by $\alpha = k_2/k_1$ into (11) results the characteristic equation of natural pulsations in respect with ($m, J, L, \eta, \alpha, k_1$) parameters

$$\omega_n^4 - \frac{k_1 L^2}{J}\left[(\xi-1)^2\alpha + \frac{J}{mL^2}(\alpha+1) + \xi^2\right]\omega_n^2 + \frac{k_1^2 L^2 \alpha}{mJ} = 0 \tag{12}$$

Taking into account the last equation and following the initial purpose of this paragraph results that the inequality between the squared values of natural pulsations can be evaluated as follows

$$\varepsilon = \omega_{n1}^2 - \omega_{n2}^2 = \sqrt{\left(\frac{k_1 L^2}{J}\left[(\xi-1)^2\alpha + \frac{J}{mL^2}(\alpha+1) + \xi^2\right]\right)^2 - 4\frac{k_1^2 L^2 \alpha}{mJ}} \tag{13}$$

Optimization of the mass geometry configuration with respect in position of the center of gravity ξ and stiffness ratio α imposes an appropriate estimation of basic parameters values. Without any significant diminishing of general character of this analysis it will be consider that the truck tip equipment have a mass range between (30000...50000) kg and a wheels rigidity having $k_1 = 100000$ N/m.

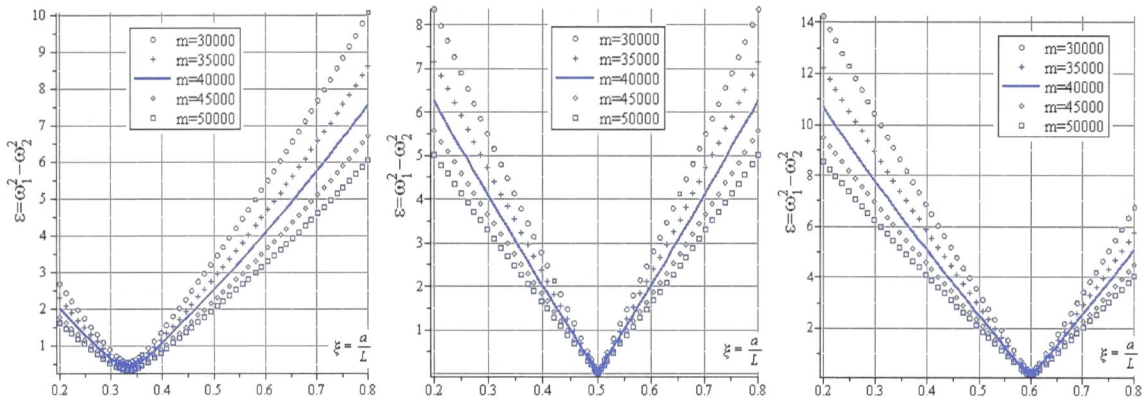

Figure 3: Evolution of natural pulsation difference in respect with ξ and m parameters for (left) $\alpha = 0.5$, (middle) $\alpha = 1.0$ and (right) $\alpha = 1.5$

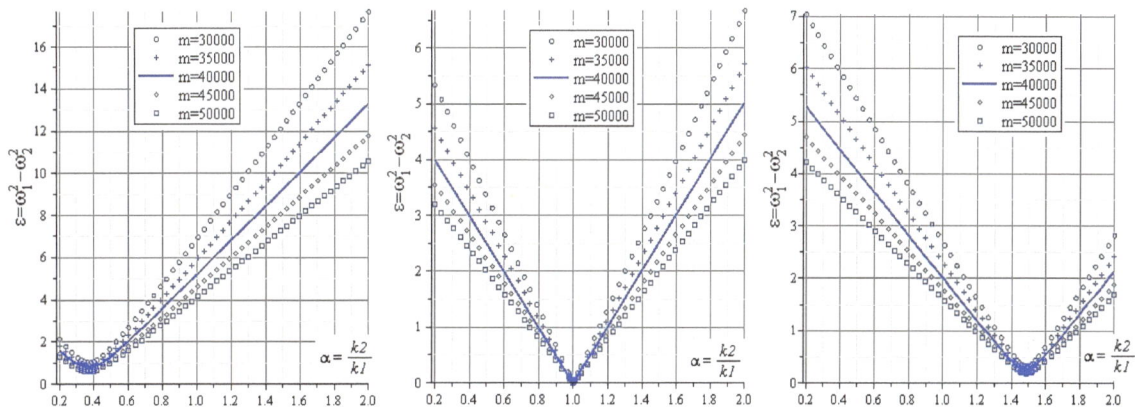

Figure 4: Evolution of natural pulsation difference in respect with α and m parameters for (left) $\xi = 0.25$, (middle) $\xi = 0.5$ and (right) $\xi = 0.6$

Simulations presented in Figs. 3...5 shows the evolution of the natural pulsation difference ε as a function of the relative position ξ, the rigidity ratio α and five main values of mass m.

The diagrams in Fig. 3 show the evolution of the ε parameter in respect with the relative position ξ, according with discrete variation of mass m and rigidity ratio α. These diagrams dignify the system behaviour into the area of small differences of ε and also reveal possible null values of natural pulsation difference.

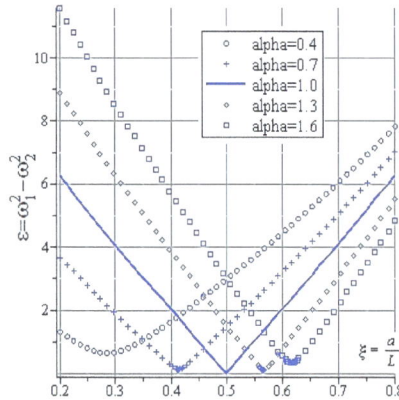

Figure 5: Evolution of natural pulsation difference in respect with (ξ, α) for $m = 40000$ kg

An appropriate analysis with previous case results from the diagrams in Fig. 4. These graphs correspond with the evolution of the natural pulsation difference in respect with the rigidity ratio α, according with discrete variation of mass m and relative position ξ.

According the previous analysis the diagram in Fig.5 presents the evolution of the natural pulsation difference ε parameter as a function of the relative position ξ and the rigidity ratio α for mass m of 40000 kg. This value was adopted according with the total mass of the truck tip used into the instrumental tests. It had to be mentioned that the entire simulation procedure supposing the rigidity ratio α into a range of (0.2 ... 2.0) and the relative position ξ having limit values of 0.2, respectively 0.8. Greater values of rigidity ratio have not a serviceable purpose.

4. RESULTS AND DISSCUSSIONS

The analysis of the natural frequencies dispersion and distribution of the excitation system is very useful to evaluate the characteristic of the perturbation type into the bridge and also the dynamic behaviour of the deck bridge and its bearings right away the impulsive load have been produced. The truck tip equipment had to working in post-resonance domain according with its main purpose into this analysis that is the excitation system. Also the impulsive test of the deck bridges have not to affects this kind of excitation system.

One of the basic ideas of this study supposes to restrain the spectral spilling of the natural frequencies of excitation. Hereby it will reduce the width of possible resonance area. The entire study follows the hypothesis of best minimizing of the natural pulsations difference parameter ε thus that for a certain reduction of the reference pulsation it will be simply left-side shifted the entire resonance area. In presented case the rigidity k_1 is the appropriate element to do this intention through different technical means.

According these hypotheses lets briefly evaluate the presented results. From the diagrams depicted in Figs. 3 to 5 results that unitary ratio of stiffness and symmetrical center of gravity provides a superposition of both natural pulsations. It also results than greater masses imply reduced values of ε on the entire simulated domain. The direct linkage between stiffness ratio and relative position of the center of gravity (CG) of truck tip equipment is mainly sustained by the graphs in Fig. 5. It can be observed that in the same time with stiffness ratio α reduction, results also a reduction of the CG relative position ξ for the minimum value of ε parameter. But the absolute minimum value of ε increases in respect with dispersion of α parameter from unitary value, regardless the dispersion direction. Analysis of the (α, ξ) dependences depicted in Figs. 3 and 4 intensify the previous observations.

5. CONCLUSION

The main purpose of this study consisted by presentation and validation of some practical possibilities to adjust the characteristic of the excitation system usually used in bridge bearings health monitoring procedures without reducing its main capabilities but helping it to work into the appropriate domain regarding initial requirements.

Taking into account the remarks in previous paragraphs it had to be concluded that optimization of main configuration of the masses geometry provides a serviceable way to control dynamic effects induced into a deck bridge and its bearings during the experimental tests and analysis of the damage level.

REFERENCES

85. Bratu P. (2008). Theoretical Mechanics, Impuls House Publishing, Romania.
86. Debeleac C. (2006). Dynamic Behaviour Analysis of Speedy Frontal Loader on View of Qualities Performances Established, Ph.D. Dissertation, Dunarea de Jos University of Galati, Romania.
87. Leopa, A., (2011). The Impulsive Loading Influence on Dynamic Response Parameters of the Viaduct   Type
Structure, The Annals of "Dunarea de Jos" University of Galati, Fascicle XIV Mechanichal Engineering, No2, 2011, ISSN 1224-5615, pp.65-67.
88. Leopa, A., Nastac, S., Dragan, N., Debeleac, C., (2011). Considerations on the Influence of Viscoelastic Behavior of Nonlinear Systems Bearing on the Dynamic Constructions Response, Proceedings of the International Conference on Structural Engineering Dynamics, Tavira, Portugalia, 2011, ISBN 978-989-96276-1-1.
89. Nastac S. (2004). Computational Engineering with Applications, Impuls House Publishing, Bucharest, Romania.
90. Nastac, S. (2008). Analysis of Working Performances at Damaged Vibration Isolation Devices (2008). 0949-0950, Annals of DAAAM for 2008 & Proceedings of the 19th International DAAAM Symposium, pp 475, Editor B. Katalinic, Published by DAAAM International, Vienna, Austria 2008.
91. Nastac, Silviu (2008). Performances Evaluation at Damaged Vibration Isolation Devices, Chapter 47 in DAAAM International Scientific Book 2008, pp.551-564, B. Katalinic (Ed.), Published by DAAAM International, Vienna, Austria.

RHEOCASTING PROCESS APPLIED TO ATSi5Cu1 ALLOY

V. Geamăn

University Transilvania of Braşov, ROMANIA, e-mail: geaman.v@unitbv.ro

Abstract. *Rheocasting process (RC), has been developed for manufacturing near-net shape components of high integrity directly from liquid alloys. The rheocasting process innovatively adapts the well-established high shear dispersive mixing action to the task of in situ creation of semisolid slurry followed by direct shaping of the semisolid slurry into a component using the existing cold chamber die-casting process.*

The rheocast component has close to zero porosity, fine and uniform microstructure throughout the entire component. Compared with those produced by conventional high-pressure die-casting, rheocasting samples have much improved tensile strength and ductility.

In the paper are presented the rheocasting process, the microstructure and mechanical properties of rheocast samples made from AtSi5Cu1 alloy.

Keywords: semisolid slurry, rheocasting process, aluminum alloy.

1. INTRODUCTION

Aluminum alloys, as lightweight structural materials, are playing an important role in achieving vehicle weight reduction and improving fuel economy in the automotive industry. Since 1990, the use of Al has been doubled in cars and tripled in the light truck market. Currently, 85% of all Al alloy castings are used by the automotive and mass transport industry, and a large proportion of such castings are produced by high-pressure die-casting (HPDC) process.

The new processes need to be capable of producing components of high integrity and improved performance while being comparable with the HPDC process in terms of production cost and efficiency.

Porosity due to turbulent mould filling could be reduced or even eliminated if the viscosity of the melt could be increased to reduce the Reynolds number sufficiently so that trapped air is minimized [1, 2]. This is the concept of semisolid metal (SSM) processing.

Since early 1970s, a number of SSM processing techniques have been proposed [2]. One of the most popular SSM processes is thixocasting, in which non-dendrite alloys are pre-processed by electromagnetic stirring and reheated to the semisolid region prior to the shaping process.

As a new processing technique, thixocasting does improve component integrity and performance, but proves to be cost intensive, low efficiency and less flexible. After 30 years of extensive R&D, thixocasting is currently experiencing a decline in acceptance as a viable production technology [3, 4]. Under such circumstances, the new processing concept - rheocasting process, has been developed.

2. EXPERIMENTAL DATA

The process innovatively adapts the well-established high shear dispersive mixing action of the twin-screw extruder (originally developed for polymer processing) to the task of *in situ* creation of SSM slurry with fine and spherical solid particles followed by direct shaping of the slurry into a near-net shape component using the existing cold chamber die-casting process.

The rheocasting process starts from feeding predetermined dose of liquid metal from the melting furnace into the slurry maker where it is rapidly cooled to the SSM processing temperature while being mechanically sheared by a pair of closely intermeshing screws converting the liquid into a semisolid slurry with a pre-determined volume fraction of the solid phase dictated by the barrel temperature. The semisolid slurry is then transferred to the shot chamber of the HPDC machine for component shaping. In order to prevent Al-alloy from oxidation, nitrogen gas is used as the protective environment during the slurry-making process.

2.1. Microstructures

A commercially aluminum alloy ATSi5Cu1 was used in this work. Alloy was melted at 700°C and fed into a slurry maker at the temperature of 40°C above their melting point. A laboratory cold chamber die-casting machine was used for casting the standard tensile test samples. Figure 1 shows the typical microstructures of rheocasted alloy versus the die-casting procedure.

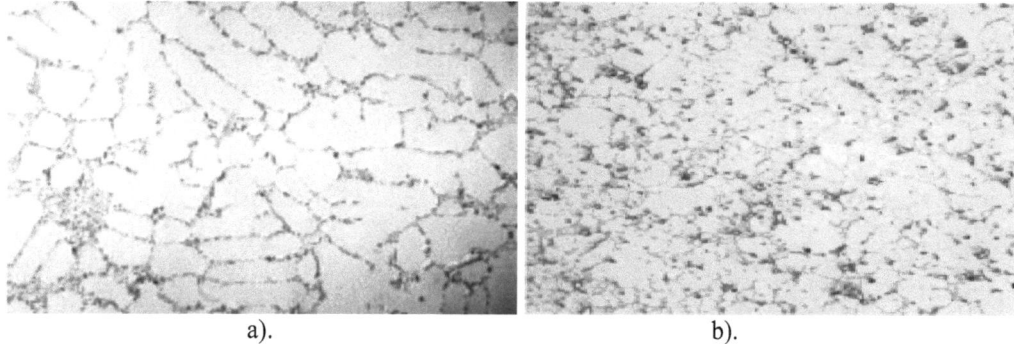

a).   b).

Fig. 1. Microstructures of experimental ATSi5Cu1 alloy,
after die-casting (a) and rheocasting respectively (b).

Detailed micro structural characterization of various rheocast samples has revealed the following micro structural characteristics:
 • Porosity is well below 0.4-0.5 vol.%. pores are rarely observed in the rheocast samples. Occasionally observed pores are small in size.
 • Primary particles have a fine size, spherical morphology and uniform distribution throughout the entire casting.
 • The remaining liquid in the SSM slurry solidifies under high cooling rate in the die resulting in the formation of extremely fine Al-phase (<10μm).
 • Oxide particles are fine (few μm), spherical and well dispersed and uniformly distributed, reducing the harmfulness of oxide particle clusters and oxide film in cast components.
 • Si-rich phases can be dispersed uniformly without any macro-segregation.

2.2. Mechanical properties

A special die was made to cast standard tensile test samples for mechanical testing. Processing parameters, such as screw rotation speed, shearing time, shot velocity, shot pressure, intensifying pressure and die temperature, were systematically varied. The effects of such processing parameters were assessed against sample quality in terms of microstructure and mechanical properties [5].
Table 1 summarizes the mechanical properties of the rheocast ATSi5Cu1 alloy in comparison with those of the same alloys produced by die-casting and sand cast processes. Rheocast ATSi5Cu1 alloy has much improved tensile strength and acceptable ductility.

Table 1. Experimental data with tensile strength, yield strength and elongation results for ATSi5Cu1 alloy.

ATSi5Cu1 – alloy			
Casting process	Tensile strength [MPa]	Yield strength [MPa]	Elongation [%]
Rheocast	252	125	7.8
Die-cast	212	115	6.3
Sand cast	172	92	5.0

3. CONCLUSIONS

The semisolid metal processing technology, rheocasting, has been developed for the production of aluminum alloy components with high integrity.

The rheocast samples have close to zero porosity, fine and uniform microstructure and are free from other casting defects.

Compared with high pressure die-casting or any available semisolid processing techniques, rheocasting offers components with improved strength and ductility, which can be attributed to micro structural refinement and uniformity, much reduced or eliminated porosity and refined and dispersed oxide particles.

Rheocasting process is particularly suitable for production of high safety, airtight and highly stressed components in the automotive industry.

REFERENCES

[1] .Metz S. A. and Flemings M. C. - *AFS Trans.,* Nr. 78 (1970), p. 453.

[2] . Kirkwood D.H. and Kapranos P. - *Proc. 4th Int. Conf. on the Semi-Solid Processing of Alloys and Composites*, Sheffild, England, 1996.

[3]. Chayong S., Atkinson H.V. and Kapranos P. – *Thixoforming of 7075 aluminum alloys*, Materials Science and Engineering, Series A 390 – 2005, p. 3÷12.

[4]. Basner T. – *Rheocasting of Semi-Solid A357 Aluminum,* Delphi Automotive Systems, 2006.

[5]. Zaharia I.I. and Geamăn V. - *Practical aspects regarding to thixoforming process applied to aluminum alloys.* in Advanced Materials Research – Switzerland, Vol. 23/2007, p. 161-164.

CONTRIBUTIONS ON THE INFLUENCE OF MECHANICAL PARAMETERS TO THE QUALITY OF WELDED JOINTS OF SMART COMPOSITES IN ULTRASONIC FIELD

Z. Apostolescu[1], G. Amza[1], S.L. Pais[1], M. Dragomir Groza[1]

Polytechnic University of Bucharest, Romania, amza@camis.pub.ro , zoia@camis.pub.ro.

Abstract: Quality of welded joints in ultrasonic field depends on three parameters: technological, mechanical and acoustic. The paper presents the main mechanical parameters with special influence on the quality of welded joints in ultrasonic field experimental results obtained in welding stops intelligent auto protection of smart bars.
Key words: *parameters, technological, mechanical, acoustic.*

1. INTRODUCTION

Ultrasonic welding process depends on a smart composites multitude of parameters grouped into three categories: technological parameters, mechanical parameters and acoustic parameters. Mechanical parameters are the result of technological scheme used in the design and construction of equipment used in welding. In general, ultrasonic welding machine built allows more welding technological schemes, depending on the shape and size ultrasound system that can be fixed or changed. The main mechanical parameters with significant influence on the quality of welded joints are static pressing force, contact local static pressure, during activation with ultrasound sonotrode shape, form anvil acoustic form factor the concentrator sonotrode and others.

2. THE INFLUENCE OF MECHANICAL PARAMETERS ON THE QUALITY OF THE WELDED JOINT

Mechanical parameters with significant influence on the welding process in the field ultrasonic are static pressing force, local contact pressure surfaces combined ultrasound and activation time.

2.1. Influence of static force push on quality welds with ultrasound smart composites

Theoretical and experimental research revealed that the size of static force push decisively influence the value of the average breaking strength of ultrasonic welds. A_s shown in figure 1, the results obtained in welding ABS matrix smart composite materials there is pressure optimal static force, which depends on the optimal activation time t_s and I sonotrode amplitude (fig. 2) and the density ultrasonic energy./1/, /2/, 17/, /18/.

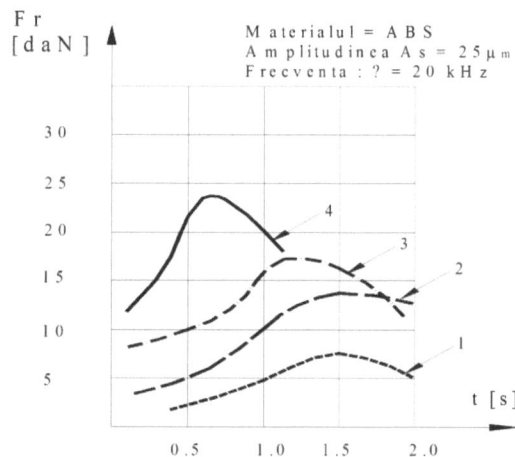

Figure: 1. Influence of static pressure force and duration of ultrasonic activation on resistance welded under the conditions:1 - PS = 5daN, 2 - PS = 10daN, 3 - PS = 15daN, 4 - PS = 20daN.

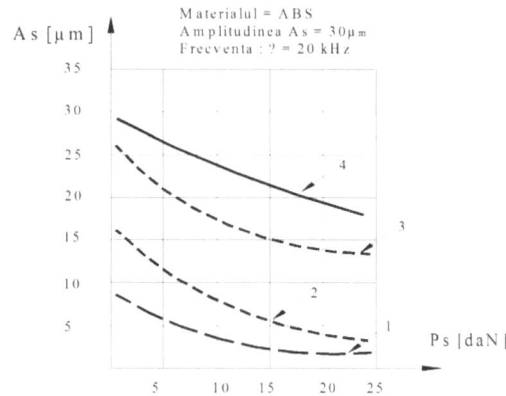

Figure 2: Ps static pressure force influence on the amplitude As, acoustic energy at different densities: 1 to 2 W/cm2, 2 to 4 W/cm2, 3-6 W/cm2, 4-8 W/cm2.

As shown in figure 2 there is a close connection between the pressure and the magnitude of static force active part of all As, in the sense that, as the static pressure force increases, the active part of sonotrode amplitude decreases and overcome the optimal value static pressing force, joint strength decreases substantially. Force static pressure and has a great influence on local contact static pressure (fig. 3), increasing as the power increases static pressure at different densities of acoustic energy./1/. /2/, /4/, /5/, /6/.

The static pressing force is chosen and the thickness of the parts to be joined, with a Fr optimal joint strength (fig. 4) depending on thickness and static power pressing.

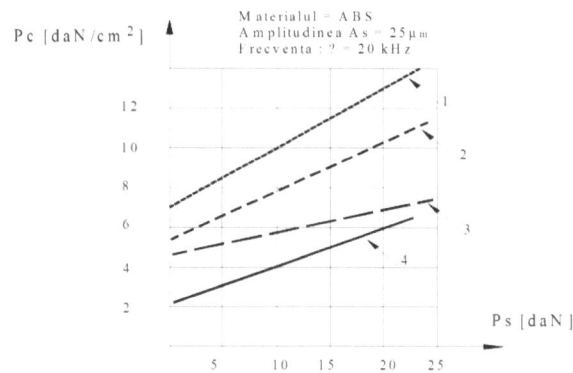

Figure 3: P_s static pressure force influence on local contact pressure P_c at different acoustic energy densities: 1 to 2 W/cm2, 2 to 4 W/cm2, 3-6 W/cm2, 4-8 W/cm2.

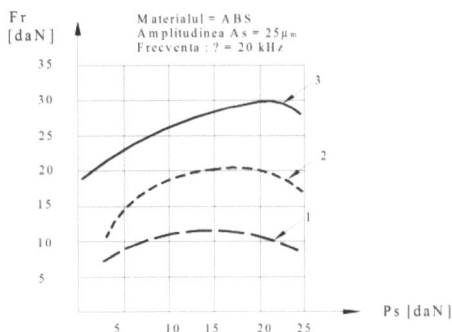

Figure 4: Influence of static pressure P_s force, the joint resistance F_r, in various thicknesses:
1 - s = 1.0+ 1.0, 2 - 2.0 + 2.0 s, 3 - s = 3.0 + 3.0.

2.2. Contact local static pressure influence on the quality of welded joints ultrasonic smart composites

The materials brought to the experimental results obtained revealed that the contact pressure static local variable and static force depends not only on the pressure but also the thickness of the parts to be joined, the size geometric configuration of the combined area. /1/, /16/.

It appears that as they move from stage one to stage two of the welding process, contact the local static pressure increases and decreases contact (fig. 5) and therefore decreases joint strength, above which there is a combination can not be achieved.

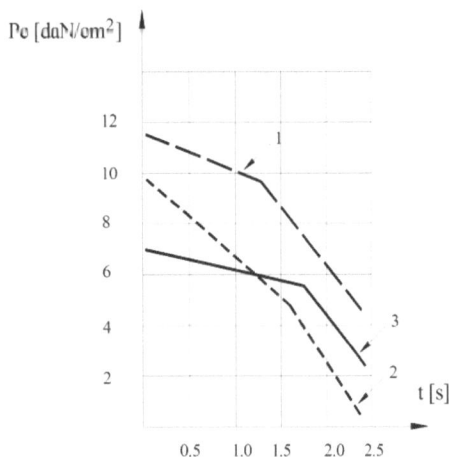

Figure 5: Influence the ultrasonic welding process on the contact pressure pc ts for different plastics:
1 - ABS 2 - High density polyethylene, 3 - polycarbonate.

Contact local static pressure depends on the configuration of the contact area and therefore to be possible to merge the contact area has some form in the first favorable thick pieces welded joint area and the geometric configuration of the functional role of piece (fig. 6). It appears that those acoustic energy concentrators required to be processed on the contact not only is a rapid concentration of heat in local areas but also combining static contact pressure much higher in these areas, pressure decreases as concentrators plastic melt or deform, increasing the contact area. /1/, /12/, /7/, /8/, /13/.

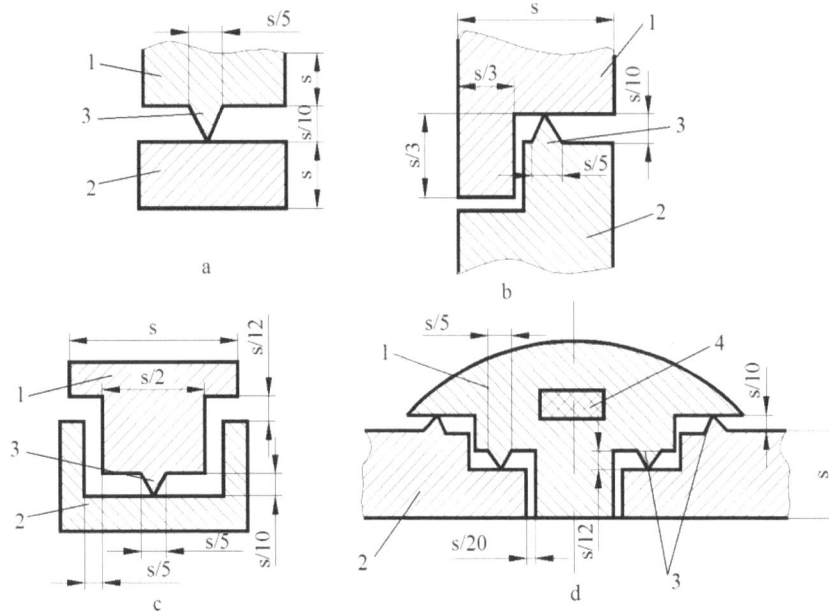

Figure 6: Types of Training contact area depending on the thickness of the welded parts and the local static pressure required contact: a - the interval b - in steps c - channel and wedge d - specifically, 1 - piece top, 2 - play less, 3 - acoustic energy concentrator, 4 - piezoceramic plates

Number, size and distances that put the concentrate acoustic energy and heat that are based on the geometric configuration of the joint and combined nature of the material. Training module shown in figure 6, d   is especially determined to achieve arrest Smart Car Smart Car or a bar, which serves to acoustic and optical signal near the vehicle remotely dangerous./8/, /9/, /10/, /11/, /12/.

2.3. Influence the quality of welded joints welding ultrasonic intelligent composite materials

During welding, which is the ultrasound waves drive the contact area has a great influence not only on the possibility of achieving joint but also the quality and the strength of the welded joint.

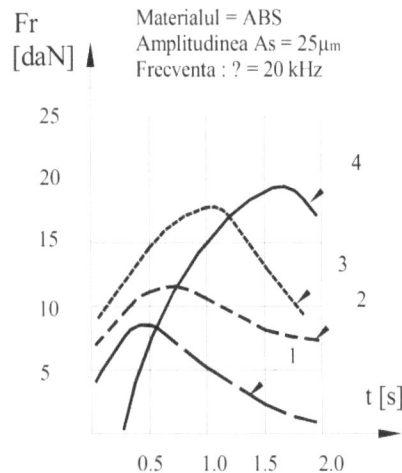

Figure 7: Influence of duration t_s welding, the joint resistance F_r, depending on the thickness of the parts to be joined:1 to 1.0 +1.0, 2 to 2.0+ 2.0, 3 to 3.0+ 3.0, 4 to 4.0+ 4.0.

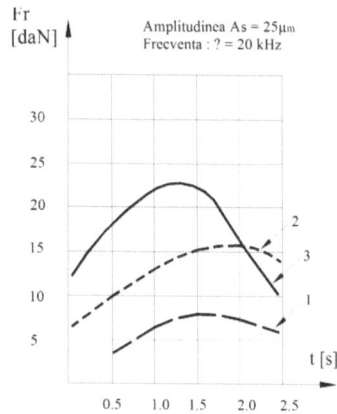

Figure 8. Influence of duration ts welding, the joint resistance F_r, depending on the nature of the materials to be joined: 1 - High density polyethylene, 2 - ABS, 3 - polycarbonate.

Experimental results have shown that the optimal value depending on the length of welding Welding thickness (Fig. 7), the maximum joint strength moving according to the combined thickness reduction. In the determination of optimal welding time should be taken into account the nature of the materials to be welded, activation time is very different depending on the material properties combined (Fig. 8).

Operation during ultrasonic contact area depends not only on the material to be welded and welding parts but thick and static pressure force (Fig. 9) and section sonotrode form in the contact area (Fig. 10).

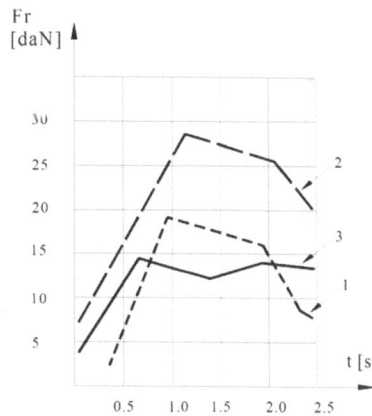

Figure 9: Influence of duration t_s welding, the joint resistance F_r, according to the static pressure force: 1 - P_S = 15daN, 2 - P_S = 25daN, 3 - P_S = 35daN.

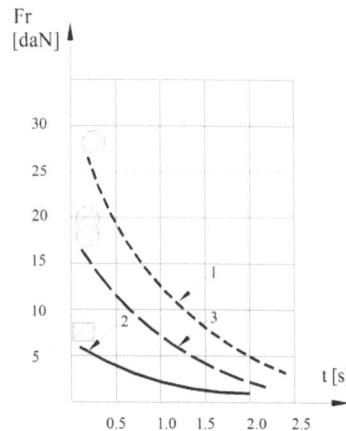

Figure 10: Influence of duration ts welding, the joint resistance F$_r$, depending on the final part of the sonotrode section:1 - Section Round 2 - the square, 3 - section some.

In direct correlation with the speed of flow during welding plastic contact areas as increasing plastic flow velocity in a short time based on the application of acoustic energy in the joint, creating favorable conditions for joint deployment process./11/, /2/, /12/, /13/, /14/, /15/.

Experiments were carried out on machines at different businesses or facilities constructed in the department TMS as prototype models in various composite materials. or different products.

Note! In addition to the parameters mentioned above and treated, the quality of welded joints in composite materials smart ultrasound also depends on other parameters which also include: nature of the material it is made sonotrode, condition and quality of the contact surface sonotrode-piece, state combined cleaning area; nature of the material anvil acoustic condition and lower anvil surface; physical limit of the area of contact of the two materials to be joined, the nature of the environment in which the process of welding and others.

3. CONCLUSIONS

Experimental results in the field ultrasonic welding of automotive taillights smart and intelligent auto bar leads to the following conclusions:

- Development of intelligent composite parts through assembly by welding. Ultrasound technology is very complex because it must take into account a number of factors related to the intelligent processing of composite materials and the mechanical and technological parameters influencing differently acoustic welding process;
- Size static pressure force decisive influence average resistance value. Breaking the welded joints in ultrasonic field there is a optimal value according to nature of the materials to be joined and ultrasonic activation time;
- Experimental results obtained on various smart composite materials were removed shows that local pressures-pressure static pressure but also the thickness of the parts to be joined, dimensions and geometric configuration combined area;
- During ultrasonic activation in the contact area has a great influence not only on joint implementation possibilities but the quality and on resistance welded joint.

REFERENCES

[1]   Amza Gheorghe, Ultrasound of high energy - Ed. Academiei, Bucureşti, 1984.

[2]   Amza Gheorghe, Systems ultraacustice – Ed. Tehnică, Bucureşti, 1989.

[3].   Amza, Gh. Borda C. , Marinescu M., Arsene D., Design of Ring – Type Ultrasonic Motor – Scientific Session of the University "Petru Maior" - Targu Mures, 27-28 oct. 2001, vol.6, pa.7-   14, Petru Maior University Press - Targu Mures , 2001, ISBN 973-8084-10-5, vol.6ISBN 973-   8084-19-0.

[4]   Berlin, A.A., et al. – Principles of Polymer Composite, Ed. Springer – Verlag, New York 1986 (Polymer – Properties and Aplications, vol. 10).

[5]   Dry M. Carolyn, Sottos, R. Nancy, Passive smart self-repair inpolamer matrix composite materials-Smart Structures abd Materials 1993, Albuquerque, N.M. USA- Proceedings of   SPIE – The International Society for Optical Engineering v 1916 1993, Publ by Society of Photo Optical Instrumentation Engineers, Belligham, WA USA, p- 438-444.

[6]   Gandhi M.V. and Thompson B.S. , A New Genertaion of Revolutionary Ultra Advanced Compostes Materials Faturing Electro-Reological Fluide, U.S Army Research Office   Workshop on Smart Materials Structures and Mathematical Issues, 1988.

[7]   Kim, J., Varadan V.V., Varadan V.K. and Bao X.Q., Finite elemente modelling of a smart cantilever plate and comparison with experiments, Smart Materials and Structures, vol. 5, 1996, pag. 165-170

[8]   Lawrence C.M., Nelson D.V., Spingarn J.R. , Measurement of process-induced strain in   composite materials using embedded fiber-optic sensors, SPIE Vol. 2718, San Diego, 1996, pag.60-68.

[9].   M. Surgeon, Continuous damage monitoring technique for laminated CFRP composite   material, Ph. D. Thesis, KU Leuven, 1999.

[10]. Macosko Ch. W., Rheological Changes During Cosslinking- British Polymer Journal, 1985, 17, No.2 pag. 239-245.

[11] Mall S., Dosede S.B. and Holl M., Performance of graphite/epoxy composite with embedded optical fibres under compression. Smart Mater Struct. No 209-15, 1996.

[12] Schwartz, M.M.., Fabrication of Composite materials , Source Books, ASTM, Metals Park, OH, 1985.

[13] Senturia S.D. et al., In-sit measurement of the properties of curing systems with microdielectometry, J Adhesion, vol. 15, 1982, pag. 69.

[14] Shapery R.A., Stress analysis of viscoelastic composite materials, Journal of Composites Materials 1967, 1 : 228-67.

[15] Singh D.SA. and Vizzini A,J. , Structural integrity of composite laminates with interacted actuators, Smart materials and Structures No 3, pag. 71-9, 1994.

[16] Stockmayer, W.W. , Journal Chemistry Physics, 11, 45 (1943): 12, 125 (1944).

[17] Tomycawa V., Nishitsuka N., Takano T., Sensors and Materials , vol. 1,1989.

[18] Tsai S.W. and Hahn, H.T. , Introduction to Composite Materials, Technomic, Lancaster, PA, 1980.

A STUDY ON THE PARAMETERS THAT INFLUENCE THE PERFORMANCE OF SANDWICH PANELS WITH CHIRAL TOPOLOGY CORES

D.M. Constantinescu[1], Şt. Sorohan[1], A. Sandu[1], M. Sandu[1], D.A. Apostol[1]

[1] University POLITEHNICA, Bucharest, ROMANIA,

e-mail: dan.constantinescu@upb.ro, fanes777@yahoo.com, agsandu@yahoo.com, marin.sandu@upb.ro, apostolda@yahoo.com

Abstract: The sandwich plates can be ideal components for large and lightweight structures with increased strength, stiffness and stability. The use of cores with chiral cellular geometry will lead to the development of structural components with superior elastic and impact resilient properties. This paper proposes the design of lightweight sandwich panels with aluminum skins and a core made with a chiral geometry network having circular nodes. As a reference construction, the panel with the skins joined only by means of circular bushes (without supplementary stiffeners) is considered. Finite element analyses (FEAs) are undertaken in order to characterize the behaviour of the considered panels as having supported edges, and loaded under lateral pressure.

Keywords: lightweight panels, chiral topology cores, FEA

1. INTRODUCTION

There is a relatively wide range of sandwich panels that are used as components of advanced lightweight structures as automotives, aircrafts, ships, containers, and modern buildings. The typical sandwich structure consists of two relatively thin high strength face sheets separated by and bonded to a relatively thick, low density, low strength core. Thus, the sandwich structure is characterized by a high flexural strength with reduced weight.

Different types of cores as polymeric and metallic rigid foams, honeycomb structures made from different materials, corrugated plates, lattice type components and others are currently used. In aerospace the most extensively used technology is the honeycomb core sandwich structure with aluminium face sheets and aluminium or titanium honeycomb core. In this structure, the honeycomb cell generatrix is perpendicular to the face sheet and, therefore, the bonding between the honeycomb core and the face sheet can be achieved only by line contact. This is the major drawback of this type of sandwich structure, because the line-contact bonding between the honeycomb cross section and the face sheet can easily lose its bonding integrity as a result of corrosion. Also, honeycomb structures have a high strength to mass ratio in the through-plane direction, but have a low strength in in-plane directions.

Classical composite panels form naturally anticlastic surfaces rather synclastic ones. Consequently, their use to form synclastic surfaces (domes) is therefore limited by the need for complex manufacturing techniques to form those shapes. Chiral honeycomb is a particular form of honeycomb structure in which the ligaments are joined at chiral nodes. A chiral node is one which cannot be superimposed on its mirror image. Fig. 1 shows some shapes of structures with chiral nodes having three, four and six ligaments. The most studied configuration is the hexagonal chiral system that may be considered as being construct from units (highlighted in bold in Fig. 1, e) consisting of a central bush with six attached ligaments. Chiral honeycombs are conventionally manufactured by injection moulding, by bonding together preformed strips or by cutting the honeycomb from solid material.

The peculiar properties of the new developed material are largely due to their auxetic geometry. The term auxetic refers to a novel class of materials characterized by negative Poisson's ratio, that induce beneficial effects as: increased resistance to indentation, improved acoustic properties and a natural tendency to form dome-shaped surfaces [1]-[4]. The auxetic behaviour is a scale independent property and therefore the same mechanism can operate at macro, micro and nano level.

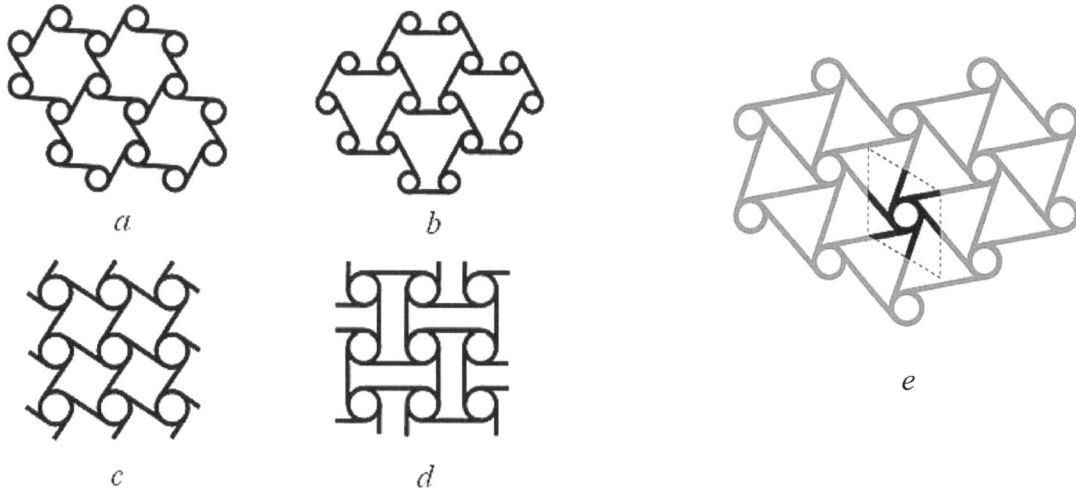

Figure 1: Some chiral structures: a) trichiral, b) anti-trichiral, c) tetrachiral, d) anti-terachiral, e) hexachiral

There are many papers which comment on the geometries and the properties of auxetic materials [1]-[7], but there is little information on the behaviour of the sandwich panels with chiral configured cores. The objective of our paper is to evaluate and to compare the performances of the four sandwich panels presented in Figs. 2-4, by using the finite element analysis.

2. THE MODELLING OF SANDWICH PANELS

The panels that are studied will be denoted as follows: SPP – the panel with the grid components parallel to the edges (Fig. 2), SPD – the sandwich panel with diagonal grid type core (Fig. 3), SPTC – the sandwich panel with a tetrachiral core (Fig. 4), SPAT-1 – the sandwich panel with a anti-tetrachiral core and an aluminium border (Fig. 5), SPAT-2 – the sandwich panel with the same geometry, but having a border form PVC. Additionally was analyzed an associated panel SPP-O obtained from the SPP structure by removing the strips.

These structures, assembled by adhesive bonding, were considered simply supported on the edges of the bottom face sheet and loaded with a lateral pressure $p = 0.07$ MPa applied on the upper face sheet. This is the mean value of differential pressure usually taken into account for aircraft panels.

The physical properties of the materials that are involved in the analysis are given in Table 1. Because the PVC and the araldite AV 119 (produced by Huntsman) have very close values of the elastic moduli and of the Poisson's ratios, the adhesive will be not emphasized explicitly in the numerical model, being included into the polypropylene.

Table 1: Properties of materials used in the finite element modelling of sandwich panels

Component	Face sheets and Borders	Bushes, Strips and Borders	Adhesive
Material	Aluminium 2024 T3	Rigid PVC	Araldite AV 119
Young's modulus [MPa]	72000	3200	3100
Poisson's ratio	0.33	0.35	0.34
Allowable stress [MPa]	300	40	45
Mass density [kg/m^3]	2700	1400	1380

In order to compare the five sandwich panel variants, linear and geometrically nonlinear finite element analyses were done. Each structure was discretized in shell finite elements and static, buckling and modal analyses were undertaken using ANSYS Code [8].

The geometric parameters taken into account were the following: $a = 600$ mm, thickness of face sheets $t_f = 1$ mm, thickness of the strips $t_1 = 2$ mm, thickness of the bushes wall $t_2 = 1.8$ mm, thickness of borders $t_3 = 2$ mm, mean radius of bushes $r = 19.1$ mm, total thickness of the sandwich panel $t = 25$ mm, $b = 75$ mm, $d = 50$ mm, $c = b/2$, $e = d/2$.

A comparison between the results of linear and geometrically nonlinear calculus was shown that the last one is a more suitable approach.

The main results of this study are presented in Table 2; the stresses which exceed the corresponding allowable values are bolded. It is to observe that the strength condition $\sigma_{eq} \leq \sigma_a$ is not accomplished in the case of panel SPP-O, that is the most flexible from the all structures which were analyzed.

Table 2: Results of finite element analyses

Sandwich panel type	Maximum deflection w_{max} [mm]	Maximum equivalent stresses into the panel components [MPa]				Buckling safety coefficient	Fundamental eigenfrequency [Hz]	Mass of the panel [kg]
		faces	strips	bushes	border			
SPP	7.508	298	27.4	35.4	90.6	1.064	238	3.01
SPD	7.533	287	27.4	28.0	69.6	1.064	237	3.02
SPTC	7.663	293	31.1	27.1	97.7	1.212	227	3.18
SPAT-1	6.932	**310**	32.0	26.2	79.2	2.022	224	3.36
SPAT-2	7.708	262	34.2	26.2	**47.8**	1.728	213	2.83
SPP-O	15.32	**667**	-	**59.2**	186.5	1.024	124	2.71

The responses of variants SPP, SPD and SPTC give similar results, while the panels SPAT-1 and SPAT-2 present an increased buckling safety coefficient. The allowable stresses are slightly exceeded in the face sheet and in the border in the case of panels SPAT-1 and SPAT-2, respectively.

It is interesting to observe that the maximum deflection is of 15 mm in the case of panel SPP-O and of 7.5 mm in case of structure SPP, i.e. by adding the strips the rigidity of the panel is doubled. Also, the fundamental eigenfrequency is increasing from 124 Hz to 238 Hz.

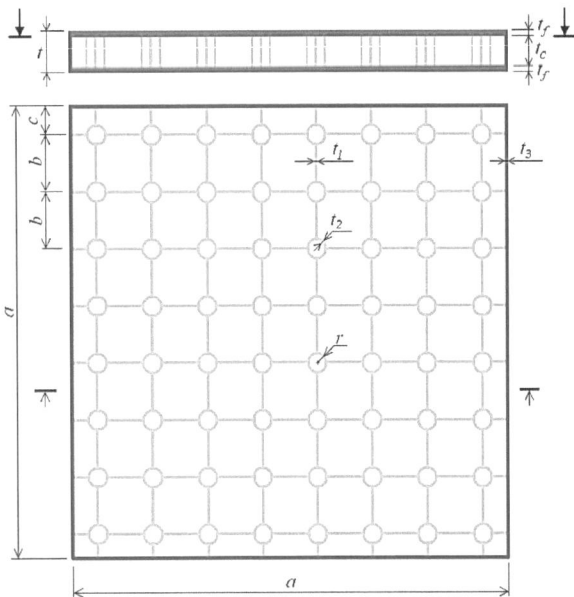

Figure 2: Configuration of panel SPP   **Figure 3**: Configuration of panel SPD

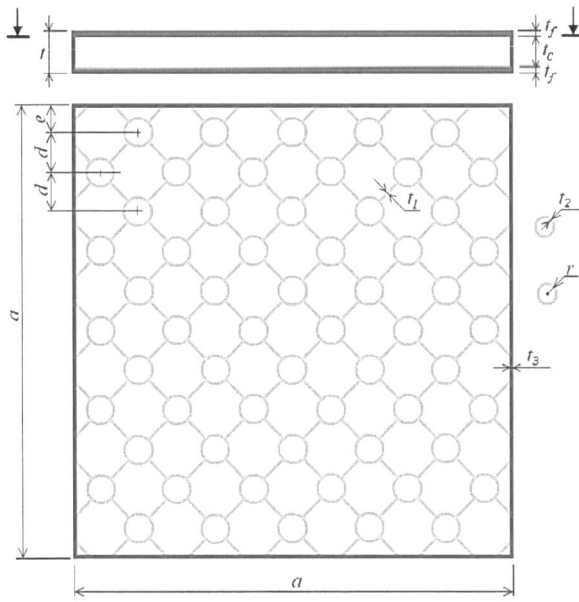

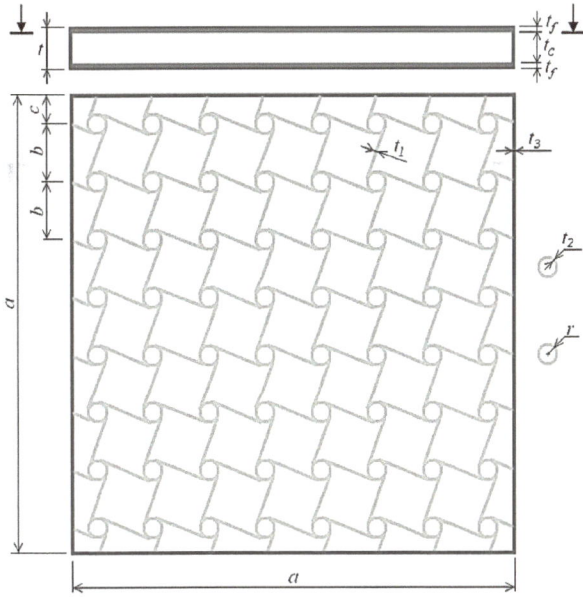

Figure 4: Configuration of panel SPTC

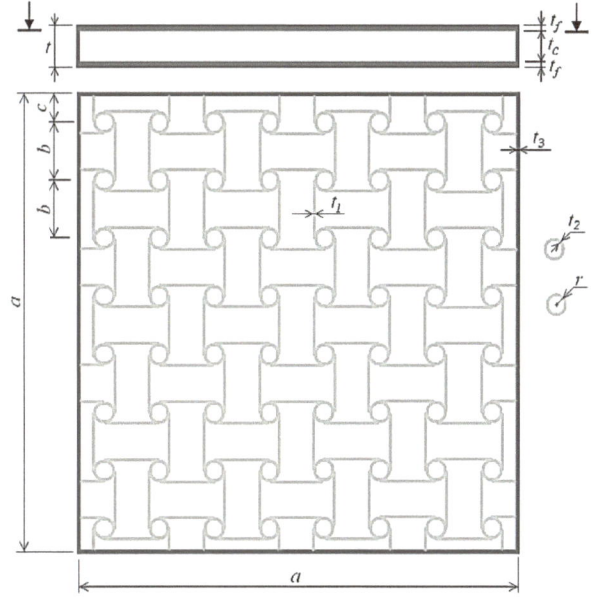

Figure 5: Configuration of panel SPAT

The values from the last but one column are referring to the first mode of local buckling that can appear in the upper sheets of the analyzed panels.

Some results obtained for the structure SPAT-1 are presented in the Figs. 6 to 12.

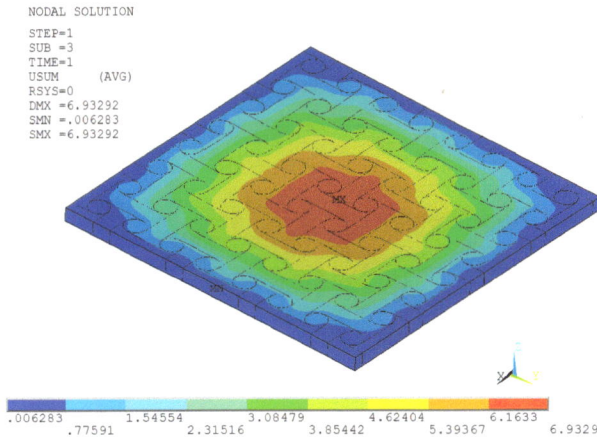

Figure 6: Normal displacements in the upper face sheet

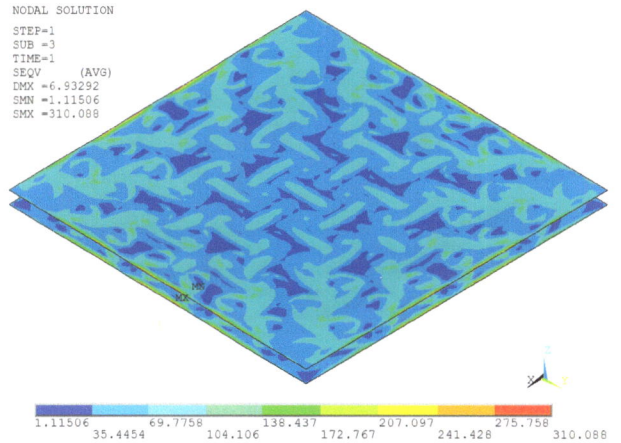

Figure 7: Equivalent stresses in the face sheets

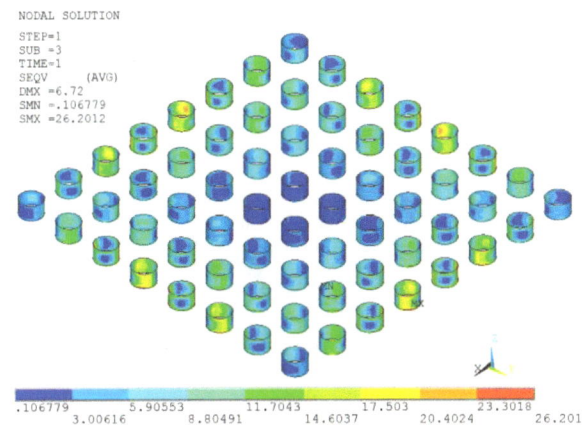

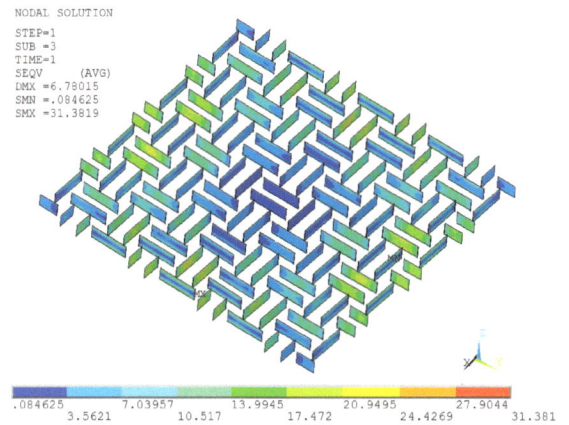

Figure 8: Equivalent stresses in the bushes   **Figure 9**: Equivalent stresses in the strips

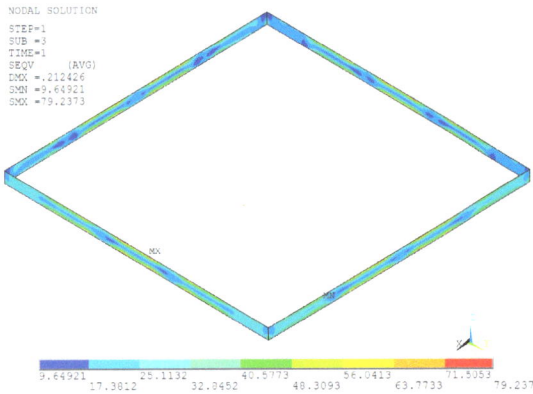

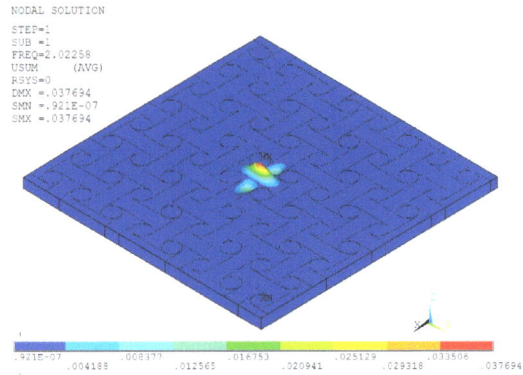

Figure 10: Equivalent stresses in the border   **Figure 11**: First mode of local buckling

The panel SPAT-2 can be considered the most convenient because its strength, rigidity and stability and, first of all, due to its reduced weight. However, a reduction of the maximum equivalent stress in the border by thickening this component is required.

The reaction forces have a different distribution on the panel contour when the coupling with the support is considered as to stop the upper tendency of movement (Fig. 13,a) comparatively to the case of an unilateral restriction (Fig. 13,b), because in the last situation the corners of the panel tend to be lifted.

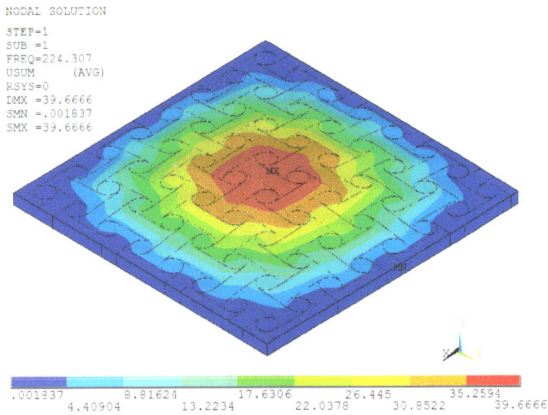

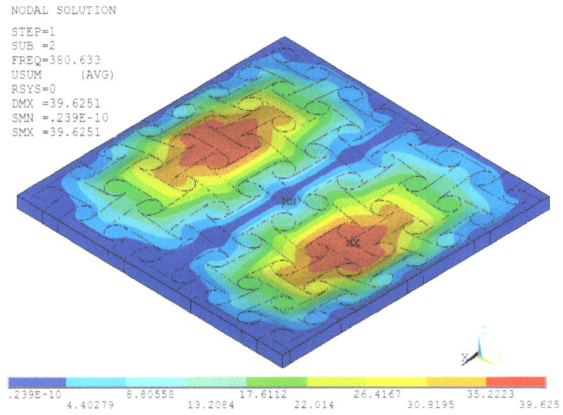

SPAT-1: First vibration mode (224 Hz)   SPAT-1: Second vibration mode (380 Hz)

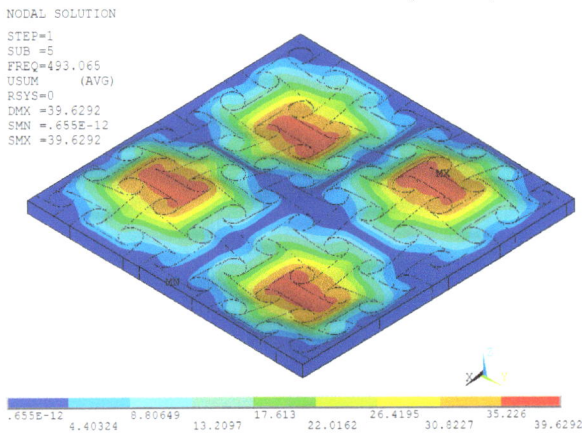

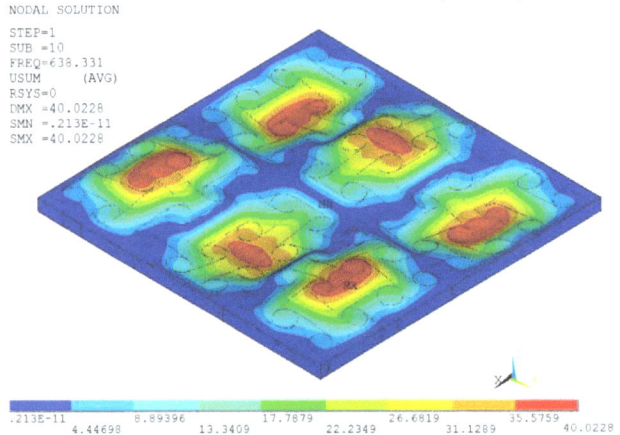

SPAT-1: The fifth vibration mode (493 Hz)   SPAT-1: The tenth vibration mode (638 Hz)

Figure 12: Four vibration modes in the case of panel SPAT-1

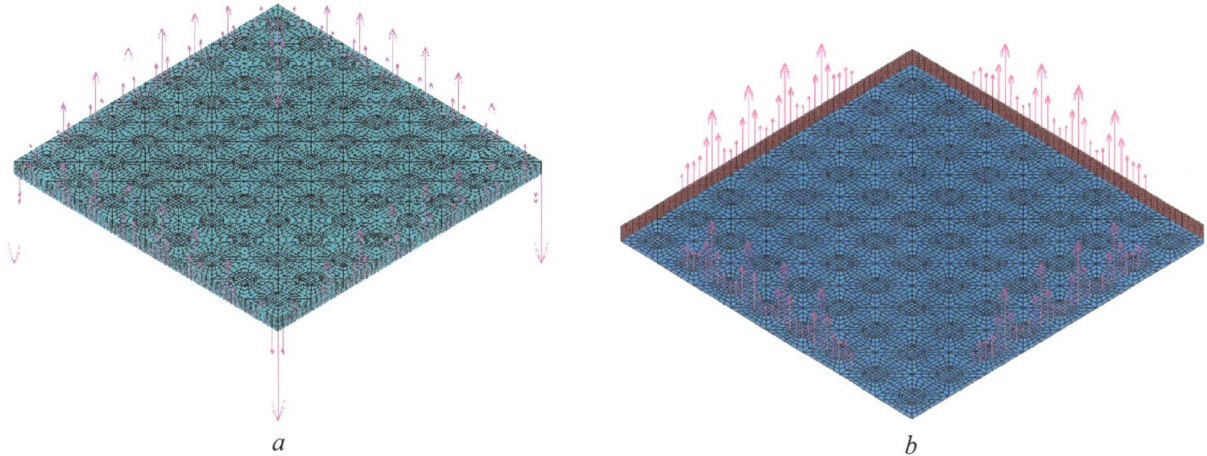

a   *b*

Figure 13: The distribution of reaction forces on the supported contour when the stand up is: a) stopped, b) free

3. CONCLUSIONS

The traditional honeycomb and different types of cores act mostly as spacer while the novel chiral cores can withstand loads very well, having also a structural function. The sandwich panels with chiral geometry honeycombs cores have more convenient mechanical properties comparatively to other similar structures, but are less analysed in the literature.

A correct dimensioning of the components and an adequate choice of the used materials can lead to enhanced properties of this kind of composite structures. Also, recycled materials can be used to manufacture the cores.

The presented study can be the start point in an extended research regarding the design of low cost sandwich panels with increased strength and stability.

ACKNOWLEDGEMENT

This work was supported by a grant of the Romanian National Authority for Scientific Research, CNDI– UEFISCDI, Project number PN-II-PT-PCCA-2011-3.2-0068.

REFERENCES

[1]   Grima J.N., Gatt R., Farrugia P.S., On the properties of auxetic meta-tetrachiral structures, Phys.Stat.Sol., vol. 245, 2008, pp. 511-520

[2]   Grima J., Alderson A., Evans K., Negative Poisson's ratios from rotating rectangles, Computational Methods in Science and Technology, 10, 2, 2004, pp. 137-145

[3]   Spadoni A., Ruzzene M., Numerical and experimental analysis of the static compliance of chiral truss-   core airfoils, J. of Mechanics of Materials and Structures, 2, 5, 2007, pp. 965-981

[4]   Dirrenberger J., Frest S., Jeulin D., Effective elastic properties of auxetic microstructures: anisotropy   and structural applications, Int. J. Mech. Mater. Des.   (http://link.springer.com/aticle/10.1007%2Fs10999-012-9192-8)

[5]   Meraghni F., Desrumaux F., Benzeggagh M. L., Mechanical behavior of cellular core for structural sandwich panels, Composites: Part A, vol 30, 1999, pp. 767-779

[6]   Liu T., Deng Z.C., Lu T.J., Structural modeling of sandwich structures with lightweight cellular cores,   Acta Mechanica Sinica, vol. 23, no. 5, 2007, pp. 545-559

[7]   Ruzzene M., Applications of chiral cellular structures for the design of innovative structural assemblies,   ARO Project no. 45518-EG, Final report, Georgia Institute of Technology, Atlanta, 2008

[8]   * * *   ANSYS Program, Swanson Analysis System, 1990

STRUCTURAL OPTIMIZATION FOR COST

K. Jármai[1]

[1] University of Miskolc, Miskolc, HUNGARY, altjar@uni-miskolc.hu

Abstract: *The paper deals with the cost of composite materials. It shows, that there are several benefits using composite materials, but to select the proper composite one should know its properties better. The factors governing fibre selection include; density, cost, strength and modulus. An example shows the cost calculation of a composite beam with prepreg.*

Keywords: *composite materials, cost calculation, fibre selection*

1. INTRODUCTION

Composites offer engineers a new freedom to optimize structural design and performance. Composites have several advantages over conventional metallic structures. The most significant of these are:

- Low density leads to high specific strength and modulus. Very strong and stiff structures can be designed, with substantial weight savings.
- Fibre can be orientated with the direction of principle stresses, increasing structural efficiency.
- Exceptional environmental and corrosion resistance.
- Improved vibration and damping properties.
- The ability to manufacture complex shapes and one offs from low cost tooling.
- Very low and controllable thermal expansion.
- Excellent fatigue resistance, carbon fibre composites can be designed to be essentially fatigue free.
- Potential for energy absorbing safety structures.
- Damaged structures can be easily repaired.

The costs of the composites are very different. Some of them are relatively cheap, some are expensive. The aim of this study is to show some information about cost calculation, to help designers to choose the proper material. For comparison the main metals, the wood and the concrete is compared to composites through density, tensile modulus and tensile strength [1]. Figure 1-3 show these comparisons.

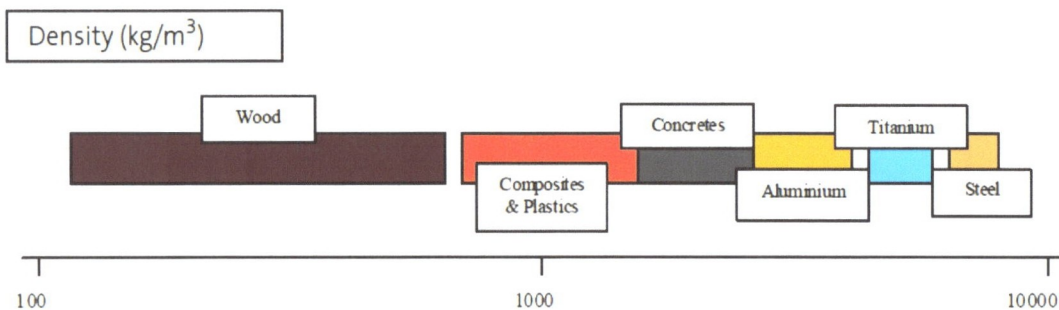

Figure 1. Comparison of the density of different materials

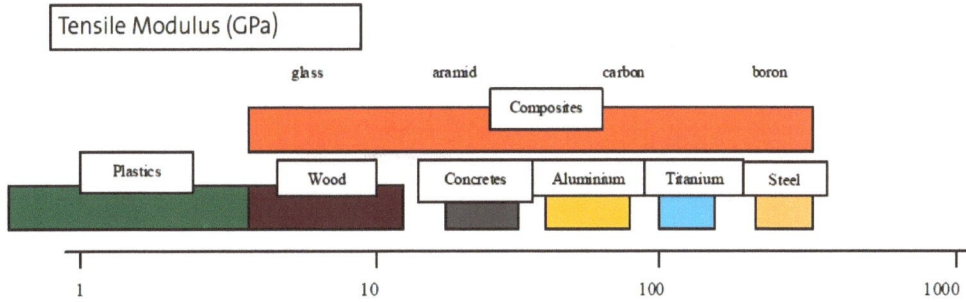

Figure 2. Comparison of the tensile modulus of different materials

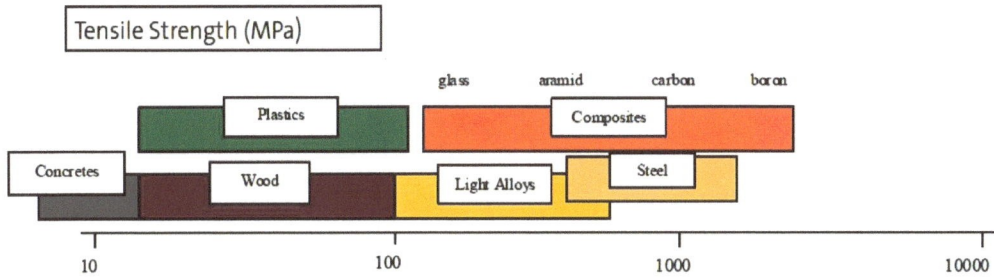

Figure 3. Comparison of the tensile strength of different materials

2. KEY FIBRE SELECTION CRITERIA

Within the composite materials there is still a great difference between the properties. Factors governing fibre selection include; density, cost, strength and modulus. Figures 4 to 7 give comparisons of these factors for a range of fibre types.

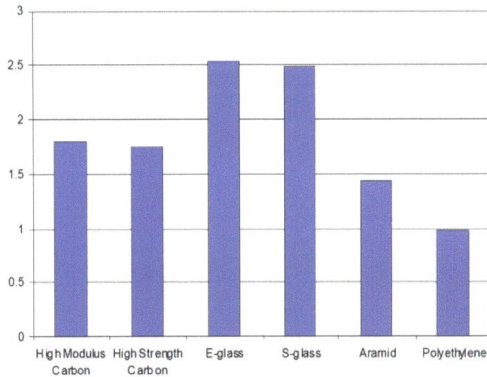

Figure 4. Relative Properties – Density

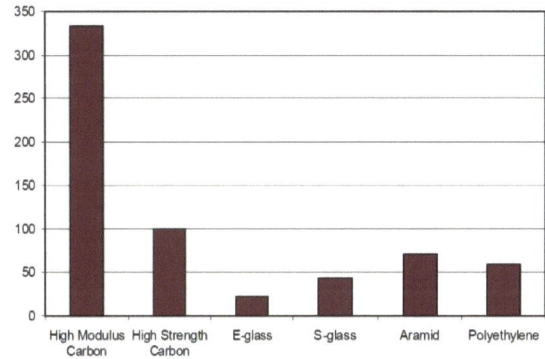

Figure 5. Relative Properties – Cost Ratio

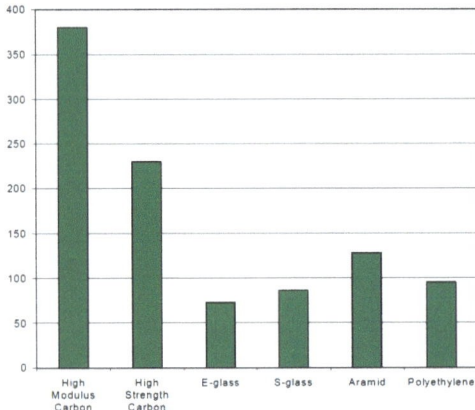

Figure 6. Relative Properties – Modulus GPa

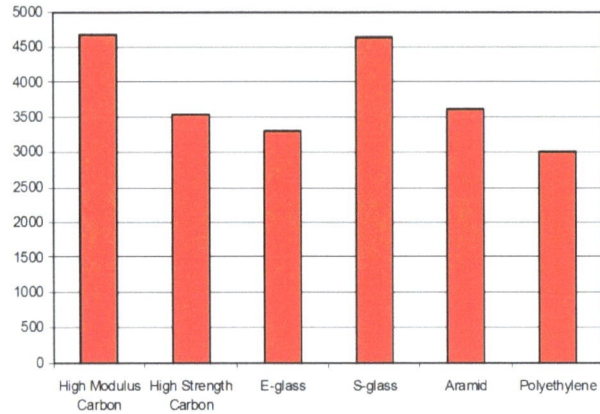

Figure 7. Relative Properties - Tensile Strength MPa

3. THE COST FUNCTION

Calculate different structures the cost function may include the cost of material, assembly, the different fabrication costs such as welding, surface preparation, painting and cutting, edge grinding, forming the geometry and are formulated according to the fabrication sequence. Not too much research has been done in this field, but we have to refer to the work of Klansek & Kravanja [3,4], Jalkanen [5], Tímár et al. [6], Farkas & Jármai [7,8,9]. For composites the calculation is very different and there are some good information available on the internet [10, 11].

3.1. The cost of material

$$K_M = k_M \rho V \,, \qquad (1)$$

for steel the specific material cost can be $k_M = 1.0$ \$/kg, for glass fibre 20-30 \$/m^2 depending on the thickness.

where K_M [kg] is the fabrication cost, k_M [\$/kg] is the corresponding material cost factor, V [mm^3] is the volume of the structure, ρ is the density of the material. For steel it is 7.85×10^{-6} kg/mm^3. If several different materials are used, then it is possible to use different material cost factors simultaneously in Eq. (1).

3.2. The fabrication cost in general

$$K_f = k_f \sum_i T_i \,, \qquad (2)$$

where K_f [\$] is the fabrication cost, k_f [\$/min] is the corresponding fabrication cost factor, T_i [min] are production times. It is assumed that the value of k_f is constant for a given manufacturer. If not, it is possible to apply different fabrication cost factors simultaneously in Eq. (2).

4. GENERIC ATL PROCESS LOOKS AS FOLLOWS

       Tooling Manufacture
       Clean Mould
[1] Tooling Preparation
       Pre-preg
[2] ATL pre-preg tape   n layers
       Consumables
[3]
       Thermo form
[4] Curing

[5] Remove part & debag

[6] Part Finishing

[7] Non Destructive Testing

[8] Part Transfer

Table 3 Mechanical properties and feedstock cost for typical prepreg laminates (fibre fraction-0.6, QI lay-up) [2]

Fibre	Resin	Young's modulus (GPa)	Shear modulus (GPa)	Design tensile strength (MPa)	Laminate density (kg/m^3)	Feed-stock costb (€/kg)
E-glass	Epoxy	22	8.7	110	1980	65
Aramid	Epoxy	30	11	112	1382	95
HS carbon	Epoxy	55	22	220	1560	100
IM carbon	Epoxy	68	27	280	1560	220

4.1. Fixed capital investments and manufacturing cost estimation for higher capacities of carbon fibre plant

Fixed capital investment estimation for similar kind of plant:

$$C_{FC,b} = C_{FC,a} \left(r_{mb}/r_{ma}\right)^{0.7} \tag{3}$$

where,

r_{ma} = monthly production rate of plant a

r_{mb} = monthly production rate of plant b

$C_{FC,a}$ = Fixed capital investment of plant a

$C_{FC,b}$ = Fixed capital investment of plant b

This method is an adaptation of the six-tenth-factor rule, which applies for use in estimation of equipment cost. A similar rule is applied to fixed capital investment except that the absolute value of the power term is governed by following conditions:

- For the average chemical process, the power term will be 0.7 as shown in equation
- For very small installation or for processes employing extreme conditions of temperature or pressure, the value of power term will be from 0 3 to 0 5
- For plant achieving higher capacities through the employment of a high proportion of multiple units rather than large-sized equipment, the term will be 0 8

4.2. Manufacturing cost estimation for carbon fiber plant:

$$A_p = 0.09*C_{FC} + 16200*C_L*N + A_U \tag{4}$$

where

A_p = Annual processing cost

C_{FC} = Fixed capital investment

C_L = Labour charges (€/hr)

N = Number of persons working per shift

A_U = Annual utility and raw material cost

A_p = 0 09*125000000 + 16200* (300/24)*25 + 150000*300

= 3100 / year for 20000 kg of carbon fibres

= 3100 €/kg of carbon fibre

The annual processing cost for A_{p2} for a similar plant of a different size designed for annual production rate R_2 can be approximately calculated by

$$A_{p2} = 0.09*C_{FC1} \left(R_2/R_1\right)^{0.7} + 16200*C_L *N_1 \left(R_2/R_1\right)^{0.25} + A_{u1} \left(R_2/R_1\right) \tag{5}$$

A similar approach for estimating manufacturing cost with a power factor of 0.8 for utilities is as

$$A_{p2} = 0.09*C_{FC1} \left(R_2/R_1\right)^{0.7} + 16200*C_L *N_1 \left(R_2/R_1\right)^{0.25} + A_{u1} \left(R_2/R_1\right)^{0.8} \tag{6}$$

Table 2. Estimated fixed capital investment (excluding land, building and fire hydrant system)

Plant capacity of carbon fibres tons/ year	Estimated fixed capital investment (Crores)	Estimated manufacturing cost of carbon fibres €/kg Eq.5	Estimated manufacturing cost of carbon fibres €/kg Eq.6
20	12.5	45.8	45.8
100	35	39.1	30.3
300	65	37.4	23.6
600	100	36.5	20.4
1000	150	35.9	18.5

5. SIMPLE EXAMPLE FOR COST CALCULATION AT FRP

We have made a calculation of the composite beam cost considering the recurring and non-recurring cost system.
Part Dimensions and Features (Figure 8)

Part Length	10 m	Flange Width	0.13 m
Web Height	0.9 m	Flange Thickness	0.02 m

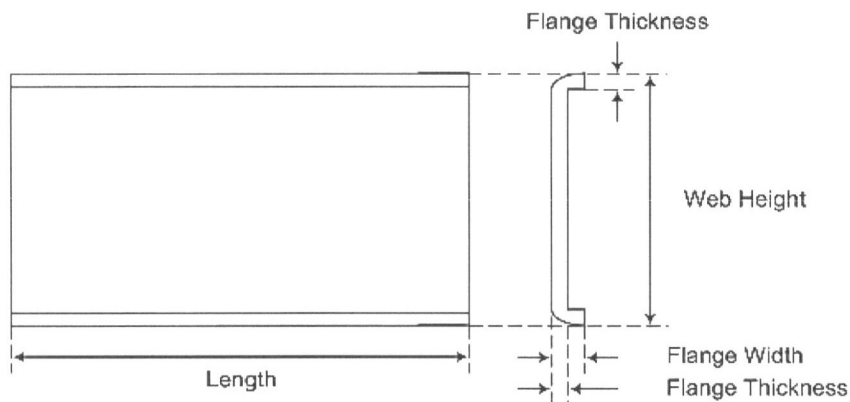

Figure 8. The composite beam cross section

5.1. Recurring Cost Summary Sheet

Labour Recurring Costs [2]

Labour	Manufacturing Hours	Charge Rate €/hour	Cost €
Clean Mould Tooling	1.99	64.3	127.9
ATL Part	5.26	138.5	728.5
Forming	5.68	80.4	456.7
Autoclave Cure Part	0.50	80.4	40.2
Debag Part	2.71	80.4	217.9
"finishing" (machining)	3.72	138.5	515.2
Non Destructive Inspection	5.30	80.4	426.1
Part Transfer	n/a	n/a	0.00
		Sub total	2512.5
Machine Rib Posts and de-burr on bench	3.00	86.6	259.9
Assembly metal Rib Posts to Part	4.00	77.9	311.7
Total Labour Recurring Costs	32.16		3084.0
Total Material Recurring Costs			9490.3
Total Recurring Costs			12574.3 per Part

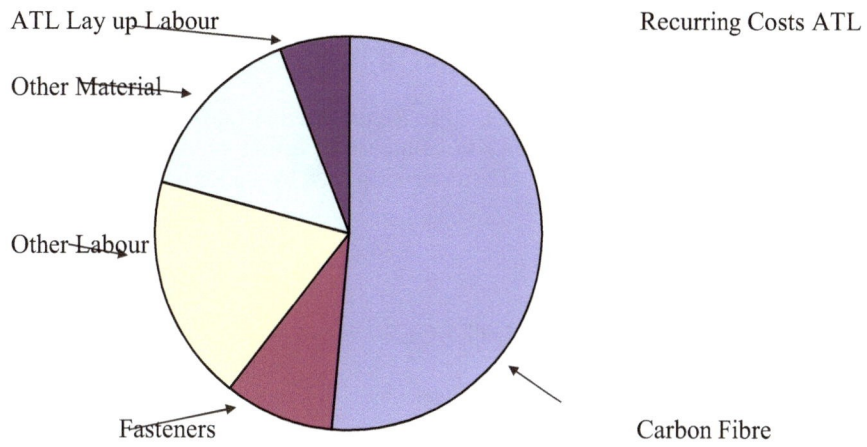

Figure 9. Distribution of the recurring costs at CFRP beam

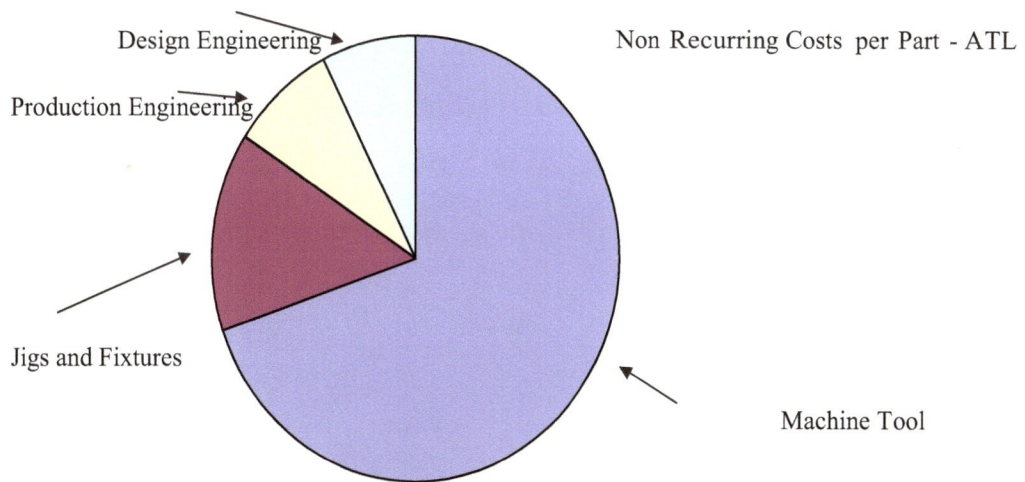

Figure 10. Distribution of the non-recurring costs at CFRP beam

6. CONCLUSION

The composites offer engineers new opportunities to optimize structural design and performance. Composites have several advantages over conventional metallic structures. The paper deals with the cost of composite materials. It shows, that to select the proper composite one should know its properties better. The factors governing fibre selection like density, cost, strength and modulus have an important role. An example shows the cost calculation of a composite beam.

ACKNOWLEDGEMENTS

The research was supported by the Hungarian Scientific Research Fund OTKA T 75678 and by the TÁMOP 4.2.1.B-10/2/KONV-2010-0001 entitled "Increasing the quality of higher education through the development of research - development and innovation program at the University of Miskolc supported by the European Union, co-financed by the European Social Fund."

REFERENCES

[1]   http://www.advanced-composites.co.uk/data_catalogue/catalogue%20files/sm/SM1010- INTRO%20TO%20 ADV%20COMPS-Rev06.pdf

[2]   http://www.acoste.org.uk/uploads/EMC_seminars/COST-STUDIO-example.pdf

[3]   Klansek,U. & Kravanja,S. (2006a) Cost estimation, optimization and competitiveness of different composite floor systems – Part 1. Self manufacturing cost estimation of composite and steel structures,   *Journal   of Constructional Steel Research*, **62** No. 5, pp. 434-448.

[4]   Klansek,U. & Kravanja,S. (2006b) Cost estimation, optimization and competitiveness of different composite floor systems – Part 2. Optimization based competitiveness between the composite I beams, channel-section and hollow-secsion, *Journal of Constructional Steel Research*, **62** No. 5, pp. 449-462.

[5]   Jalkanen,J. (2007) *Tubular truss optimization using heuristic algorithms*, PhD. Thesis, Tampere   University of Technology, Finland. 104 p.

[6]   Tímár,I., Horváth,P. & Borbély,T. (2003) Optimierung von profilierten Sandwichbalken, *Stahlbau*, **72**   No. 2. 109-113.

[7]   Farkas,J. & Jármai,K. (1997) *Analysis and Optimum Design of Metal Structures*. Balkema Publishers, Rotterdam, Brookfield,

[8]   Farkas J. & Jármai K. (2003) *Economic design of metal structures*, Millpress Science Publisher,   Rotterdam, 340 p. ISBN 90 77017 99 2

[9]   Farkas,J.,Jármai,K.: Design and optimization of metal structures, Horwood Publishers, Chichester, UK, 2008. 328 p. ISBN: 978-1-904275-29-9

[10]   R. G. Boeman, N. L. Johnson, "Development of a Cost Competitive, Composite Intensive, Body-in-White," Proceedings of 2002 Future Car Congress, Arlington, Virginia, June 3-5, 2002

[11]   Michael G. Bader, Selection of composite materials and manufacturing routes for costeffective performance, Composites: Part A 33 (2002) 913–934.

NEW TYPES OF COMPOSITE

M. Barbuta[1], M. Harja[2*]

[1] "Gheorghe Asachi" Technical University of Iasi, Iasi, ROMANIA,
barbuta31bmc@yahoo.com, maria_harja06@yahoo.com

Abstract: *The polymer concrete properties are influenced by the components and also by the type of additions. The experimental studies realized on polymer concrete obtained of epoxy resin and aggregates have shown that components dosages, especially resin dosage, have an important contribution in obtaining high performance concretes. Different types of materials (waste, by-products, natural powders, etc.) can be introduced in the polymer concrete mixture for improving the properties or for obtaining new materials. The experimental results presented in the paper analyzed new composite polymers obtained by introducing in the mix different additions such as: silica fume, fly ash and tire powder. The resin dosage and addition quantity were maintained the same for all compositions (12.4% epoxy resin and 12.8% addition). The polymer concrete with fly ash shown good workability, a good microstructural compactity, the density of about 2,300kg/m³ and the highest value of compressive strength.*

Keywords: *composite, polymer concrete, compressive strength, fly ash, silica fume*

1. INTRODUCTION

The construction material industry had developed in the last decades new materials on the base of different types of polymer. Polymer composites are competitive in the competition with cement concrete because they offer some advantages such as: great resistance to the chemical aggressions, fast hardening, good adhesion to other types of materials, high mechanical strengths, resistance to frost thaw cycles, etc. [1, 2, 3].

The use of polymer in concrete is a new possibility of obtaining construction performance materials [4. 5. 6], considering the strengths and/or durability properties. In the same time, as in the case of cement concrete, in the polymer concrete composition a lot of types of addition can be used, especially for improving the properties or just for consuming some wastes [7. 8, 9, 10].

Very large quantities of waste are produced in the world. The usual method of managing wastes is through their disposal in landfills. The most important wastes are: fly ash, silica fume, steel slag, tires etc. which occupy great area of land and in time they create a severe social and environmental problem [11, 12, 13]. In this situation, waste recycling alternatives are gaining increasing importance.

The purpose of the study was to analyze new types of polymer composite concretes obtained by introducing some additions in the mixture.

2. EXPERIMENTAL PROGRAM

The materials used were: epoxy resin and crushed aggregates of two grades 0-4 mm (Sort I) and 4-8 mm (Sort II). Epoxy resin used was type ROPOXID, made in Romania [14]. The hardener was type ROMANID 407 [14]. The epoxy resin in combination with the hardener forms the binder of the polymer concrete. A minimum resin content of 12.4% was adopted from the workability condition [15].

The additions used in the study was type powder, silica fume (SUF) and fly ash (FA) were used as they resulted as wastes. The powder of used tires was mechanically obtained firstly, by cutting the tires in small pieces, after that the steel insertions were eliminated and the cut pieces were transformed in balls and finally in powder.

For the study of polymer concrete properties all compositions were prepared with the same dosages of components: 12.4% epoxy resin, 12.8% filler, 37.4% sort I and 37.4% sort II. In the case of polymer concrete used as witness (PCW) the resin dosage was maintained at 12.4% and the aggregates were 43.8% for both sorts. In function of additions the samples was named: PCS (with SUF), PCFA (with fly ash) and PCTP (with tire powder). The test samples (cubes of 70 mm sizes) for determining the compressive strength were poured according to Romanian standards [16].

3. RESULTS AND DISCUSSION

3.1. Additions characterization

The **SUF** is a by-product that results from ferrosilicon production having the following characteristics: Particle sizes of 0.01…0.5 µm; Shape of particles is spherical and angular, Figure 1
Specific surface is 130,000 m²/kg; Density is 2.250 kg/m³ [17]. For chemical analysis it was used the non-destructive method (EDX). Experimental results obtained for elementary composition are presented in Figure 1 and Table 1.
The principal elements contained by the SUF are: Si, O, Al, K, and Na. These elements appear in crystalline form, Figure 1.

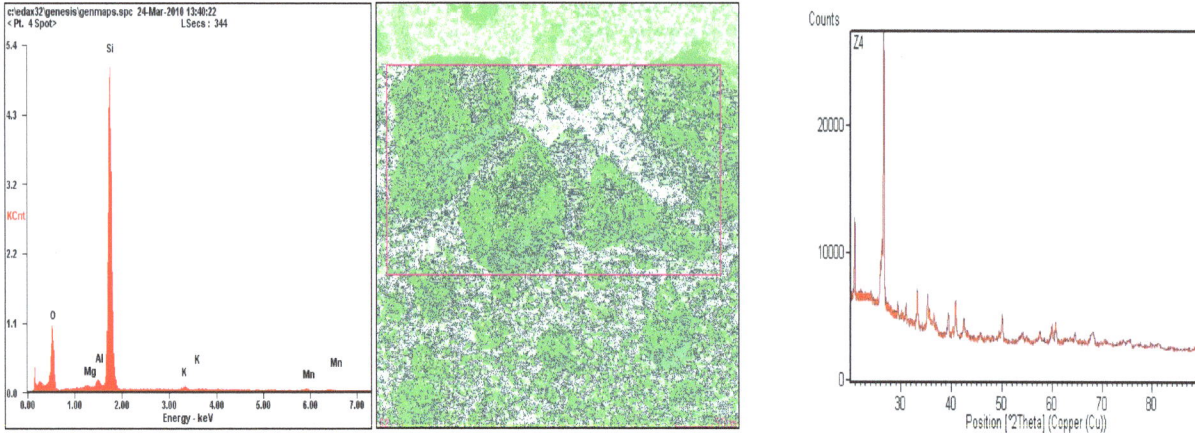

Figure 1: SEM, EDAX and XRD characterization of SUF

XRD as seen in Figure 2 shows that the SUF contains principal crystalline phase quartz and small quantities of kaolinite, mullite, muscovite and rutile.

Table 1: Chemical composition of additions

Element	SUF	Fly Ash
O	36.62	26.61
Mg	0.4	1.49
Al	1.64	17.54
Si	60.75	36.19
K	0.78	3.38
Ca	0.89	4.86
Fe	0.52	6.73

The **fly ash** is a by-product of power plants and it has particles with spherical shape and elementary composition contains [18]: Si, O, Al, Mg, C, K, Fe, and Ca (Figure 2 and Table 1).

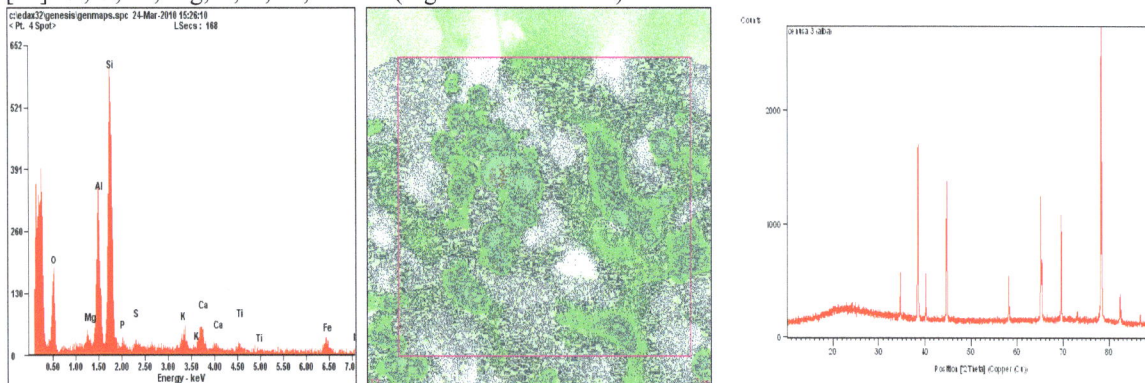

Figure 2: SEM, EDAX and XRD characterization of power plant ash

The elements from fly ash appear in crystalline form, XRD diagram is presented in Figure 2.
XRD as seen in Figure 2 shows that the ash contains crystalline phases: quartz, illite, kaolinite, mullite, hematite, muscovite, rutile and a glassy phase.
The SEM imagine for tire powder is presented in Figure 3.

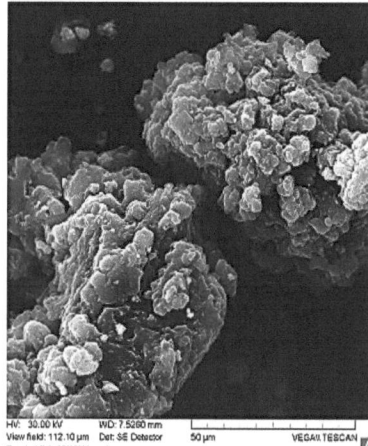

Figure 3: SEM of tire powder

3.2. Properties of polymer concretes

Workability of fresh concrete

The workability was one of the important conditions in preparing polymer concrete [19], because it imposed the maximum dosage of addition and the minimum dosage of resin (which is the most expensive component). Each type of polymer concrete was prepared with the same dosage of resin and addition.

- Polymer concrete with silica fume was a robust concrete, the silica fume made to decrease the workability.
- In the case of tire powder the workability of the concrete was the lowest in comparison with the other types of polymer concrete;
- The polymer concrete with fly ash had a good workability compared with the other compositions, approximately the same with that of the witness.

Density of polymer concretes was determined for each type of addition, Table 2

Table 2: Density of polymer concrete

Mixture	Density, kg/m^3	
	Fresh	Hardened
PCW	2086	2128
PCS	2255	2098
PCFA	2352	2120
PCTP	1692	1678

The experimental results obtained for density on polymer concrete with different types of additions are presented in Table 2. For comparison in the table are presented also the experimental results for polymer concrete without addition (PCW).

- The density of polymer concrete with tire powder indicated that this type of concrete is a lightweight composite;
- The highest density was obtained for polymer concrete with fly ash.

By introducing addition such as silica fume and fly ash the density of polymer concrete increase, compared with that of a polymer concrete without additions.

Microstructure of polymer composite concrete

For analyzing the microstructure of polymer concrete morphological [18, 19] studies were done. Scanning electron microscope (SEM) images were realized with Vega Tescan, for composite without filler the image is presented in Figure 4.

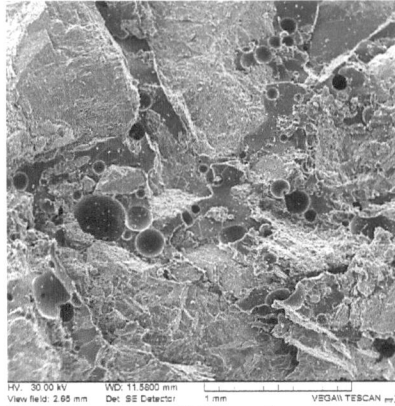

Figure 4: Polymer concrete with 12.4% epoxy resin (EP) and aggregates without filler

From the figures it can observe the microstructure of witness, Figure 4, the presence of voids and also microcracks to the interface between resin and aggregates. In the case of polymer concrete with silica fume, Figure 5, the number of voids is reduced, but a lot of resin sinterings are among aggregates.

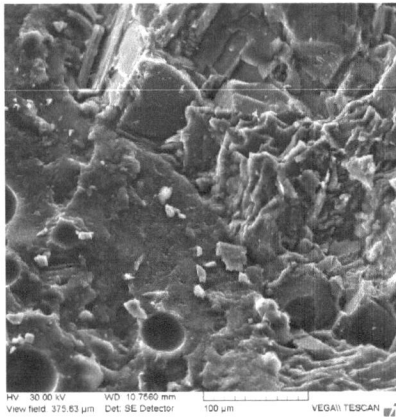

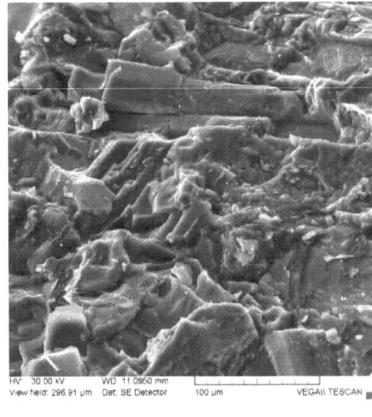

Figure 5: Polymer concrete with 12.4% epoxy resin (EP) and SUF

In the case of polymer concrete with fly ash, Figure 6 the presence of additions produced a more compact structure, also with small voids, but without resin sinterings.

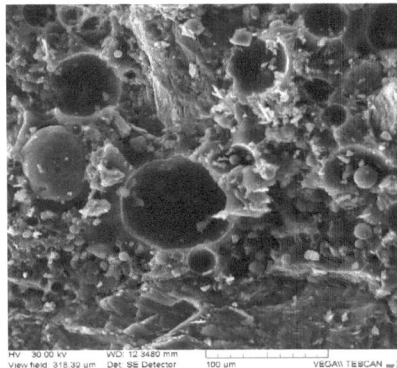

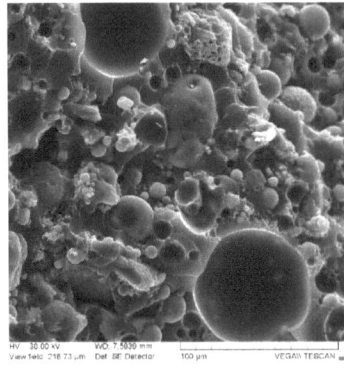

Figure 6: Polymeric concrete BPF2 with 12.8% fly ash and 12.4 % resin

In the case of polymer concrete with tire powder, Figure 7, it can observe that there are much more voids and microcracks.

Figure 7: Polymer concrete with tire powder

Compressive strength

The experimental results, presented in Figure 8 have shown that the polymer concrete with fly ash presented the highest value, 69. 82 MPa, very closed to that of the witness, 69.92 MPa. The smallest value was obtained for PCTP, of about 25.75 MPa (with about 60% smaller than the strength of witness). In the case of polymer concrete with silica fume the compressive strength was only with 10% smaller than the strength of the witness [20, 21, 22].

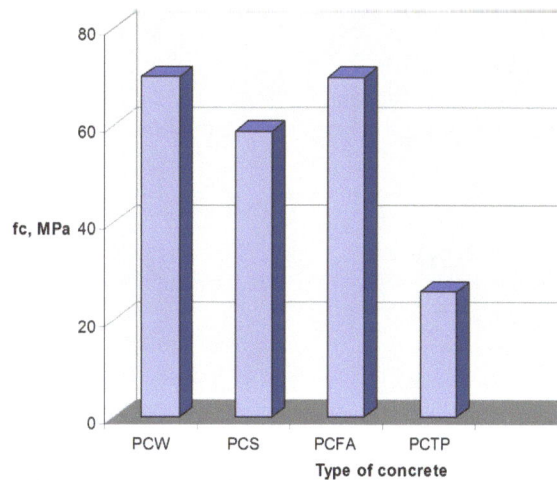

Figure 8: Compressive strength of polymer concrete

3. CONCLUSION

In the paper are presented new composite polymer concretes obtained with different additions such as silica fume, fly ash and tire powder. The purpose was to study the influence of additions type on physical-mechanical characteristics of polymer concrete. A good workability is obtained if in the polymer concrete is introduced fly ash. In the case of tire powder the polymer concrete is very robust, compared with all other types of polymer concretes. The densities have shown that by introducing the additions the densities increase in the case of silica fume and fly ash and it decreases in the case of tire powder. The microstructure of analyzed polymer concretes had shown a more compact structure in the case of addition of fly ash.

The compressive strength was near the value of the witness in the case of addition type fly ash, with about 10% smaller in the case of silica fume and lower in the case of tire powder (with about 60%).

In conclusion, even these additions are not improving the characteristics of fresh concrete and the compressive strength, it is possible that they can improve other mechanical or durability properties, which were not studied yet. Also, the use of

311

these wastes as components in the composition of composite polymer concrete contributes to the superior capitalization, on the one hand, and on the other hand storage pollution is reduced.

REFERENCES

[1]   Blaga A., Beaudoin J. J., Polymer concrete, Canadian building digest; CBD-242, NRC-IRC   Publication, Canada, 1985.

[2]   Fowler D. W., Polymers in concrete: a vision for 21st century, Cement and Concrete Composite, 21,   449-452, 1999.

[3]   Reis J. M. L., Ferreira A. J.M., Assessment of fracture properties of epoxy polymer concrete reinforced   with short carbon and glass fibers, Construction and Building Materials, 18, 523-528, 2004.

[4]   Barbuta M., Harja M., Baran I., Comparison of Mechanical Properties for Polymer Concrete with Different Types of Filler, Journal of Materials in Civil Engineering, 22, 696-701, 2010.

[5]   Magureanu C., Sosa I., Negrutiu C., Heghes B., Mechanical Properties and Durability of Ultra-High-Performance Concrete, ACI Materials Journal, 109, 177-184, 2012.

[6]   Pendhari S., Kant T., Desai Y., Application of polymer composites in civil construction: A general review. Composites and Structure, 84(2), 114-124, 2008.

[7]   Gorninski J. P., Dal Molin D. C., Kazmierczak C. S., Comparative assessment of isophtalic and   orthophtalic polyester polymer concrete: Different costs, similar mechanical properties and durability, Construction   and   Building Materials, 21(3), 546-555, 2007.

[8]   Muthukumar M., Mohan D., Studies on Furan Polymer Concrete, Journal of Polymer Research, 12,   231–241, 2005.

[9]   Jo B. W., Park S. K., Kim D.K., Mechanical properties of nano-MMT reinforced polymer composite   and polymer concrete, Construction and Building Materials, 22, 14–20, 2008.

[10]   Barbuta M., Harja M., Cretescu I., Soreanu G., Influence of wastes content on properties of polymer concrete. 7th International Symposium Cement Based Mater Sustainable Agriculture, Canada, 18-21 September 2011, 358-364.

[11]   Gorninski J. P., Dal Molin D. C., Kazmierczak C. S., Strength degradation of polymer concrete in acidic environments., Cement and Concrete Composite, 29, 637-645, 2007.

[12]   Benazzouk A., Douzane O., Mezreb K., Laidoudi B., Queneudec M., Thermal conductivity of cement composites containing rubber waste particles: Experimental study and modeling, Construction and Building Materials, 22, 573-579, 2008.

[13]   Lin C., Hong Yu-J., Huc A., Using a composite material containing waste tire powder and polypropylene fiber cut end to recover spilled oil, Waste Management, 30, 263–267, 2010.

[14]   Harja M., Barbuta M., Rusu L., Obtaining and Characterization of Polymer Concrete with Fly Ash, Journal of Applied Science, 9, 88-96, 2009.

[15]   Barbuta M., Lepadatu D., Mechanical characteristics investigation of polymer concrete using mixture design of experiments and response surface method, Journal of Applied Science, 8, 2242–2249, 2008.

[16]   SR EN 12390-3:2005 – Testing hardened concrete. Part 3: Compressive strength of test specimens.

[17]   Barbuta M., Taranu N., Harja M., Wastes used in obtaining polymer composite, Environmental   Engineering and Management Journal, 8, 1145-1150, 2009.

[18]   Barbuta M., Harja M., Babor D., Concrete polymer with fly ash. Morphologic analysis based on scanning electron microscopic observations, Romanian Journal of Materials, 40, 3-14, 2010.

[19]   Harja M., Barbuta M., Gavrilescu M., Study of morphology for geopolymer materials obtained from   fly ash, Environmental Engineering and Management Journal, 8(5), 1021-1027, 2009.

[20]   Barbuta M., Diaconescu R.M., Harja M., Using Neural Networks for Prediction of Properties of Polymer Concrete with Fly Ash, Journal of Materials in Civil Engineering, 24, 523-529, 2012.

[21]   Magureanu C., Negrutiu C., Heghes B., Chiorean A., Creep and shrinkage of high performance   concrete. World Scientific an Engineering Acad Soc editor, Proceedings 7th International Conference   Heat   Transfer, Thermal Engineering Environmental, 204-208, 2009.

[22]   Corobceanu V., Giusca R., The durability of prestressed concrete elements reinforced with fibers, Bulletin   of the polish academy of Sciences Technical Sciences, 60(1), 165-170, 2012.

OIL INFLUENCE OVER DYNAMIC STABILITY OF ROLLING BEARING AND HYDRODYNAMIC BEARING TURBOCHARGERS

C.C. Boricean[1], I.C. Rosca[1], P. Grigore[1]
[1]University Trasilvania Brasov, Brasov, Romania,
cosmin.boricean@unitbv.ro, icrosca@unitbv.ro, paul.grigore@unitbv.ro

Abstract: The dynamic stability of rotating machines can be influenced by some functional parameters than can in some functional conditions, rise or decrease the running performances of the rotating machines. One of the main parameters that can majorly influence the dynamic stability of the rotating machines is represented by the oil film. The objective of this study is to highlight some aspects related to oil film influence over the dynamic performances of diesel engines turbochargers rotors sustained by rolling bearings or hydrodynamic bearings. The study brings into audience attention some experimental data related to turbocharger rotordynamic stability, data gathered from test rig testing of modern turbochargers that can reach functioning speeds beyond 250000 rpm.
Keywords: rotordynamics, vibration, turbocharger, oil film, unbalance.

1. INTRODUCTION

Modern turbochargers use as sustaining elements rolling bearings which have the major advantage that in comparison with classic hydrodynamic bearings can reach speeds that in some functioning cases reach 300000 rpm. The other major advantage that results from rolling bearing usage is related to bearing durability. In comparison with hydrodynamic bearings the usage life time of rolling bearings is much higher with approximately 25% [3], [4].

In order to obtain functioning performances at high speeds the turbocharger rotor has to be perfectly balanced and the oil film parameters have to be at the indicated scale in order to supply cooling agent (oil) at the level of the turbocharger rotor and also in order to supply additional damping to the turbocharger rotor system.
The oil film parameters are also important because they influence the self-centering phenomena that occurs at high speed at the level of all functioning rotating machines that have rotors sustained on hydrodynamic bearings [5], [6].

The study of the oil film parameters can provide properly information regarding the functioning conditions of turbochargers, showing the bearing wear scale of the bearings.

2. CONTENT

In order to accomplish the objective of the study there had been developed several test in order to identify the level of vibration amplitudes, tests performed on GARRETT GTV 2600 hydrodynamic bearing turbocharger and on GARRETT GTB 2260V rolling bearing turbocharger.

The tests were accomplished using the Bruel&Kjaer Pulse 12 vibration platform. In figure 1 it is presented the 6 channel acquisition platform and accelerometer connections to the platform.

Figure 1. Acquisition platform and channel set up

In order to have a sufficient precision of the measurements there were used 6 accelerometers mounted on the turbocharger in the following order:

- on channel 1 accelerometer no.1 mounted on the intermediate turbocharger housing near to the compressor wheel on X axis;
- on channel 2 accelerometer no.2 mounted on the intermediate turbocharger housing near to the compressor wheel on X axis with 90^0 phase shift form accelerometer no.1;
- on channel 3 accelerometer no.3 mounted on the intermediate turbocharger housing near to the turbine wheel on X axis;
- on channel 4 accelerometer no.4 mounted on the intermediate turbocharger housing near to the turbine wheel on X axis with 90^0 phase shift form accelerometer no.3;
- on channel 5 accelerometer no.5 mounted near to the bearing outer ring;
- on channel 6 accelerometer no.6 mounted near to the bearing outer ring with 90^0 phase shift form accelerometer no.5;

The placement of the accelerometers on the GTV 2260V turbocharger is presented in figure 2.

Figure 2. Accelerometer montage on to the GTV 2260 V turbocharger

The two turbocharger rotors subjected to study are presented in figure 3 a and b.

Figure 3a. Turbocharger rotor sustained by hydrodynamic bearing

Figure 3b. Turbocharger rotor sustained by rolling bearing

In order to measure the amplitude given by the oil film whirl it had been performed several test at stabilized, accelerated and decelerated regimes at different rotor revolution speed like: 30000, 55000 and 90000 rpm considering these revolution speed to all of the above mentioned regimes.

The measurements performed on the rolling bearing turbocharger GTB 2260 V highlighted the fact that the oil film frequency is obtained at the frequency of 550 Hz, frequency obtained at measurements performed at 55000 rpm at stabilized regime.

In figure 4 there are presented the frequency obtained at 55000 rpm stabilized regime where we can observe the oil film frequency.

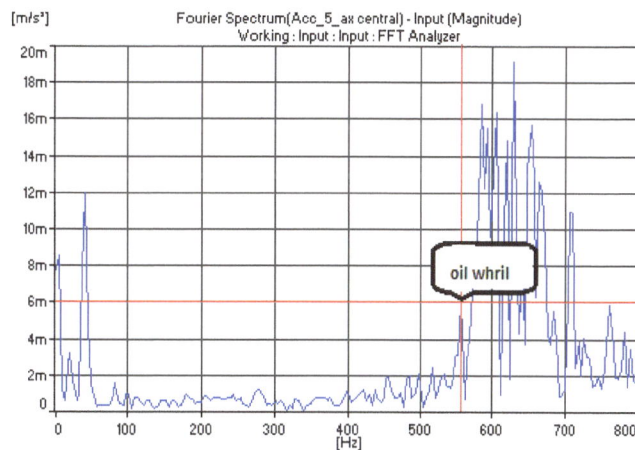

Figure 4. Oil film frequency at 55000 rpm stabilized regime for the rolling bearing turbocharger

Generally speaking the oil film frequency is obtained in measurements at 0.55X Hz from the fundamental frequency f_n given by the formula [1], [2]:

$$f_n = \frac{n}{60} = \frac{55000}{60} = 916.66 \text{Hz} \tag{1}$$

In case of the measurements the oil film frequency is obtained at 0.6X from the fundamental frequency given by the revolution speed.

In order to determine the oil film frequency of the hydrodynamic bearing turbocharger rotor the measurements and the accelerometer montage on to the turbocharger where maintained the same with the ones considered for the rolling bearing turbocharger.

For the case of the hydrodynamic rolling turbocharger rotor the oil film frequency is presented in figure 5.

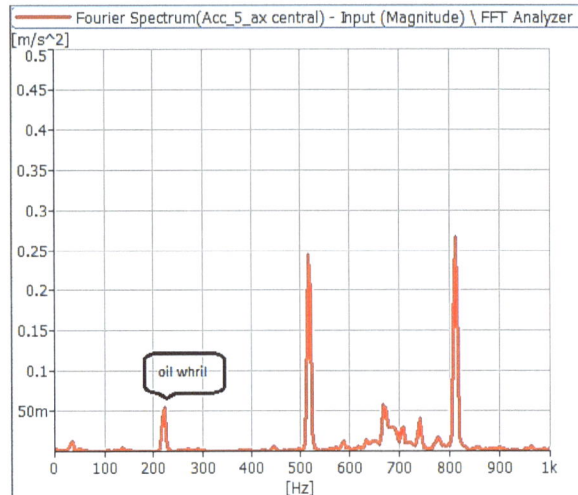

Figure 5. Oil film frequency at 30000 rpm stabilized regime for the hydrodynamic bearing turbocharger

In the case of the hydrodynamic bearing turbocharger the obtained oil film frequency was found at the value of 0.45X Hz from the fundamental frequency given by formula 1.

The measurements performed by specialists in the field of turbocharger rotordynamics found that oil film frequency is obtained at 0.55X Hz [7] from the fundamental frequency.

It is to be mentioned that the two turbocharger rotors subjected for analysis in this study where considered for the similarity between the two regarding mass, turbocharging properties and rotors geometric data.

The tests accomplished on both turbocharger rotors highlight the fact at only at certain functioning regimes the oil film frequency can be observed, because of the multitude of the vibration sources that can be tricky in real vibration source identification.

3. CONCLUSION

Oil film frequency can be found only on certain functioning regimes. It is very important to correctly identify the oil film frequency in order not to confuse the vibration sources that can be found during monitoring the rotating machines in this case the turbocharger. High vibration amplitudes denote some aspects related to malfunctioning of turbocharger rotor assembly including bearings (hydrodynamic or rolling bearings).

The measurements accomplished in this study are accurate, fact highlighted by the comparison of the gathered data with data sets obtained by other studies performed by specialist in the field of rotordynamics [7].

The oil film has major relevance in the self-centering phenomena that occurs at high revolution speed that turbochargers reach during functioning.

The self-centering phenomena is important for the hydrodynamic turbocharger because if the self-centering phenomena does not occur the rotor could function on semi viscous friction an unwanted friction domain because of the damages that this phenomena induces to the life time of the turbocharger rotor in general.

If we refer to rolling bearing turbocharger the oil film is important because the oil film influences the contact type between the ball and the outer/inner ring of the bearing. The contact type between these two elements is important because if the oil film does not exist the friction is dry and at the level of the bearing at high speeds can appear the micro welding phenomena when the bearing is rapidly and permanently damaged.

The study highlighted the fact that level of amplitudes of the oil film are strongly different for the two considered turbocharger rotors, fact highlighted in figures 4 and 5.

ACKNOWLEDGEMENT

This paper is supported by the Sectoral Operational Programme Human Resources Development (SOP HRD), financed from the European Social Fund and by the Romanian Government under the contract number POSDRU/88/1.5/S/59321

REFERENCES

[1]   Gafiţanu, M., Creţu, S.P, Drăgan, B.: Diagnosticarea vibroacustică a maşinilor şi utilajelor, Editura Tehnică Bucureşti, Bucureşti, 1989.

[2]   Gafiţanu, M., Năstase, D., Creţu, S.P.: Rulmenţi. Proiectare şi tehnologie,vol.I,II, Editura Tehnică Bucureşti, Bucureşti, 1985.

[3]   Guangchi, Y., Meng. G., Jing, J.: Turbocharger Rotor Dynamics with Foundation Excitation, Science Direct, 2008.

[4]   Gupta, K.K.: Formulation of Numerical Procedures for Dynamic Analysis of Spinning Structures, International Journal for Numerical Methods in Engineering, vol.23, 1986.

[5]   Lalanne, M., Ferraris, G.: Rotordynamics Prediction in Engineering 2th Edition, Editura John Wiley & Sons Inc., USA, 1997.

[6]   Nicoară, D.: Optimizarea sistemelor mecanice Aplicaţii la sistemele rotor-lagăre, Editura Universităţii Transilvania din Braşov, Braşov, 2003.

[7]   Nyugen, H-S.: Rotordynamics of automotive turbochargers, Editura Springer, Berlin, 2012.

CHARACTERIZATION OF PERFORMANCE COATINGS USING THE MULTI-CRITERIA ANALYSIS METHOD

I.S. Radu

Transilvania University of Brasov, ROMANIA, serban.radu@unitbv.ro

Abstract: *The paper presents a method for evaluating surface properties of various materials and coatings, based on multi-criteria analysis. Newly developed materials, have a complex surface topology due to the manufacturing processes enabling strict surface parameter control.*
Evaluating these materials is a costs and time consuming process. The main objective of the presented method is to offer a good characterization of the surface properties, based on basic laboratory measurements and to enable an effective comparison tool between two or more materials with similar properties.
The method was developed by analyzing the basic surface parameters of three performance coatings. The measured surface values were transposed into specific parameters of the multi-criteria analysis method and a final comparison was done in order to determine the best material. In order to confirm the method, more complex investigations (endurance tests on a specific test rig and scanning electron-microscope investigations) were done in parallel. The results of these investigations confirmed the results obtained by using the multi-criteria analysis method.
Keywords: *performance materials, coatings, surface, multi-criteria analysis, tribology*

1. INTRODUCTION

For evaluating a series of performance coatings with different surface profile parameters, a method was developed, enabling an effective characterization of the coatings by using only basic laboratory measurements. The method is based on the multi-criteria analysis principle and basically enables the fast evaluation of data collected from specific tests like roughness and friction measurements. By using this method, time consuming endurance tests and costly scanning electron-microscope investigations can be eliminated.

2. DESCRIBING THE PERFORMANCE COATINGS

Three types of performance coatings were analyzed. Their specific parameters are presented in Table 1. The coating is an alloy of chrome and has a granular surface shape, enabling the depositing of lubrication material [3]. As seen in Table 1, the main differences between the coatings, consists in their roughness profile and structure type. Coatings are considered having a closed structure, when a certain degree of overlapping exists between its spheres [5].

Table 1: Main characteristics of the performance layers

	Coating 1	Coating 2	Coating 3
Structure Type	Open	Open	Closed
Average Maximum Profile Height Rz [µm]	9,5	32,9	17,84
Density of Spheres [Sph./cm]	227	87	132
Hardness HV0,1	1000	1000	1000
Layer Thickness [µm]	20	70	70

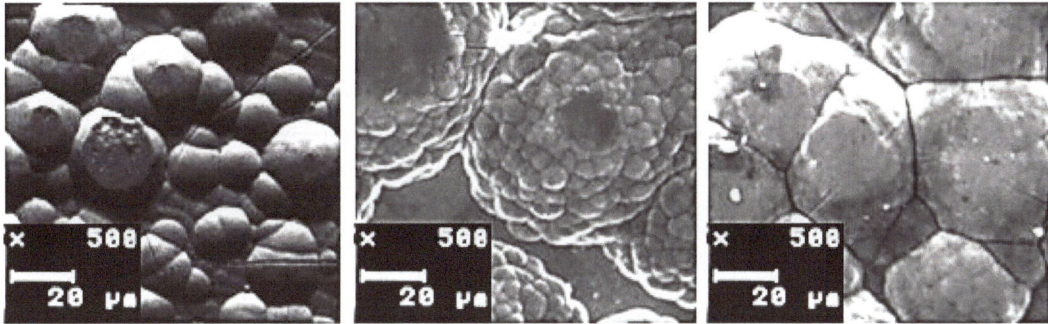

Figure 1: Surface view of the three performance coatings

3. DETERMINING THE MECHANICAL PROPERTIES OF THE COATINGS

Mechanical properties of the coatings were organized in four main categories:
- Bearing area (based on surface roughness)
- Friction (at rotary and oscillatory movement)
- Wear (linear and volumetric)

Figure 2: Determining the roughness profile   **Figure 3:** Determining friction and wear

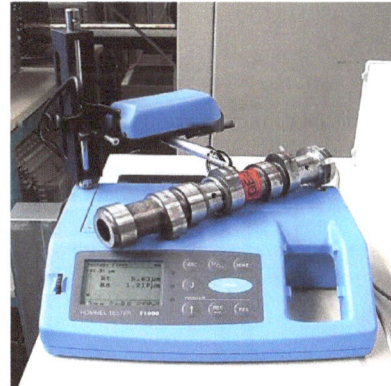

The bearing area was determined by measuring the roughness profile of coatings using a basic roughness meter, as shown in Figure 2. Friction and wear were measured using a tribometer, pictured in Figure 3, capable of rotary and oscillatory movement [2,5]. Measurement results were only organized as data in specific folders. Their evaluation will be done by using a multi-criteria analysis method.

4. ON THE MULTI-CRITERIA ANALYSIS METHOD

In order to evaluate and rank the three coatings based on the experimental results, various models and methods of decision taking were analyzed. Considering the efficiency to complexity ratio, the "Electric" model developed by the General Electric Company and the multi-criteria analysis method were chosen [1].

These analysis techniques offer very accurate and objective results, for the following reasons [4]:
- the contribution of each criteria is established by comparing them to each other
- the relative position between two criteria can be of equality, superiority or inferiority
- marks are awarded separately for each criteria
- the comparative analysis of variant is done separately, considering each criteria

5. ESTABLISHING THE CRITERIA OF COMPARISON

The chosen criteria have to define essential properties and characteristics in order to individualize the variants to be analyzed. Starting from this concept, the chosen criteria and their abbreviations are shown in Table 2.

Table 2: The criteria chosen for the analysis

Criterion	Symbol
1 – Fabrication costs	**FC**
2 – Bearing area	**BA**
3 – Friction coefficient (rotary movement)	**FCR**
4 - Friction coefficient (oscillatory movement)	**FCO**
5 – Volumetric wear (rotary movement)	**VWR**
6 - Volumetric wear (oscillatory movement)	**VWO**
7 – Linear wear (rotary movement)	**LWR**
8 - Linear wear (oscillatory movement)	**LWO**
9 – Sensitivity at lubricant quality	**SLQ**
10 – Sensitivity at movement type	**SMT**

6. ESTABLISHING THE LEVEL OF EACH CRITERION

In order to establish a level of each criterion, a square table containing the ten criteria was generated, see Table 3. The contribution of the criteria was done by using the "Latin grid" with three values [1,6]. The criterion of one row is compared to the corresponding criterion of the column. If it is more important its value will be 1. If it is less important its value will be 0. If they are equally important the value will be 0,5.

Table 3: The table of levels

	FC	BA	FCR	FCO	VWR	VWO	LWR	LWO	SLQ	SMT	Score p	Level	Contribution coefficient γ_i
FC	0,5	0	0	0	0	0	0	0	1	1	2,5	**8**	0,58
BA	1	0,5	0	0	0	0	0	0	1	1	3,5	**7**	0,9
FCR	1	1	0,5	0	0	0	0	0	1	1	4,5	**6**	1,3
FCO	1	1	1	0,5	0	0	0	0	1	1	5,5	**5**	1,77
VWR	1	1	1	1	0,5	0	1	1	1	1	8,5	**2**	4,16
VWO	1	1	1	1	1	0,5	1	1	1	1	9,5	**1**	5,6
LWR	1	1	1	1	0	0	0,5	0	1	1	6,5	**4**	2,37
LWO	1	1	1	1	0	0	1	0,5	1	1	7,5	**3**	3,14
SLQ	0	0	0	0	0	0	0	0	0,5	0	0,5	**10**	0,07
SMT	0	0	0	0	0	0	0	0	1	0,5	1,5	**9**	0,3

By adding the values in the columns, the score "p", of each criterion is obtained (in Table 3, the antepenultimate column). The sum of all points is always equal to half the square of the criteria number: for the present case, for a number of 10 criteria, the sum of all points will be 50.

The score accumulated by each criterion, is used to determine its level and thus its rank (in Table 3, the penultimate column).

7. DETERMINING THE CONTRIBUTION COEFFICIENTS OF EACH CRITERION

For calculating the contribution coefficients γ_i, the Frisco method was used [1,6]. It is one of the most well known procedure used nowadays, and is given by the following equation:

$$\gamma_i = \frac{p + \Delta p + m + 0,5}{-\Delta p' + 0,5N}$$

(1)

Where:

p – score of each criterion
Δp – difference between the actual criterion and the criterion corresponding to the lowest level
m – number of criteria surpassed by the actual criterion
$\Delta p'$ – difference between the score of the actual criterion and the criterion corresponding to the highest level
N – number of criteria

The contribution coefficients of each criterion are shown in the last column of Table 3.

8. GRANTING OF MARKS TO EACH CRITERION

The procedure of granting marks to each criterion, gives the presented method a certain character of subjectivity. The German VDI 2225 standards, recommend only five marks [1,6]. For the present project, a notation with marks from 1 to 10 was considered. The granted notes are shown in Table 4. Except criterion FC (fabrication costs), the marks Ni, highlighted in Table 4, are actually averages of the marks granted to the respective criterion. For example, the VWO (volumetric wear at oscillatory movement) criterion was granted a mark for its behavior at mixed lubrication, as well as hydrodynamic lubrication.

Table 4: The table of marks

Criterion		Coating 1	Coating 2	Coating 3
Level	Symbol	N_i	N_i	N_i
1	FC	8	7,5	10
2	BA	9,25	8,5	9,5
3	FCR	9	7	9,5
4	FCO	7,5	8	9,75
5	VWR	8	9	9,5
6	VWO	7,5	8,5	9,5
7	LWR	7	8	10
8	LWO	10	8	9
9	SLQ	8,75	8,25	9
10	SMT	9,5	9,5	8

9. DETERMINING THE RANKING OF VARIANTS

The final ranking (also called the raking of variants), is calculated tabular, using the matrix of consequences, shown in Table 5. In this table, the weighted score of each criterion is calculated. This is the product of the marks granted the criterion, amplified by the contribution of the respective criterion. By adding for each of the variants the weighted score, the final score is obtained.

According to the values shown in Table 5, it can be concluded that the final ranking of the analyzed coatings is: coating 3, coating 1 and coating 2 ; meaning that the coating with the best properties is coating 3 by achieving a score of 195,09.

Table 5: The table of scores

Criterion	γ_i	Coating 1		Coating 2		Coating 3	
		N_i	$N_i\,\gamma_i$	N_i	$N_i\,\gamma_i$	N_i	$N_i\,\gamma_i$
FC	5,6	8	44,8	7,5	42	10	56
BA	4,16	9,25	38,48	8,5	35,36	9,5	39,52
FCR	3,14	9	28,26	7	21,98	9,5	29,83
FCO	2,37	7,5	17,77	8	18,96	9,75	23,1
VWR	1,77	8	14,16	9	15,93	9,5	16,81
VWO	1,3	7,5	9,75	8,5	11,05	9,5	12,35
LWR	0,9	7	6,3	8	7,2	10	9
LWO	0,58	10	5,8	8	4,64	9	5,22
SLQ	0,3	8,75	2,62	8,25	2,47	9	2,7
SMT	0,07	9,5	0,66	9,5	0,66	8	0,56
Final score		-	**168,6**	-	**160,25**	-	**195,09**

10. CONCLUSION

A procedure, based on the multi-criteria analysis, for ranking performance coatings was developed. It enables a fast evaluation of laboratory measurements. Three different coatings were used for this application. After preliminary laboratory measurements, consisting in surface and tribological measurements, ten evaluation criteria were determined. By using a "latin grid" of three values, the level and the contribution coefficient of each criteria was calculated. By granting marks, the final ranking of the three coatings was established.

ACKNOWLEDGEMENT

The project supported by the Sectorial Operational Programme Human Resources Development (SOP HRD), financed from the European Social Fund and by the Romanian Government under the contract number POSDRU/89/1.5/S/59323.

REFERENCES

[1]   Bobancu, Ş., Cioc, V., Inovare inginerească în design, Editura Infomarket, Braşov, 2002.
[2]   Bobancu, Ş., Cozma, R., Tribologie. Universitatea Transilvania, Braşov, 1995.
[3]   Bogatzky, T., Gümpel, P., Die Anpassung von Hochleistungsschichten an Komponenten von Verbrennungsmotoren, HTWG Forum, Ausgabe 2006/2007.
[4]   Pomerol. J., C., Multicriterion Decision in Management, Kluwer Academic Publisher, 2000.
[5]   Radu, S., Cercetări privind utilizarea straturilor de înaltă performanţă în cuplele de frecare ale   mecanismului motor, Teza de doctorat, Universitatea Transilvania, HTWG Konstanz, 2003.
[6]   Statnikov, R., B., Multicriteria Analysis in Engineering, Springer, 2002.

EXPERIMENTAL MEASURES FOR MEETING STANDARDS OF DIESEL ENGINE VISIBLE EMISSIONS

V. Sandu[1], C. Bejan[2]

[1] Transylvania University,Brasov, ROMANIA, e-mail sandu@unitbv.ro
[2] Road Vehicle Institute, Brasov, ROMANIA, e-mail cvbejan@yahoo.com

Abstract: *The paper presents research work done on a turbocharged inter-cooled direct injection diesel engine for the reduction of smoke emission. The engine type D2156MTN8R having a rated power of 280 HP at 2100 rpm, manufactured by Roman Truck Company, was tested on the dynamometric bench being measured performances and emissions. According to standards there were measured performance parameters such as corrected power (P_{ec}), corrected torque (M_{ec}), hourly fuel consumption (C_h), specific fuel consumption (c) as well as emissions –in terms of smoke number (N_s)expressed in Hartridge smoke units (HSU). The opacity of exhaust gas was compared to limits imposed by European Regulation ECE 24.Further experimental adjustments were done to correct smoke emissions at lower speeds,in four steps: variation of the injection timing, variation of fuel flow rate injected in the combustion chamber, axial rotation of the injector nozzles and variation of the injected fuel flow rate according to charged air pressure.*

Keywords: *diesel engine, visible emissions, experimental adjustments*

1. INTRODUCTION

Diesel engine applications are dominant in freight and passenger transportation due to fuel economy and reliability, being also redesigned to reduce exhaust gas emissions. Although major abatements were done, visible exhaust gas can cause air pollution and health problems. Heavy duty diesel engines must meet simultaneously severe requirements both on performance and emissions, regulated by European standards. Among the most spread emission standards, ECE R 24 [1] defines provisions for approval with regard to "visible pollutants" which is called, in lay terms, smoke. The regulation defines smoke limits which must be checked by testing engine on the dynamometric bench. If the engine smoke emissions are lower than imposed limits, the engine type is considered certified according to Regulation 24.

2. ENGINE REQUIREMENTS

The D2156MTN8R engine having the block series no.1226 was tested on the dynamometric test bench at Road Vehicle Institute (INAR). The engine performance depends on atmospheric conditions and fitted equipment. During the tests the atmospheric conditions were as follows [2]:

Pressure : 710 - 721 mm Hg;

Air temperature : 9-15 ° C;

The engine performance is corrected according to pressure and temperature with coefficient α, with f_a atmospheric factor and f_m engine factor, with the formula:

$$\alpha = f_a^{fm} \tag{1}$$

$$f_a = \left(\frac{99}{p_s}\right)^{0.7}\left(\frac{T}{298}\right)^{1.5} \tag{2}$$

p_s - atmospheric dry pressure expressed in kilopascals, T - atmospheric temperature in Kelvin.

The air temperature in absolute Kelvin scale is measured at the engine inlet at 0.15 m upstream the air filter.

$$f_m = 0.036 \cdot q_c - 1.14 \tag{3}$$

$$q_c = \frac{q}{r} \tag{4}$$

with q - fuel flow in milligramme per cycle per liter of total swept volume (mg/(l.cycle) and r - pressure ratio of compressor outlet and compressor inlet. The calculated values of α range in 1.04-1.06 the interval being included in the requirement of 0.9-1.1.

As a cooling agent it was used the dedurised water from the cooling system of the test bench, the temperature being kept in 75-80°C. For the engine operation it was used diesel fuel according to standard EN 590.

The engine was fit with a 6 blade Φ680 fan, an aluminum made charge air cooler, an (unloaded) alternator and without air compressor.

3. ECE 24 REQUIREMENTS AND TEST RESULTS

There are two dissimilar tests when are determined the emission of visible pollutants, the measurements at steady speeds (A) and at free acceleration (B)[1].

The measurements were performed with opacimeter AVL 465 in which the gas is measured in a confined enclosure with a non-reflecting internal surface; the effective length of light path is determined avoiding the influence of the protective devices of light source and photoelectric cell. The opacimeter has two scales, a linear scale ranging from 0 to 100 and a logarithmic scale in absolute units of light absorbtion ranging from 0 to 4 unit m^{-1}.

(A) Test at steady speeds over the full-load curve

The test is carried out on an engine being measured the opacity of the exhaust gases running under full-load and at steady speed, between minimum and maximum rated speed, in this case 1000 - 2100 rpm.

For each of the engine speeds at which the absorption coefficient is measured, the nominal gas flow shall be calculated by means of the following formula, for four-stroke engines:

$$G = \frac{V \cdot n}{120} \tag{5}$$

in which:

G - nominal gas flow, in liters per second (l/s), V - cylinder capacity of the engine, in liters (1), n - engine speed, in revolutions per minute (rpm).

In all the tests which were represented below, there were calculated corrected power (P_{ec}), corrected torque (M_{ec}), hourly fuel consumption (C_h), specific fuel consumption (c) as well as emissions –in terms of smoke number (N_s) expressed in Hartridge Smoke Units (HSU).

The initial tests were performed with injection timing at 25°CR (Crankshaft Rotation), illustrated in figure 1, with continuous line. The measured engine performance in terms of power, torque and specific fuel consumption complies with the engine type declaration, as stated in Romanian standard STAS 6635. The allure of performance curves is typical for heavy duty diesel engine, having the lowest specific fuel consumption in the range of speeds corresponding to the highest torque.

As it can be noticed, the smoke number expressed in Hartridge units exceeds the limit required by Regulation 24 for lower speeds, in this case lower than 1200 rpm.

The first step to meet the limits was done by variation of injection timing around the value of 25°CR which is imposed by engine standard. So injection timing was modified with ±2° at 27° and 23°. The higher injection timing corresponding to 27° is represented in figure 1 with dashed line and the smaller injection timing corresponding to 23° with dotted line.

It was noticed that from the three measurements, the one performed for injection timing of 27°CR has the power, torque and specific fuel consumption improved; the hourly fuel consumption was even and the smoke number was the best only until the speed of maximum torque of 1400 rpm. At lower speeds the measured smoke was higher than admitted limit.

When the engine run with 23° injection timing the performances were worse at speeds higher than 1400 rpm, but at at lower speeds, critical from the point of view of meeting regulation, the smoke number was a little bit better. Also the specific fuel consumption increased, the lowest pole being deviated in the area of smaller speeds, which are not so frequently operated during engine life span.

The influence of injection timing of only ±2° is significant in terms of power difference (around 20 kW) and specific fuel consumption difference (around 15 g/kWh) while the hourly fuel consumption was not sensitive at all.

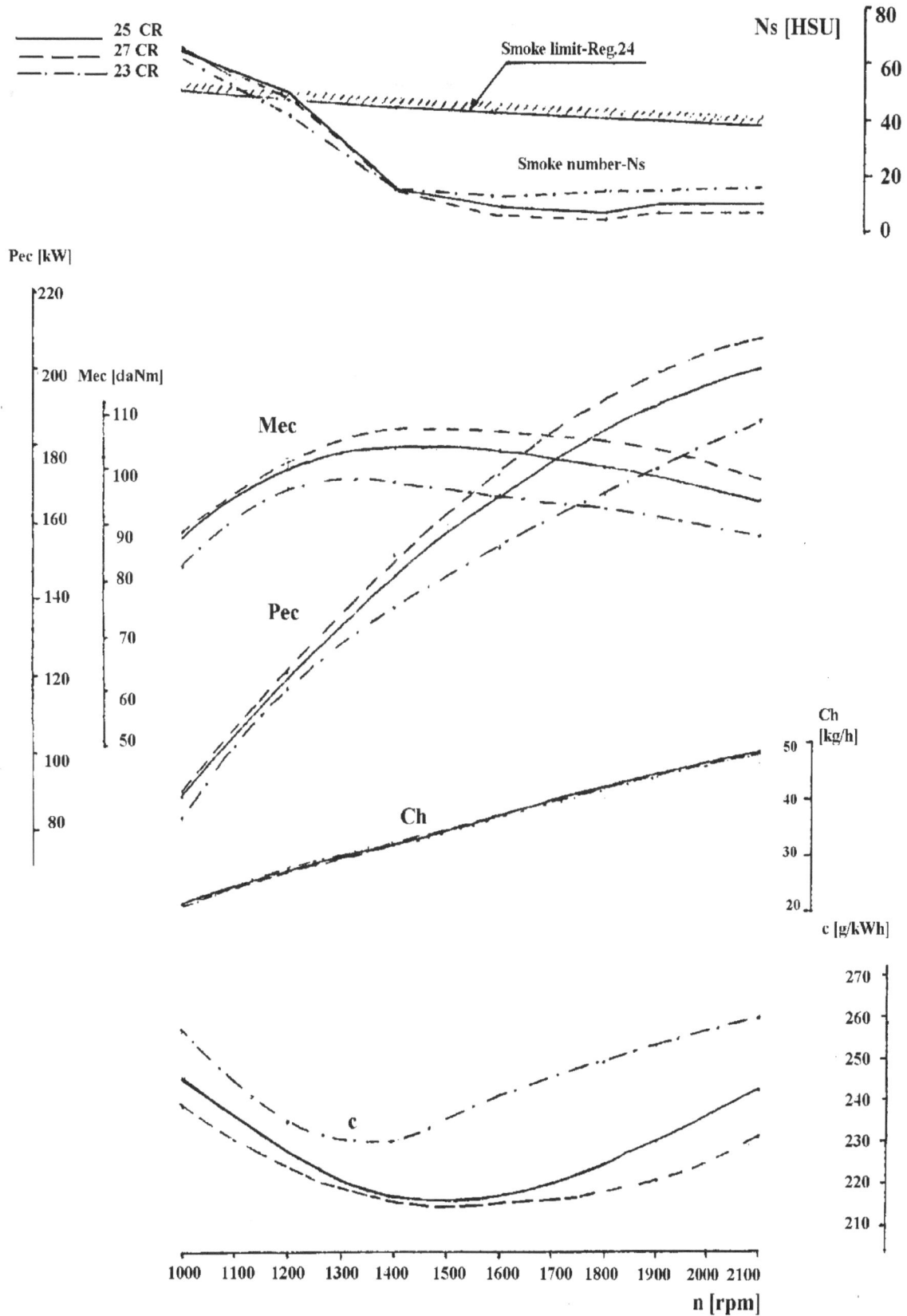

Figure 1: Engine performance versus injection timing (crankshaft rotation - CR) in degrees

The second step was to study the influence of fuel flow rate in the injection pump being represented in figure 2. The injection pump was adjusted for maximum rated power at 280 HP and for a lower power of 240 HP. For the fuel flow rate adjusted for 240 HP the smoke number was higher than for 280 HP on the whole range of

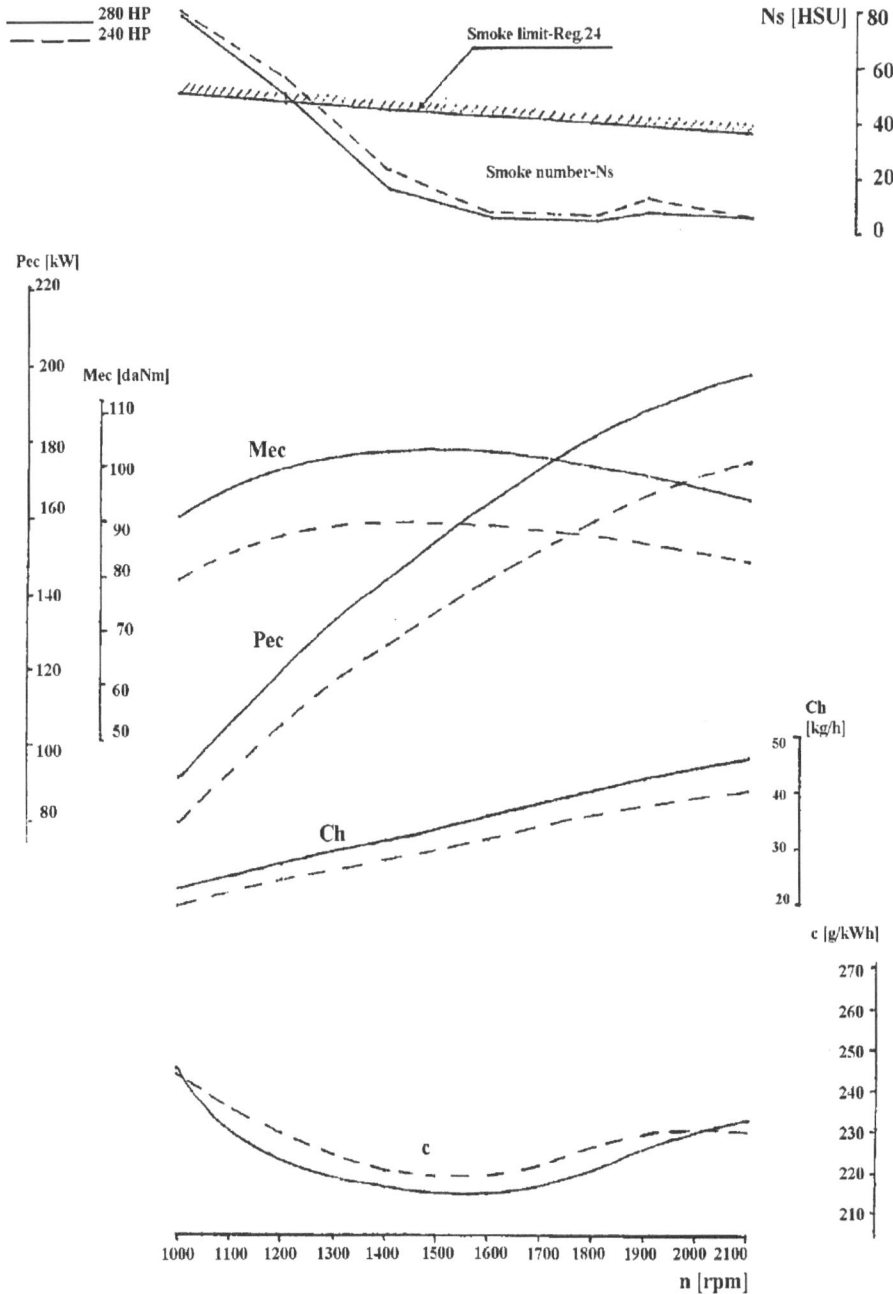

Figure 2: Engine performance versus fuel flow rate

The third step investigated the influence of the axial position of the fuel sprays in the combustion chamber. The engine combustion chamber is a sperical Meurer type, being very sensitive to the position of injectors in the combustion chamber, due to the asymmetrical shape of the walls which influences the air-fuel mixture evolution. From the reference position indicated on the documentation the injector nozzles were rotated with ±5° (forward rotation -counter clockwise and backward rotation –clockwise).The performances are presented in figure 3. It can be observed that the backward rotation produced an increase of power, torque and a smoke reduction; even so, the smoke limit was exceeded for speeds lower than 1100 rpm. This new position of the injector nozzles was mantained in the next tests.

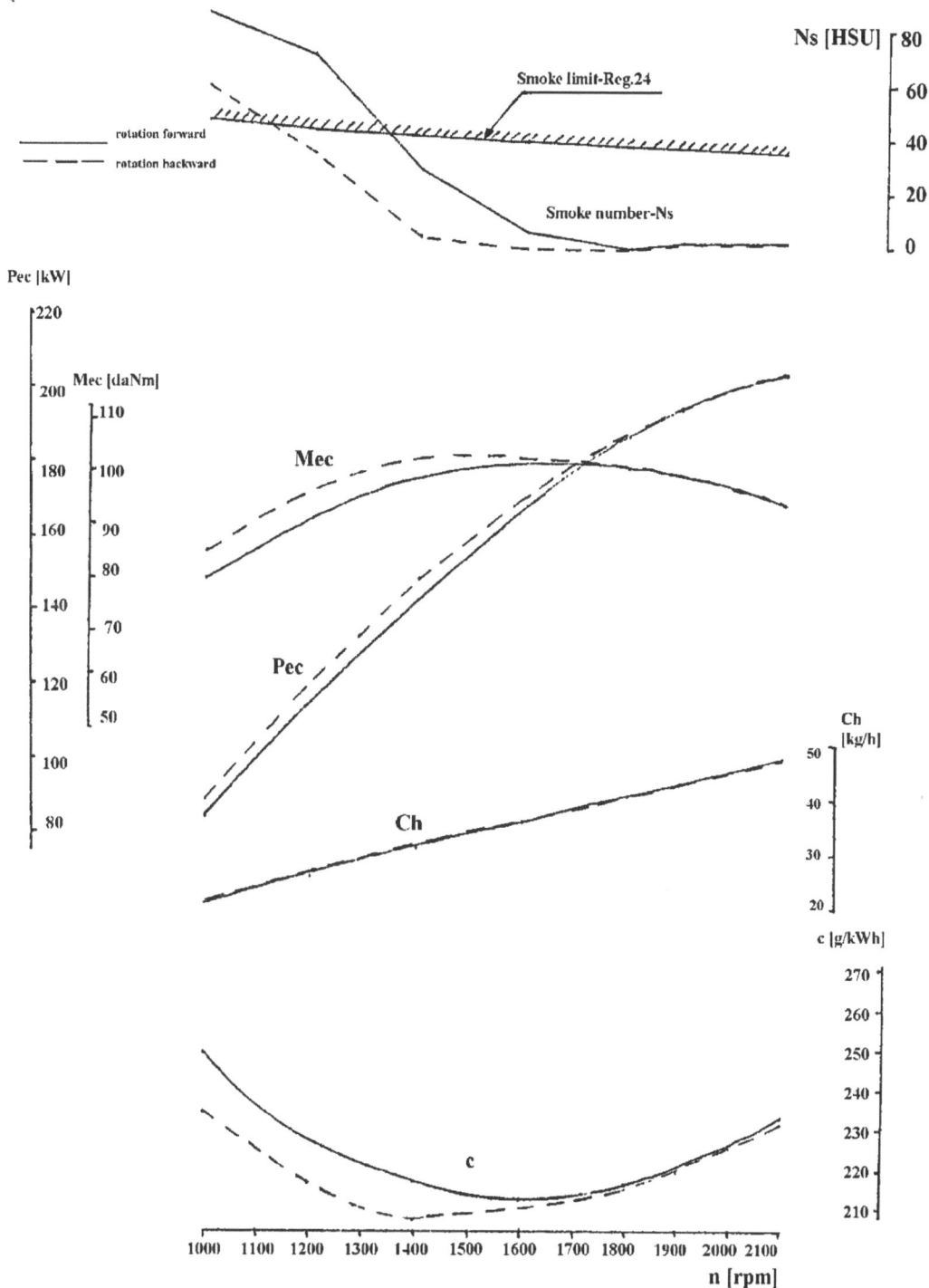

Figure 3: Engine performance versus injector rotation 5°CR forward and backward

The fourth step was to reduce the smoke number under the limits in the range of 1000-1100 rpm by means of adjusting the injection pump fuel flow rate by means of a device called LDA (abbreviation from German term Ladedruckabhängigkeit). That device, specific to in-line pumps, can correct the injected fuel flow rate according to charged air pressure. The law of variation of the measures aforementioned can be simply adjusted passing from the „series value" to a lower value by modifying the charge air pressure corresponding to that speed and reducing the fuel flow rate.The effect of the modification can be seen in figure 4, the smoke number was decreased under the limit as the fuel flow rate was reduced from 140 mm^3/cycle to 110 mm^3/cycle. At 1000 rpm the smoke value was 50 HSU and the imposed limit was 50.4 HSU.

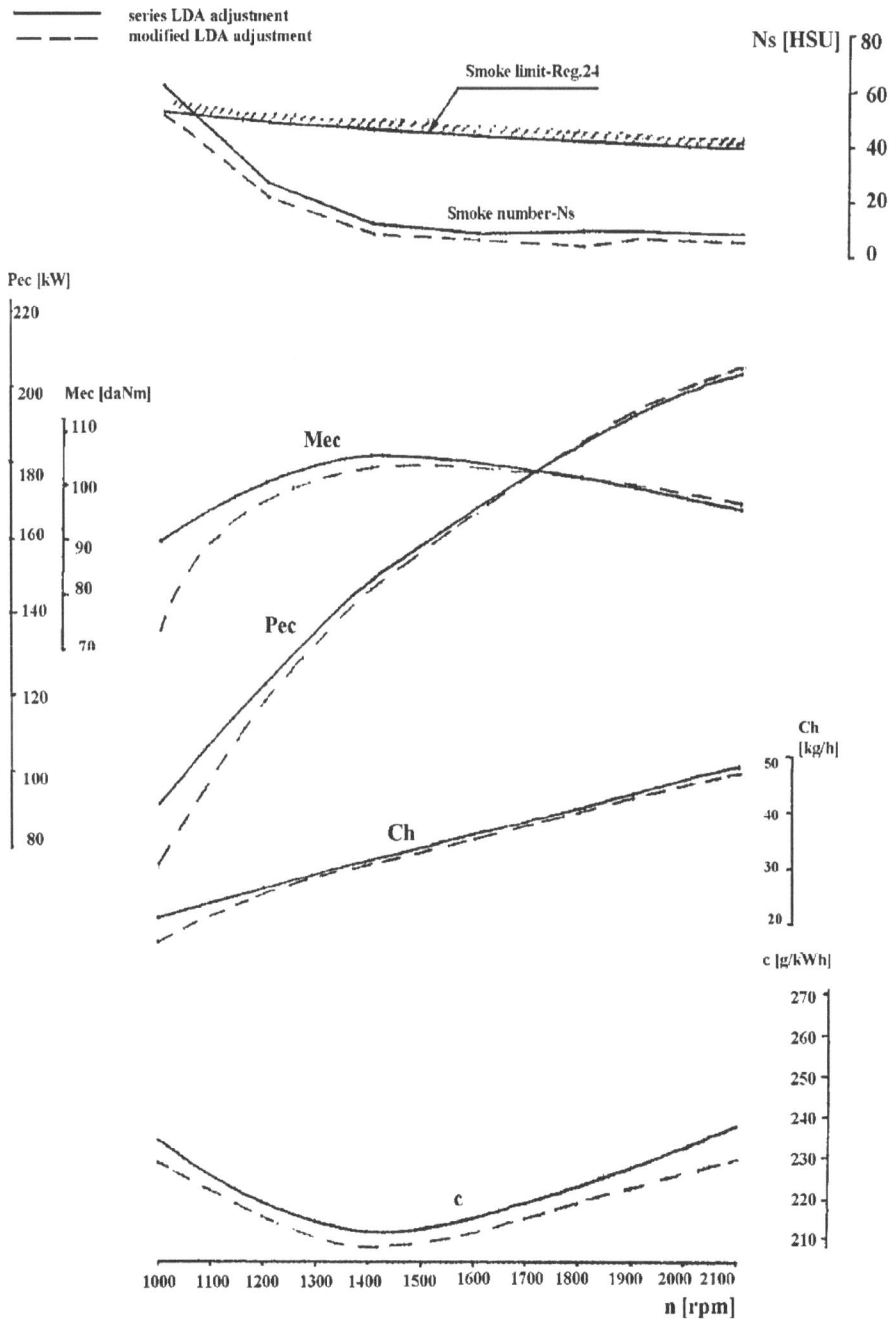

Figure 4: Engine performance versus injection pump correction

In order to verify the engine operation there were measured the mechanical efficiency through the method of power loss in motoring operation, the values ranging from 0.79 to 0.86. It was verified the compression ratio by means of measuring combustion chamber volume being calculated the value of 15.3:1.

(B) Free acceleration test

The visible pollutants are measured in free acceleration on the engine running in the maximum rated speed and maximum power. The absorption coefficient X_M is the mean value of four readings which should not differ with more than 0.15 m^{-1}.

For turbocharged engines the absorption coefficient measured under free acceleration shall not exceed a limit depending on the nominal flow rate corresponding to the maximum absorption coefficient measured during the tests at steady speeds, plus 0.5 m^{-1}.

For the present test, free acceleration smoke measurement were performed according to requirement from Appendix no. 5 of 24 ECE Regulation. The measured value of opacity index in free acceleration is $X_M = 2.0$ m^{-1}. The corrected value X_C of opacity index in free acceleration is: $X_C = 2.5$ m^{-1}. Being the smallest value calculated from: $X_C = X_M (S_L / S_M)$ and $X_C = X_M + 0.5$, in which $S_M = 0.796$ m^{-1} – the measured opacity in stabilised operation mode which is the closest to the prescribed limit and $S_L = 1.90$ m^{-1} – the limited value of opacity corresponding to the speed at which S_M is measured.

4. CONCLUSIONS

The paper summarizes research work performed in four steps in order to obtain the approval on the engine type on visible emissions. The procedure was done step by step proving that experimental adjustments may lead to the meeting of the requirements.

The test results of the engine D2156MTN8R indicated that the engine complies with the values of smoke limits according to ECE 24 Regulation in the following conditions:

- the injection timing is 27° crankshaft rotation;
- injector nozzle position is rotated with 5° backwards (clockwise) from the initial position prescribed in documentation;
- injection pump correction device should allow a fuel flow reduction at 1000 rpm of 20% (30 mm^3/cycle).

ACKNOWLEDGEMENTS

The authors thank to research engineers A. Gal and N. Lungu for useful support and discussions. The support of the Department of Engine Testing of Road Vehicle Institute Brasov is gratefully acknowledged.

REFERENCES

[1]   *** ECE-R 24.03 – Uniform provisions concerning the approval of the compression ignition engines   with regard to the visible pollutant emissions of the engine.

[2]   *** Test performed for engine D2156MTN8R approval according to ECE 24 regulation, Test bulletin 184M, INAR Brasov.

COMPOSITES ACRYLIC COPOLYMERS –WOOD WASTE

L. Dumitrescu[1], A. Matei[1,2], I. Manciulea[2]

[1]Transilvania University of Brasov, , Romania, lucia.d@unitbv.ro
[2]National Institute for Research and Development in Microtehnologies IMT-Bucharest, Romania,

Abstract: *Wood is a three dimensional polymeric composite made primarily of cellulose, hemicelluloses and lignin. Various composite wood products are now obtained, which are preferred as engineering materials, because they are economical, low in processing energy, renewable, displaying superior mechanical properties, dimensional stability, greater resistance to chemical and biological degradation, and less moisture absorption.*

Our research was focused on synthesis and characterization of new composites materials based on acrylic copolymers and wood waste sawdust and calcium lignosulfonate.

Improved properties of the new composites are obtained as the result of complex interactions, by esterification and etherification reactions, between the hydroxyl groups of wood waste sawdust and calcium lignosulfonate and the ester, carboxyl and carboxylate functional groups from the acrylic copolymer (the matrix). The chemistry behind the interactions of acrylic copolymer with lignocellulosic waste sawdust and calcium lignosulfonate was demonstrated by FT-IR analysis and AFM.

The proposed new ecological composites materials are in agreement with nowadays research in the field of recycling wood waste to obtain new ecological materials used as coatings or landscape pavers.

1. INTRODUCTION

Wood has always been an important and versatile material with many uses because of its very aesthetically pleasing character. But wood has some drawbacks such as high moisture uptake, biodegradation, and physical and mechanical property change with environmental factors [1]. These negative inherent properties of wood can be minimized by appropriate chemical treatment such as the structural modification or formation of wood polymer composites [2].

Nowadays the production of composites based on lignocellulosic materials and synthetic polymers has become an important way for recovering, reusing and recycling biomass/wood waste [3].

Lignocellulosic materials contain lignocellulosic materials, important natural renewable resources, contain polymers cellulose, hemicelluloses and lignin, which possess many active functional groups susceptible to chemical reactions, such as: primary and secondary hydroxyls, carbonyls, carboxyls, esters, ether etc. Based on the variety of functional groups, etherification, esterification, alkylation, hydroxyalkylation, graft copolymerization, crosslinking and oxidation reactions have been conducted to different lignocellulosic materials to produce a series of products with many practical applications [3, 4, 5, 6].

The composite materials based mainly on natural polymer, e.g. natural rubber latex, cellulose and starch have a much lower undesirable impact on the environment since they are made from renewable resources. Sawdust and wood flour is the most common wood filler used in wood-plastic composites and other wood-alternative material composites has made major advances in material performance. It is typically a postindustrial source consisting of wood shavings, chips, and sawdust produced by secondary wood product [7, 8].

Sawdust, obtained from wood industry is an abundant by-product which is easily available in the countryside at negligible price. It contains various organic compounds (lignin, cellulose and hemicelluloses) with polyphenolic groups that could bind heavy metal ions through different mechanisms [9, 10]. The performance and stability of sawdust-reinforced composite materials depends on the development of coherent interfacial bonding between sawdust and matrix. The general components of sawdust are cellulose, hemicelluloses, lignin, pectin, waxes, and water-soluble substances [4], [5]. Cellulose, hemicelluloses, and lignin are the main components contributed the strength, flexural, and impact properties of the composites. Moreover the bonding between sawdust and the hydrophobic matrix has effected to the mechanical properties of the composite material [9, 6].

The aim of the present work was to synthesize and characterize new ecological composites materials based on acrylic copolymers (as matrix) and wood waste sawdust and calcium lignosulfonate (as fillers). The sawdust was mixed with calcium lignosulfonate then dispersed into the acrylic copolymer.

2. EXPERIMENTAL

There are several possibilities available when considering the synthesis of wood-polymers composites. Based on some previous experiences [10] we select to mix the aqueous dispersion of the experimental acrylic copolymer with 20% of sawdust and lignin derivative-calcium lignosulfonate (5% in CPa and 10% in CPb).

The matrix of the new composite materials consists on the water based dispersion of the acrylic copolymer obtained by copolymerization of acrylic monomers: ethyl acrylate, butyl acrylate, acrylonitrile and acrylic acid.

Calcium lignosulfonate, was obtained, as waste, from wod and paper industry and characterized in order to put into evidence its chemical reactive potential able to combine/react with natural and synthetic polymers or reagents. The complex analysis [11, 12] of the chemical structure of the calcium lignosulfonate put into evidence the presence of the functional groups: alcoholic hydroxyl (15.00 %), phenolic hydroxyl (14.75 %) carbonyl (10.50%) and carboxyl (11.25 %) in his structure [4]. The sawdust was provided by a furniture plant and has the following characteristics: 44.50% cellulose, 19.00% hemicelluloses, 27.15% lignin, 4.35% organic solvents extractives, 5.00% hot water extractives. The composites

3. RESULTS AND DISCUSSIONS

The new synthesized composites based on acrylic copolymer (as matrix) and wood waste sawdust and calcium lignosulfonate (as fillers) were characterized as follows:

a. The morphology of the proposed composite materials was analysed with an AFM NT-MDT model BL 222 RNTE. The AFM images (Fig.1) detail the surface morphology showing a continuous and uniform distribution of wood waste sawdust and calcium lignosulfonate in the polymeric matrix. The acrylic copolymer –wood waste composites present a dense structure, able to assure good properties, having average roughness value 80.60 nm (CPa), respectively, average roughness value 91.20 nm (CPb).

a)   b)

Figure 1: AFM images of (a) composite CPa and (b) composite CPb

b. The interphase characterization of the composites synthesised was performed by FT-IR spectrometry with a spectrometer Spectrum BX Perkin Elmer, in reflectance mode, in the range of 500-4500 cm^{-1}, after four scans, with 4 cm^{-1} resolution.

The bonding process between the matrx - acrylic copolymer and fillers - sawdust and calcium lignosulfonate can be considered to develop mainly due to the presence of the hydroxyl groups in the structure of sawdust and calcium lignosulfonate which can participate to the esterification reaction with carboxyl groups from the acylic copolymer.

The FT-IR spectra (Fig. 2) performed on the surface of the composites CPa and CPb, confirm the interaction of the acylic copolymer segments with the sawdust and calcium lignosulfonate. Infrared absorption bands of acrylic copolymer, sawdust and lignosulfonate show specific peaks which explain the interactions between the composite matrix (acrylic copolymer) and wood waste fillers - sawdust and calcium lignosulfonate, as follows:

-The absorption bands corresponding to alcoholic hydroxyl (1020-1050 cm^{-1}) and to phenolic hydroxyl (1240-1265 cm^{-1}) are characteristic for lignocellulosic waste (sawdust and lignosulfonate);

-Absorption band at 1085 cm^{-1} was attributed to –SO$_3$H group from calcium lignosulfonate bonded on the acylic copolymer structure;

-Absorption band at 1425 cm^{-1} is characteristic to methylene (–CH$_2$-) groups from lignin, respectively lignosulfonates and sawdust;

- Absorption bands at 1510-1600 cm^{-1} are characteristic to the aromatic nucleus from sawdust and lignosulfonate as lignin structure;

-Absorption band at 1580-1600 cm^{-1} is attributable to the aromatic nucleus from sawdust and lignosulfonate

-Absorption band located at 1724-1729 cm^{-1} is characteristic for carboxyl groups from carboxylic acids and esters, certifying the formation of esters by esterification of carboxyl groups present in the matrix of acrylic copolymer with hydroxyl groups from both fillers - sawdust and lignosulfonate.

-The absorption band at 2981-2958 cm^{-1} indicates the presence of metoxy group (- OCH$_3$) characteristic to lignin structure, respectively to sawdust and calcium lignosulfonate.

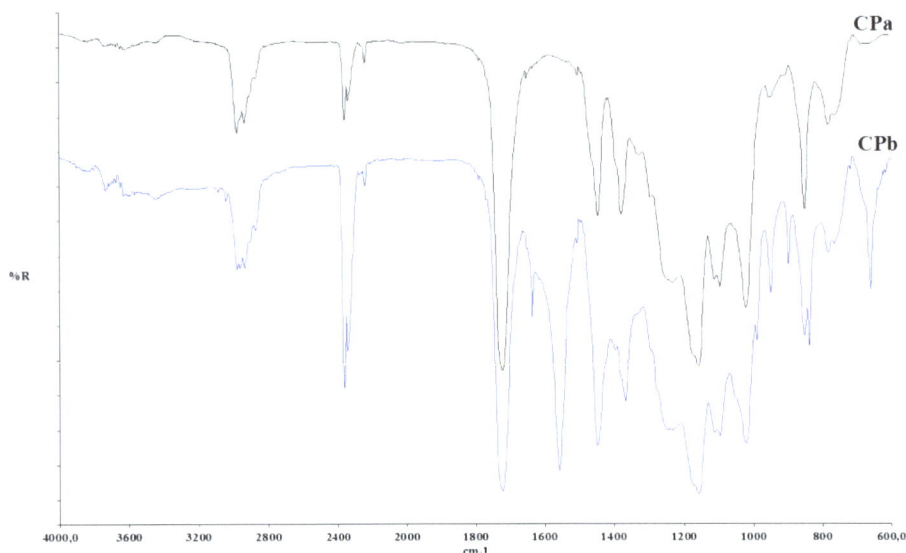

Figure 2: FTIR spectra of the acrylic copolymer-wood waste composites CPa and CPb

c. Investigation of biological durability of composites

Investigation of biological durability of composites wood-polymers is very important for outdoor applications. *The biocide activity investigation* have been performed (according to STAS 8022/91) in order to identify the composites resistance against biological attack of the microorganism from soil. Considering the biocide activity of acrylic copolymers [10] and of the component calcium lignosulfonate, the obtained composites were biologically investigated by insertion in soil for a period of 28 days. After testing, the samples were visually examined by optical microscopy in order to establish the attack level. The results of the biological testing are presented in the Table 1. The fungal growth was classified between 0 and 4, as following:

Table 1:. The results of the biological testing of the copolymer-wood waste composites

Treatment type	Degree of attack	Note	Preservation Efficiency
Wood reference sample	90% of surface	4	According to STAS 8022/91
Composite CPa	14% of surface	2	slight growth (good)
Composite CPb	9% of surface	2	slight growth (good)

The fungal growth was classified between 0 and 4, as following:

0 - no growth;
1 - trace of growth detected visually;
2 - slight growth or 5-20% coverage of total area;
3 - moderate growth or 20-50% coverage;
4 - plenty of growth or above 50% coverage.

d. Determination of the mass losses of composites

The mass losses of the composites after 28 days testing in soil was determined gravimetrically and releved that low mass losses were observed for both composites samples exposed to decay testing in soil, respectively 3.0% for composite CPa and 2.0% for composite CPb, comparing with 28% mass loss for pine sapwood control sample.

4. CONCLUSIONS

The research was focussed on synthesis and characterization of two new ecological composites based on an experimental acrylic copolymer as matrix with wood waste sawdust and calcium lignosulfonate, as fillers.

FTIR spectra show that the presence of the hydroxyl, carbonyl, carboxyl groups in the structure of both sawdust and calcium lignosulfonate represents anchoring points between the acrylic copolymer chains and lignocellulosic materials and explains the improved properties of the new composites acrylic copolymer-wood waste sawdust and calcium lignosulfonate.

The proposed composites materials submitted to the standard testing procedures for wood preservatives exhibited low moisture absorption and biocide activity against the microorganisms from soil. It was observed that increasing the proportion of calcium lignosulfonate increase also the biocide activity of composite CPb.

REFERENCES

[1]   Hill CAS. Modifying the properties of wood. In: Hill CAS, editor. "Wood modification. Chemical, thermal and other processes", John Wiley & Sons Ltd., 175–90, 2006.

[2]   Kamdem DP, Pizzi A, Jermannaud A. "Durability of heat treated wood", Holz als Roh-Werkstoff 60,1–6, 2002.

[3]   Dumitrescu L., Perniu D., Manciulea I. "Nanocomposites based on acrylic copolymers, iron lignosulfonate and ZnO nanoparticles used as wood preservatives", Solid State Phenomena, vol. 151, ISSN 1012-0394, 139-144, 2009.

[4]   Hon, D.N.S., "Chemical Modification of Lignocellulosic Materials", Mark Dekker, Publishers, New York, NY, 1996.

[5]   Zakis G. F., "Functional analysis of lignins and their derivatives", TAPPI Press, Atlanta, Ga, 1994.

[6]   Dumitrescu L., Timar M. C, Proceedings of the Baltic Polymer Symposium, Riga, RTU, ISBN, 9984-32-535-0, 2003, p. 69-73.

[7]   Winandy, J.E., Stark, N.M. and Clemons, C.M., "Considerations in recycling of wood-plastic composites". *In* Proceedings of the 5th International Conference on Global Wood and Natural Fibre Composites, Germany. 27-28 April 2004. P A6-1-9, 2004.

[8]   A. Prompunjai, W. Sridach, World Academy of Science, "Preparation and Some Mechanical Properties   of Composite Materials Made from Sawdust, Cassava Starch and Natural Rubber Latex", World   Academy of Science, Engineering and Technology, 48 ,
        928-932, 2010.

[9]   Li, X., Tabil, L.G. and Panigrahi, S., "Chemical Treatments of Natural Fiber for Use in Natural Fiber-
        Reinforced Composites", A Review. J. Polym. Environ. 15: 25–33, 2007.

[10]   Dumitrescu L., Manciulea I., "New ecomaterials for wood preservation". Environmental Engineering   and Management Journal, Vol.8, No.4, 793-796, 2009.

[11]   Dumitrescu L., Petrovici V., Tica R., Manciulea I., "Reducing the environment hazard using the
        lignosulfonates as copolymerization partners" , Environmental Engineering and Management Journal,   **1**,
437-443, 2002.

[12]   Zakis G.F., "Functional analysis of lignins and their derivatives", TAPPI Press, Atlanta, Ga., 1994.

METALLOGRAPHIC AND MICROSCOPIC ANALYSIS OF FIBERGLASS COMPOSITE MATERIAL, AFTER TENSILE TEST

A.E. Stanciu[1], R. Purcarea[1]

[1] *Transilvania* University of Brasov, Brasov, ROMANIA, ancastanciu77@yahoo.com, r.purcarea@unitbv.ro

Abstract: *This metallographic analysis technology can get information about manufacturing multilayer structure type, warp and weft. Microscope meets the needs to reduce the time for evaluation and quality improvement were seen in the image entirely in real time just by turning the adjustment while song notes studied.*
Keyword: *fiberglass, 2D and 3D view, rupture in material*

1. INTRODUCTION

Metallographic analysis is examining the overall naked eye, a magnifying glass or by stereomicroscope at low magnification (below 100x) of blanks, parts, tools or samples specially prepared for this purpose (breaks, sections). Through this analysis you can get information on manufacturing technology as well as details on the operating conditions (breaking the static or fatigue, etc. Through this analysis you can get information on manufacturing technology as well as details on the operating conditions (breaking the static or fatigue, etc.

Microscopic analysis can be both an interim control method and final. Its importance is even greater as this method is simple and requires no special equipment and can be controlled a large number of parts. Application can be made: breaking the surface or form (solidification, condensation, etc.).

Different measurements can be made in 3D: volume measurement, measurement profile section, measuring the distance between two parallel planes, measuring plane angles.

2. MATERIAL STRUCTURE ANALYSIS

2.1. Metallographic analysis

To view the structure we made an metallographic analysis and we could see layers, product homogeneity, and resin.It should be emphasized that the internal structure of typical grain requires no resemblance to the real structure that can be isolated in a section of material. The notion of granule feature is a hypothetical construction, conveniently placed to separate the effects of orientation, the influence of composition and geometry of reinforcement (as reflected by the specification of behavior typical grain).

Macroscopic analysis was performed in this work, through a final control method. The study was conducted on specially prepared surfaces (ground and attacked with a reagent). By analyzing macroscopic defects can be identified and determined which were formed at different stages of manufacturing technology.

It generally means multilayer composite as a constructive element made of two layers showing minimum for a given fixed direction, different values of the elastic properties (modulus of elasticity longitudinal, transverse, transverse contraction coefficients). It deals specifically with multilayer composites whose individual layers are unidirectional armed; all the fibers are linear and parallel to each other.

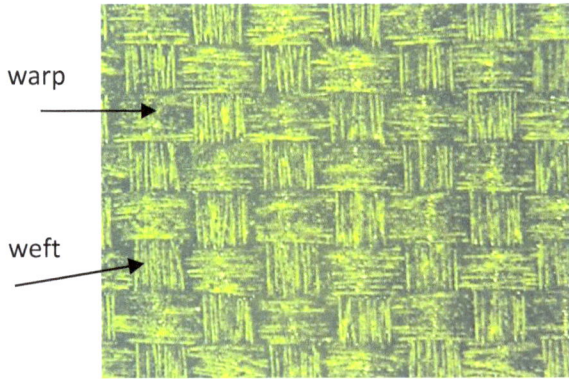

Figure 1 Dull face of compozite material
MAT-Roving

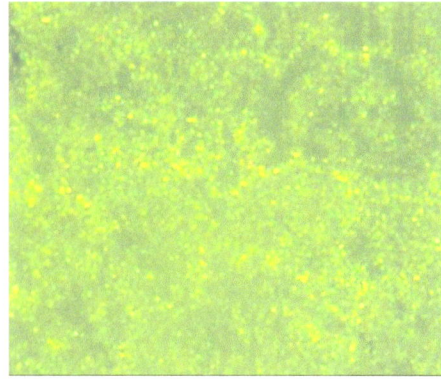

Figure 2 Bright face of compozite material
MAT-Roving

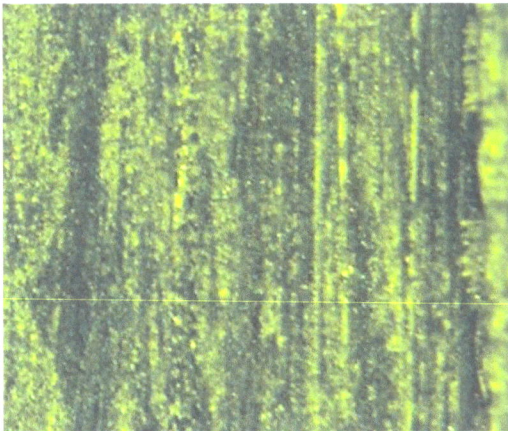

Figure 3 Longitudinal view

Figure 4 Transversal view

Appropriate fiber reinforced composite tubes required by internal or external pressure. If there are preferential directions of fiber orientation when there are preferential directions of elastic properties. If fiber arrangement is symmetrical about two axes perpendicular to each other then the special case of anisotropy is orthogonal or orthotropic. Symmetry axes are in this case orthotropic axes.

2.2. Microscopic analysis

To study changes in the composite material were used two powerful devices: camera and microscope VHX. After tensile and bending test the specimens have been studied in the breaking area.

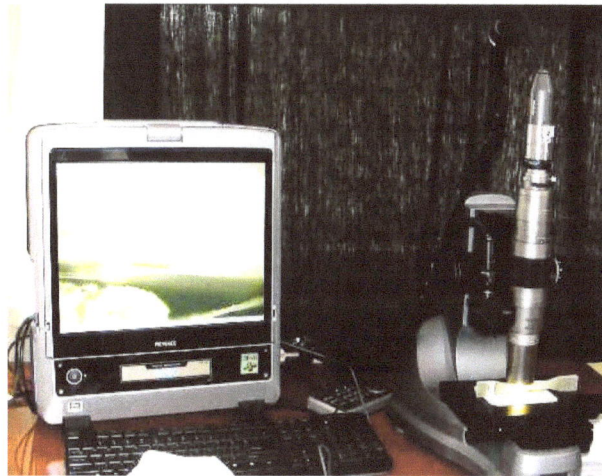

Figure 5 Microscop VHX

VHX 500 integrates all steps from zoom to measure 3D profiles with the ability to zoom the object between 500 and 2000 times.

After tensile test , we studied the specimens with the camcorder which was increased by 5 times in areas where breaking and microscope VHX were increased by 500 times.

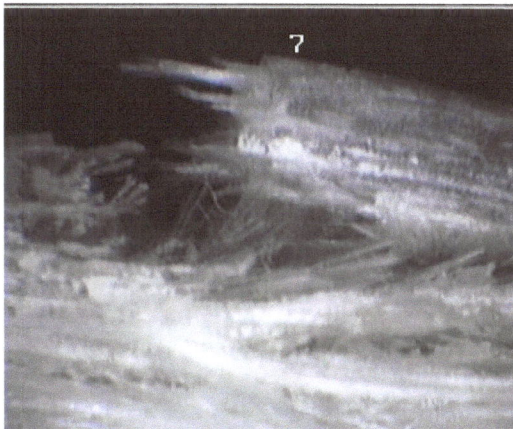

Figure 6 Specimen 8 MAT-Roving increase 5 times with camcorder

Figure 7 Specimen8 MAT-Roving increased 500 times with the microscop

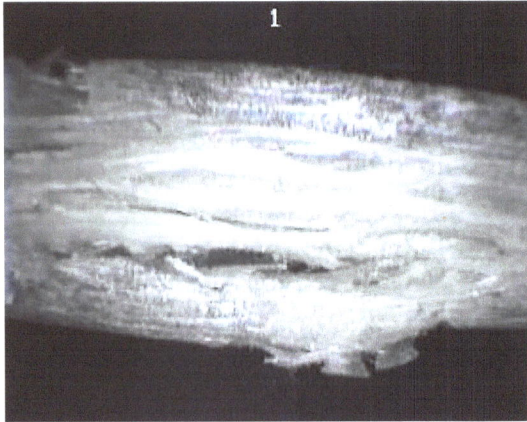

Figure 8 Specimen2 MAT-Roving increase 5 times with camcorder

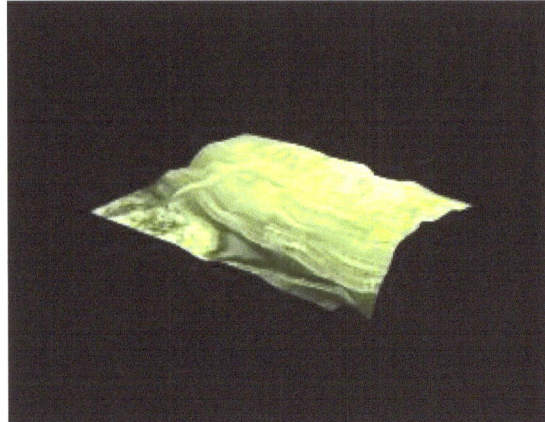

Figure 9 Specimen 2 MAT-Roving increased 500 times with the microscop

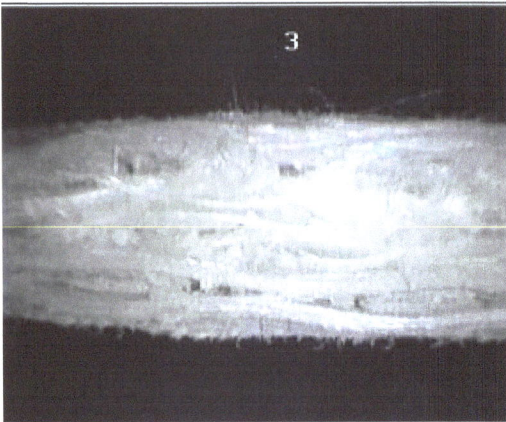

Figure 10 Specimen b3 Roving increase 5 times with camcorder

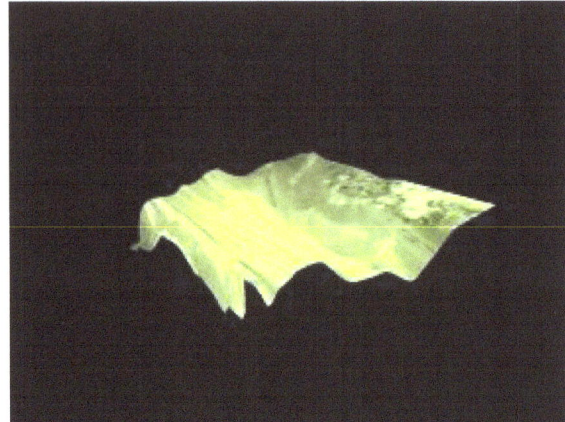

Figure 11 Specimen b3 Roving increased 500 times with the microscop

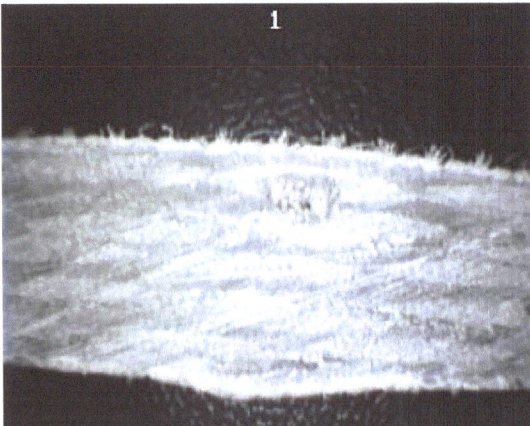

Figure 12 Specimen u3 Roving increase 5 times with camcorder

Figure 13 Specimen u3 Roving increased 500 times with the microscop, 2D

3. CONCLUSION

Conducting research on macroscopic composite samples revealed stratified and inhomogeneous structure of the material. The analysis of micrographs for specimens studied was observed distributions and glass fiber orientation of the material MAT and those of roving, distributions of fillings and holes, uneven, material damage, vacuoles and existing inclusions in the matrix resin.

Extreme complexity of products, continuous emergence of new scientific models and theories that changed the approach to technological act itself makes anything requiring a high concentration of material and conceptual forces. View in detail was done with a powerful microscope that can increase 500 times the area studied up to 2000 times, and the images were as clear and even in the depth of the material. Results were obtained with a microscope both 2D and 3D, noting also how to change the whole structure of these materials.

ACKNOWLEDGEMENTS

This paper is supported by the Sectoral Operational Programme Human Resources Development (SOP HRD), financed from the European Social Fund and by the Romanian Government under the contract number POSDRU/89/1.5/S/59323.

REFERENCES

[1]   Stanciu A., Teodorescu Draghicescu H., Candea I., Secara E., Experimental Approaches Regarding the Elastic Properties of a Composite Laminate Subjected to Static Loads, The 3rd International Conference   on International Conference *Computational Mechanics and Virtual Engineering COMEC 2009*, 29 – 30   October 2009, Brasov, Romania.

[2]   Stanciu A., Cotoros D., Study Concerning the Mechanical Tests of MAT&ROVING Fiber Reinforced Laminated Composites, WSEAS, Puerto de la Cruz, Spain, 2009

[6]   Donaldson, R.L. & Miracle, D.B. (2001). ASM Handbook Volume 21: Composites, ASM   International, ISBN: 978-0871707031.

THEORETICAL AND EXPERIMENTAL DETERMINATION OF PROPERTIES FOR COMPOSITE MATERIAL TYPE ROVING SUBJECTED TO BENDING

A.E. Stanciu[1], R. Purcarea[1], M.V. Munteanu[1], S. Vlase[1], H. Teodorescu -Draghicescu[1]

[1] *Transilvania* University of Braşov, ROMANIA, ancastanciu77@yahoo.com, r.purcarea@unitbv.ro,

Abstract: *Finite element modeling has enabled reinforcement solution optimization of welded frame in order to stress reduction in critical points based on strain and stress solutions. Validation of reinforcement solutions of welded frames obtained by finite element modeling was done experimentally. Thus, the first model is a physical model, very often called analytical model and it is derived from the actual structure by abstraction or idealization of the geometry, boundaries and boundary conditions etc.*
Keyword: *Finite element method, fiberglass composite, roving*

1. INTRODUCTION

As far analytical techniques have developed significantly and although still widely used shell discontinuity analysis in structural analysis, it is increasingly replaced by numerical computing methods. Most widely used technique in contemporary design of complex mechanical systems is the finite element method, a powerful technique that allows detailed modeling of complex systems.

Method is used to determine the flexural behavior of test specimens and determination of flexural strength, modulus of elasticity in bending and other aspects of relationship stress / strain under the circumstances. Apply a lever simply supported, loaded in three or four point bending. How to place and test specimen is chosen so as to limit deformation to shear and interlaminar shear avoid breakage.

If the materials have physical properties, such as elasticity, dependent on direction, specimens should be chosen so that during the test, voltage sag is applied in the same direction as that in which the products are requested service.

Experimental determination of the specimens was performed according to standard EN ISO 14125 *Compozite de materiale-plastice armate cu fibre, Determinarea proprietăţilor de încovoiere.*
Specimen supported as a lever is subjected to bending constant speed until rupture or deformation reaches a predetermined value. During the test measured the force applied to the specimen and arrow.

2. THEORETICAL METHOD (FEM) AND EXPERIMENTAL METHOD (ETM)

The material behavior was studied during pure bending, when subjected to a force of 600N and the postprocessor program used was MSC Nastran.

Most widely used technique in contemporary design of complex mechanical systems is the finite element method, a powerful technique that allows detailed modeling of complex systems.

This idealization requires very often and assumptions regarding the characteristics of the material or constitutive laws, which are unknown for real structural.

The study was made for the following types of composite materials:
• RT 800, structured on 4 layers, weft disposition, fiberglass composite (fabric) in the matrix of epoxy, figure 1;
• RT 800, structured on 4 layers, warp disposition, fiberglass composite (fabric) in the matrix of epoxy, figure 2.

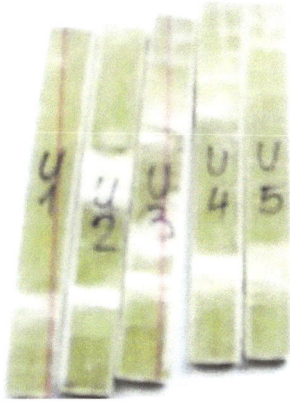

Figure 14 Roving specimens with 4 Layers on warp

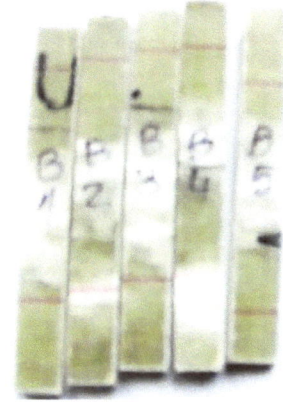

Figure 15 Roving specimens with 4 Layers on weft

Table 1 Load values of the parameters for roving specimen

	E U1	E U2	E U3	E U4	E U5	E B1	E B2	E B3	E B4	E B5
Calibrated part length *mm*	60	60	60	60	60	60	60	60	60	60
Load speed *mm / min*	5	5	5	5	5	5	5	5	5	5
Test-piece width *mm*	10.4	10.8	11	10.5	10.6	10.5	10.8	10.5	10.4	10.6
Test-piece thick-ness *mm*	4	4.2	4,2	4.1	4.5	4.5	4.4	4.5	4.4	4.4
Area *mm²*	41.6	45.36	46.2	43.05	47.7	47.25	47.52	47.25	45.76	46.64

Table 2 Mean values of mechanical bending

Mechanical properties for Roving subjected at bending	Roving average values for roving at warp	Roving average values for roving at weft
Stiffness, *N/m*	207560	247850
Young's modulus, *MPa*	11397	10404
Tensile Strength, *Nm²*	0,69286	0,82738
Extension from preload at Minimum Extension, *MPa*	288,02	285,86
Strain at Maximum Extension	0,031726	0,030984
Deformația specifică a epruvetei de la valoarea inițială până la valoarea maximă, *mm*	7,9326	7,4885
Work to Minimum Extension,*Nmm*	4967	5652
Load at Minimum Extension, *kN*	0,97834	1,1098
Stress at Minimum Extension, *MPa*	261,39	258,36
Elongation at Fracture	0,040164	0,040205

For material RT 800 (warp fabric) the specific deformation experimental using 600 N force, was determined between values -0.016164 și 0.017041 and theoretical was determined between values -0.014 și 0.014, figure 2.

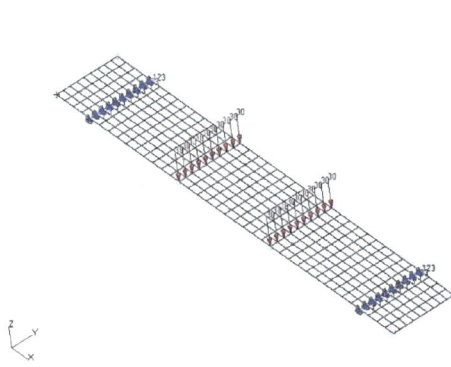

Figure 16 Specimen RT800 warp subjected to bending , at 600 N

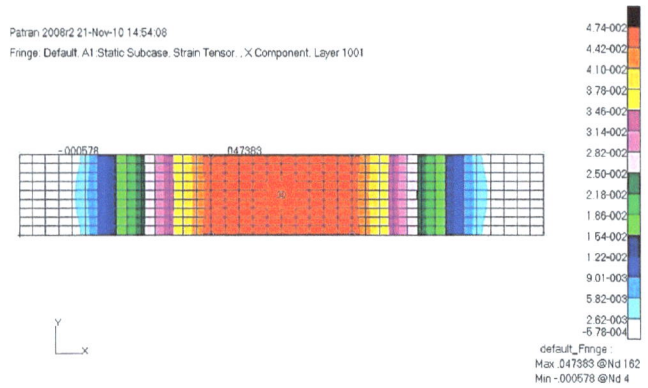

Figure 17 Specific strain distribution RT800, at warp, layer 1

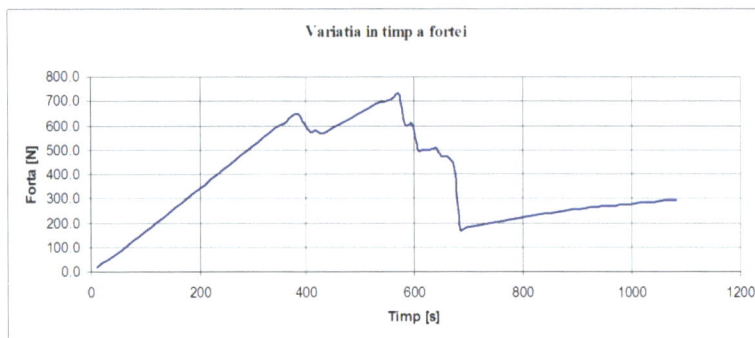

Figure 18 Time variation of force for the specimen RT800 at warp

For type of material RT 800 (weft fabric) the specific deformation experimental using 600 N force, was determined between values -0.02 și 0.02 and theoretical was determined between values -0.020 și 0.021, figure 3.

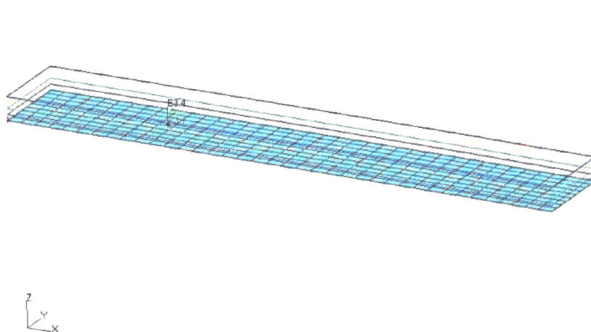

Figure 19 Discretized specimen and layout layers RT800 at weft

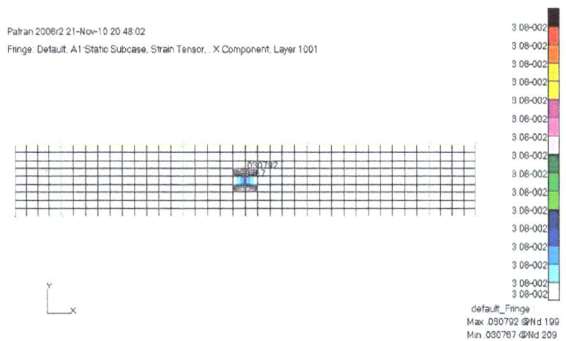

Figure 20 Specific strain distribution for specimen RT800, at weft, layer 1, TER zone

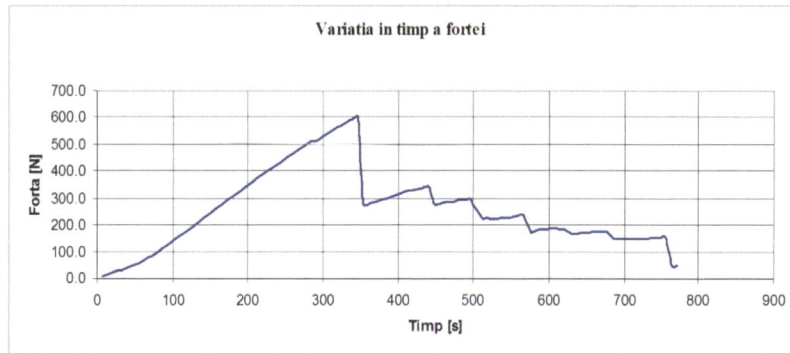

Figure 21 Time variation of the force for specimen RT800 at weft

3. CONCLUSION

Validation of reinforced solutions obtained by finite element modeling was experimentally done by electrical resistive tensometry. The measurement results were correlated with results from finite element modeling.

We find that using the finite element method for the study of anisotropic material such as the composite materials we are able to easily change the load, boundary conditions, way of application, having the opportunity of selecting the optimum choice, the dimensions and required characteristics of materials. Each part of the material can be assessed and the results are checked out by help of the experimental determination.

For reinforcement materials tensile test results depend essentially on the size of the specimen used, temperature of polymers specific influence on the strength and deformation at break them upward.

ACKNOWLEDGEMENTS

This paper is supported by the Sectoral Operational Programme Human Resources Development (SOP HRD), financed from the European Social Fund and by the Romanian Government under the contract number POSDRU/89/1.5/S/59323.

REFERENCES

[1]   Stanciu A., Teodorescu Draghicescu H., Candea I., Secara E., Experimental Approaches Regarding the Elastic Properties of a Composite Laminate Subjected to Static Loads, The 3rd International Conference on International Conference Computational Mechanics and Virtual Engineering COMEC 2009, 29 – 30 October 2009, Brasov, Romania.

[2]   Stanciu A., Curtu I., Bencze A., Method to Obtain the Specific Deformations Theoretical and   Experimental, COMEC 2011, Brasov.

[3]   Stanciu A., Study Concerning the Mechanical Tests of MAT&ROVING Fiber Reinforced Laminated Composites,

CHARACTERIZATION OF MICRO-STRUCTURES OF COMPOSITE OF SMALL WIND TURBINE BLADE FOLLOWING STRUCTURAL TESTING UP TO FAILURE

Č. Mitrović[1], N. Petrović[1], D. Bekrić[1], V.Dragović[2], I. Mileusnić[3],

[1]Aeronautical Department, Faculty of Mechanical Engineering, University of Belgrade, Belgrade, Serbia
[2]WING d.o.o., Belgrade, Serbia
[3]NanoLab, Biomedical Engineering, Faculty of Mechanical Engineering, University of Belgrade, Serbia

Abstract. *This paper presents the results of researchof the microstructure of the composite blade W55RBVS for the wind turbine of up to 6kW power after structural testing up to failure.The first part of the testing consists of the static testing of the structure up to the moment of the blade failure.The aim of the first part of the test was to definerigidityof the blade W55RBVS, to determine the maximum force which leads to faliure and the relative span of the blade failure. Blade testing is performed in the Aerotechnics Laboratoryof the Faculty of Mechanical Engineering, Belgrade University.The second part of the testing consists of comparing the criticallyloaded part with sub-critically loaded part. This test was carried out by atomic force microscopy (Eng. Atomic Force Microscopy-AFM). All the results and analysis are presented in this paper.The test result will be used to redesign the blades.*

Keywords: *structural testing, small wind turbine, composite blade,atomic force microscopy, microstructure.*

1. INTRODUCTION

The program for testing rigidity of the composite rotor blade W55RBVS of the wind turbine Scirocco is defined by standard IEC 61400-2. The rotor blade testing is done in an Aerotechnics Laboratory at the Faculty of Mechanical Engineering, Belgrade University. The purpose of this testing is to define the rigidity of the rotor blade and to determine the maximum force that causes the failure of the rotor blade as well as the spot of the failure.

This research uses atomic and magnetic force microscopy to determine differences of the surface topography and magnetic properties between parts of the wind turbine blade that were under different load during the mechanical testing of rigidity. This may lead to new insight in the structure of the materials used for wind turbine blades and potentially bring improvements to their mechanical properties.

2.STRUCTURAL TESTING

The purpose of this research is to test the established technology of designing the composite rotor bladeW55RBVS of wind turbine Scirocco.
Scirocco is a 5.6 m diameter rotor, 6 kW output professional-scale wind turbine ideal for remote sites with medium power needs, such as small farms, houses, or large homesteads and especially rural or village electrification, sea water desalination, direct electrical water wind-driven pumping.Thiswind turbine is a common practice andit is builtin several countriesin Europeand America (Figure 1&2).

Unlike most of its competitors, which use furling technique to decrease acquisition cost at the detriment of performances and reliability, Scirocco WT is designed and realized without compromise, using state of the art components, centrifugal pitch overspeed regulation device, slewing rings for blades attachment and yawing, 100% sealed direct drive permanent magnet generator. Design and calculations have been made according to IEC 61400-2 design rules.

Scirocco is characterized by low rotation speed, a very high efficiency, especiallyin the lower to medium wind speed range which is prevailing most of the time.

Figure 1 – Wind turbine Eoltec Scirocco

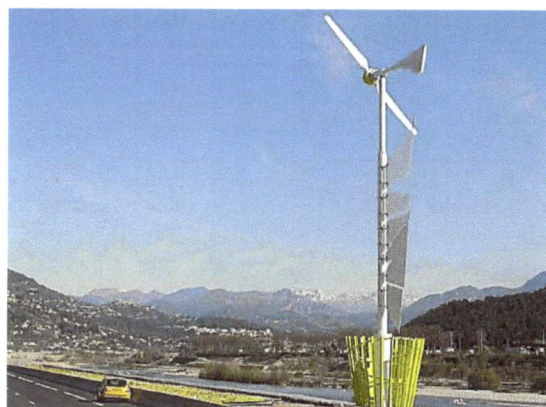

Figure 2 – EoltecSciroccowind turbine along the highwayin France

This is the result of the conjunction of an optimized 2 blades rotor, an efficient direct drive PM generator, and variable speed management combined with the maximal power point tracking control following the constantly varying wind speed. Composite rotor blades W55RBVS are produced in the WING.d.o.o company (Figure3).

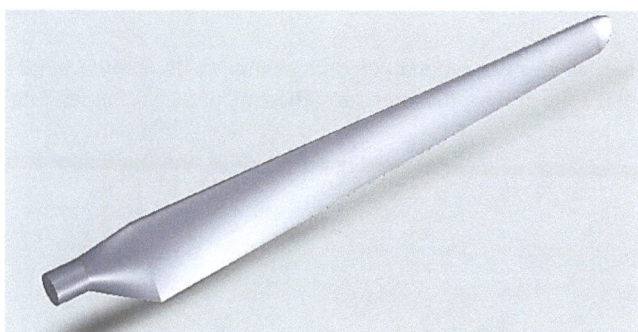

Figure 3 – Designof compositebladesW55RBVS

Figure 4 – Built-composite blades W55RBVS

The design of rotor blades is a balanced integration of aerodynamics, structural analysis and dynamics, choice of different type of composite materials, production technology and overall economy. The design of modern rotor includes choice of blade number, airfoil, chord and twist distribution. Additional criteria are reliability, noise and aesthetic considerations. The blade design process is multi-objective and generally include two optimization functions, maximum annual energy production or minimum cost of energy

To optimize minimum cost of energy requires a multi-disciplinary method that includes an aerodynamic model, structural model, along with cost models for the blades and all major wind turbine components.

2.1 Program for testing

The program for testing rigidity of the composite rotor blade W55RBVS, made in WING.d.o.o. company, of the wind turbine Scirocco is defined by standard IEC 61400-2. The rotor blade testing is done in AerotechnicsLaboratory at the Faculty of Mechanical Engineering, Belgrade University.

Figure 5 Rotor blade acceptance

The purpose of this testing is to define rigidity of the rotor blade and to determine the maximum force that causes failure of the rotor blade as well as the spot of the failure.

Methods used for this purpose and the way of performing this testing are common practice of aeronautical reseach institutes.

2.2 Rotor blade acceptance

Rotor blade acceptance provides a holding link between the wind turbine blade and the test table grid. The rotor blade is mounted on the structure at the angle of attack.

A special rotor blade holder is prepared for this testing.This holder constitutes of extremely rigid specious grid made of steel C and L profiles that are mutually connected by bolt links. The link between the rotor blade and the holder is accomplished through 30mm board and by bolts M8 of 8.8 quality.

2.3 Measuring equipment

The central acquisition unit HBM SPIDER 8 represents a multifunctional model of receiving analogue and digital signals with parallel tracing of the flow of input units, by means of integrated microcomputer, using higher level system to relieve the acquisition routeand to provide the flow of signals from certain "smart sensors" directly to the control unit.

SPIDER 8 is a multichannel acquisition unit designed for dynamic parallel measuring. Thanks to integration with personal computer as a higher level system, the process of measuring is remarkably simple and the total acquisition system is compact and of small dimensions.

This eight-channel acquisition unit provides 9600 measurings per second per each channel with resolution of 16 bits. All 8 A/D convertors work simultaneously and in a real time follow the transformation of physical values into a digital signal.

Two modules are used for this experiment in this way providing that 16 measuring values are at a disposal for the experiment.

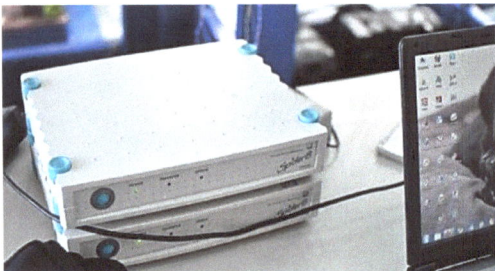

Figure 6 – Two modules of SPIDER 8

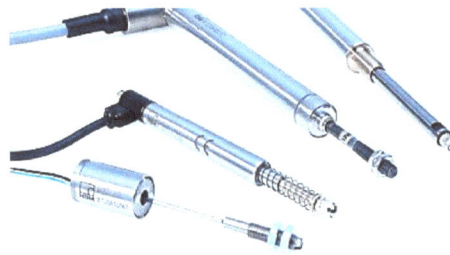

Figure 7 –Inductivedisplacementsensor **LVDT**- HBM

Measuring of chosen spot movement on the wind turbine blade is performed by means of standard inductive movement measurer**LVDT**brand HBM accuracy class ±0,2% and measuring scale from ±1 to ±500mm. Four different spots are considered in measurement performing.

Six force sensors specially made for this purpose are used for measuring the force. Each force sensor consists of four measuring tapes glued on the body of the sensor in a way shown on the Figure 8.The function of the two measuring tapes glued in axial direction is to measure the deformation of the sensor body during its load, while the function of the other two measuring tapes glued crosswise is to compensate the deformations caused by sensor body bending and in this way to provide that the measuring value should be proportional only to axial load.

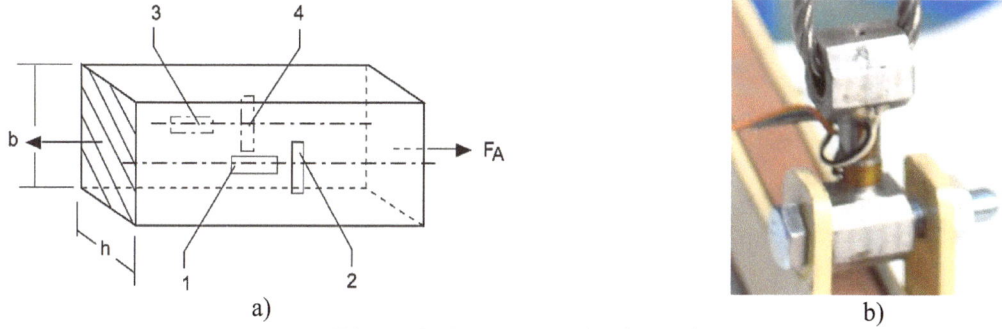

Figure 8 –Force sensora) scheme; b) on theblade

Each of the force sensors is connected to Vitston full measuring bridge. The measuring bridge is supplied with direct input voltage V_I, while output voltage V_O depends on resistance change of the measuring tapes R_1, R_2, R_3 and R_4 caused by their deformations and it is proportional to axial force value that influences the sensor.The method of bonding the measuring tapes into measuring bridge provides above mentioned compensation of bending moment influences on the output voltage.

Introducing force is performed by means of specially made system that consists of supporting structure, pulley, reductor unit and engine with frequent control. The maximum force possible that can be achieved with this system is 2500 daN.

Figure 9 –Constrictionforforce distribution

2.4 Realization of the experiment

The rotor blade is loaded in six sections by lyres and measured from the place of the blade clamp. The rotor blade mass is 11.9kg.

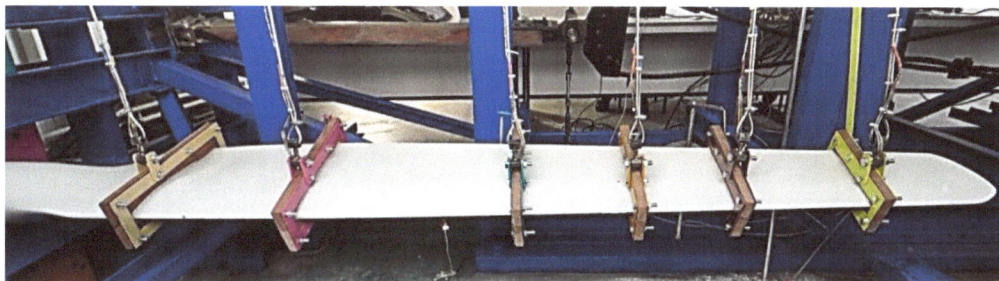

Figure10 –Lyresfor the introductionforce

The experiment is performed in the accordance with IEC 61400-2 standard. The rotor blade failure is marked with the resulting force of 490.76 daNunder maximum deflection of27.999mm.After the testing is completed, some separations are recorded on the following distances (mm):

Leading edge			Trailing edge		
	Beginning	56.5		Beginning	57
	End	106.5		End	103

Figure 11 – The position of the bladefracture

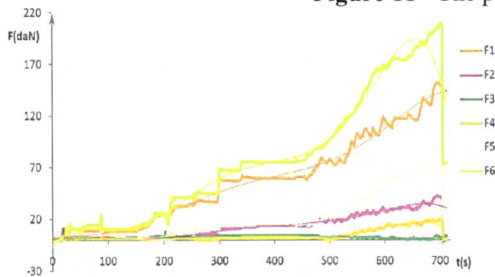

Figure 12 – The load distribution of each section in function of time

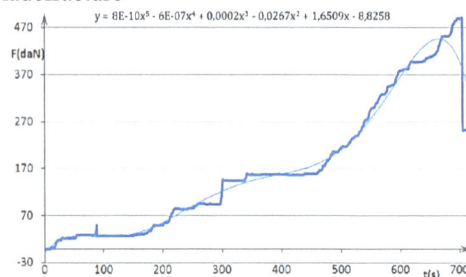

Figure 12 – The resulting force distribution in function of time

3. TESTING BY AFM/MFM MODE

The second part of the testing consists of comparing the critically loaded part with sub-critically loaded part. The composite material is micro-glass fibers reinforced with epoxide resin. The impellers were exposed to critical load until failure and parts from critically and sub-critically loaded regions of wind turbine blade are investigated by Magnetic Force Microscopy (MFM). Magnetic force microscopy is able to provide characterization of surface and internal structure near surface of the sample, which is used in this paper to give analysis of structural change of material with different loads. Results and comparison of microstructure of two parts of wind turbine blade are analyzed and presented

3.1 Measuring equipment

The scanning probe microscope used in this study is a SPM-5200 from JEOL, Japan. The SPM-5200 is a multipurpose, high resolution scanning probe microscope offering ease of use with diverse measurement and sample environments.

3.2 Realization of the testing

AFM mode. The AFM mode of SPM consists of a cantilever with a sharp tip (probe) at its end that is used to scan the specimen surface. The cantilever is typically silicon or silicon nitride with a tip radius of curvature on the order of nanometers. When the tip is brought into proximity of a sample surface, forces between the tip and the sample lead to a deflection of the cantilever according to Hooke's law.

Figure 13 – JEOL SPM 5200

Typically, the deflection is measured using a laser spot reflected from the top surface of the cantilever into an array of photodiodes. TappingMode AFM is a more recent development in which the imaging probe is vertically oscillated at or near the resonant frequency of the cantilever.

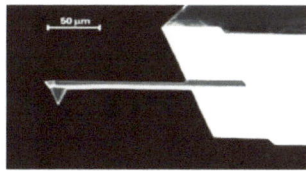

Figure 14 – Cantilever with a sharp tip (probe)

Electromechanical feedback maintains the oscillation at a constant amplitude during scanning. The image is produced by mapping the distance when the scanner moves vertically, to maintain the constant oscillation amplitude at each lateral data point.

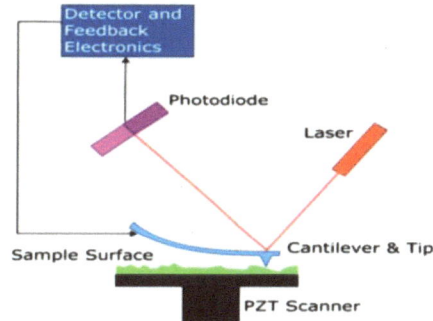

Figure 15 – Scheme ofmeasurementsusing a laser

The key advantage of TappingMode is the elimination of the lateral shear forces present in contact mode, which, on many specimens, can damage the structure being imaged. TappingMode AFM can be conducted in an air or liquid environment.

MFM mode. Magnetic Force Microscopy (MFM) is a secondary imaging mode derived from TappingMode mode that maps magnetic force gradient above the sample surface. Unlike AFM, the MFM uses a probe tip coated with a thin film of ferromagnetic material. This kind of probe tip will react to the magnetic domains on the sample surface. The image of the sample is obtained with two-pass technique in which initial scan is used to obtain the topography of the sample. In the second scan, the tip-sample distance is increased and the biased tip is scanned along the topography line obtained from the first scan. The topographical line maintains constant tip sample distance, which equals the line of the constant van der Waals force.

So, when the tip follows the topography line in the second scan, the van der Waals forces acting on the tip are kept constant. Thus, the only change in force affecting the signal is the change of the magnetic force. Magnetic property evaluation is based on the magnetic force gradient image that shows the qualitative distribution of the magnetic field within the scanned area, pointing to the local distribution of magnetic properties. The cantilever used in this study is produced by MikroMasch (Estonia) by trade name NCS18 Co-Cr. MFM probe is silicon etched probe tip that has conical shape. It is coated with Co and Cr layers, so resulting tip radius with the coating is 90nm. Full tip cone angle is 40°.

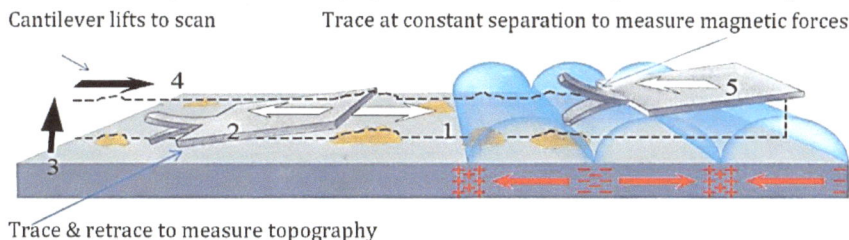

Figure 15 – LiftMode to obtain both topography and magnetic force gradient

3.3 Results

Average of roughness:

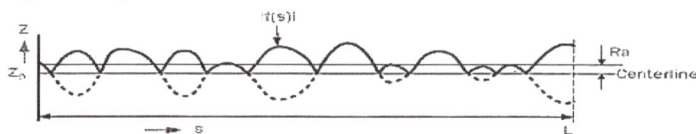

$$R_a = \frac{1}{L}\int_0^L |f(s) - Z_0|\, ds \tag{1}$$

Root mean square roughness:

$$Rq = \sqrt{\frac{1}{L}\int_0^L (f(s) - Z_0)^2\, ds} \tag{2}$$

The 10-point average roughness is defined as the "sum of the average value of the absolute values of the deviation from the centerline between the largest deviation and the fifth deviation, and the average value of the absolute values of the deviation between the smallest deviation and the fifth deviation.

$$Rzijs = \frac{\left|z_{p1} + z_{p2} + z_{p3} + z_{p4} + z_{p5}\right| + \left|z_{v1} + z_{v2} + z_{v3} + z_{v4} + z_{v5}\right|}{10}$$

(3)

The height z at the position d along the line is defined as: z=f(s). The height Zo of the centerline is defined as:

$$Z_0 = \frac{1}{L}\int_0^L f(s)ds$$

(4)

The maximum difference between high (maximum value Z) and low (minimum max value Zmin) (Rz) : Rz= Rmax-Rmin

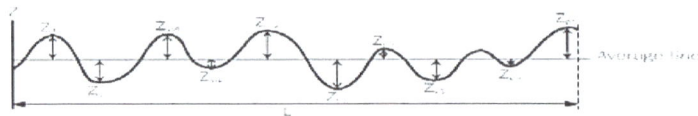

Sample 1- Lessloaded

It is expected for the part of material that is less loaded (Sample 1) to have less surface roughness as layers of silicone fibers held by epoxy will be more evenly aligned. AFM research shows that Sample 1 have smaller average surface roughness (Ra)by 200 nm than in Sample 2 which had higher load during the breaking of wind turbine. This could mean that inner layers of the material were deformed and may not be aligned as in their original position.

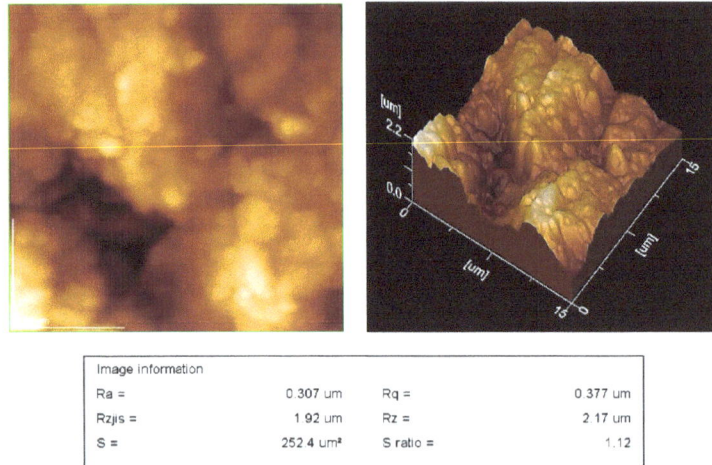

Image information			
Ra =	0.307 um	Rq =	0.377 um
Rzjis =	1.92 um	Rz =	2.17 um
S =	252.4 um²	S ratio =	1.12

Figure 16 – Sample 1a

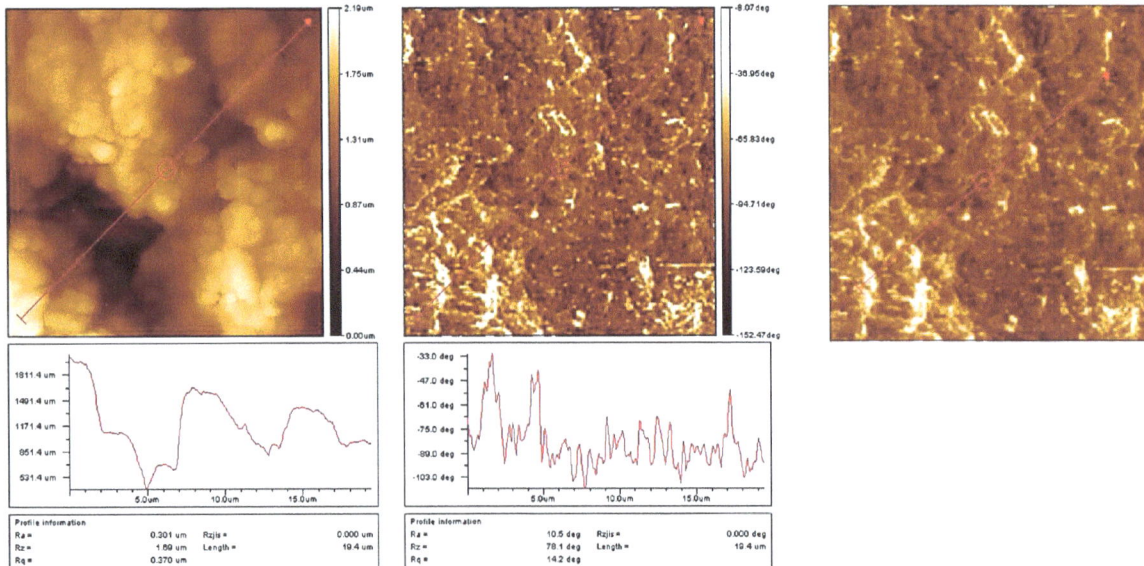

Profile information			
Ra =	0.301 um	Rzjis =	0.000 um
Rz =	1.59 um	Length =	19.4 um
Rq =	0.370 um		

Profile information			
Ra =	10.5 deg	Rzjis =	0.000 deg
Rz =	78.1 deg	Length =	19.4 um
Rq =	14.2 deg		

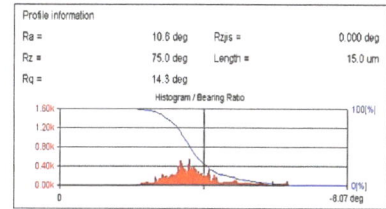

Figure 17 – Sample 1b

Sample 2- Higher load

From the figures representing magnetic characteristics of the material it is apparent that sample 1 has relatively uniform distribution of magnetic properties while sample 2 has ''black dots'' which represent abrupt change in sample magnetization i.e. change in magnetic field gradient. The existence of so called ''black dots'' on the figure of higher loaded sample (sample 2) indicates on possible changes in material structure that is on its stratification upon loading.

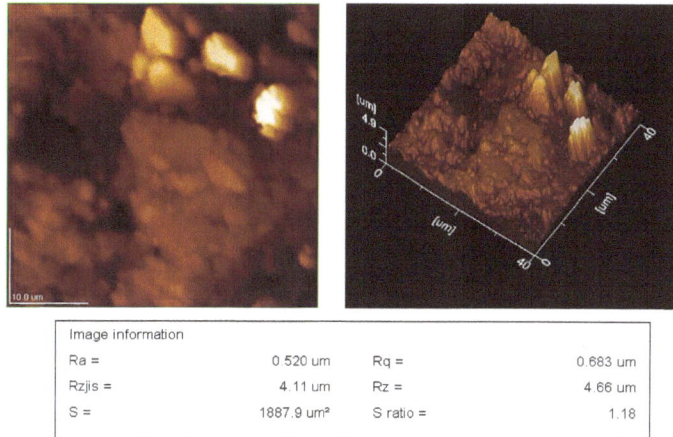

Image information			
Ra =	0.520 um	Rq =	0.683 um
Rzjis =	4.11 um	Rz =	4.66 um
S =	1887.9 um²	S ratio =	1.18

Figure 18 – Sample 2a

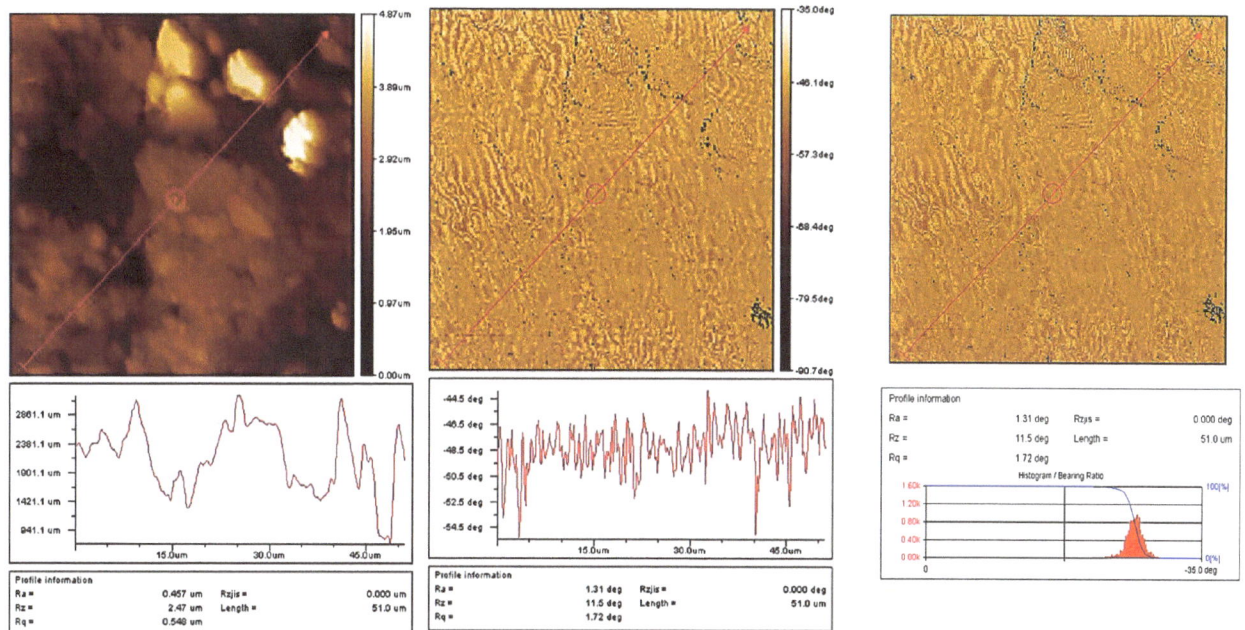

Figure 18 – Sample 2b

4. CONCLUSION

This experiment confirmed that the wind turbine blade W55RBVS has very high rigidity.
Research by Magnetic Force Microsopcy shows that sample taken from part with higher load have largest surface roughness and height differences as it is expected. The existence of ''black dots'' (''magnetic holes'') indicates on stratification inside the composite.

351

ACKNOWLEDGMENTS

This research was supported by the Ministry of Science and Technological Development, Republic of Serbia projects: TR35035, TR36001

REFERENCES

[1]   Č. Mitrović, N. Petrović, A. Bengin, D. Bekrić, V.Dragović, A. Simonović, G. Vorotović, S. Radojević, D. Stamenković,*Structural Testing of Small Wind Turbine Blade up to Failure*, International Conference on Innovative Technologies. Conference IN-TECH 2011, Bratislava,

[2]   R.Backers , Eoltec Shirocco E5.6-6, *Instalation&Maintanence Manual*, 2009-20010, Solacity Inc.

[3]   International standard IEC 61400-1, *Wind turbines – Part 1: Design requirements*, Switzerland.

[4]   International standard IEC 61400-2, *Wind turbines – Part 2: Design requirements for small wind turbines*, Switzerland.

.

RESEARCH ON MECHANICAL ALLOYING EFFECTS ON MAGNETIC PROPERTIES OF BARIUM FERRITE TYPE M

I. Ştefan[1], A. Olei[2], C. Nicolicescu[3]

[1] University of Craiova, Department of IMST, Dr.Tr. Severin, ROMANIA, stefan_iuly@yahoo.com, adrian_olei@yahoo.com, nicolicescu_claudiu@yahoo.com

Abstract: The paper presents the effects of mechanical alloying process of $BaCO_3$ and αFe_2O_3 powders on the magnetic properties of barium ferrite type M.

For preparation of the magnetic ferrite powders was used a physical method as mechanical ball milling method. The powders magnetic property was characterized with a Vibrating Sample Magnetometer. The phase's structure was presented by an X-ray diffractometer. The morphology of the powder was observed with a particle size analyzer and a scan electron microscope.

There was observed a significant increase in remanence magnetization and coercive field of barium ferrite samples obtained by mechanical alloying.

Keywords: barium ferrite, mechanical alloying, remanence, coercive field.

1. INTRODUCTION

Barium ferrite type M is widely used in the fabrication of commercial permanent magnets, computer data storage, high density perpendicular magnetic and magneto-optic recording, magnetic fluids and microwave devices [1–3].

This type of ferrite is termed as hard ferrite due to their high coercive force, high saturation magnetization, high Curie temperature, chemically inert and mechanically resilient [4].

Mechanical alloying is one of the suitable processing techniques for synthesis of ferrite powders.

Homogenous distribution of the powder particles after mechanical alloying is an important factor which affects the ferritization process and the magnetic properties of ferrites. From this point of view, milling in water should be beneficial for the material properties, because distribution of particles size is narrower than in the case of milling in air [5]. For this reason, all the experiments of this work were carried out in water.

This present work aims to study the mechanical alloying effects of iron oxide Fe_2O_3 and barium carbonate $BaCO_3$ powders on magnetic properties of barium ferrite type M. Were studied three powder mixtures.

The first mixture, also known as reference mixture (denoted RM) was performed by mixing the Fe_2O_3 and $BaCO_3$ powders for 30 minutes in a planetary mill. The other two mixtures were carried out by mechanical alloying process for 4 to 12 hours (denoted MA4 and MA12). All the mixtures were heated to the same temperature and we pursued the influence of the mechanically alloyed powders on their magnetic properties.

2. MATERIALS AND EXPERIMENTAL PROCEDURES

For synthesis of barium ferrite $BaFe_{12}O_{19}$, mixture of iron oxide Fe_2O_3 (99% purity) and barium carbonate $BaCO_3$ (99% purity) powders was used with composition $BaCO_3 + 6Fe_2O_3$ (Figure1).

| a) Fe_2O_3 | b) $BaCO_3$ |

Figure 1: Fe_2O_3 and $BaCO_3$ powders used for mechanical alloying process

The used powders were weighed out in stoichiometric proportion in order to form the desired composition.
These starting materials were mixed by a mechanical activation process. Ball milling process was carried out in a vibratory mill type Pulverisette 4 for 4 and 12 hours in water medium. A mass ratio ball to powder was 4:1. The mill generated vibrations of balls and the material inside the container during which their collisions occur [6,7].

After milling process the initial powders were analyzed by thermal analysis in air atmosphere. The phases structure were presented by an X-ray diffractometer.
The morphology of the powder was observed with a particle size analyzer and a scan electron microscope.

The magnetic hysteresis loops of obtained powder material were measured by a Vibrating Sample Magnetometer.

3. RESULTS AND DISCUSSIONS

There were analyzed from morphological point of view the initial materials used to form barium ferrite type M and the mixtures resulting from mechanical alloying too. Thus, in figure 2 and 3 are presented the particle size distribution of $BaCO_3$ and Fe_2O_3 powders and in figures 4 and 5 are shown the mechanically alloyed mixtures of $BaCO_3x6Fe_2O_3$ powders for 4 and 12 hours.

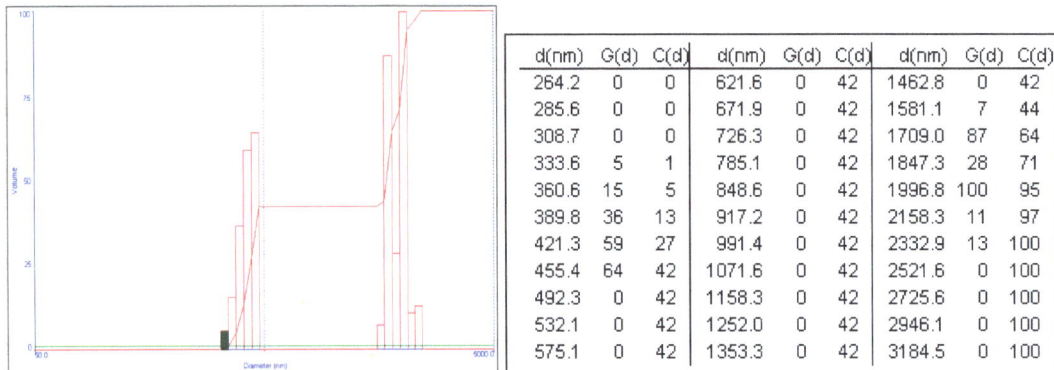

d(nm)	G(d)	C(d)	d(nm)	G(d)	C(d)	d(nm)	G(d)	C(d)
264.2	0	0	621.6	0	42	1462.8	0	42
285.6	0	0	671.9	0	42	1581.1	7	44
308.7	0	0	726.3	0	42	1709.0	87	64
333.6	5	1	785.1	0	42	1847.3	28	71
360.6	15	5	848.6	0	42	1996.8	100	95
389.8	36	13	917.2	0	42	2158.3	11	97
421.3	59	27	991.4	0	42	2332.9	13	100
455.4	64	42	1071.6	0	42	2521.6	0	100
492.3	0	42	1158.3	0	42	2725.6	0	100
532.1	0	42	1252.0	0	42	2946.1	0	100
575.1	0	42	1353.3	0	42	3184.5	0	100

Figure 2: Particle size distribution of $BaCO_3$ initial powders

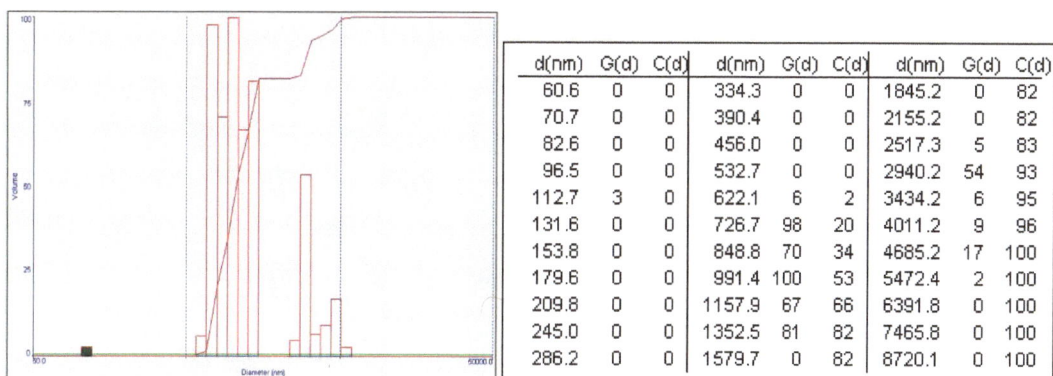

d(nm)	G(d)	C(d)	d(nm)	G(d)	C(d)	d(nm)	G(d)	C(d)
60.6	0	0	334.3	0	0	1845.2	0	82
70.7	0	0	390.4	0	0	2155.2	0	82
82.6	0	0	456.0	0	0	2517.3	5	83
96.5	0	0	532.7	0	0	2940.2	54	93
112.7	3	0	622.1	6	2	3434.2	6	95
131.6	0	0	726.7	98	20	4011.2	9	96
153.8	0	0	848.8	70	34	4685.2	17	100
179.6	0	0	991.4	100	53	5472.4	2	100
209.8	0	0	1157.9	67	66	6391.8	0	100
245.0	0	0	1352.5	81	82	7465.8	0	100
286.2	0	0	1579.7	0	82	8720.1	0	100

Figure 3: Particle size distribution of Fe_2O_3 initial powders

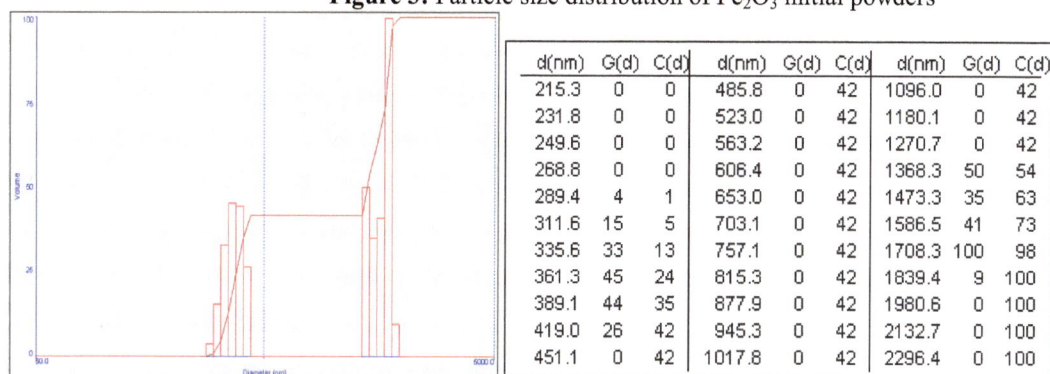

d(nm)	G(d)	C(d)	d(nm)	G(d)	C(d)	d(nm)	G(d)	C(d)
215.3	0	0	485.8	0	42	1096.0	0	42
231.8	0	0	523.0	0	42	1180.1	0	42
249.6	0	0	563.2	0	42	1270.7	0	42
268.8	0	0	606.4	0	42	1368.3	50	54
289.4	4	1	653.0	0	42	1473.3	35	63
311.6	15	5	703.1	0	42	1586.5	41	73
335.6	33	13	757.1	0	42	1708.3	100	98
361.3	45	24	815.3	0	42	1839.4	9	100
389.1	44	35	877.9	0	42	1980.6	0	100
419.0	26	42	945.3	0	42	2132.7	0	100
451.1	0	42	1017.8	0	42	2296.4	0	100

Figure 4: Particle size distribution of $BaCO_3xFe_2O_3$ powders after 4 hours of milling

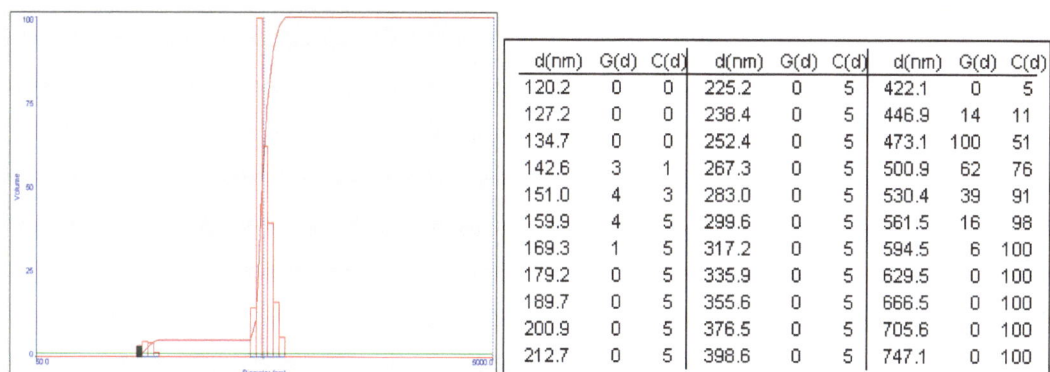

d(nm)	G(d)	C(d)	d(nm)	G(d)	C(d)	d(nm)	G(d)	C(d)
120.2	0	0	225.2	0	5	422.1	0	5
127.2	0	0	238.4	0	5	446.9	14	11
134.7	0	0	252.4	0	5	473.1	100	51
142.6	3	1	267.3	0	5	500.9	62	76
151.0	4	3	283.0	0	5	530.4	39	91
159.9	4	5	299.6	0	5	561.5	16	98
169.3	1	5	317.2	0	5	594.5	6	100
179.2	0	5	335.9	0	5	629.5	0	100
189.7	0	5	355.6	0	5	666.5	0	100
200.9	0	5	376.5	0	5	705.6	0	100
212.7	0	5	398.6	0	5	747.1	0	100

Figure 5: Particle size distribution of $BaCO_3xFe_2O_3$ powders after 12 hours of milling

Particle size ranges and maximum size of analyzed powder particles were systematized in table 1 and in figure 6 are shown the SEM micrographs of the $BaCO_3xFe_2O_3$ samples mechanically alloyed and unalloyed.

Table 1: Particle size ranges and maximum particle size of materials

Sample No.	Initial materials used in mixtures of BFM	Granulometric ranges [nm]	Maximum size of particles [nm]
1	$BaCO_3$	[333.6 – 455,6] [1581,1 - 2332,9]	1996,8
2	Fe_2O_3 unmilled	[622,1 – 1352,5] [2517,3 – 5472,4]	991,4
3	$BaCO_3xFe_2O_3$ milled 4 hours	[289,4 – 419,0] [1368,3 – 1839,4]	1708,3
4	$BaCO_3xFe_2O_3$ milled 12 hours	[142,6 – 169,3] [446,9 – 594,5]	473,1

a)　　　　　　　　　　b)　　　　　　　　　　c)

Figure 6: SEM micrographs of the $BaCO_3xFe_2O_3$ samples: **a)** unmilled; **b)** milled for 4 h; **c)** milled for 12 h

From the obtained data, it can be seen that powders particle size decreases with increasing of mechanical alloying time. The SEM micrograph of the milled sample for 12 hours, shows that the size of powder particles belongs to the submicron field.

In order to determine the heating temperature of $BaCO_3$ and Fe_2O_3 powders was performed the thermogravimetric analysis using a MOM derivatograph device. Heating process was performed with 10°C/min. The diagram of the $BaCO_3x6Fe_2O_3$ mixture is shown in figure 7.

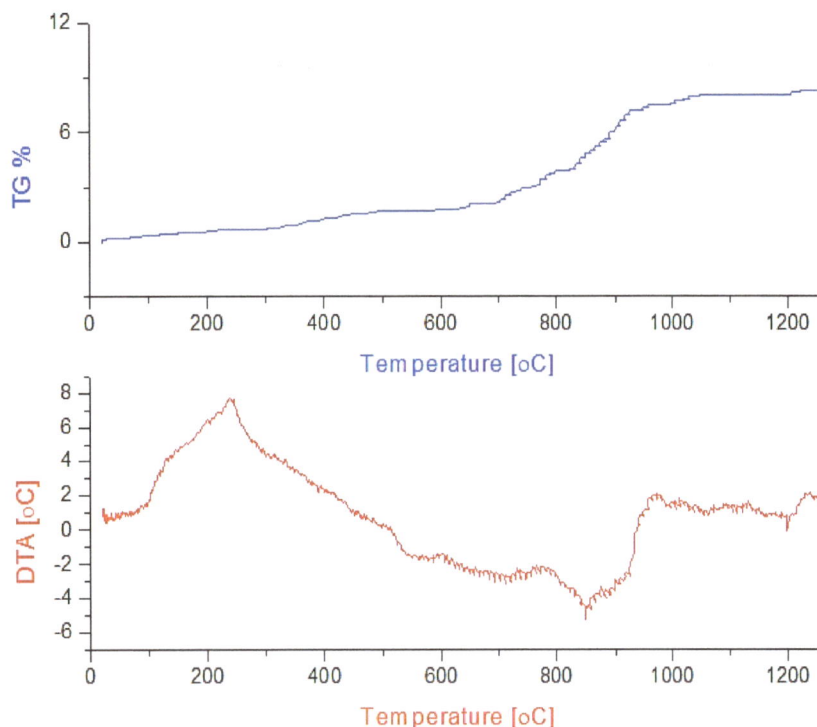

Figure 7: Diagram of $BaCO_3 x6Fe_2O_3$ mixture: cuves TG and DTA

Analyzing the TG and DTA curves obtained from heating the mixture of $BaCO_3 x6Fe_2O_3$ powders it can be observed two pronounced endothermic peaks at temperatures of 850 and 1100°C. The peak from temperature of 850oC placed on the DTA curve is related to weight loss slope from TG curve at the same temperature where a reaction product is formed. In order to identify the reaction product, a mixture of $BaCO_3 x6Fe_2O_3$ was heated up to a temperature of 850oC, then was suddenly cooled in water and was subjected to X-ray diffraction, and is observed in the diffractogram from figure 8 that at this temperature is born barium monoferrite, $BaFe_2O_4$.

On the DTA curve is also observed another peak at 1200oC corresponding to the formation of a new phase which was identified by X-ray diffraction (figure 9) as barium ferrite (BFM).

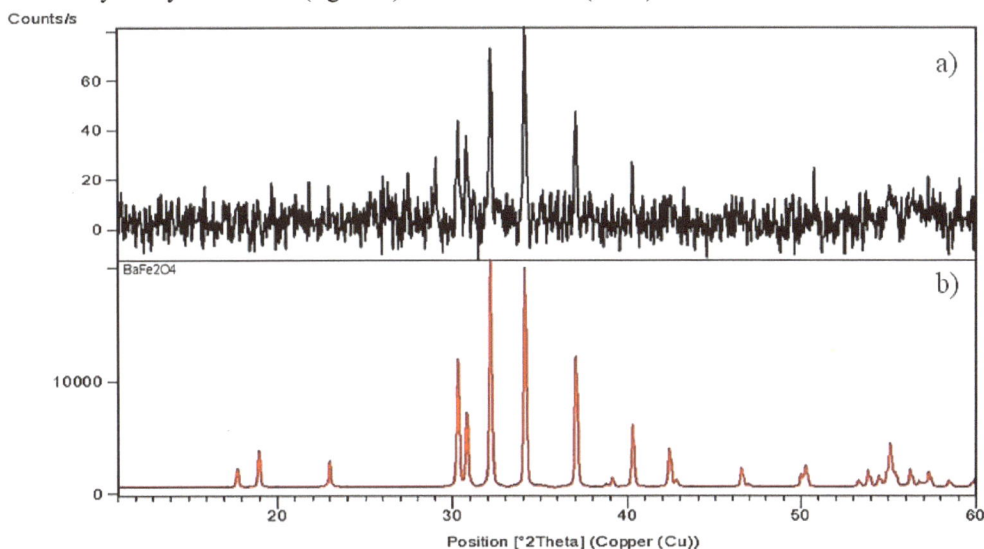

Figure 8: X-ray diffraction of $BaCO_3 x6Fe_2O_3$ sample heated at 850°C: a) sample; b) monoferrite phase

357

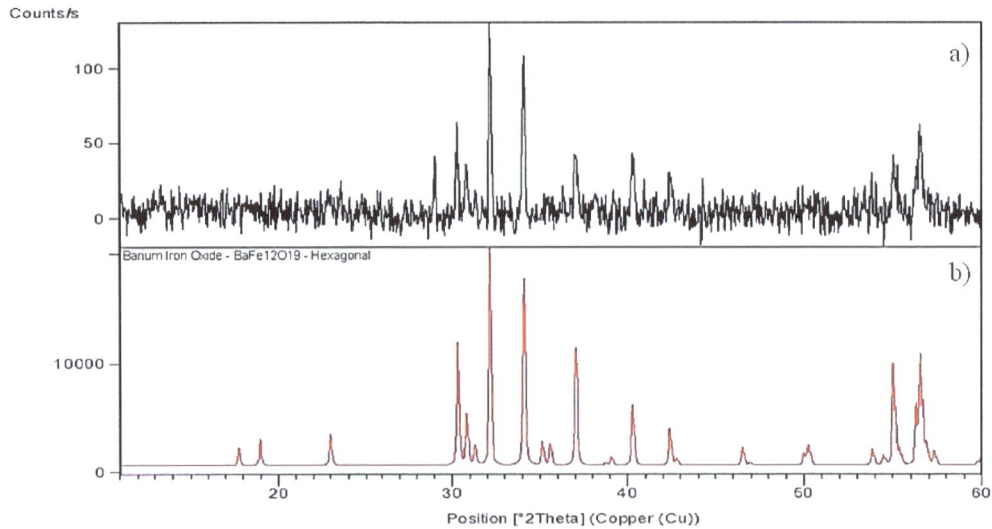

Figure 9: X-ray diffraction of $BaCO_3x6Fe_2O_3$ sample heated at 1200^oC: a) sample; b) ferrite phase

The magnetic hysteresis loops of obtained barium ferrite were measured by a Vibrating Magnetometer and there are presented in figure 10.

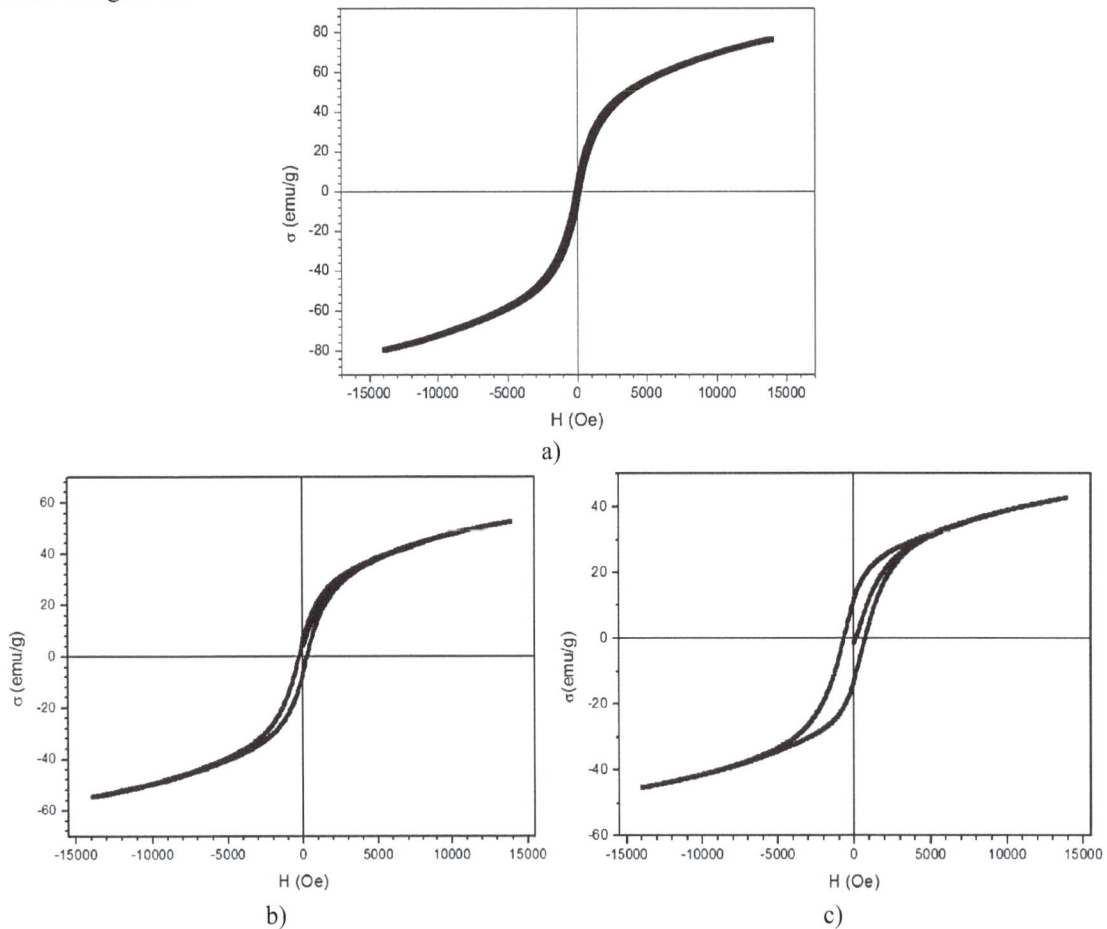

Figure 9: Magnetic hysteresis loops of $BaCO_3xFe_2O_3$ samples: **a)** unmilled; **b)** milled for 4 h; **c)** milled for 12 h

The magnetic characteristics of obtained ferrite are presented in table 2. From hysteresis loops and from table 2 is observed an increase in magnetic properties once with increasing of mechanical alloying time. Remanent magnetization σr reaches the value of 8 emu/g and coercive field has a value of 250 Oe for $BaCO_3x6Fe_2O_3$ sample mechanically alloyed 4 hours.

The mechanical alloyed sample for 12 hours shows much higher values of remanent magnetization and the coercive field, respectively 12 emu/g and 800 Oe.

Table 2: The magnetic characteristics of barium ferrite samples

Sample / Mag. Param.	MA4	MA12
σ_r /emu/g	8	12
Hc/Oe	250	800

4. CONCLUSIONS

Analyzing the results from experimental research can conclude the following:

- increasing of mechanical alloying time for $BaCO_3$ and Fe_2O_3 mixtures lead to a reduction of powders particle size starting from micron field and reaching in the submicron field after 12 hours of mechanical alloying;

- mechanical alloying time influence the magnetic characteristics of barium ferrite type M samples;

- it can be seen that ferrite derived from mechanically alloyed mixture for 12 hours has the highest amount of remanent magnetization, namely σr =12 emu/g;

- regarding to the coercive field value it is found that this increases to a value of 800 Oe for ferrite obtained by mechanical alloying for a longer time. As it is known, the coercive field is influenced by the particle size of the magnetic powder.

The more the mechanical alloying time increases, or the more particle size of $BaCO_3$ and Fe_2O_3 powders tend to the nanometer field, the better the magnetic properties are.

REFERENCES

[1] http://www.aacg.bham.ac.uk/magnetic materials/.

[2] V. Pillai, P. Kumar, M.S. Multani, D.O. Shah, Structure and magnetic properties of nanoparticles of barium ferrite synthesized using microemulsion processing, Colloids and Surfaces A: Physicochemical and Engineering Aspects 80 (1993) 69.

[3] S. Castro, M. Gayoso, C. Rodríguez, J. Mira, J. Rivas, S. Paz, J.M. Grenèche, Synthesis and characterization of small $BaFe_{12}O_{19}$ particles, Journal of Magnetism and Magnetic Materials, 140–144 (1995) 2097.

[4] M.M. Hessien, M. Radwan, M.M. Rashad, Enhancement of magnetic properties for the barium hexaferrite prepared through ceramic route, J. Anal. Appl. Pyrolysis 78 (2007) 282–287.

[5] A. Witkowski, M. Leonowicz and W. Kaszuwara // Inzynieria Materialowa XXV (2000) 140.

[6] J. Konieczny, L.A. Dobrzański, A. Przybył, J.J. Wysłocki, Structure and magnetic properties of powder soft magnetic materials, Journal of Achievements in Materials and Manufacturing Engineering 20 (2007) 139-142.

[7] P. Gramatyka P., A. Kolano-Burian, R. Kolano, M. Polak, Nanocrystalline iron based powder cores for high frequency applications, Journal of Achievements in Materials and Manufacturing Engineering 18 (2006) 99-102.

PREDICTING GRAIN REFINEMENT BY COLD SEVERE PLASTIC DEFORMATION PROCESSES IN ALLOYS BY USING THE METHOD OF VOLUME AVERAGED DISLOCATION GENERATION

I. Popa [1], V. Geamăn [2]

[1] Universitatea *Transilvania* din Braşov, Romania, ioan.popa@unitbv.ro, geaman.v@unitbv.ro

Abstract: *The grain refinement during severe plastic deformation processes (SPD) is predicted using volume averaged amount of dislocations generated. The model incorporates a new expansion of a model for hardening in the parabolic hardening regime, in which the work hardening depends on the effective dislocation free path related to the presence of non shearable particles and solute-solute nearest neighbour interactions. These two mechanisms give rise to dislocation multiplication in the form of generation of geometrically necessary dislocations and dislocations induced by local bond energies. The model predicts the volume averaged amount of dislocations generated and considers that they distribute to create cell walls and move to existing cell walls/grain boundaries where they increase in the grain boundary misorientation.*
Keywords: *Severe Plastic Deformation (SPD); Equal Channel Angular Pressing (ECAP); Dislocation Mobility; Aluminium Alloys; Grain refinement.*

1. INTRODUCTION

The main purpose of severe plastic deformation (SPD) processes, such as equal channel angular pressing (ECAP, see Fig.1) [28,29,40] and high pressure torsion (HPT, see Fig.2) [6,13,46], is to achieve a refined grain structure. Predicting the grain size attainable through SPD processes is important, because the grain size will determine the strengthening achievable through grain boundary strengthening and because the grain size, especially that related to the stable high angle grain boundaries, has a major influence on superplastic forming properties [4,37]. These SPD processes are mostly conducted under cold deformation conditions, where dynamic recrystallisation is suppressed whilst strain rate dependence is very limited [17]. It is generally thought that FCC metals with grain size 50-100 nm deform predominantly via the slip of lattice dislocations, and for grains larger than 100nm they deform exclusively through this mechanism [47]. Various works have shown that in nanostructured materials (grain size < 50 nm) alternative deformation mechanisms involving deformation twins [20,22,42], and stacking faults (SFs) [8,21] can occur. In some cases these types of defects have even been observed in grains with size up to 100 nm [22], but it is not evident that they make a significant contribution to deformation in these SPD processed materials.

Many researchers (see e.g. [36]) consider that the structure evolution during SDP processing broadly follows the classic mechanisms and concepts on structural changes occurring during conventional processes as shown in classic works [1,2,27,30,45]. At the early stage of deformation, a very high dislocation density is introduced, which leads to the formation of an intragranular structure consisting of cells with thick cell walls and low angles of misorientation. As the strain increases, the thickness of the cell walls decreases. These walls evolve into grain boundaries, and ultimately an array of ultra fine grains with high-angle non-equilibrium grain boundaries (GBs) [7,38,39] are formed. (Nonequilibrium grain boundaries may be present where there are non-geometrically necessary dislocations i.e. excess dislocations that do not contribute to the formation of misorientation at a grain boundary.)

In broad terms we may term this the classic model for microstructure evolution in highly deformed metals. Xu et al. and Langdon [19,44] noted that the classic model would predict a gradually increasing refinement of the microstructure as a result of the continuous introduction of dislocation during the straining process. However, these researchers considered this to be inconsistent with some experimental observations and an alternative model based on an inter-relationship between the formation of subgrain boundaries and shear deformation during ECAP was proposed [19]. The model incorporates the geometries relevant to repeated ECAP passes (see Fig.1).

The original grains become elongated to a band shape subgrain when the billet passes the corner in the first pass. In the second pass, the elongated subgrain is either further elongated (route C) because the shear plane remains in the same direction or sheared (route A and B_C) because the shear plane is changed to another direction. Especially when route B_C is used, several intersecting slip systems lead to a high density of dislocations and then these dislocations re-arrange and become subsumed in the grain boundary (some researchers consider the latter „ annihilation" of dislocations, as will become clear below we prefer to avoid that term). As a result, for route B_C, a reasonably equiaxed array of grains is formed. Cell wall and grain boundary evolution involves various complex processes [14,17]. Cell wall formation of low angle boundaries by dynamic recovery during the deformation. Low misorientation (1°) boundaries form a 3-d or 2-d cell structure which tends to remain equiaxed during deformation and are sometimes referred to as "incidental dislocation boundaries" [17]. Subgrain size tends to be constant at larger strains, and this is interpreted in terms of a dynamic equilibrium between dislocation generation and annihilation. But also other types of low angle boundaries have been identified. High angle boundaries can form by deformation banding in which grains may split on a coarse scale into several sections which then follow different orientation paths during subsequent deformation, or by increase in misorientation of the persistent boundaries discussed above. However, high angle boundaries may also evolve from low angle boundaries by assimilation of dislocations. Further rigid body rotation during deformation of high angle boundaries, will tend to align them with the rolling plane, thus forming a lamellar microstructure [23].

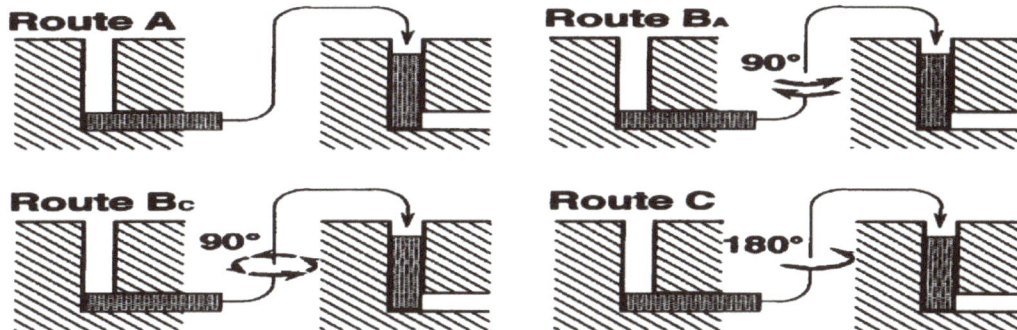

Figure 1:
Schematic illustration showing the billet and ECAP die, with the billet rotations during subsequent passes for the 4 basic routes of ECAP. (from [25])

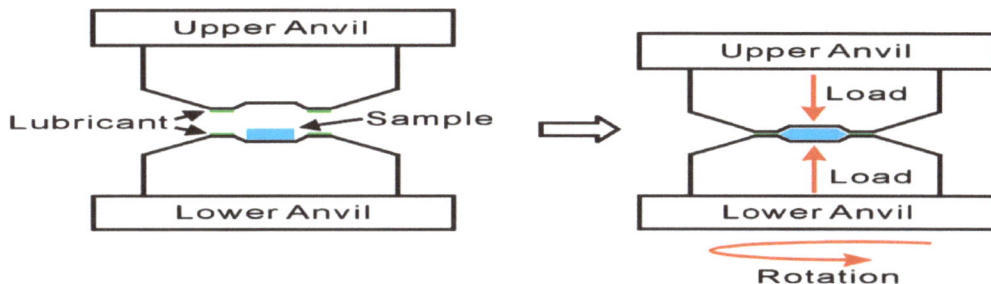

Figure 2: The principle of high pressure torsion (HPT) (from [13])

Grain refinement during SPD is significantly influenced by the presence of non-shearable particles in the alloy, which can produce a factor 2 difference in grain size between alloys with different content of non-shearable particles [24]. Also elements dissolved in the matrix phase have a significant influence on grains refinement, with 3wt%Mg additions to Al causing about a factor 3 reduction in grains size. A model that quantitatively explains these effects has as yet not been available. It has been noted (Beyerlein et.al. [5] and Signorelli et.al. [31]) that prediction of grain size evolution is principle possible in a visco-plastic self-consistent (VPSC) polycrystal model when combined with an (empirical) criterion for the grain subdivision process during ECAP. However, no reports on successful predictions of grain sizes have as yet appeared.

The aim of the present work is to derive a model for grain size in the SPD regime incorporating especially the effect of non-shearable particles and dissolved alloying elements. Our new model will employ some of the concepts from the models and papers reviewed above. However, to make the problem tractable, we will use a substantial simplification by treating only averaged dislocation densities as caused by averaged particle spacings and averaged solute contents, effectively reducing a 3D problem to volume averaged properties. This will effectively by-pass much of the details of the cell wall and boundary evolution processes in favour of a model describing a volume averaged behaviour.

The model will be tested against microstructure data of several SPD processed Al based alloys. It proved possible to construct a simplified model with good predictive properties in closed form solutions, and hence we will provide a computationally efficient model. The structure of the present paper is as follows.We will first introduce the materials, processing and experimental techniques applied in the work. Next we will present a model for grain size evolution illustrating several aspects of it with selected results from transmission electron microscopy (TEM) and electron backscatter diffraction (EBSD) work performed. Finally we will present a full description of results and a literature survey on published data for a critical comparison of model predictions with data.

2. EXPERIMENTAL: MATERIALS, PROCESSING AND MICROSTRUCTURE ANALYSIS.

Microstructure development during SPD of 6 alloys was analysed; the compositions of these alloys are given in Table 1. Prior to SPD the Al-7034 and Al-1050 alloys had been extruded into a rod. The Al-2024 alloy was cast, hot rolled to plate and subsequently heat treated to T3 condition. The Al-3Mg-0.2Mn was cast, hot rolled, solution treated and subsequently cold rolled. The other alloys were not thermomechanically processed prior to SPD, they were in as-cast and homogenized condition.

The average grain size was 2.1 μm for the 7034 alloy and 45 μm for the 1050 alloy. The grain size for the Al-Zr and Al-Zr-Si-Fe alloys was very large, about 690 μm and 540 μm respectively. The grains in the rolled Al-2024 alloy were pancake shaped with sizes about 100 by 500 μm. The content of second phase particles is very different between these alloys. The 7034 has the highest amount of second phase particles by virtue of the high content of η phase particles present [41]. The amount of second phase particles is intermediate for the 1050 alloy, with the Al-Zr alloy having a very low amount of second phase particles.

Five alloys were processed by ECAP and 2 were processed by HPT (see Table 1). ECAP was conducted using a solid die with 9.7 mm diameter channel, with a 90° channel intersection angle (Φ) and a 20° curvature on the outer side of channel intersection (Ψ) (see Fig. 3). Specimens were lubricated with a suspension of MoS_2 in mineral oil ('ASO oil' supplied by Rocol) in order to reduce the friction between the plunger, specimen and the die. A careful alignment of the plunger and upper channel of the die was carried out. A plunger pushing speed of 0.5 mm/s was employed. After one pass of ECAP, another specimen was put in the die to push out the first specimen. Processing of Al-7034 and Al-2024 and selected microstructure analysis results on these alloys was reported before [11,41,43]. All SPD processing was carried out at room temperature, with the exception of ECAP on the Al-7034 alloy which was carried out at 200°C. Processing by HPT was conducted using disks having diameters of 10 mm and thicknesses of 0.9 mm. All processing by HPT was conducted at room temperature under pressures of 34.0 GPa and with torsional straining between 1 and 5 turns. No lubrication was applied on the sample. In additional tests slippage was found to be negligible. Electron backscattered diffraction (EBSD) was used to characterize the microstructure as well as grain and subgrain boundary misorientation distribution in billets of all ECAP-processed alloys. Samples of 10 mm length used for EBSD analysis were machined from the middle of ECAP-processed billets.

For sample preparation, the surface of cross section was first mechanically ground up to 4000-grit SiC paper, then electropolished employing an electrolyte composed of 33 vol% nitric acid and 67 vol% methanol. The electropolishing was carried out with a DC voltage of 20-30 V for 30 seconds. The electrolyte was cooled to and maintained at a temperature of -30°C using liquid nitrogen. The equipment used was a JEOL JSM6500F thermal field emission gun scanning electron microscope (FEG-SEM) equipped with an HKL EBSD detector and HKL Channel 5 software.

The SEM accelerating voltage was set to 15 kV. Step size is reported with the results; in most cases it was between 0.1 and 0.5 μm. Orientation imaging microscopy (OIM) maps were obtained from the cross section perpendicular to the longitudinal direction of ECAP-processed billets. Intercept lengths were determined using an automated procedure. For misorientation angle distributions the lowest cut off angle was set at 2°. TEM was conducted on Al-1050 subjected to ECAP and HPT, the Al-Zr alloy subjected to ECAP, and the 2024 alloy subjected to HPT. For TEM disks of 3mm in diameter were punched out from slices cut from the processed billets, ground to about 0.20 mm in thickness and then electropolished using a solution of HNO3 and methanol (1:3 in volume).

TEM foils were examined using a JEOL 3010 microscope operating at 300 kV. EBSD analysis of grain size of an HPT processed (5 turns) Al-1Mg-0.3Mn alloy proved unreliable due to very low pattern identification rates, and it was decided to not use EBSD for HPT processed alloys. Instead, to supplement the TEM work on HPT processed samples, the grain size of selected HPT processed samples of an Al-1050 and Al-3Mg-0.3Mn alloys were studied through SEM. For this analysis samples were sectioned, ground and polished, and subsequently etched in Keller's reagent for 3 min. to reveal grain boundaries. SEM was conducted on a JEOL JSM6500F FEG-SEM in secondary electron mode.

Grain sizes reported and quoted in this paper were determined through analysis of TEM, EBSD and SEM data. For EBSD analysis of grain sizes on our SPD processed alloys grain boundary intercept lengths, L, were determined using an automated procedure, with lower cut off angle set at 2°.

For reliable TEM grain size measurements it needs to be considered that not all grain boundaries in a sample will show a detectable contrast, and also finite sample thickness (causing overlap of grains in the TEM image) needs

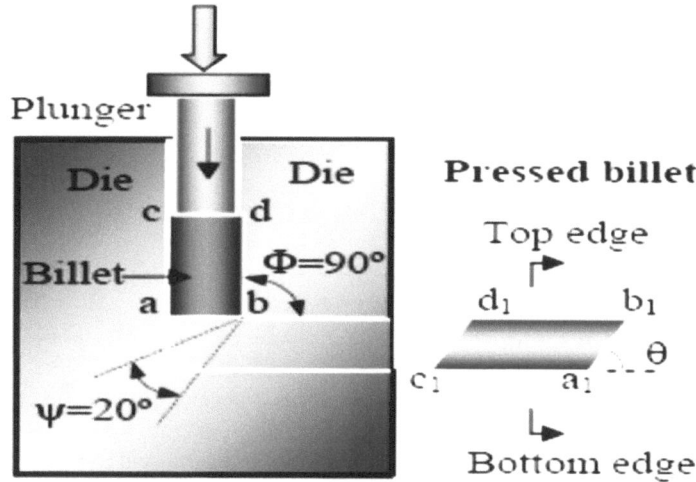

Figure 3: Schematic illustration of the ECAP die.[32]

Table 1. Compositions and grain size for the alloys studied subjected to SPD, with process that the alloys were subjected to.[32]

Alloy	Composition [wt%]							grain size [μm] before SPD	SPD	
	Zn	Mg	Cu	Zr	Si	Fe	Mn		ECAP	HPT
Al-7034	11.5	2.5	0.9	0.2	-	-	-	2.1	✓	-
Al-1050 A	0.006	0.01	0.009	-	0.12	0.18	0.007	45	✓	-
Al-1050 B	0.006	0	0.008	-	0.14	0.26	0.005	45	-	✓
Al-2024	-	1.4	4.4	-	0.1	0.1	0.4	~200	✓	✓
Al-Zr	<0.01	<0.01	<0.01	0.15	<0.01	<0.01	-	690	✓	-
Al-Zr-Si-Fe	0.01	<0.01	<0.01	0.16	0.17	0.19	-	540	✓	-
Al-3Mg-0.2Mn	0.003	2.95	-	-	0.15	0.20	0.24	30×100	-	✓

to be taken into account. For TEM micrographs obtained for the alloys investigated by us and most TEM micrographs published in the literature, grain size was determined by us by considering only the 20-30% of the micrograph with the lowest beam intensity (i.e. darkest grains) and subsequently eliminating all areas with aspect apparent aspect ratio larger than 2 (because these are likely to be due to overlapping grains). Subsequently intercept lengths were determined on random lines.

A correction was made for TEM foil thickness, which is assumed to be 120±50nm.(Note that much of the literature in the field does not seem to consider this correction.) When this procedure produced less than 5 grains, more grains up to at least 5 grains were included. In selected other cases grain size data used, was obtained directly from grain sizes reported in published work (e.g. Ref [33]).

We will report accuracies of determinations of average grain size which take into account the uncertainty concerning foil thickness and distribution of sizes. For analysis of SEM micrographs of etched samples, a line intercept method was used. Care was taken to avoid areas where overetching was obscuring grain boundaries.

There is a range of ways in which an 'average' grain size can be defined (see e.g. [35]). Although sometimes L has been taken as 'grain size' it is actually an underestimate of most realistic definitions of the grain size, and in this paper we will report the grain size, d, consistently as $d = 1.455 L$. This is based on $d = V^{1/3}$, V being the average cell volume, with the assumption of a Poisson-Voronoi size/shape distribution. Accuracy of experimental d determinations is typically ±8% (1 standard deviation), rising to ±15% for cases where less than 10 grains can be reliably detected.

To elucidate elements of the model, tensile tests were performed on 15 further alloys. The first 9 low Cu Al- Mg-(Cu)-Mn alloys were direct chill (DC) cast, the cast ingots were preheated and homogenised at 540°C, and subsequently hot rolled down to 5 mm in thickness. After that, the hot rolled and cold rolled to required reduction, and subsequently solution treated at 500°C.

The Al-Cu-Mg and Al-Li-Cu-Mg alloys were produced by conventional casting followed by hot rolling. They were all solution treated at 495°C. The tensile testing was conducted according to the ASTM-E8M standard. The tensile axis is taken in the longitudinal (L) direction (i.e. the rolling direction).

For each condition usually two tests were performed. Tensile tests were performed using an 8800 series Instron machine at a constant strain rate of 0.001 s-1. Selected processing and microstructure data on the Al-Mg-(Cu)-Mn, Al-Cu-Mg and Al-Li-Cu-Mg alloys was reported before [12,18,48].

3. THE MODEL AND SUPPORTING MICROSTRUCTURAL DATA.

3.1 General model structure and main assumptions.

Most available models on grain refinement during SPD follow the strategy of considering the gradual evolution of the microstructure starting from low strains up to the very high strains typical of SPD [3,5,9,17,34]. An important factor in these models is the details of cell wall and subgrain boundary formation. Whilst recognising the value of those models, especially in elucidating the processes occurring at low and medium strains in 3D, we note that these approaches have not led to a model that is successful in predicting grain sizes of alloys subjected to SPD.

In this work will depart from this approach and aim to produce a model that predicts grain sizes in the regime of SPD (effective strains in excess of 3). The model does not include any qualitative predictions on cell and subgrain formation, we just acknowledge that it does occur and is dominant at low to medium strains. We will introduce a substantial simplification in treating only averaged dislocation densities as caused by averaged particle spacings and averaged solute contents, effectively reducing the 3D dislocation movement and cell wall creation case to a volume averaged model.

As a consequence the spatial pattern of the grain boundaries has no direct relation to the spatial arrangement of particles and the original orientation of the crystal lattice. We will consider SPD processing routes that lead to grain structures that are close to being equiaxed. Within the model we view a dislocation as being the border of a surface where slip has occurred. It is the cumulative effect of this deformation on a range of slipped surfaces that determines changes in the CW/GB misorientation angle. When a dislocation is assimilated in a CW/GB it may appear „annihilated" in the sense that it can not be discerned in TEM, but in the sense of being the border of a 2D surface where slip has occurred it has not disappeared: it is present in the grain boundary at the intersection of the slipped surface and the grain boundary. (These views are in many ways comparable and compatible to the more elaborate treatments by Estrin, Toth and co-workers [9,34] which incorporates low angle grain boundaries/cell walls of finite thickness and a finite volume density of dislocations. In terms of the concepts these latter works, we are here considering an infinitely thin grain boundary, and an area density of dislocations, and we expand that treatment by extending it to higher angle grain boundaries.). The grain size can be predicted well by considering the total amount of dislocations formed in the straining process, without regard to the detailed geometry and mechanisms of cell wall formation. To simplify terminology we will term any feature that is either a cell wall, low angle grain boundary (LAGB) or high angle grain boundary (HAGB) a „CW/GB" (a cell wall and/or grain boundary). The derivation of the model is described in two parts: dislocation generation formation and evolution of cell walls. The grain size development as a function of the accumulated strain determined in this work and other works [10,15,26] of Al alloys between 99.5 and 99.99wt% purity, and low purity (97wt%) Al (Total equivalent strain during ECAP is determined using the equation described by Iwahashi et.al. [16] and the equivalent strain is determined using the approximation for large strains recommended by Zhilyaev and Langdon [46].). At strains 1 to 3 the CW/GB size is very different between the different alloys.

4. RESULTS AND ANALYSIS: GRAIN SIZE DURING SPD

We have tested the present model for grain size evolution by comparison of its predictions to a range of data on grain size of SPD Al alloys. New data was obtained by performing ECAP on Al-1050, Al-Zr and Al-Zr-Si-Fe alloys and HPT on Al-2024 and Al-3Mg-0.2Mn, size data for a range of other commercial and experimental alloys was obtained from [11,24,33]. The database contains a total of 21 alloys, in a total of 37 alloy-processing combinations, with strains ranging from 1 to 17 and with resulting grain sizes between 2 μm and 50 nm.

5. DISCUSSION

As grain size gets progressively refined, the dislocations created due to the presence of non shearable particles will start to arrive at grain boundaries in ever greater numbers. These dislocations are geometrically necessary in the sense that they are required to create deformation at the non-shearable particle, but they are in general not geometrically necessary in terms of the misorientation angle of the grain boundary. Thus a part of these dislocations will not contribute to misorientation of the grain boundary, and a constellation sometimes described as non equilibrium boundary will result. Thus nonequilibrium boundaries, often observed in SPD metals, are a natural consequence of SPD in the present model. [32]

6. SUMMARY

The work hardening behaviours at strains up to 0.05 and the grain refinement during SPD up to a strain of 16 at room temperature of a wide range of alloys was investigated. A model was presented for the grain refinement and the model for hardening in the parabolic regime was expanded. The work hardening analysis showed: It is confirmed that the work hardening factor depends on the dislocation free path related to the presence of non shearable particles. These particles give rise to dislocation multiplication in the form of generation of geometrically necessary dislocations. [32]

ACKNOWLEDGMENTS

This paper is supported by the Sectoral Operational Programme Human Resources Development (SOP HRD), financed from the European Social Fund and by the Romanian Government under the contract number POSDRU/107/1.5/S/76945.

REFERENCES

[1]   Argon, A.S., Haasen, P., Acta. Metall. Mater., 1993, 41:3289.

[2]   Ashby, M.F,. In: Kelly, K., Nicholson, R.B., editors. *Strengthening Methods in Crystals*, Elsevier, Amsterdam, The Netherlands, 1971., p.137.

[3]   Baik, S.C., et. al., Mater. Sci. Eng., A, 2003, 351:86.

[4]   Baretzky, B., et.al., Rev. Adv. Mater. Sci., 2005, 9:45.

[5]   Beyerlein, I.J., Lebensohn, R.A., Tome, C.N., Mater. Sci. Eng., A, 2003, 345:122.

[6]   Bridgman, P.W., *Studies in large plastic flow and fracture*, McGraw-Hill, 1952, New York, USA.

[7]   Chang, C.P., Sun, P.L., Kao, P.W., Acta Mater., 2000, 48:3377

[8]   Chen, M.W., et.al., Science, 2003, 300:1275.

[9]   Estrin, Y., et. al., Acta Mater., 1998, 46:5509.

[10]   Field, D.P., Weiland, H., In: Schwartz, A.J., Kumar, M., Adams, B.L., editors. *Electron Backscatter Diffraction in Materials Science*, Kluwer Acad., New York, 2000, p. 199.

[11]   Gao, N., et. al., Mater. Sci. Eng., A, 2005, 410-411:303.

[12]   Gao, N., et. al., Mater. Sci. Tech., 2005, 21:1010.

[13]   Horita, Z., Langdon, T.G., Mater. Sci. Eng. A, 2005, 410-411:422.

[14]   Hughes, D.A., Mater. Sci. Eng., A, 2001, 319:46.

[15]   Iwahashi, Y., et.al., Acta Mater., 1998, 46:3317.

[16]   Iwahashi Y, et. al., Scripta Mater., 1996, 35:143.

[17]   Jazaeri, H., Humphreys, F.J., Acta Mater., 2004, 52:3239 .

[18]   Kamp, N., et. al., Int. J. Fatigue, 2007, 29:869.

[19]   Langdon, T.G., Mater. Sci. Eng. A, 2007, 462:3.

[20]   Liao, X.Z., et.al., Appl. Phys. Lett., 2004, 84:592.

[21]   Liu, M.P., et. al., Mater. Sci. Eng., A, 2008, 483-484:59.

[22]   Liu, M.P., et. al., Mater. Sci. Eng., A, 2009, 503:122.

[23]   Liu, Q., et. al., Acta Mater., 2002, 50:3789.

[24]   Munoz Morris, M.A., Gutierrez-Urrutia, I., Morris, D.G., Mater. Sci. Eng., A, 2008, 493:141.

[25]   Nakashima K., et. al., Mater. Sci. Eng., 2000, A281:82.

[26]   Reihanian, M., et. al., Mater Char., 2008, 59:1312.

[27]   Rollett, A.D., Kocks, U.F., Solid State Phenomena, 1994, 35-36:1.

[28] Segal V.M, et. al., Russian Metall, 1981, 1:99.
[29] Segal V.M., Mater. Sci. Eng. A, 1995, 97:57.
[30] Sevillano, J.G., Aernoudt, E., Mater. Sci. Eng., A, 1987, 86:35.
[31] Signorelli, J.W., et. al., Scripta Mater., 2006, 55:1099.
[32] Starink, M. J., et.al., *Predicting grain refinement by cold severe plastic deformation in alloys using volume averaged dislocation generation,* Acta Mater., 2009, SO171BJ, UK.
[33] Stolyarov, V.V., et. al., Mater. Sci. Eng., A, 2003, 357:159.
[34] Toth, L., Molinari, A., Estrin, Y. J., Eng. Mater. Tech., 2002, 124:71.
[35] Underwood, E., Quantitative Stereology, Addison-Wesley, Cambridge Massachusetts, 1970.
[36] Valiev, R.Z., et.al., Acta. Mater., 1996, 44:4705.
[37] Valiev, R.Z., et.al., Scripta Mater. 1997, 37:1945.
[38] Valiev, R.Z., Islamgaliev, R.K., Alexandrov, I.V., Prog. Mater. Sci., 2000, 45:103.
[39] Valiev, R.Z., Nat. Mater., 2004, 3:511.
[40] Valiev R.Z., Langdon, T.G., Prog. Mater. Sci., 2006, 51:881.
[41] Wang, S.C., et. al., Rev. Adv. Mater. Sci., 2005, 10:249.
[42] Wu, X.L., Ma, E., Zhu, Y.T.J., Mater. Sci., 2007, 42:1427.
[43] Xu, C., et. al., Mater. Lett., 2003, 57:3588.
[44] Xu, C., et. al., Mater. Sci. Eng., A, 2005, 398:66.
[45] Zehetbauer, M., Seumer, V., Acta Metall Mater., 1993, 41:577.
[46] Zhilyaev, A.P., Langdon, T.G., Prog. Mater. Sci., 2008, 53:893.
[47] Zhu, Y.T., Langdon, T.G., Mater. Sci. Eng. A, 2005, 409:242.
[48] Zhu, Z., Starink, M.J., Mater. Sci. Eng., A, 2008, 489:138.

THE STRUCTURAL TRANSFORMATION OF THE CERAMIC LAYER ZrO$_2$/20%Y$_2$O$_3$, OBTAINED BY THERMAL SPRAYING AFTER HEAT TREATMENT

L.G. Pintilei[1], F. Branza[2], G.N. Basescu[1], S.C. Iacob Strugariu[1], B. Istrate[1], C. Munteanu[1*]

[1] "Gheorghe Asachi" Technical University of Iasi, Faculty of Mechanical Engineering, 61-63 Prof. dr. doc. D. Mangeron Blvd, 700050, Iasi, Romania, laura_rares082008@yahoo.com

[2] Faculty of Physics, Alexandru Ioan Cuza University, Blvd. Carol I no. 11, 700506, Iasi, Romania, fbrinza@uaic.ro

Abstract: *In this paper presents a new concept of thermal barrier coating which consist of a ceramic layer of ZrO$_2$/20%Y$_2$O$_3$ and an adherent layer of Ni-22wt% Cr-10wt% Al-1wt% Y, deposited by atmospheric plasma spraying (APS), on samples of Ni super alloy used for the manufacturing of turbine blades. In this paper a heat treatment is applied on the layer obtained by atmospheric plasma spraying, by cooling it with a jet of air at a pressure of 100 bar after maintaining it a a temperature of 1000°C for intervals of 5, 10 and 15 hours. After the heat treatment the surfaces were analyzed microstructurally and morphologically using electronic microscopy and in terms of the phase composition by X-ray diffraction. After the heat treatment an increase in grain size on the surface layer can be observed which is caused by the formation of "bridges" between neighboring splats that had the effect of closing the initial micro cracks produced during spraying.*
Keywords: *SEM, X-ray, heat treatment, ZrO$_2$/20%Y$_2$O$_3$*

1. INTRODUCTION

Plasma spray is one of the most important achievements in the field of surface engineering, a field in which remarkable progress has been made in recent years. Since gas turbine power units used in aeronautics work at very high temperatures, the thermal barrier coatings have become essential for increasing the lifetime of the gas turbine. This coating is designed to isolate components also protecting them from corrosion, erosion, thermal and mechanical stresses that occur on their surfaces in operating conditions. Using these thermal barrier coatings allows an increase in the maximum operating temperature of the gases without increasing the temperature of the material and reducing the amount of air necessary for cooling while keeping the temperature of the turbine.

Heat treatments applied on the superficial layers obtained by plasma spray jet are aimed at structural changes and removal of defects caused during spraying. Through heat treatment is aimed at acquiring physical-mechanical, chemical or technological imposed by the terms of use.

2. EXPERIMENTAL PROCEDURE

Thermal barrier coatings (TBCs) used on turbine blades consist of a two-layer system. First deposited layer is relatively thin and is called adherent layer and the second layer is sprayed over the first layer and called thermal barrier. The intermediate layer is of a MCrAlY alloy. The purpose of this layer is to protect the substrate from oxidation and corrosion and also to improve the adhesion between the substrate and insulation. The standard material used in aerospace industry as thermal insulator is zirconia stabilized with yttrium. The ceramic powder which was sprayed on the samples was ZrO$_2$/20% Y$_2$O$_3$ using the Sulzer Metco 7MB type installation produced by METCO.

The spray parameters used for the plasma jet deposition are presented in Table 1.

Table 1: Technical parameters

Powder	NiCrAlY	YSZ (202NS)
Spraying distance (mm)	120	120
Injector	1,8	1,8
The intensity of the gas Plasma, (A)	600	600
Electrode voltage (U)	62	65
Velocity of rotation (rot/min)	55	55
Composition of plasma (m^3/h)	50%Ar 13,51%H	40%Ar 13,51%H

3. EXPERIMENTAL RESULTS

This paper presents the thermal treatment applied to the layers obtained by plasma jet spraying. They were cooled with a jet of air at a pressure of 100 bar, after keeping them at a temperature of 1000 °C at intervals of 5, 10 and 15 hours. Average furnace heating speed in the first 10 minutes is 100 °C. After the first 10 minutes the oven heats up with 20 °C at every 10 minutes until the temperature reaches 1000 °C.

The heat treatment goal is to analyze the effect of sintering of the layer in morphology and phase transformations terms. After the heat treatment the microstructure and surface morphology were examined by electron microscopy and in terms of the phase composition by X-ray diffraction.

3.1. Structural analysis by electron microscopy of the heat treated layers

In the following figures are presented secondary electron images obtained by scanning electron microscopy on the Quanta 200 3D microscope in the Low Vacuum module which is working at pressures ranging from 50 to 60 Pa and using the LFD (Large Field Detector) type detector. The voltage used to accelerate the electron beam has the value of 30kV and the working distance of the beam cannon varies from 12 to 15 mm.

From the taken SEM images of the surface layer and from the line analysis (Figure 1) it can be seen that the layer structure became more rough, peaks increasing in intensity. The same can be concluded from diffraction and X-ray diffractometry following also the performed scan and the profile of the surfaces.

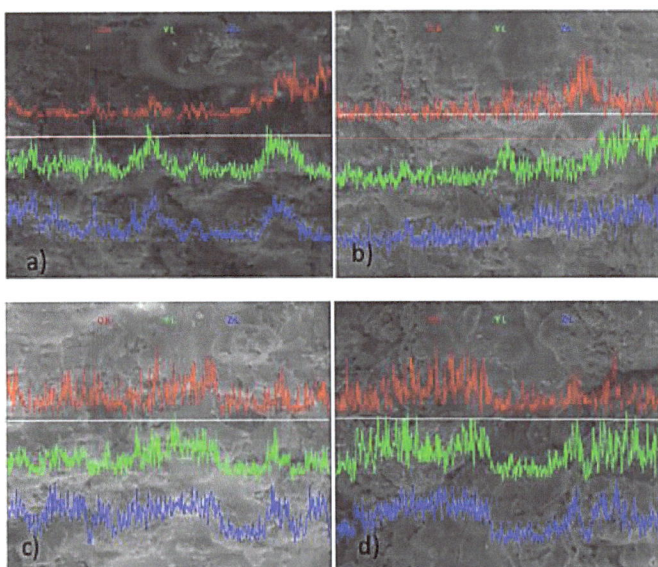

Figure 1: The SEM images show the microstructure of the ZrO_2/20%Y_2O_3 deposited layer:
   a) before the heat treatment;
   b) after the heat treatment at 1000°C for 5 h;
   c) after the heat treatment at 1000°C for 10 h; d) after the heat treatment at 1000°C for 15 h.

In Figure 2 are shown SEM images of the ceramic layer with the thickness of 100 μm and the adherent layer thickness of 30μm. The adherent layer has a filiform layout and a very good adherence to the base material and makes a good relation with the ceramic layer which consists of successive splats with regular geometric shape, a porous structure [1]. Also intergranular corrosion is observed due to the porous ceramic layer combined with its high temperature undergoing heat treatment (Figure 2-c). Black particles are located at the adhesion layer/ceramic layer interface and have a high content of Al and Ni from the oxidation process of the elements which compose the adherence layer during the deposition and they rise during the heat treatment (Figure 2-d).

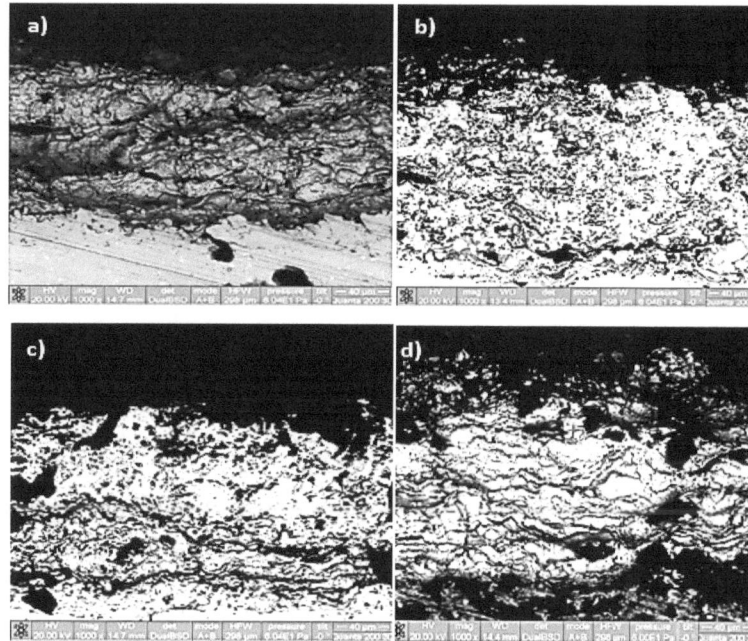

Figure 2: The SEM cross-section images present the microstructure of $ZrO_2/20\%Y_2O_3$ deposited layer in contrast Z:
a) without thermal treatment;
b) after the thermal treatment at 1000°C for 5 h;
c) after the thermal treatment at 1000°C for 10 h; d) after the thermal treatment at 1000°C for 15 h

In Figure 3 are presented the limits between neighboring splats with interlamellar cracks. The heat treatment caused an increase in grain size on the layer surface by the formation of "bridges" between neighboring splats (Figures 3-b, c, d), with the effect of closing the initial microcracks [1, 2]. Large pores are maintained after prolonged heat treatment. There were no qualitative differences between the surface layer heat treated for 5 h and the one treated for 10 h (Figure 3-b, c). For the treatment carried at the temperature of 1000 °C for 15h a splitting of the grains is observed and a tendency of disappearance of the successive arrangement of the splats (Figure 3-d).

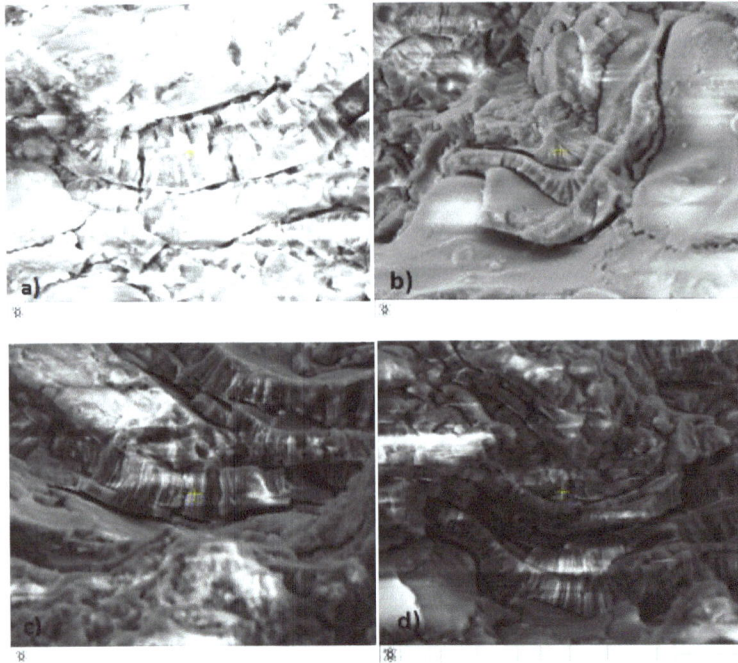

Figure 3: The SEM cross-section images which present the microstructure of the $ZrO_2/20\%Y_2O_3$ deposited layer:
- a) without heat treatment;
- b) after the heat treatment at 1000 °C for 5 h;
- c) after the heat treatment at 1000 °C for 10 h; d) after the heat treatment at 1000 °C for 15 h.

3.2. Structural analyses using X-ray diffraction

The X-ray diffractometer is equipped with a X-ray anode tube made of Cu kα, λ = 1.54 Å, to which has been applied voltage of 45 kV with a current of 40 A, at an angle of diffraction (2θ) ranging from 25 to 130°. X-ray diffraction analysis was performed in order to observe and highlight the structural changes depending on the number of hours applied to each treatment compared with the phases that were highlighted on the untreated samples (Figure 4). [1]

Wavelengths are: K Alpha 1 [A]: 1.54060, K-Alpha 2.1,54443, K-Beta [A]: 1.39225, K-A2 / K-A1 Ratio: 0.50000.

Structural analysis were performed using a dedicated software (Xpert High Score Plus) thought which the crystallographic parameters were identified (lattice type, network constant values a, b and c and angles elementary cell alpha, beta and gamma elementary cell volume, density) and the possible compositional parameters. For the precise choosing of the stoichiometric compositions.

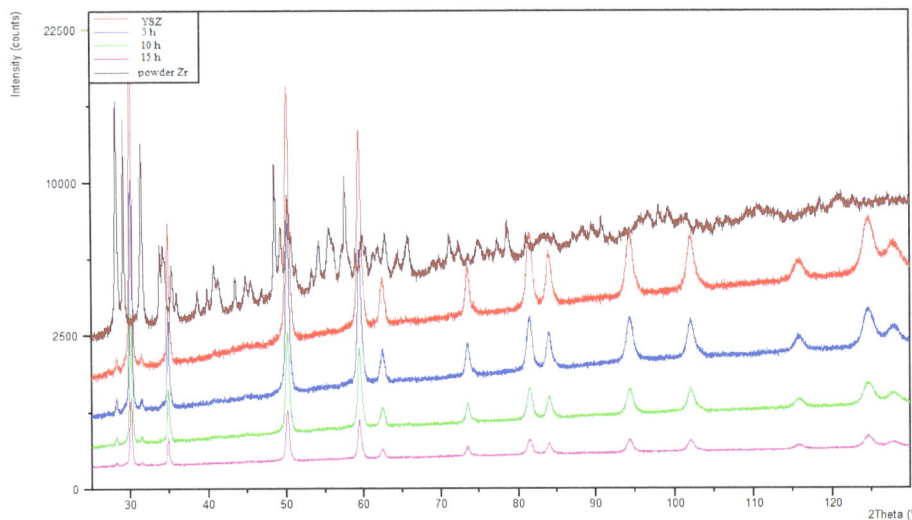

Figure 4: The X-ray diffractogram for the $ZrO_2/20\%Y_2O_3$ deposited layer, at an diffration angle of 2θ = 25...130°

On the cumulative diffractogram was also inserted the specific one for the initial powder. Since the initial powder is a mixture of Zr, Y, O, peaks for all the specific crystallographic planes are present. The XRD patterns of the samples have a smaller number of peaks, on one hand due to the formation of new compounds with new types of crystallographic networks, on the other hand due to the reduced exposure of some crystallographic planes, mainly due to the influence of the substrate and texturing[3].

The determination of the elementary cell parameters is showing the appearance on the substrate of a modified network from the original zirconium oxide. The changes are due to the inclusion in the network of yttrium atoms of the stabilizer.

The original tetragonal monocrystalline lattice of the zirconium oxide has the following parameters a = 5.1240 A and c = 5.1770 A. For the sample with the coating thickness of 100 μm, the new network has the following parameters a = 3.6120 and c = 5.2120. All the networks are monoclinic, but after 15 hours of heat treatment the networks relocate presumably by the migration of the yttrium atoms in the place of the missing oxygen atoms, so the network is tetragonal, with monoclinic precipitates. The new tetragonal network has different parameters from the original zirconium oxide, namely a = 3.6260 and c = 5.2350.

Table 2: Stoichiometric structures occurring for different types of heat treatment

Parametre	$ZrO_2/20\%Y_2O_3$	$ZrO_2/20\%Y_2O_3/$ 5 h heat treatment	$ZrO_2/20\%Y_2O_3/$ 10 h heat treatment	$ZrO_2/20\%Y_2O_3/$ 15 h heat treatment
a	5,1484	5,1507	5,1507	3,6260
b	5,2067	5,2028	5,2028	3,6260
c	5,3154	5,3156	5,3156	5,2350
alfa	90,0000	90,0000	90,0000	90,0000
beta	99,2290	99,1960	99,1960	90,0000
gama	90,0000	90,0000	90,0000	90,0000

The XRD patterns with indexed crystallographic planes are shown in the following figure:

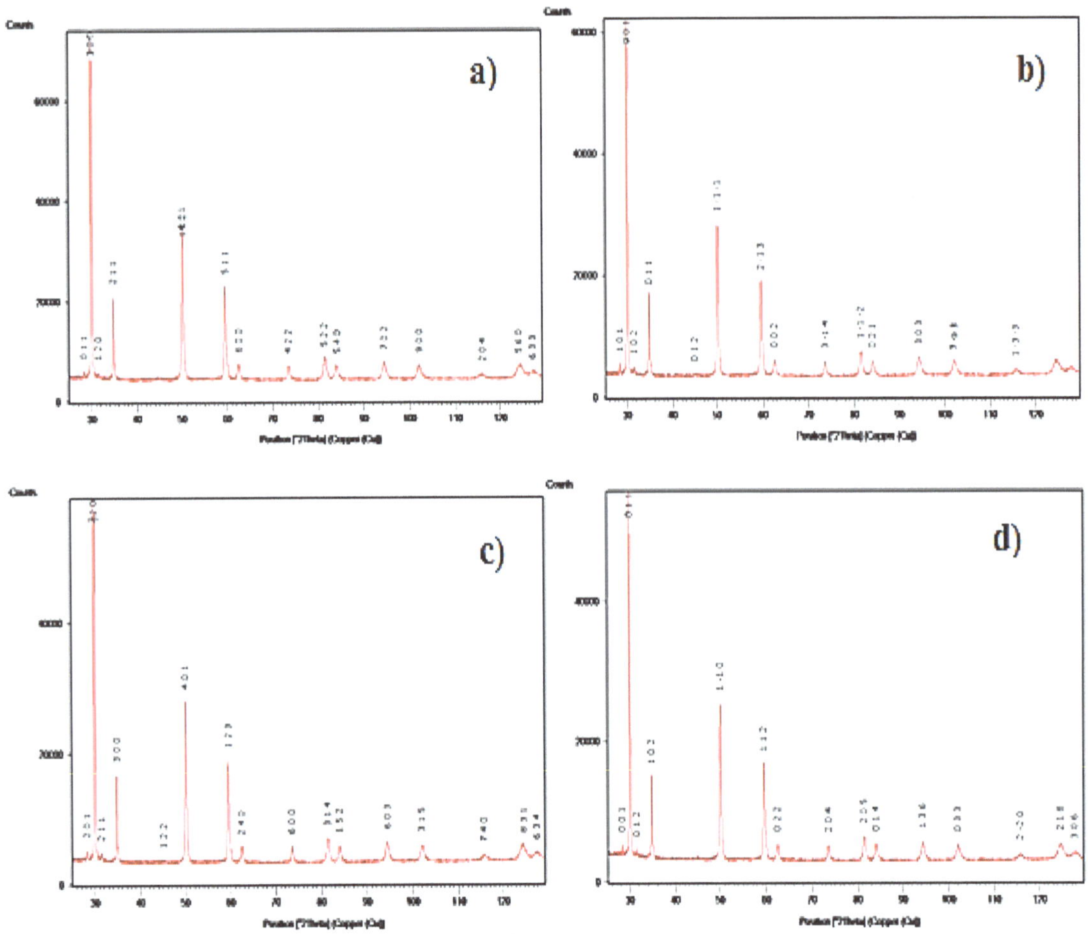

Figure 5: The crystallographic planes for the ZrO2/20%Y2O3 deposited layer, with a diffraction angle of: $2\theta = 25...130$:
a) without heat treatment;
b) after the heat treatment at 1000°C for 5 h;
c) after the heat treatment at1000°C for 10 h; d) after the heat treatment at 1000°C for 15 h.

3.3. The determination of the surface roughness

The surface roughness measurements for the layer sprayed with the $ZrO_2/20\%Y_2O_3$ ceramic powder were made using the Form Talysurf Intra system.

In Figure 7 is presented the roughness analysis on the four profiles using the LS-Line module. The arithmetical mean deviation of the profile roughness measured for the sample with the layer thickness of 100 μm without heat treatment (Figure 7-a) is: $R_a = 6,4238$ μm, the standard deviation of the assessed profile roughness $R_q = 8,1743$ μm and the average height roughness $R_z = 40,1180$μ.

Arithmetical mean deviation of the roughness profile measured for the sample with the layer thickness of 100 μm after a 5 hours heat treatment (Figure 7-b) is: $R_a = 5,9197$ μm. The standard deviation of the assessed profile roughness $R_q = 7,6334$ μm and the average height of roughness profile measured $R_z = 38,4407$ μm.

In the case of the roughness profile measured on the sample with the layer thickness of 100 μm subjected to heat treatment for 10 h (Figure 7-c) the parameters are: $R_a = 6,0821$ μm, the standard deviation of the assessed profile roughness $R_q = 7,4465$ μm and the average height roughness $R_z = 32,464$.

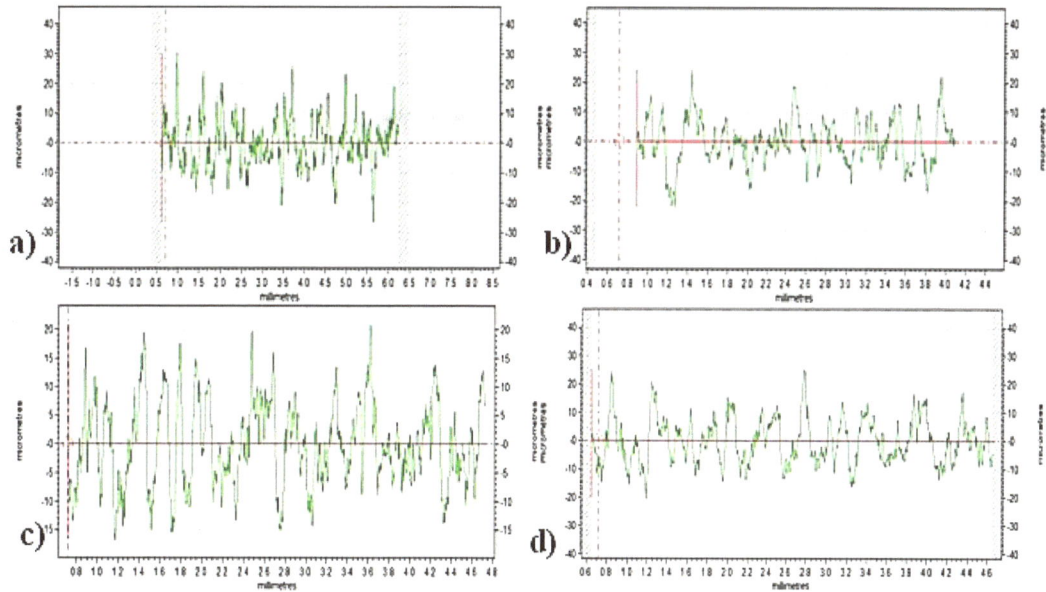

Figure 7: The roughness values:
a) without heat treatment;
b) for the samples subjected to heat treatment for 5 h;
c) for the samples subjected to heat treatment for 10h;
d) for the samples subjected to heat treatment for 15 h.

The roughness profile measured on the samples with the layer thickness of 100 μm and subject to a 15h heat treatment (Figure 7-d) is: $R_a = 6,5547$ μm, the standard deviation of the assessed profile roughness $R_q = 7,9397$ μm and the average height of roughness profile measured is $R_z = 34,9909$ μm.

3. CONCLUSIONS

After the heat treatment at a temperature of 1000 °C the matrix becomes more compact. The grain size increases due to the formation on the surface layer of "bridges" between neighboring splats.
Following the heat treatment carried out at a temperature of 1000 °C with a 15 h maintenance, a division of the grains appears and a tendency of disappearance of the successive arrangement of the splats.
Following the X-ray diffraction investigations it is revealed that the spraying of the ceramic powder under high temperature, a multiphase structure coating is obtained. The structure consists of elementary monoclinic cell particles for the heat treatment from 5 to 10 hours.
For the 15 hour heat treatment, multiphase changes were revealed in the assembly. Particles with a tetragonal crystalline structure appear, stable at room temperature and which, according to the literature, show superior thermo-mechanical properties for technological applications.

ACKNOWLEDGEMENTS

We acknowledge that this paper was realized with the support of **EURODOC** "Doctoral Scholarship for Research Performance at European level" project financed by the European Social Found and Romanian Government.
We thank to **SC Plasma Jet SA** for the support in the plasma jet deposition of the analyzed layer from this paper.
We thank **SC AEROSTAR SA** from Bacau for providing the super alloy substrates.

REFERENCES

[1]   SONG Xiwen, XIE Min, ZHOU Fen, JIA Guixiao, HAO Xihong, AN Shengli, *High-temperature thermal properties of yttria fully stabilized zirconia ceramics,* JOURNAL OF RARE EARTHS, Vol.   29,   No.   2,
Feb.   011, p. 155, 2010.
[2]   Nicholas Curry, Nicolaie Markocsan, Xin-Hai Li, Aure′lien Tricoire, and Mitch Dorfman, Next   Generation Thermal Barrier Coatings for the Gas Turbine Industry, ASM International, 2010;
[3]   Chin-Guo Kuo [a], Hong-Hsin Huang [b], Cheng-Fu Yang [c,*], Effects of the oxygen pressure on the   crystalline orientation and strains of YSZ thin films prepared by E-beam PVD, Ceramics International   37 (2011) 2037–2041, 2011.

USING POLYMERIC GRIDS IN MASONRY RETROFITTINGS

D. Stoica[1], R. Sofronie[1]

[1] Technical University of Civil Engineering, Bucharest, ROMANIA, danielstoica2001@yahoo.com

Abstract: *The use of composite materials in religious buildings and historical monuments, include two key principles, supported by the Venice Chart (1964), namely: structural interventions are less visible and does not alter the appearance of the building; action is reversible, with the possibility of disassembling initially established in the event of non performance. Composite building efficiency is obtained only by ensuring adequate anchorage length at the ends of the material.*

Keywords: *FRP, polymeric, grid, strength, capacity*

1. INTRODUCTION

In the entire world there is a huge number of existing masonry buildings with masonry structural walls system made of: stones, solid or cored bricks.

Because of their emplacement in the earthquake or mining subsidence areas for a lot of them a retrofitting solution must be adopted as quick is possible.

Generally speaking there are only several retrofitting possibilities such as:

1. A new structural system insertion which to cover the lack of strength capacity and which may be:
a. A confining system with RC small columns and belts
b. A frame system: RC or steel
c. RC structural walls system
2. Retrofitting of the existing structural system using jacketing:
a. In classical solutions: RC coat with steel reinforcements
b. In modern solutions: using FRP and epoxy resins or using polymeric grids and lime mortar

2. MASONRY STRUCTURES MAIN ASPECTS

Normally each type of structures may satisfy two very important demands to offer safety for gravitational and exceptional actions: stiffness through the lateral displacements and strength capacities. From this point of view the buildings with masonry walls structural systems present enough stiffness. As a measure of it the fundamental period of vibration may rise to T=0.05n seconds (where n represent the number of oscillating building levels).

The main characteristic of the existing stock of masonry buildings (without adequate design codes) is that usually the masonry blocks have high strength capacities and the mortars are with a low quality and strength. Thinking to these we may say that the main problem of the existing masonry structures is the lack of strength capacities and not the stiffness. But the masonry as a construction material is a composite material consists in blocks and mortar layers and its strength depend of both components.

Each seismic or mining subsidence zone have proper accelerations and displacements spectra (in function of the recorded data) but almost each one have a period of resonance T_r (highlighted as an exemplification in Figure 1).

In function of it, the fundamental period of vibration T of a masonry building may be like in the table 1.

The existent masonry buildings stock consist of ordinary buildings (dwellings for instance) but also from cultural heritage buildings (churches, historical monuments).

Their retrofitting solution is a complex decision because of the functionality and architectural aspects but also for the structural safety. Mostly the RC solution seems to become traumatizing for the buildings. The decision must be in accordance of all these.

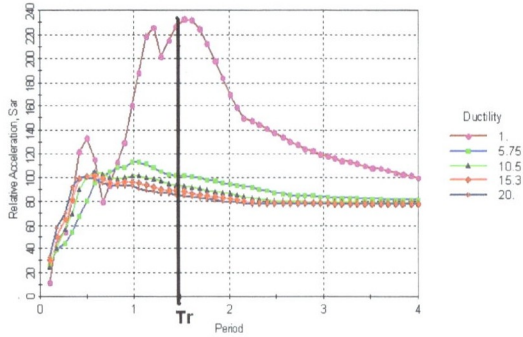

Figure 1 - Equal ductility spectra exemplification

Table 1 Spectral position.

Case	Comparison	Acceleration Spectra value	Stiffness	Drifts
1	T<Tr	minimal	great	small
2	T=Tr	maximal	enough	normal
3	T>Tr	medium	low	large

Usually the architectural point of view is more or less take into consideration by the civil engineer designers but for the cultural heritage buildings, the architects have the principal role. If RC solution are normally chosen for the first category, for the second normally remain just one possibility – to use the modern materials and techniques (FRP or polymeric grids solutions).

We shall ask which one is better to use?

From our opinion, after a lot of tests made for FRP solutions we observed that in the majority of the cases (even it was about stones or clay bricks) an exfoliation of FRP from the masonry surface appear (and this happened into laboratory conditions, well supervised and controlled). Some of these are presented in the figure 2.

Figure 2 - FRP solution after the tests

Whatever, the FRP materials and their application system is expansive and if the final result conduct to this behavior shows that something is not match. A lot of explanations can be but we think about a big difference between the materials characteristics (including Young modulus and strengths).

To change the system is possible after this, but is rather expansive too.

On another hand these are specimens, simply peers easy to retrofit and to manage all the connections between the FRP and the main structure. But there are a lot of places into a real building case where the entire systems become complex and difficult for application.

But is not the case of our article so in the followings we'll try just to highlight why a polymeric grid solution may be better first than a classical solution and second than the modern FRP one.

3. POLYMERIC GRIDS RETROFITTING SOLUTION FOR MASONRY STRUCTURES

More than 90% from the human loss became from their building collapses and huge economic losses appear after major earthquakes.

We shall expect than during a building life a lot of small earthquakes may occur in an year, several moderate or medium earthquake during the exploitation life and the severe earthquakes 2-3 times or never. In the following figures the European seismic (Figure 3), geological (Figure 4) and world (Figure 5) seismic maps are presented.

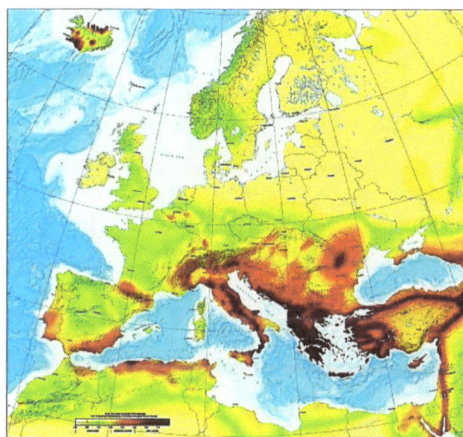

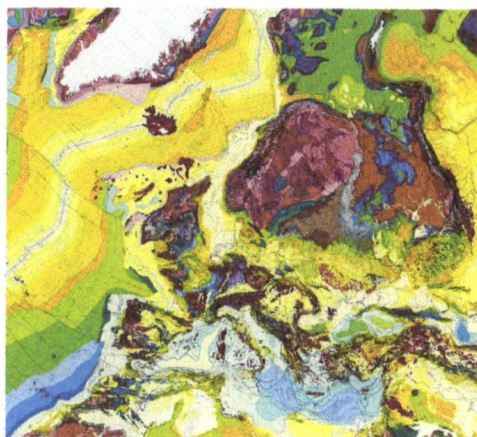

Figure 3   - European seismic hazard map   **Figure 4** - European geological map

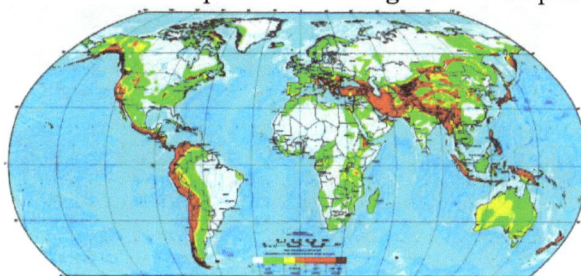

Figure 5 - World seismic hazard map

An alternative solution for the steel nets reinforcement is to use the polymeric grids. For the railways, highways or airports infrastructure the geo-grids were used for over 20 years and if the distance between Brussels and Paris today is cover in a couple of ours this performance is tributary to geo-grids. In the present, for masonry structures three types of grids are available: RG20, RG30 and RG4 (RichterGard system) with the strength capacities of 20, 30 and 40 kN/m. These products are fabricated by former Tensar International Limited from England (named RichterGard in the present) ant having the initial index SS20, SS30 and SS40. It is not possible to use any other type of synthetic product for reinforcement as polymeric grids.

The performances of the grids with integrated joints (made from high density and strength polyethylene) consist in variable and self-reliable strengths ant the capacity for adaptation in case of local stress concentrations resulting energy dissipation inducted in excess. This is coming from the geometrical conformation of those grids and from the macro-molecular structure of the used polymers. All those performances were experimentally granted on the shaking tables or seismic laboratories (Bergamo-Italy, Lisbon-Portugal, Ispra-Italy and Iasi-Romania) for natural scale models of masonry elements or buildings (1D, 2D and 3D-presented in Figure 10-23). In all the cases the results were over expectations and highlighted new and innovative constructive solutions, meaning the homogenization of the reinforced masonry constructions after the elimination of RC structural elements.

The RG new system concept consist in participation of all the masonry strength resources and avoiding the local unitary stress concentrations. In this aim the masonry buildings will be jacketing only on the perimeters, at the external envelopes, with polymeric grids inserted in a simple lime or even lime-cement mortar layer. In some cases also some masonry structural elements may be confined.

Using the RG system the masonry confinement conduct to a reduction of masonry anisotropy and non-homogenously switched by continuous geometrical deflections and to triaxial compression unitary stresses.

The strength capacity of the confined masonry increasing about 5 times as the experimentally tests proves. On another hand, in comparison with the unconfined masonry, for the instantaneously or short term dynamic actions the confined masonry respond in elastic stage, but with a higher dissipative capacity through for the inducted energy.

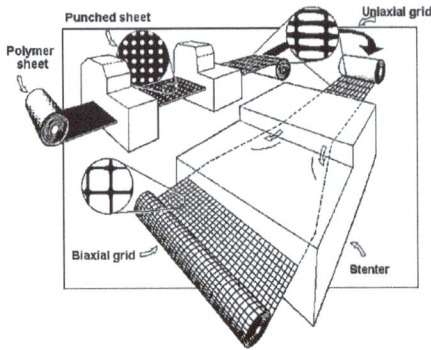

Figure 6 - Tensar® **process**

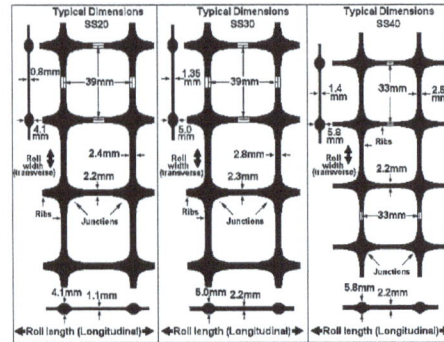

Figure 7 - Geometric Characteristics of *Tensar* Grids

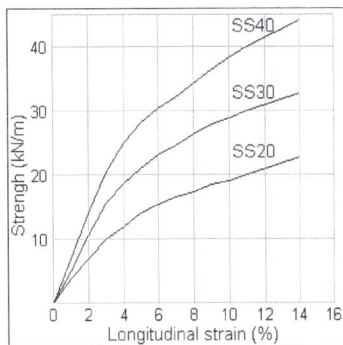

Figure 8 - Longitudinal stress-strain diagram

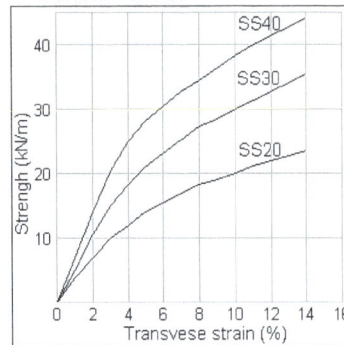

Figure 9 - Transverse stress-strain diagram

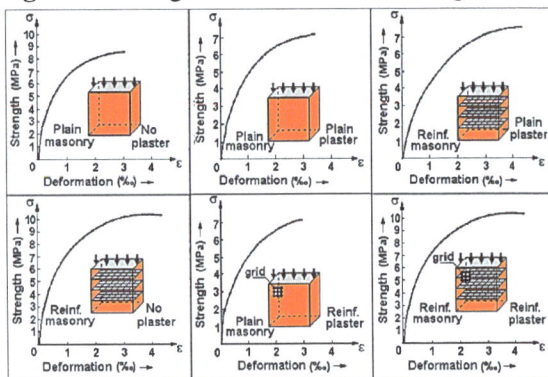

Figure 10 - Wall panels of plain and reinforced masonry/plaster submitted to axial compression

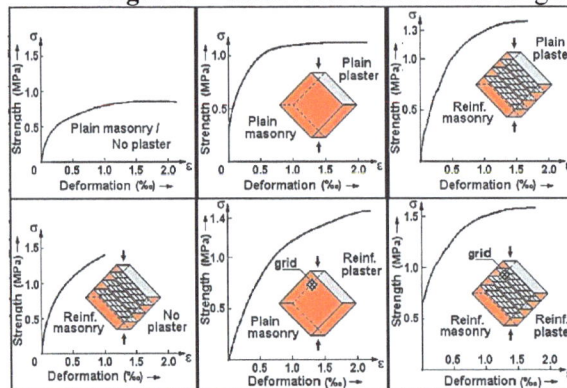

Figure 11 - Wall panels of plain and reinforced masonry/plaster submitted to diagonal tension

All those jacketing solutions previously described are totally integrated to the masonry and no additional structural components which to doubling the columns or the masonry walls. The increasing strengths capacities coming only from the grids and the spatiality of the unitary stresses in the confined masonry.

The RG system keep the original masonry integrity without any other traumas (like cuttings, drillings) and because of its reversibility anytime become easily to take off without to produce any injuries to the initial masonry structures or elements.

Using the polymeric grids jacketing or confinements substitute the RC or steel structural components role. The buildings with confined masonry structures become more homogeneously without any supplementary unequilibrate masses and local stress concentrators, being in concordance with the EC8 (Annex A) requests. In plus, the confinement state for a unitary behavior of the building parts and reduce as called "whiplash" during the seismic actions.

There is a totally difference between the steel reinforcement net fixation and the polymeric grids, in their case the joints anchorage is used. The stress transfer mechanism from the mortar to polymeric grids take place only in the rigid joints through normal stresses (σ), without any tangential stresses (τ) contribution and only the tension efforts are transferred from the mortar to the polymeric grids.

Looking to figures 18 and 19 it is easily to understand that for the solid bricks model with confinement the bricks remain integers and because of the energy dissipation the polymeric grids exfoliate and for the second model with cored bricks the cored brick crushed and the grid remain integer.

Both models stand up after a lot of tests near PGA=1.96g. For the first model, with solid bricks it is easily to remove and replace the grids in safety conditions for the solid bricks masonry but for the second model with cored bricks it is explicit that even the model stand-up this is only from confinement and the crushed bricks cannot be remove and replace. [1] [2] [3] [4] [5]

Figure 12 - Damage patterns for ordinary masonry

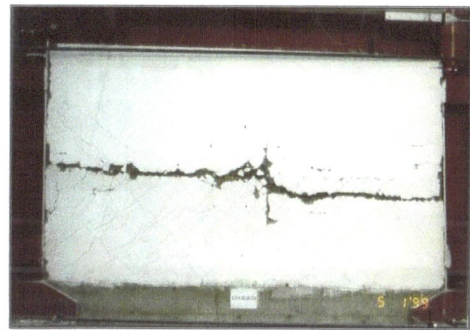

Figure 13 - Damage patterns for confined masonry

Figure 14 - Damage patterns for ordinary masonry

Figure 15 - Damage patterns for confined masonry

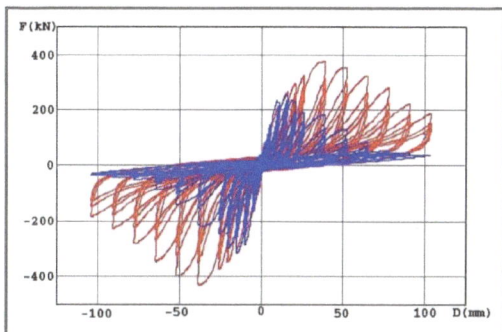

Figure 16 - Comparative hysteretic diagrams for wall panels without openings of plain and confined masonry

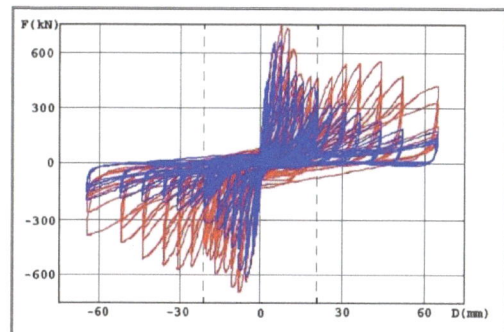

Figure 17 - Comparative hysteretic diagrams for wall panels with openings of plain and confined masonry

Responses on the tested models with confinements

Figure 18 - Failure pattern in Tensar SS20 after 18 dB = 1.96 g

Figure 19 - Failure pattern in Tensar SS20 after 18 dB = 1.96 g

3D – Bergamo Shaking Table tests

Figure 20 - Model 1 – model without confinement

Figure 21 - Model 2 – model without confinement

Figure 22 - Model 1 – model with confinement

Figure 23 - Model 2 – model with confinement

REFERENCES

[1]   **Sofronie, R.**, Bergamo G., **Stoica, D.**, Toanchina, M., „Civil structures of reinforced masonry without rc-structural members". *Proceedings of the National Convention on Structural Failures and Reliability of Civil Structures*. University of Architecture, Venice 6-7 December 2001, pp.347-358.

[2]   **Sofronie, R.,** Bergamo G., **Stoica, D.,** Toanchină, M., „Seismic response of masonry buildings reinforced with polymer grids." *Third World Conference on Structural Control* Como – Italy 7-12 April 2002;

[3]   **Sofronie, R.,** Bergamo G., **Stoica, D.,** Toanchină, M., „Buildings of composed masonry with cored or solid bricks?" *Proceedings of the XXX IAHS Congress on Housing.* 9-13 September 2002, University of Coimbra – Portugal

[4]   **Sofronie, R.,** Franchioni, G., Bergamo G., **Stoica, D.,** Toanchină, M., „Masonry irregular buildings reinforced with polymer grids*" Proceedings of the 3rd European Workshop on Seismic Behavior of Irregular and Complex Structures.* 17-18 September 2002, Florence, Italy

[5]   **Stoica, D.,** Tragakis, P., Plumier, A., **Sofronie, R.,** Majewski, S. – "Masonry structures retrofitting with polymeric grids", *FRPRCS – 8, University of Patras, Greece, July 16-18, 2007*

THE FLY-ASH ADMIXTURE INFLUENCE CONCERNING THE PERMEABILITY OF THE BELITIC CEMENT CONCRETE

M. Rujanu[1], D.C. Covatariu[1]

[1]„Gheorghe Asachi" Technical University from Iaşi, Faculty of Civil Engineering and Building Services, ROMANIA,
mrujanu@yahoo.com, covadan@ce.tuiasi.ro

Abstract: The actual research involving fly-ashes resulted as a residue from the thermo power station processes point out the fact that the admixture of a certain quantity in the fresh concrete has as an effect the considerable increasing of the permeability. Also, the admixture of the fly-ashes could considerable decrease the cement quantity in the concrete mixture, so that the price of the concrete will be more decreased. In this case it is important to analyse the influence of the fly-ashes addition over the impermeability rate of the belitic cement concrete.
Keywords: fly-ash, belitic concrete

1. INTRODUCTION

The concrete works assume the realization of some performant features from all points of view. In this way, a concrete with high permeability requirements can be accomplish using active additions, in case where we do not neglect the strength characteristic. The permeability characteristic it is performed through the structural characteristics - compactness, porosity. If on the compactness it can interfer in the way of enlargement, by using active additives, experimentally was observed that the ensemble of technical characteristics obtained on these concrete were close to those of concrete without addition. So it is impose that for every mixture of concrete made with belitic cement to analyze the permeability characteristic comparative with others technical characteristics important for favorable behavior of concrete and to choose the variants with the best overall behavior.

2. GENERAL ASPECTS CONCERNING OF THE OF FLY-ASH ADMIXTURE' INFLUENCE ON SOME CONCRETE CHARACTERISTICS

Cement concrete use cement as binder, and this influences the formation of concrete structure through its nature and through the dosage used for realization of the mixture. In case of hydrotechnical cements used as a binder for concrete, in the category of wich it fits the one that is the subject of the present study, it has in the composition 6-20 % basic slag of furnace that improves the structural characteristics.

From the point of view of the acting mechanism of the addition, this type of addition produces some changes in the structure of the cement stone and on the formed structural characteristics and so on the behavior of the binder used in the concrete mixtures. So, the slag is hydrated in the presence of calcium hydroxide solution which rises from the reaction with water of the cement clinker. The hardening takes place in time, the cement paste gets rigid, the viscosity around the particles of slag increases, leading to a slowing in the process of difusion. In this period there are formed the saturated solutions of gypsum, calcium hydroxide and small quantities of SiO_2 and Al_2O_3. The speed that dissolve and it hydrates Al_2O_3 is smaller than the speed that it separate later to hidroaluminat. SiO_2 that exist in the solution implies the hydration of tricalcium silicate. Later there will be the reactions between water, calcium hydroxide and calcium sulphate with the active compounds of the slag. In a first phase, the slag undergoes a superficial coloidation (coloidare) and afterwards it will take place the formation of hydrosilicates, hydroaluminates and complex hydrocompounds.The formed hydrosilicates will encourage the increase of volume of new gelic formations, which after a process of hardening lead to an intens process of microcracks. In this situation is recomended to maintain till the concrete hardens in a wet environment or underwater, to prevent the contractions and in the same time helping increase its compactness. The previous studies revealed the fact that the mechanical strenght although are growing slowly, in the end the values are close to those of the cement without addition, and at longer periods of hardening may exceed the strength of cement without addition.

From the research to date and those who used as addition thermo plant ash and belitic cement with addition of bazic furnace slag (the same type of addition), did not exceed the total proportion of addition 30-40%. In these conditions there is the possibility to occure some consequences regarding:

4. a greater content of water mixing imposed by addition, that is the smooth part with a big surface area, determins the dendency of movement of the pores dimension having the dimension 0,5-1 mm, to greater dimensions;

5. growth volume of capillary pores through increasing the volume of addition, from the concrete mixture.

These changes are theoretic unfavorable in forming the concrete and they can have negative influences on the structural characteristics of the concrete, that why it is imposed a very strict corelation on the content of smooth part from the mixture. This presume that the ash dosage must be corelated so that the smooth part do not exceed some boundaries and in the same time it must be considered the fact that the water dosage in the mixt and the ratio that is between water and the smooth part must be in lower limits or reduced using tensioactiv addition.

3. EXPERIMENTAL TESTS

The experiment was prepared taken into account the standards NE 012-1/2007 in conjunction with NE 012/2010 that reffers to the realization of concrete in different working conditions and who specify the limits for composition factors (minimum cement dosage, ratio A/C maxim etc.) and in the same time recommended in regard to the exposure class of the concrete, even the type of cement that is recommended to be used.

Given the experience in the domain and the restrictions imposed by the standards, we create 3 recipes of concrete using aggregate from rivers with maxim dimension of 16mm, composite cement type H II/A-S 32.5 and as addition fly-ash collected by dry way, all those for different values of component dosage and we realise the following compositional characteristics, presented in table 1:

Table 1: Compositional characteristics of the fresh concrete

Index recipe	Component dosage			A/C+Ce	Consistency class – cm
	Cement-C kg/m^3	Ash-Ce kg/m^3	Water-A l/m^3		
B1	100	200	223	0,74	C3(8,5)
B2	200	200	215	0,55	C3(8,5)
B3	300	150	216	0,49	C3(8,5)

Were made cubic specimens with the side of 14,1 cm, three from every recipe were retained in standard conditions to be tested for permeability, and other three specimens were also retained in standard conditions and tested for compression after 28 days.

4. EXPERIMENTAL RESULTS

Corresponding to the compositional characteristics of the 4 recipes, after subjecting the samples to test, was recorded the following results presented in table 2:

Table 2: Experimental results

Simbol samples	Content of cimentation material C+Ce	Strengh at 28 days N/mm^2	Specific growth at 28 days $N/mm^2/kg \times 1000$	Degree of impermeability
B1	100+200	6.0	20	P4
B2	200+200	18.0	45	P8
B3	300+150	25.2	50	P12

A graphical transposition of the experimental results, would allow a clearer appreciacion of the permeability characteristics, as follows:

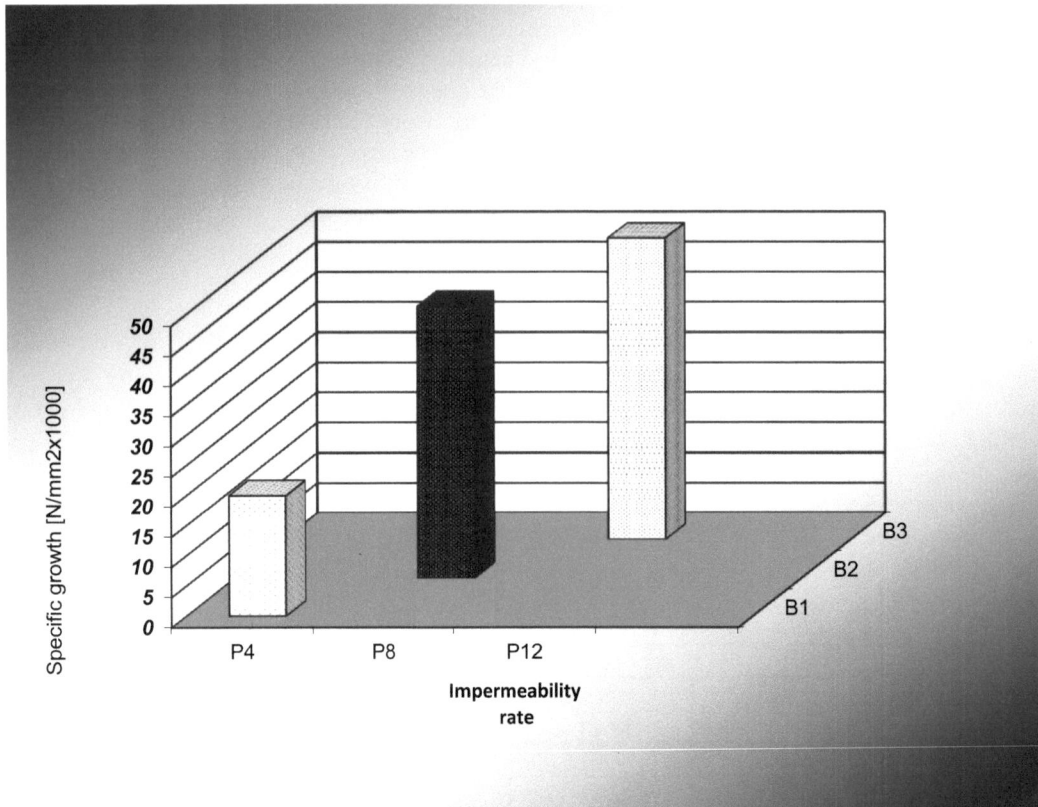

Figure 1: Permeability variation depending on the specific growth

5. CONCLUSIONS

Analysis of the experimental values supplemented with the interpretation of the chart allows the following conclusions:

5. The recipes 1, 2 and 3 have the cement dosage of 100, 200 and 300 kg/m^3, were the dosage of ash is 200, 200 respectively 150kg/m^3, present a degree of impermeability rising in accordance with specific growth, in conditions were in general were kept practically the same consistency;

6. Recipe 1 present a reduced impermeability characteristic, in conditions were the dosage of cement is just 100kg/m^3, and the dosage of ash is 200% in regard the dosage of cement. In these conditions the strenth of concrete is reduced in relation with the dosage of cement used, except that the duration of hardening over 90 days its value will be in concordancy with the dosage of cement used. It is worth to mention the fact the degree of impermeability in this case is reduce, only P4, but it stands the reserve that at 90 days will present specific improvements to this type of cement;

7. It is considered that recipe B3, presents very good characteristics in cured state at 28 days from casting, witch recommends from point of view of the strength and also from the degree of impermeability;

8. From economical efficency point of view, in terms of impermeability characteristic, I would notice recipe B2. Even though in this case, for a dosage of cement, of 200 kg/m^3 and ash 200 kg/m^3, it is obtained a specific growth of 10% smaller than in case of recipe B3 in conditions of a dosage of cement increased with 33% and in a dosage of ash smaller with 25%, consider that this recipe (B2) reflects the improvementof the behavior from impermeability point of view, in conditions of introduction of some increased dosage of fly-ash, if it makes the comparaison with recipe B1.

REFERENCES

[1]   A.M. Neville, *"Proprietățile betonului"*, ediția a IV-a, Editura Tehnică București, 2003;

[2]   Rujanu, M., Teza de doctorat, Facultatea de Construcții și Arhitectură Iași, 1993

[3]   Koichi Maekawa ș.a., *Modelling of concrete performance*, Routledge, London, 1999;

[4]   Marta, L., ș.a., Adaosuri de impermeabilizare pentru mortare și betoane, Sesiunea științifică a I.P. Iași,   Secț. B.p., pg. 346, 1981

[5]   Marta, L., ș.a., Considerațiuni asupra posibilităților de impermeabilizare a betonului, A V-a Conferință   de Geotehnică, Cluj-Napoca, vol. III, pg. 39, 1984

[6]   Vermat, M., Revista Materiale de Construcții, nr. 629, 1968

COMPARATIVE DYNAMIC ANALYSIS FOR A DOOR CAR USING DIFFERENT TYPES OF MATERIAL BASED ON VIRTUAL SIMULATION

C. Itu[1], M. L. Scutaru[1]

[1] Transilvania UNIVERSITY, Brasov, ROMANIA,

calinitu@unitbv.ro , calinitu@yahoo.com, luminitascutaru@yahoo.com

Abstract: *In this paper, we attempt to evaluate the structural behavior of the door car (stress and displacement magnitude values) based on dynamic stress simulation, for two material types (composite and steel materia)l.*

Keywords: *finite elements, stress, strain, optimization, rigidity, and composite material.*

1. INTRODUCTION

The actual tendency of the engineers and researchers people from worldwide are focused both on the research-developed-optimization activities for some components or parts of system assemblies or sub-assemblies and on the increasingly quality, reliability and durability of products in order to remain competitive and profitable or to obtain the marketplace.

The manufacturers are pressed to refresh their product models much faster than before. Shortening time to market and reducing costs are simply not good enough anymore. They have to penetrate new markets with innovative new products that support and cultivate strong brand values. This calls for a high performance development process – with high throughput and high precision. A process that delivers the right products – designed right first time. Engineering departments now more than ever are challenged to tune the functional performance characteristics of their new designs.

Creating innovative and attractive products is not just about producing visually attractive designs. Some manufacturers consider safety and reliability not just to be constraints but critical to their image.

Today, the industry evolution towards more virtual prototyping results in fewer prototypes that must be tested in greater detail.

Virtual prototyping method won it an important role in the different high industry domains, such as: automotive, aircraft, biomechanical industry. The truth virtual prototyping is created by means of FEA software, seeing obtained some products which accomplish the request imposes by market. Therefore, by means of FEA software, it can be obtained a faithful modeling, both for system components and functional conditions of the system. In this way, it can be eliminated a long stage from experimental testing which represent an expensive process.

FE analyses provide a very efficient way to represent structural flexibility for even the most complicated geometry. The method uses an advanced numerical algorithm based on displacement calculus method. The flexible behavior for any number of parts in the simulation can be represented and visualized by graphing and animating results.

This functionality allows visualizing the stress field of a part undergoing dynamic or static loading. When using the dynamic solver, the resulting stresses are the dynamic stresses that take all transients into account. In this paper, we attempt to evaluate the structural behavior of a door car under a shock loading resulting from close door. The dynamical loading that acts on the door is an inertial acceleration.

2. TECHNICAL REQUIREMENTS

The processes and phenomena that develop in mechanical system are quite complex and difficult to represent faithful in virtual models, which represent in fact an approximation of the real systems.

Theoretical interpretation of the phenomena and processes from mechanical systems gets useful in order to make a final decision with regard to adoption some optimal functional parameters.

The analyzed model was made for a 3 point bending composite cantilever, and block diagram of the debate study is presented in the figure 1.

The input values known are mass, material & geometrical properties for cantilever. The other components: supports and nose that acts on the cantilever in the middle zone of it, will be considered as rigid body parts. The external loading is applied as velocity on the nose. The supports have all DOF restraints.

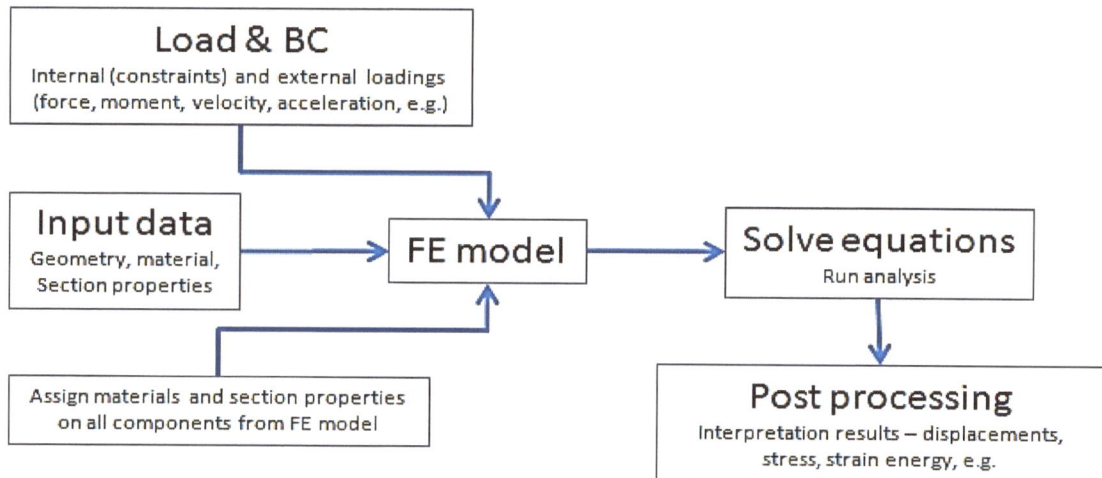

Figure 22 - Bloc schema for analysis

Based on input parameters adopted and the restraints, it can be obtained the equation solutions of the mechanical system, resulting the reactions force in the nodes that have restraints all DOF and based on them the force loading that charge the composite cantilever.

The motion equations obtained from dynamic behavior of the mechanical system along with external and internal loading applied on the study model leads on to obtained the dynamic FE model.

Dynamic stress analysis of the door car under inertial loading

In the figure 2, it is shown the FE (virtual) model of door car. According with virtual schema, the inertial loading is applied on the door and body car..

The inertial loading represents in fact the acceleration that can appear from a shock loading after a close door stage. The inertia loading imposed is $10g = 98100$ [mm/s^2].

The connection between both components (door and body car) was made using rigid elements applied in the pin connection and latch zones.

The constraints assembly was applied on the external contour of the body car included in FE model (see figure 3).

The door car was analyzed in two scenarios:
1. With door made from composite material,
2. With door made from steel material

The both analysis scenarios were made in order to appreciate the behavior of the door made from composite materials with respect of door made from steel material.

The principal output parameters that will be used for interpretation results are magnitude displacements and von Mises stress.

The material properties used for both scenarios are shown in the Table 1.

The weights of door using both materials have the next values:
- Door car from composite material = 7 [kg],
- Door car from steel material = 19 [kg]

Table 1: Material properties

	Steel mat	Composite mat
Young modulus [MPa]	202500	15000
Poison ratio	0.3	0.38
Density [tone/mm^3]	7.8e-9	2.72e-9

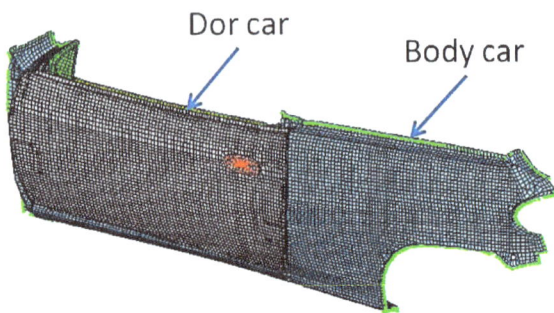

Figure 23 FE model of composite cantilever

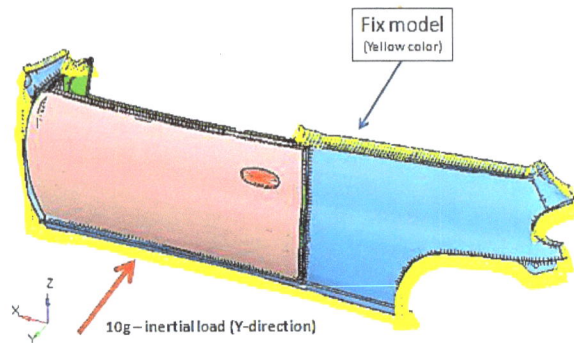

Figure 24 Load & Boundary conditions

In the figure 4 is shown the displacement magnitude field obtained on the door under 10g inertial loading applied.

External door surface

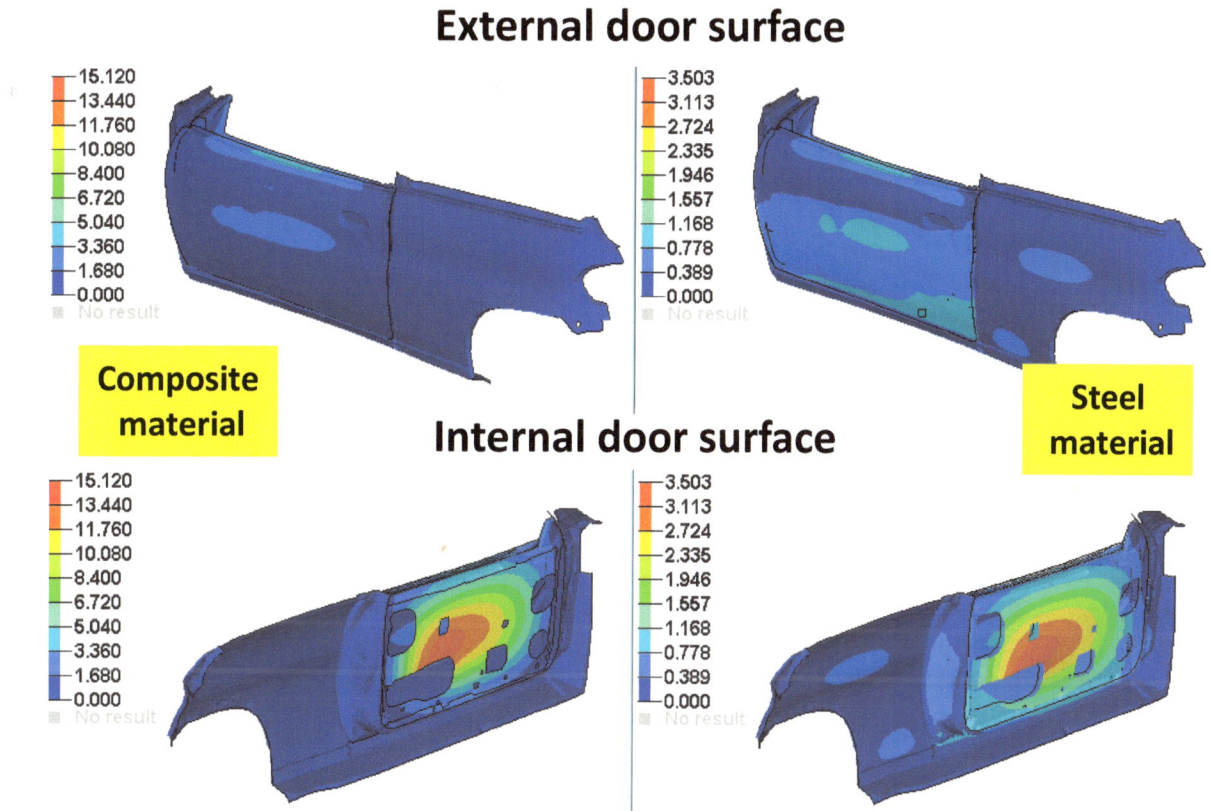

Internal door surface

Figure 25 – Plot displacement magnitude obtained on the door

In the figure 5 is shown the von Mises stress field obtained on the door under 10g inertial loading applied.

External door surface

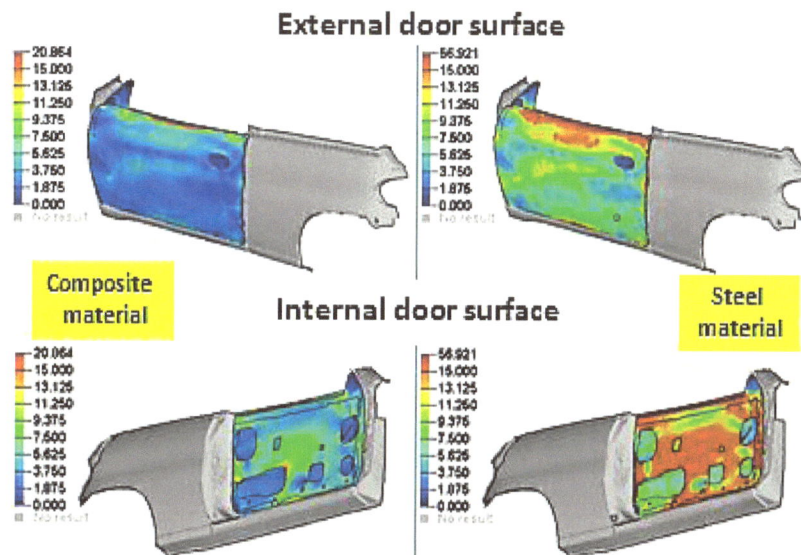

Internal door surface

Figure 26 – Plot von Mises stress field obtained on the door

Table 2: Comparison analysis Results obtained on the door car

	Composite mat	Steel mat
Maximum displacement magnitude [mm]	15.12	3.5
Maximum von mises stress [MPa]	20.86	56.92

4. CONCLUSION

In the figure 4 & 5, it can be observed the comparison results obtain from the FEA analysis for the door car analyzed.

Based on the results obtained, it can be concluded that the composite material offers a little advantage with respect to steel material for case studied. These advantages consist in: light weight and stress values lower than values obtained on steel door car.

The output parameters (displacement and stress) evaluation method based on virtual simulation offer a quickly and better information about state stress for a part under a dynamic loading comparison by evaluation method based on static analysis because the dynamical analyses with inertial loading taking into account to dynamical motion equation.

Based on dynamic FEA analysis, it can make appreciation with regards to factor of safety for a component desired on a complete duty cycle, known material properties of the part studied.

ACKNOWLEDGEMENTS

This paper is supported by the Sectoral Operational Programme Human Resources Development (SOP HRD), financed from the European Social Fund and by the Romanian Government under the contract number POSDRU/89/1.5/S/59323.

REFERENCES

[1]   Vlase, S., Elastodinamica elementelor finite. Editura LUX LIBRIS, 1996.
[2]   Vlase,S., s.a., Materiale compozite. Metode de calcul. Editura Universităţii
[3]   TRANSILVANIA, 2007. ISBN 978-973-635-890-6.
[4]   Radu N. Gh., Rezistenţa Materialelor şi Elemente de Teoria Elasticităţii, Editura Universităţii "Transilvania" Braşov, 2002
[5]   Camelia CERBU, Ioan CURTU - *Mecanica materialelor compozite;* Editura Universităţii *Transilvania* Braşov, ISBN 978-973-635-951-4, format electronic; 2007;

TENSILE TEST MECHANICAL IDENTIFICATION OF POLYMER MATRIX COMPOSITES USED IN CAR BUILDING

A. Modrea

PETRU MAIOR University, Târgu Mures

Abstract: The paper presents tensile properties of a few types of materials: low-density polyethylene (LDPE) reinforced with short fiberglass, polyester composite reinforced with fiberglass RT300 (test material taken from the warp direction), polyester composite reinforced with fiberglass RT300 (test material from weft direction) as well as polyester composite type LDPE and HDPE reinforced with short fiberglass. In order to use composites in cars parts design it is necessary the knowledge of their mechanical properties, so that loads can be calculated for conditions of use.

Keywords: RT300 glass fabric, tensile test, polymer matrix composites, high-density polyethylene

1. INTRODUCTION

Fiberglass reinforced plastic materials and composite resins are frequently used in car manufacturing. Their use requires knowledge of their mechanical properties. There are different theoretical methods to calculate these values [1], [2] but, in practice, it is seen that these methods give values that differ from real values. Therefore, it is necessary to determine experimentally the characteristics of these types of composites. Tensile tests allow for the determination of these values [3-6].

2. TYPES OF SPECIMENS USED

A series of identical specimens have been carried out of high-density polyethylene (HDPE). They were grouped according to the type of test, each sample being made of ten specimens with some spares for each sample. When the grip end broke, a spare has been used to replace the specimen. For the tensile test, LDPE and HDPE specimens' sets have been used. Specimens were used in accordance with ISO 294-1 and ISO 294-3 in order to carry out tensile tests. Figure 1 shows a series of specimens of low-density polyethylene (LDPE).

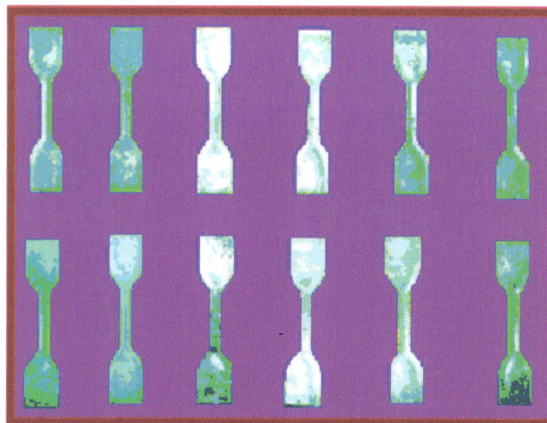

Figure 1. Specimens of glass fiber-reinforced low-density polyethylene (LDPE)

Specimens were also made of fiberglass reinforced polyester resins along the warp direction (FSU) (fig. 2) and on a weft direction (FSB) (fig. 3), for tensile tests in accordance with ISO 527-4.

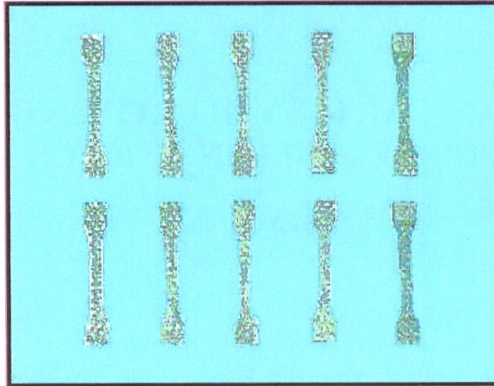

Figure 2. RT300 glass fibers-reinforced polyester resin specimens along the warp direction

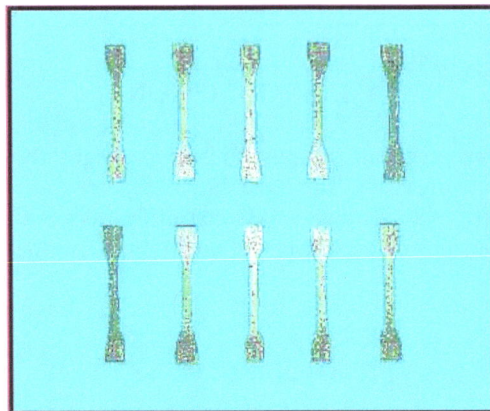

Figure 3. RT300 glass fibers-reinforced polyester resin specimens along the weft direction

Tensile tests have been carried out on specimens cut at 45° to the warp direction (fig.4).

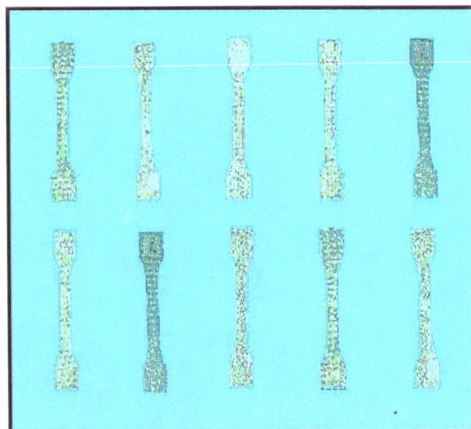

Figure 4. RT300 glass fibers-reinforced polyester resin specimens along the warp direction after tensile tests

3. RESULTS

For tensile tests, the equipment found at the Faculty of Materials Science as well as the one in the Materials Strength Lab have been used. The results of the series of tests T01 are shown in fig. 5. The mean values of the tensile tests of four types of series T01 – T04 (ten specimens each) are presented in tables 1-4.

Table 1. Basic mean mechanical properties of glass fibers-reinforced LDPE series T01 in tensile tests

Feature	Value
Load (N)	1036
Extension (mm)	9.17

Table 2. Basic mean mechanical properties of glass fibers-reinforced LDPE series T02 in tensile tests

Feature	Value
Load (N)	826
Extension (mm)	9.72

Table 3. Basic mean mechanical properties of glass fibers-reinforced LDPE series T03 in tensile tests

Feature	Value
Load (N)	887
Extension (mm)	9.28

Table 4. Basic mean mechanical properties of glass fibers-reinforced LDPE series T04 in tensile tests

Feature	Value
Load (N)	697
Extension (mm)	8.53

For the four sample trials, following values have been obtained for the Young's modulus on the length: E_{T01}=271.145MPa; E_{T02}=203.951 MPa; E_{T03} = 229.396 MPa; E_{T04} = 196.108 MPa. Specimens made of RT300 glass fibers-reinforced polyester resins, (named FSU and FSB), cut on the warp and weft respectively have been subjected to tensile tests. They have been denoted as samples T05 and T06. These specimens broke by explosion, demonstrating a good bond between the fibers and the matrix. Regarding the RT300 glass fibers-reinforced polyester specimens the failure mode is neat, which leads to the conclusion that the bond between matrix and fiber broke and the composite behavior is weaker. A detail of the tensile test is presented in fig. 6. Aspects of the breaking section are presented in fig. 7-11.

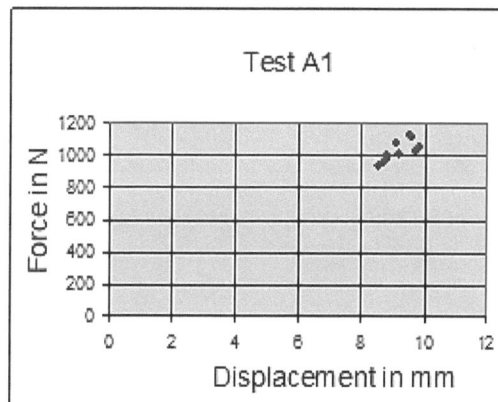

Figure 5. Experimental results for series T01

The geometric features of the test specimens are presented in table 5.

Table 5. Geometric features of test specimens

Material Type	Width b (mm)	Thickness h (mm)	Area (mm^2)
FCM	16.4	4.5	73.8
FCT	10.5	4.5	47.25
FSMB	15.5	4.0	62
FSMU	15.5	4.5	69.75
FS 45°	10	4.3	43
FSB	10	4.5	45
FSU	10	4.5	45

Figure 6. Failure mode in the center of the RT300 glass fiber-reinforced polyester specimens

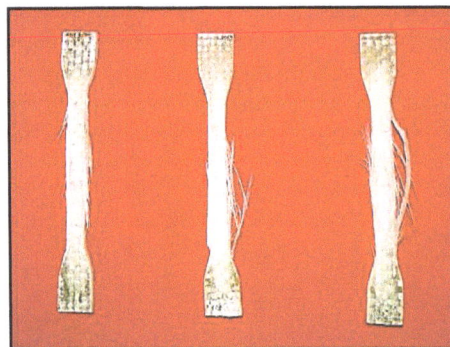

Figure 7. Broken specimens of RT300 glass fiber-reinforced polyester resins, cut on weft direction

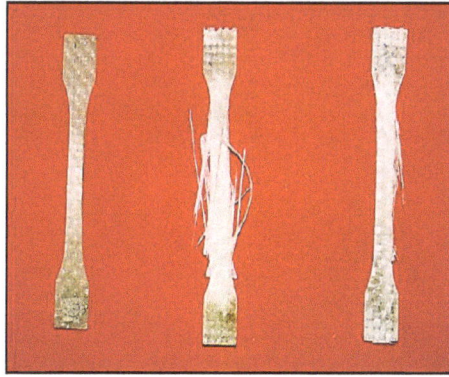

Figure 8. Broken specimens of RT300 glass fiber-reinforced polyester resins, cut on warp direction

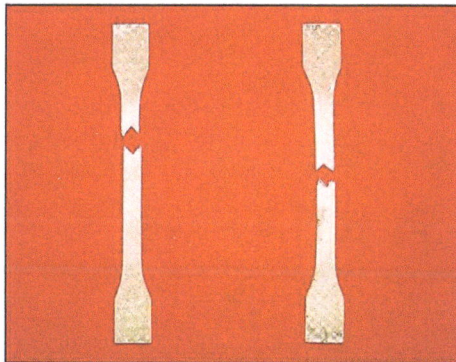

Figure 9. Broken specimens of RT300 glass fiber-reinforced polyester resins, cut at 45° to the warp

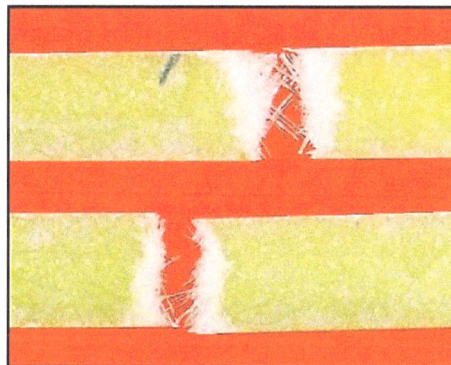

Figure 10. Broken glass fiber-reinforced LDPE composite specimens

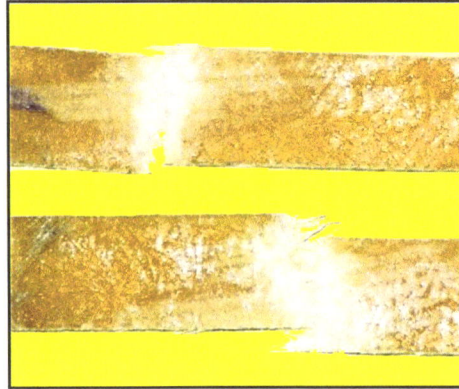

Figure 11. Broken glass fiber-reinforced HDPE composite specimens

4. DISCUSSION

Fiberglass reinforced composites type RT300 have good mechanical features and show good tensile strength and these materials are suitable for big structures. Specimens cut from weft direction show a bending strength of 290 MPa while their tensile strength is 220 MPa.

5. CONCLUSIONS

Theoretical methods will generally need to be doubled by mechanical tests in order to determine mechanical constants experimentally.

The calculations for the analyzed specimens [2] compared to experimental results show that for a series of specimens, the measured properties are outside the margins calculated in theory in ideal conditions. It can be seen that tensile tests are a relatively cheap and quick method to determine the characteristics of fiber reinforced composite materials.

REFERENCES

[1]   N.D. Cristescu, E.M. Craciun, E. Soos, Mechanics of elastic composites, Chapman & Hall/CRC, (2003).

[2]   A. Modrea, Ph.D. Thesis, TRANSILVANIA University of Braşov, 2005.

[3]   H. Teodorescu-Drăghicescu, S. Vlase, Homogenization and Averaging Methods to Predict Elastic Properties of Pre-Impregnated Composite Materials, Computational Materials Science, **50**, 4, Feb. (2011).

[4]   H. Teodorescu-Drăghicescu, S. Vlase, L. Scutaru, L. Serbina, M.R. Calin, Hysteresis Effect in a Three-Phase Polymer Matrix Composite Subjected to Static Cyclic Loadings, Optoelectronics and Advanced Materials – Rapid Communications (OAM-RC), **5**, 3, March (2011).

[5]   S. Vlase, H. Teodorescu-Drăghicescu, D.L. Motoc, M.L. Scutaru, L. Serbina, M.R. Calin, Behavior of Multiphase Fiber-Reinforced Polymers Under Short Time Cyclic Loading, Optoelectronics and   Advanced Materials – Rapid Communications (OAM-RC), **5**, 4, April (2011).

[6]   H. Teodorescu-Drăghicescu, A. Stanciu, S. Vlase, L Scutaru, M.R. Calin, L. Serbina, Finite Element Method Analysis Of Some Fibre-Reinforced Composite Laminates, Optoelectronics and Advanced Materials – Rapid Communications (OAM-RC), **5**, 7, July (2011).

FEM BASED SIMULATION OF INJECTED BONE SHAPED PP BASED COMPOSITE MATERIALS

D. Luca Motoc
"Transilvania" University of Braşov, Braşov, ROMANIA, danaluca@unitbv.ro

Abstract: *The paper aims to present a finite element simulation of polypropylene based glass fibers reinforced composite materials using an injection molding manufacturing technology. Two different configurations are used as injected locations and comparisons will be made from the fill and optimization stages of the analysis. A design of experiments (DOE) based on the Taguchi and factorial methods will be used to optimize the injection process and few specific parameters monitored along with the overall composite material behavior, both on time length or function of the melt temperature. The differences encountered based on the aforementioned process and analysis settings can be used for further developments both at the material level (e.g. constitutive selection, volume fraction, effective properties) and manufacturing process.*

Keywords: *finite element, composite, plastic, simulation*

1. INTRODUCTION

Polypropylene (PP) is one of the most widely used thermoplastic polymers in industry due to its excellent cost-to-performance ratio. The selection and use of various material combinations in structural components made from polymers naturally indicate the aforementioned as one of the promising candidate for use along with inorganic fillers as a composite materials with physical, mechanical or thermal improved properties.

Due to its intrinsic nature as semi-crystalline polymer, the plastic deformation occurs both in crystalline and amorphous phases, being a relatively complex phenomenon. Based on this, for a polymer composite subjected to an increasing load the stress transferred to the matrix material induces plastic flow in the matrix and consequently give rise to an irreversible inelastic strain within the composite.

Literature is abundant in references concerning with the inorganic fillers such as glass fibers (GF) that may contribute to the increase of the tensile modulus if embedded in polymer matrices, yet causing the decrease of the strength and toughness of the overall composite. There are several factors that may contribute to these material properties improvement/degradation, such are: poor fiber-matrix adhesion, stress concentration and fillers' aspect ratio, size, shape, surface characteristics and dispersion within the matrix, to underline the most important.

The effective material properties can be retrieved using micromechanical theoretical models, especially the ones that proved less discrepancies with the measured value, such are: Mori-Tanaka, Halpin-Tsai, Cox, etc. These predictors are suitable for arbitrary fiber orientations and different lenght values and lower volume fraction of the fillers. Most theories begin by estimating the properties for a system of fully aligned fibers and then calculating the properties of the effective composite by averaging the unidirectional properties over all directions, weighted by the orientation distribution function.

Injection molding is a common manufacturing process used in plastic industry to produce complex shapes assisted by an outstanding computer-aided technology. Nonetheless, the major problem for product designers is still present but minimized and concerns of the warpage of molded parts prediction. Warpage is caused by variations in shrinkage through the part. For fiber reinforced thermoplastic composites the anisotropic composite property caused by the fiber distribution orientation can become one of the major factors in the part shrinkage.

The process parameters such as injection speed, injection pressure, holding pressure, melting temperature, holding time, cooling time and etc. need to be optimized in order to produce finished plastic parts with good quality. Taguchi method is one of the research preferences in the design of experiments (DOE) analysis.

The present paper aims to present few issues concerning with the influence of the constitutive and injection molding related settings in case of short glass fiber reinforced thermoplastic composite samples. The fiber orientation distribution, individual material properties and different locations of the injection gates are few of the imposed entrances in the herein study, followed by a Taguchi based design of experiments analysis to determine the influence of few injection molding parameters on the composite bone-shaped parts.

2. NUMERICAL SIMULATION AND PREDICTION

It is very well acknowledged in literature that the injection molding process is generally affected by the following variables: injection speed, materials used, screw rotation speed, backpressure, coolant, cylinder and die temperatures, holding pressure transfer, manifold temperature and spear temperature, etc.

The herein bone shape geometry was investigated using a finite element based method (FEM) by selecting the dual domain mesh type from a commercial software facility (e.g. Moldflow Insight 2010 from Autodesk). The material used was a PP based 30% GF reinforced composite from Sabic Innovative Plastics (USA, trade name – MEV 1006) with the effective mechanical and thermal properties listed in Table 1. For the finite element method based a dual-domain mesh was carried out followed by a mesh diagnosis that proved the overall quality of the discretized part.

A design of experiments (DOE) analysis based on Taguchi and followed by the factorial method was used as an analysis sequence whilst the process insights were compared taking into account 1 and 2 injection locations, respectively. The injection location(s) were set at the end of the holding extremity, which is compulsory to avoid stress locators development. The effective mechanical properties of the composite part were predicted based on the Mori-Tanaka micromechanical model setting as options for the GF orientation during the injection: random distributed at core, aligned at skin.

Table 1: Effective material properties – 30% GF embedded within a PP matrix

Material property	Longitudinal	Transversal
Elastic modulus (GPa)	5.183	2.747
Shear modulus (GPa)	1.224	
Poisson ratio	0.4371	0.4617
CTE x10^{-5} (1/°C)	2.552	6.288
Thermal conductivity (W/m°C)	0.18	

In Figure 1 was represented the volumetric shrinkage at ejection for the double injection locations configuration. As it can be seen the most sensible is the central part of the bone shaped sample but the shrinkage is not accentuated comparatively with the case of a single injection location. The latter setting gives rise to a time range higher than its counterpart with approximately 1 sec.

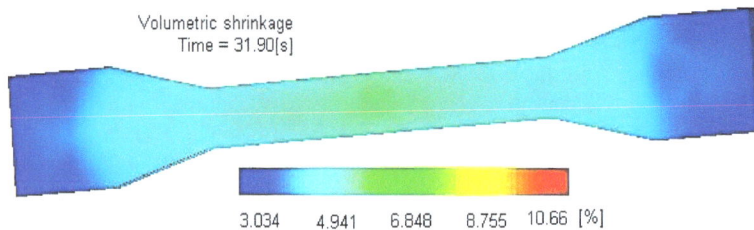

Figure 27: Composite sample's volumetric shrinkage using two injection locations

In Figure 2 (a) and (b) are being represented the tensile modulus variation encountered in the longitudinal, respectively transversal directions of the PP based glass fiber reinforced composite sample using a single injection location. Differences in the values can be regarded to the different fiber orientations within the injected parts, both at skin and core levels.

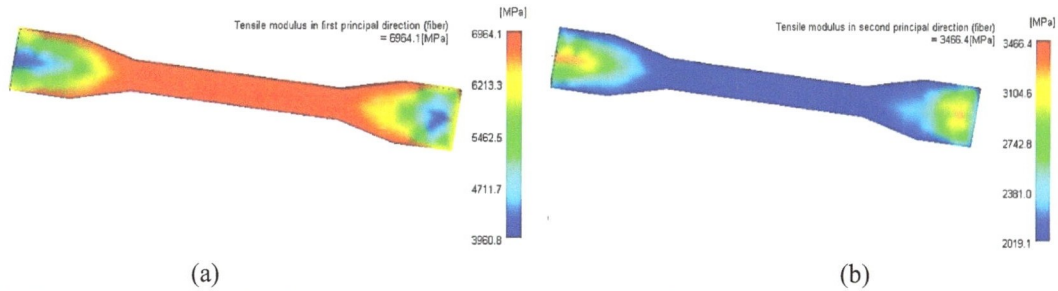

(a)   (b)

Figure 28: Tensile modulus in the first (a), respectively second (b) principal directions in case of a single injection gate Setting two different injection locations do not necessarily shortens the time until the samples reach the ejection temperature. This can be sized in the figures below, where a higher value was encountered for the double injection location configuration.

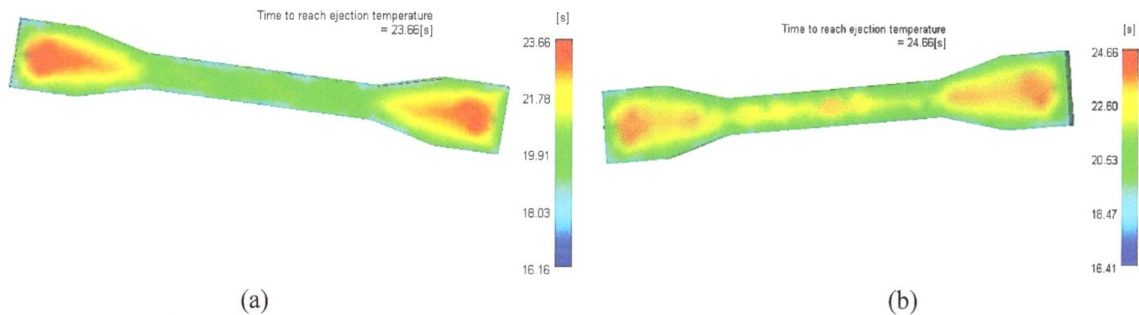

(a)   (b)

Figure 29: Time to reach ejection temperature for the composite samples using:
a single (a), respectively two injection (b) locations

In Figure 4 (a) and (b) are being plotted the optimized value distributions of the pressure at the end of fill for both configurations under the herein study. As it can be seen, the injection location influences the field distribution as well as the values recorded subsequently.

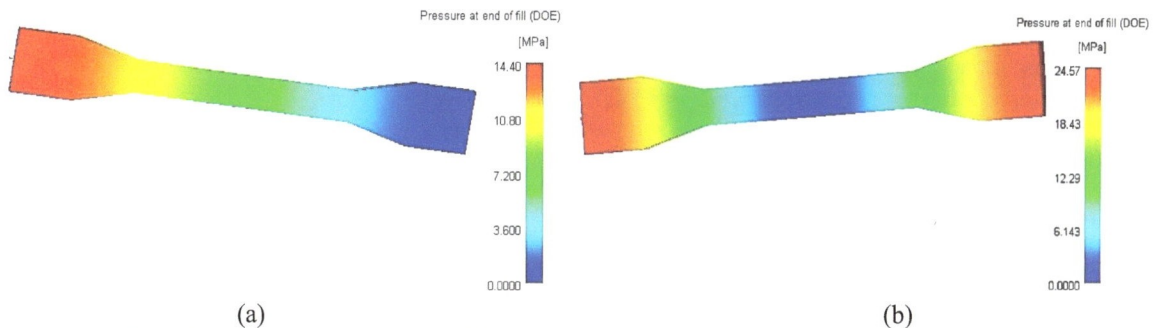

(a)   (b)

Figure 30: Pressure at the end of fill (optimized values) for the composite samples using:
a single (a), respectively two injection (b) locations

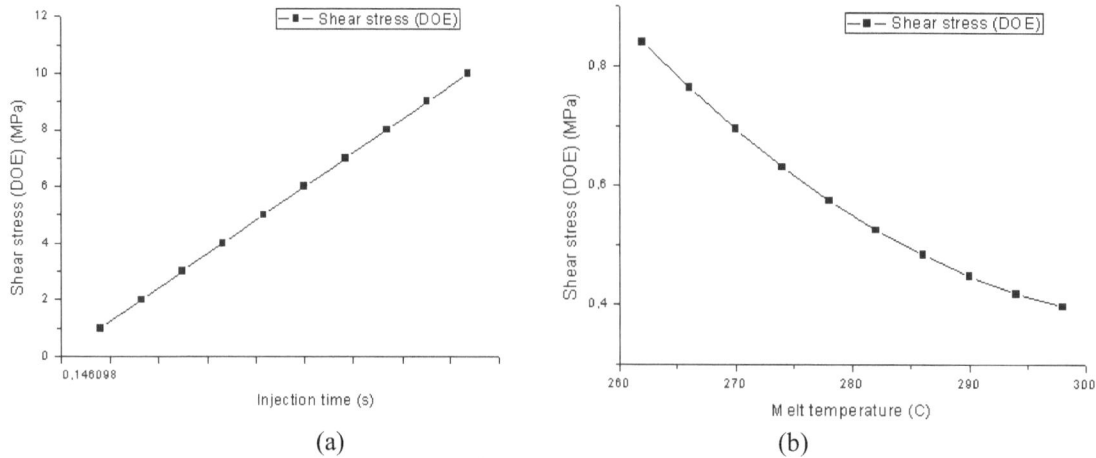

(a)   (b)

Figure 31: Optimized shear stress values variations:
(a) vs. injection time and 1 injection location (b) vs. melt temperature and 2 injection locations

Apart to the aforementioned indicators, for each analysis stage several quantities or information can be withdrawn with respect to the constitutive/entire composite material or the manufacturing process, such are: the fiber orientation at skin/at core, air traps, average velocity of the injected materials, in-cavity residual stress in both longitudinal and transversal directions, weld lines, shear stress at wall, etc.

Table 2: Process parameter ranges

Parameter	Min	Max
Mold temperature (°C)	40	80
Melt temperature (°C)	250	270
Flow rate (10^3 mm^3/s)	10	80
Packing pressure (MPa)	25	40

For the DOE based on the Taguchi method were weighted and ranked several indicators based on the following factors: melt temperature, global thickness multiplier and injection time, mold wall temperature, packing time and packing profile multiplier. The quality criterion considered herein were: flow front temperature variance, shear stress at wall, injection pressure, clamp force , volumetric shrinkage variance, sink index, total part weight and cycle time.

The parameter ranges, listed in Table 2, were chosen according to the material specification suggested from the plastic manufacturer.

3. CONCLUSIONS

Determining the optimal process parameter setting is a critical work that great influences the productivity, quality and costs of the composite based product manufacturing and production. Taguchi process parameter design method is one of the methods preferred in engineering to determine the optimal parameter settings and as is known is based upon three stages, as follows: concept design or system design, parameter design, tolerance design. Taguchi approach has potential for savings in experimental time and cost on product or process development and quality improvement. There is general agreement that off-line experiments during product or process design stage are of great value. Reducing quality loss by designing the products and processes to be insensitive to variation in noise variables is a novel concept to statisticians and quality engineers.

The results herein are revealing the fact that the melting temperature is the most significant parameter while the injection speed is the insignificant parameter. The presence of a second injection location does influence the injection manufacturing process, contributing to volumetric shrinkage of the composite sample in the most sensitive part of it. The significant factors can be determined performing an ANOVA analysis. The fiber distribution within the composite sample or at the skin surfaces represent another major influencing factor that should be address individually and in connection to the manufacturing parameters.

Further studies can be address with respect to the herein subject by integrating a supplementary methodology to the finite element based simulation, namely artificial intelligence. Neural network, genetic algorithms or fuzzy logic represents few of the approaches that can aid these optimization processes.

REFERENCES

92. Kwon, Y. W., Bang, H.: The Finite Element Method using Matlab, CRC Press, 1997.
93. Rosato, D., et. al., (ed.): Injection Molding Handbook, Kluwer Academic Publishers, 2000.
94. Zheng, R., Tanner, R., Fan, X.: Injection Molding – Integration of Theory and Modeling Method, Springer, 2011.
95. Xu, J.: Microcellular Injection Molding, Wiley, 2010.
96. Jones P.: Budgeting, Costing and Estimating for the Injection Molding Industry, Smithers, UK, 2009.
97. Wang, J. (ed.): Some Critical Issues for Injection Molding, InTech, Rijeka, Croatia, 2011.
98. Sin L. T., et. al.: Computer Aided Injection Molding Process Analysis of Polyvinyl Alcohol Starch Green Biodegradable Polymer Compound, Journal of Manufacturing Processes, 14, pp. 8-19, 2012.
99. Bergimc, B., Kampus, Z., Sustarsic, B.: The use of the Taguchi Approach to Determine the Influence of the Injection-Molding Parameters on the Properties of Green Parts, J. Achievements in Materials and Manufacturing Engineering, vol. 15, pp. 63-70, 2006.
100. Deng, W-J, et. al.: An Effective Approach for Process Parameter Optimization in Injection Molding of Plastic Housing Components, Polymer Plastic Technology and Engineering, 47:9, 910-919, 2008.
101. Kamaruddin, S., Khan, Z., Foong, S.: Application of Taguchi Method in the Optimization of Injection Molding Parameters for Manufacturing Products from Plastic Blend, IACSIT International Journal of Engineering and Technology, vol. 2, pp. 574-580, 2010.
102. Barthi, M. K, Khan, M. I.: Recent methods for optimization of plastic injection molding process – a retrospective and literature review, International Journal of Engineering Science and Technology, 2:9, 4540-4554, 2010.
103. Chang,T., Faison, E.: Optimization of weld line quality in injection molding using an experimental design approach, J. of Injection Molding Technology, 3:2, 61-66, 1999.
104. Yusuff, M., et. al.: A plastic injection molding process characterisation using experimental design technique: a case study, Jurnal Teknologi, 41(A), 1–16, 2004.
105. Thomason, J. L..: Micromechanical parameters from macro-mechanical measurements on glass-reinforced polybutyleneterephtalate, Composites, Part A, 33, 331-339, 2002.

COMPOSITE COATINGS IN ZINC MATRIX WITH SIO$_2$ IN DISPERSED PHASE OBTAINED BY ELECTRODEPOSITION

I. Constantinescu[1], E. Vasilescu[1], R.I. Novac[1]

[1]Faculty Metallurgy Material Science and Environment, "Dunărea de Jos" University of Galaţi,
111 Domnească Street,800201, email:ionel53gl@yahoo.com

Abstract: The study describes the conditions for obtaining zinc matrix composites using SiO$_2$ as dispersed phase by direct current electrodeposition. Authors made experiments for setting the best codeposition parameters. In this work are presented the results about the influence of the main technological codeposition parameters on the chemical composition, structure and properties of the composite coatings.
Keywords: electrochemical deposition, zinc matrix, coatings composite, SiO$_2$.

INTRODUCTION

Metalic coatings are often created by chemical or electrochemical plating methods, heat spraying, cladding and vacuum plating. Composites are combined heterogeneous materials created at least by two phases separated from each other by interfaces[1].The composite coatings obtained by including the dispersed-phase particles in the metallic matrix has a wide technological interest for many applications. The advantages of creating composites by using electrodeposition techniques are:

- This technology is not very expensive.
- It is possible to regulate precisely the thickness of the coatings.
- It can be obtained composite coatings with matrixes of wide ranges of metals and particles [2].

Electrodeposition is used for improving the mechanical properties such as wear, corrosion improvement and hardness of the coating surfaces. Coatings of zinc are of great practical importance because of their capacity to protect ferrous substrates against corrosion[3,4,6].In addition for pure zinc coatings, composite coatings of zinc-SiO$_2$ have attracted the attention of the researchers due to their characteristics concerning of the effect of inert oxide particles content on the crystalline structure and properties of the composites [7,9,10]. Zinc matrix composite coatings are often used for automotive, aerospace and marine parts.

2. EXPERIMENTAL CONDITIONS

The quality of the deposit depends on a good correlation between dispersed phase characteristics, bath composition, current density, pH, stirring, etc. The amount of the embedded particles plays an important role in improving the new material properties.
An SP-150 type potentiostat, a magnetic stirring machine and an electrolyte tank were used in order to obtain zinc composite coatings. The schematics of electrodeposition method are shown in Figure 1:

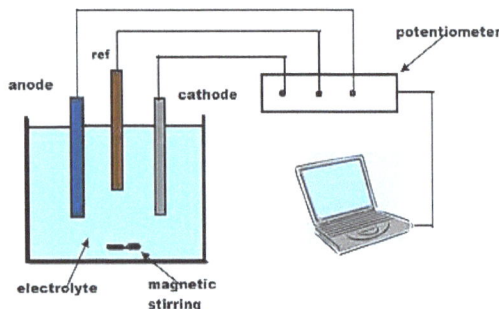

Figure 1: Schematics of the experimental setup

It was used a solution volume of 200 ml electrolyte and the experiments took place at 20° C. As for zinc layers in order to determin the optimal electrodeposition parameters were used steel samples - base metal, substances for the preparation of electrolytic solutions, SiO_2 powder (1-5μm size) as dispersed phase for the coatings.

The active electrodes surfaces were degreased and polished. Concentrations of SiO_2 particles were 10g/l, 30g/l and 50g/l in the electrolyte solution.

In this work a high purity rolled zinc anode (99,9%) and cathode of steel having an active area of 16 cm^2 were used. The cathode face that did not come in direct contact with the anode was insulated. The working parameters of the deposition are shown in Table1:

Table 1: The working parameters for electrodeposition of Zn-SiO$_2$ composite coatings

Electrolyte composition	$ZnSO_4 \cdot 7H_2O$=315g/l; $Na_2SO_4 \cdot 10\ H_2O$=75g/l; $Al_2(SO_4)_3 \cdot 18H_2O$=40g/l
pH	3 - 4
Temperature (°C)	25
Current density (A/ dm^2)	2,3,4
Magnetic stirring (rpm)	300
Electrodeposition time (min.)	60

3. EXPERIMENTAL RESULTS AND DISCUSSIONS

The EDX analysis shows that SiO_2 particles were present in layers along with zinc. Different SiO_2 percentages reported in the composite coatings are depending on particle concentrations in the bath and current densities; the re-sults are given in Figures 2and 3

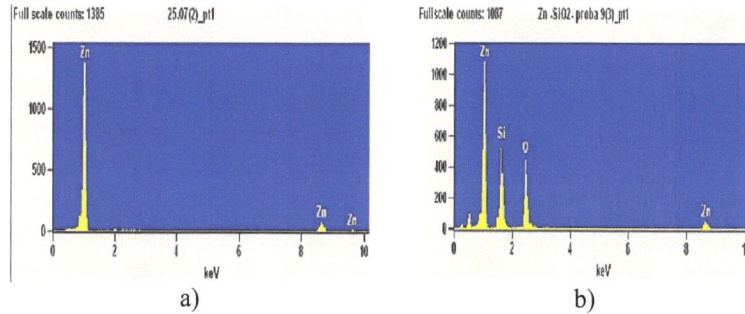

a)   b)

Figure 2: The EDX analisys for a) pure Zn and
b) Zn-SiO$_2$ composites(30g/l) obtained with the following parameters: 3A/dm^2,120min,500rpm

Figure 2 presents the difference between EDX analysis of the pure zinc layers(Figure a) and composite layers.The presence of oxide particles in zinc matrix is shown in Figure b.

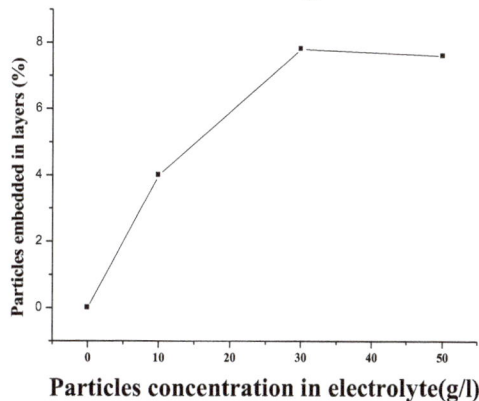

Figure 3:The variation of the particles concentration in electrolyte according to the particles embedded in layers

In Figure 3 has been observed that the amount of embedded particles increases with increasing the concentration of suspended particles in electrolyte. The highest value for the embedded phase volume is 8% obtained for 30 g/l, confirming a good adherence and uniform distribution of DP using the following deposition parameters: current density of 3 A/dm^2, time 60 min, 500 rpm stirring, temperature 20^{0}C. The percentage of the embedded particles in layer is decreasing at 50g/l DP in electrolyte.

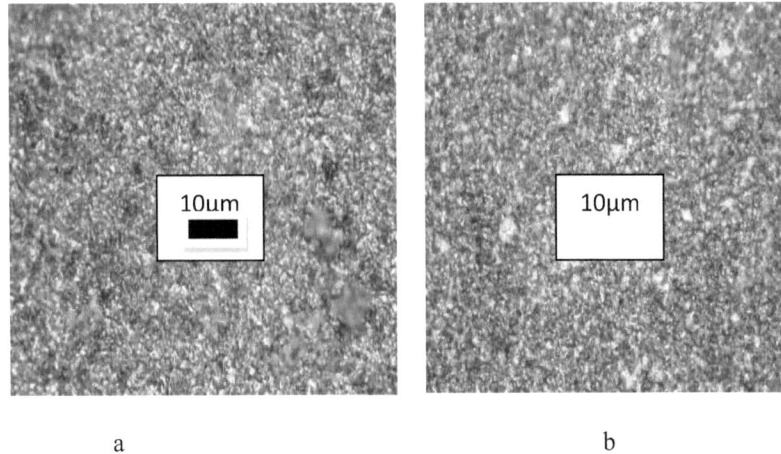

a   b

Figure 4: Surface Optical Microscopy (x400) of the coatings: a) Zn pure coating obtained at 3A/dm^2, b) Zn-SiO$_2$ composite coatings for 3 A/dm^2 and 30g/l DP in electrolyte

In Figures 4 and 5 is presented the effect of the SiO$_2$ inert particles on the structure; the structure modifications are highlighted by optical microscopy and SEM analysis.

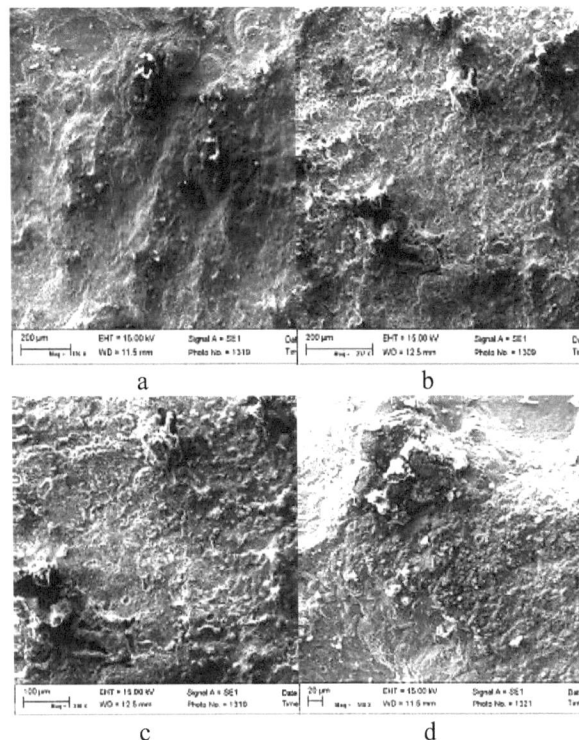

Figure 5:SEM Micrographs of Zn-SiO$_2$ coatings for different magnitudes obtained at:

3 A/dm^2, 30g/l DP ,60 min ,500rpm

Using SEM microscopy analysis and optical microscopy(OM) analysis it was observed homogeneous distribution of additional phase in zinc matrix. The presence of SiO$_2$ in zinc matrix makes modifications of structure(roughness) and mechanical properties(microhardnes).Study of microhardness is important both for estimating mechanical properties or wear resistance of the composite layers.The microhardness range of the layers is controlled by the amount of DP embedded in zinc matrix.The microhardness Vickers test was made using a CV-400DAT2 NAMICON durimeter.

The microhardness of the electrodeposited layers varies quite large; the results are shown in the chart below:

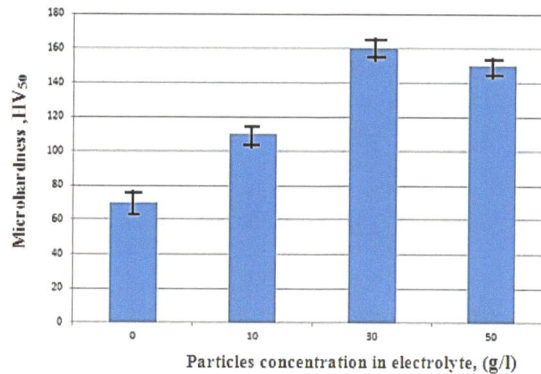

Figure 6: Microhardness (HV$_{50}$) of the samples obtained by electrodepositing according to the particles concentration in electrolyte

It is noted that the average microhardness values of the composite layers are bigger then the average microhardness values of the pure zinc. This increasing of the microhardness values can be attributed to two different phenomenons: firstly, the strengthening effect of the reinforcing phase itself and secondly of the structural modification of Zn electrodeposited crystallites.

Microstructure of Zn-SiO$_2$ presents a different morphology compared with the pure Zn layers. This depends on the particles volume in electrolyte as well as those particules embedded in composite layers. Surface morphology makes changes in roughness measured by Ra parameter.

There are quite large variations of the roughness values obtained for different current densities. The results are presented in Figure 7:

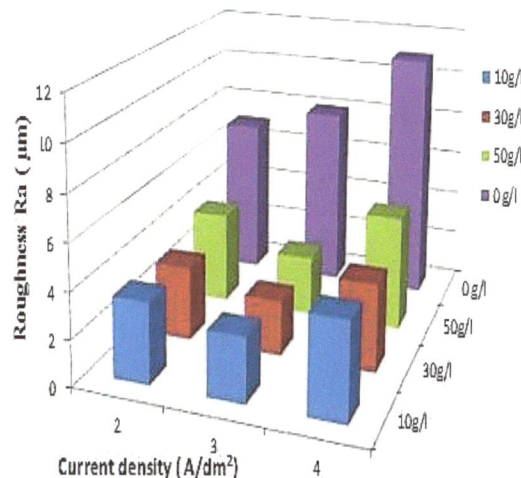

Figure 7: Roughness variation of Zn-SiO$_2$ composite layers for different current densities and different particle loadings in electrolyte

The values for roughness of the composite layers are lower then the roughness determined for pure zinc. The presence of SiO_2 particles in layers proves a change of properties comparative with simple Zn layers. The particle loading of 30g/l makes the composite surface smoother then 10g/l or 50g/l loading. The smoothest surface is obtained for 3A/dm^2 current density.

4.CONCLUSIONS

The experimental research highlights the conditions for obtaining zinc-SiO_2 composite coatings using electrodeposition method.

The composite coatings quality analysis has been made both qualitatively and quantitatively by using modern methods such as EDX and SEM analysis. These have highlighted inclusion of the dispersed phase and homogeneous distribution of the additional phase for some electrochemical parameters.

It has been stipulated experimental conditions for obtaining the best composite layers from the point of view of the best microhardness and roughness for composite layers.

REFERENCES

[1]   Schlesinger M.,:Modern Electroplating,5th Edition,ISBN: 978-0-470-16778-6,2010.

[2]   Hovestad A., Janssen L.J.J.,:Electrochemical Codeposition of Inert Particles in a Metallic Matrix,Journal   of Applied Electrochemistry, 25, pag.519,1995.

[3]   Khorsand S.,Raeissi K., Golozar M.A.: An Investigation on the Role of Texture and Surface Morphology   in the Corrosion Resistance of Zinc Electrodeposits, Corrosion Science,53,pag. 2676-2678,2011.

[4]   Binder L., Kordesch K.,:Electrodeposition of Zinc Using a Multi-Component Pulse Current, Electrochimica Acta,volume 31,Issue 2, pag.255–262,1986.

[5]   Zhang Z.,Leng W.H., Cai Q.I., Cao F.H., Zhang J.Q.,:Study of the Zinc Electroplating Process Using Electrochemical Noise Technique, Journal of Electroanalytical Chemistry, volume 578, Issue 2,pag.357–367,2005.

[6]   Zhang Q.,Hua Y.,:Kinetic Investigation of Zinc Electrodeposition from Sulfate Electrolytes in the Presence of Impurities and Ionic Liquid Additive [BMIM]HSO$_4$, Materials Chemistry and Physics,volume 134, Issue 1, pag. 333–339,2012.

[7]   Hashimoto S.,Masaki A.,:The Characterization of Electrodeposited Zn-SiO_2 Composites before   and after Corrosion Test, Corrosion Science, volume 36, Issue 12, pag.2125–2137,1994.

[8]   Fransaer J.,Celis J.P.,Roos J.R.,:Analysis of the Electrolytic Codeposition of Nonbrowian Particles with Metals., Journal of the Electrochemical Society, 139, pag.413,1992.

[9]   Kondo K., Tanaka A.O.Z.,:Electrodeposition of Zinc-SiO_2 Composite, Journal of the Electrochemical Society, 147(7): pag.2611,2000.

[10]   Khan T.R., Erbe A.,Auinger M.,:Electrodeposition of Zinc–Silica Composite Coatings:Challenges in Incorporating Functionalized Silica Particles into a Zinc Matrix,Science and Technology of Advanced Materials, 2:10.1088/1468-6996/12/5/055005, published 14 September 2011.

[11]   Low C.T.J., Wills R.G.A.,Walsh F.C.,:Electrodepositing of Composite Coatings Containing Nanoparticles in   a Metal Deposit,Surface and Coatings Technology, 201, pag.371-383,2006.

[12]   Celis J.P., Roos J.R., Buelens C.,:A Mathematical Model for the Electrolytic Codeposition of Particles with a Metallic Matrix, J. of the Electrochemical Society, 134 (6),1402–1408,1987.

[13]   Tolumoye J.,Tuaweri J., Wilcox G.D.,:Behaviour of Zn-SiO_2 Electrodeposition in the Presence of N,N-dimethyldodecylamine,Surface and Coatings Technology, 2:10.1016,pag.5921,2006.

[14]   Lin Z.F., Zhang D.,Wang Y.,Wang P.,:A Zinc/Silicon Dioxide Composite Film: Fabrication and Anti-corrosion Characterization,Materials and Corrosion, volume 63,Issue 5, pag. 416–420,2012.

[15]   Socha R.P., Fransaer J: Mechanism of Formation of Silica-Silicate Thin Films on Zinc, Thin Solid Films, 488,pag.45,2005.

Table of Contents

Proceedings of COMAT 2012